W9-BRR-589

Applied Reservoir Engineering

Volume 1

Applied Reservoir Engineering

Volume 1

Charles R. Smith

G. W. Tracy

R. Lance Farrar

OGCI Publications

Oil & Gas Consultants International, Inc.

Tulsa

Charles R. Smith, G. W. Tracy and R. Lance Farrar
4554 South Harvard
Tulsa, Oklahoma 74135

Printed in the United States of America.

Library of Congress Catalog Card Number: 92-080738

International Standard Book Number: 0-930972-15-5

CONTENTS, Volume 1

4 RESERVOIR VOLUMETRICS 4-1 — 4-14

5 GAS RESERVOIRS 5-1 — 5-62

10 MULTIZONE RESERVOIR PERFORMANCE 10-1 — 10-22

11 IMMISCIBLE FLUID DISPLACEMENT MECHANISMS 11-1 — 11-28

Nomenclature

Index

CONTENTS, Volume 2

16 PRESSURE TRANSIENT ANALYSIS 16-1 — 16-150

1 GEOLOGY

Since many of those who function for part of their professional lives as reservoir engineers do not have an earth science background, it is well to remember that the reservoir to be studied has a geologic background. Typically, the reservoir has been (although not always) defined by drilling, and some data will be available at isolated positions in the system. The data can take the form of drilling times, cuttings analysis, open hole and through casing well logs, core analysis and descriptions, SEM and x-ray diffraction analysis, sidewall cores, drillstem tests, various specialized well logs, and production tests. If some time has elapsed, production data will usually also be available. What this evidence indicates in terms of good reservoir description, will depend on the skill and insight of those doing the interpretation. Geological insights will be of great help.

SOURCES OF THE HYDROCARBONS

Although the presence of hydrocarbons will be evident in an already discovered field, it is useful to speculate on the source of the hydrocarbons in the reservoir being studied. The problem has been studied extensively.[1]

It is universally agreed, with minor objection from time to time, that hydrocarbons have an organic origin. This means that organic matter has to be synthesized by living organisms and, thereafter, it must be deposited and preserved in sediments. Depending on further geologic history, part of the organic matter may be transformed into petroleum-like compounds. Not all of the geologic history has been favorable to the preservation of organic matter.

Photosynthesis is the basis for the mass production of organic matter. About two billion years ago in Precambrian time, photosynthesis appeared as a worldwide phenomenon. The enrichment of molecular oxygen in the atmosphere is a result of photosynthesis and the mass production of organic matter.

The average preservation rate of the primary organic production, expressed as organic carbon, is estimated to be less than 0.1%. The upper limit of the preservation rate of organic carbon to be found in certain oxygen-deficient environments favorable for deposition of source rock sediments is about 4%. Deuser[2] has studied the organic carbon budget of the Black Sea with results as summarized in Figure 1-1.

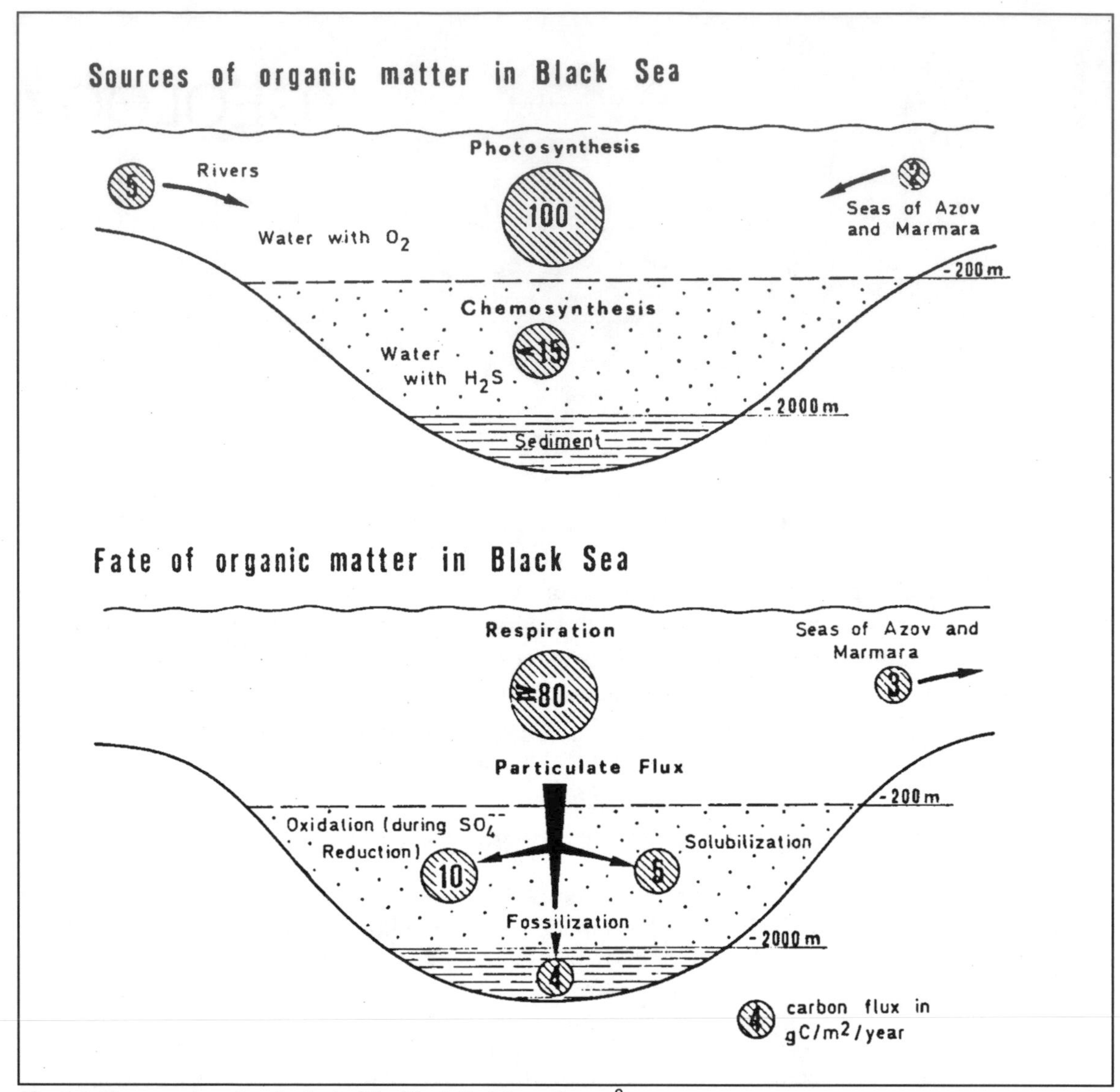

Fig. 1-1. The organic carbon budget of the Black Sea (from Deuser[2]).

Beginning in the Precambrian until the Devonian, the primary producer of organic matter was marine photoplankton. Since the Devonian, an increasing amount of organic matter has been higher terrestrial plants. Higher organized animals, such as fishes, contribute so little organic matter in sediments that they can be essentially neglected. Bacteria, phytoplankton, zooplankton, and higher plants are the main contributors of organic matter in sediments.

Favorable conditions for the depositions of sediments rich in organic matter are found on the continental shelves in areas of quiet water, such as in lagoons, estuaries, deep basins of restricted circulation, and on the continental slopes.

The three main stages of the evolution of organic matter in sediments are diagenesis, catagenesis, and metagenesis:

1. Diagenesis begins in recently deposited sediments where microbial activity occurs. At the end of diagenesis, the organic matter consists mainly of a fossilized, insoluble organic residue called kerogen.

2. Catagenesis results from an increase in temperature during burial in sedimentary basins. Thermal breakdown of kerogen is responsible for the generation of most hydrocarbons.

3. Metagenesis is reached only at great depth, where temperature and pressure are high. At this stage, organic matter is composed only of methane and a carbon residue. The constituents of residual kerogen are converted to graphitic carbon.

Figure 1-2 illustrates the general scheme of evolution of the organic matter as a function of depth. Figure 1-3 presents the sources of hydrocarbons in geological situations, this related to the evolution of organic matter. Geochemical fossils represent a first source of hydrocarbons in the subsurface, while degradation of kerogen represents a second source of hydrocarbons. The bulk of petroleum is generated at a depth where thermal degradation of kerogen becomes important. Hydrocarbons having their source in relatively recent sediments are derived from living organisms and can be regarded as geochemical fossils. Alkanes, fatty acids, terpenes, steroids, and porphyrins are the major groups of geochemical fossils.

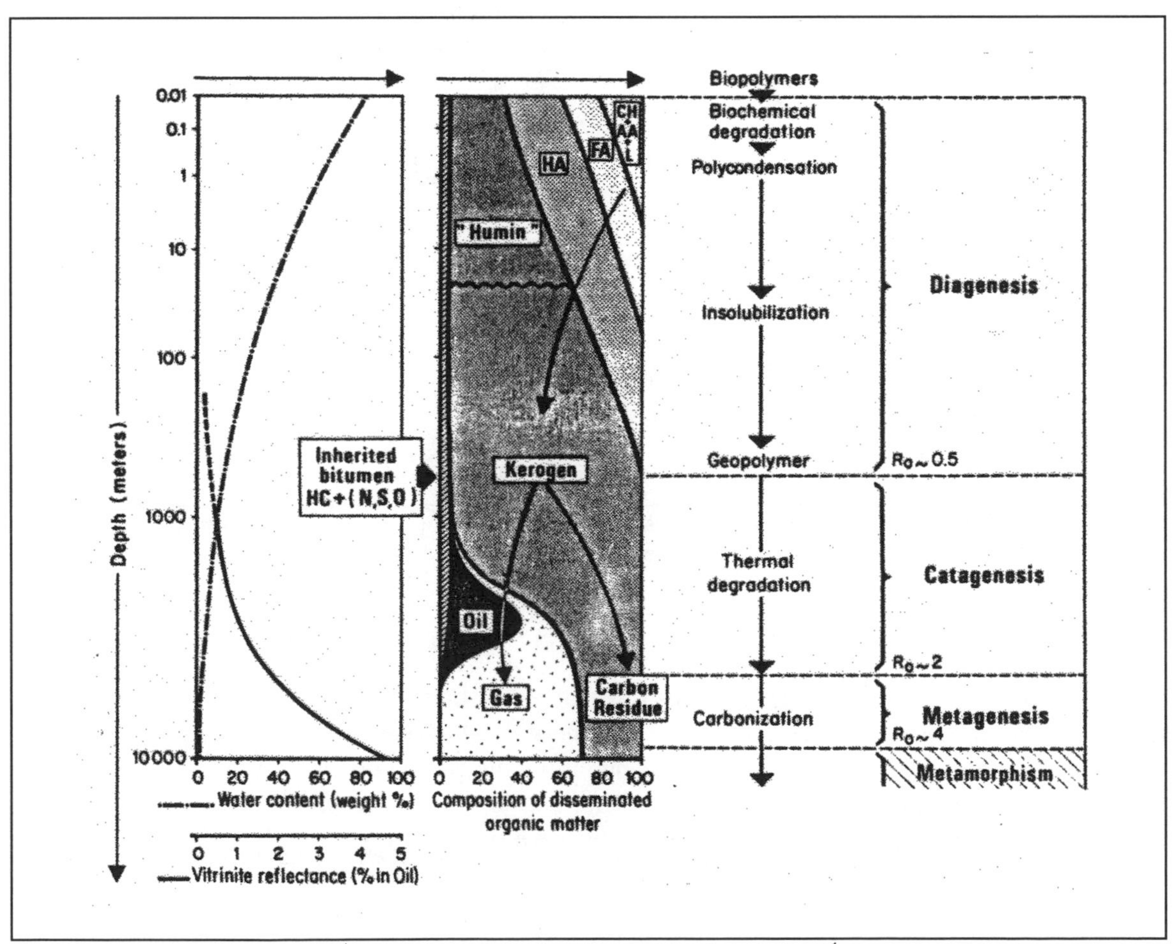

Fig. 1-2. General scheme of the organic matter related to depth (from Tissot and Welte[1]). CH: carbohydrates; AA: amino acids; FA: fulvic acids; HA: humic acids; L: lipids; HC: hydrocarbons; N, S, O compounds (non-hydrocarbons). Permission to publish by Springer-Verlag.

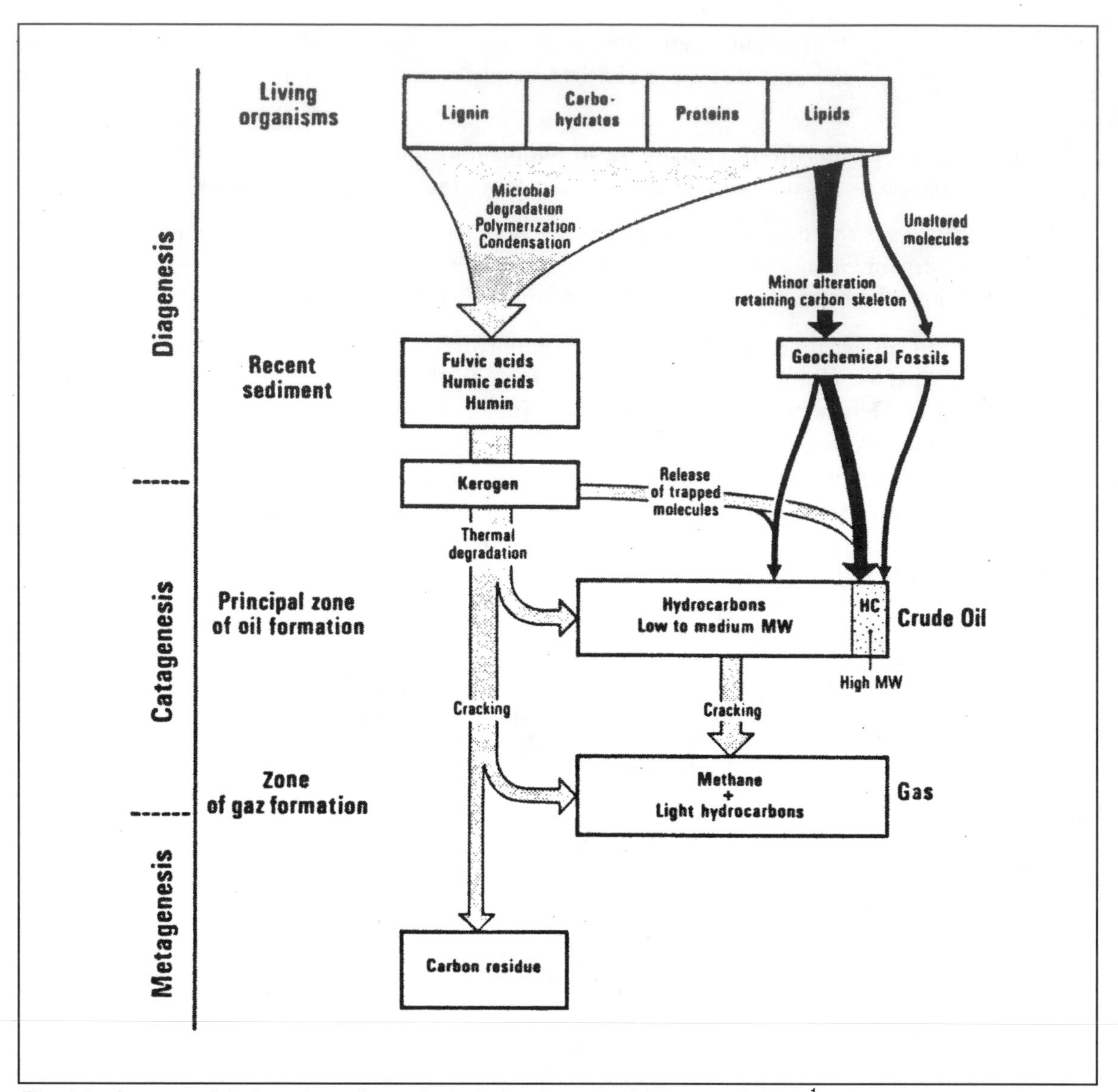

Fig. 1-3. Sources of hydrocarbons in geological situations (from Tissot and Welte[1]). Permission to publish by Springer-Verlag.

Kerogen as used here is the organic constituent of sedimentary rocks that is not soluble in aqueous alkaline solvents or in common organic solvents. That part which is extractable with organic solvents is bitumen. Kerogen is the most important form of organic carbon on earth, and is 100 times more abundant than coal plus petroleum in reservoirs, and is 50 times more abundant than bitumen. Kerogens that have a high hydrogen/carbon ratio have potential for oil and gas generation. Most source beds for liquid hydrocarbon production were deposited in marine environments with their organic matter being generally dominated by aquatic lower plants and bacteria.

OIL SHALES

It will serve here to define an oil shale as any shallow rock yielding oil in commercial amount upon pyrolysis (heating). The organic matter contained in the oil shales is mainly kerogen. Whether an oil shale is of economic interest can be decided rather easily. The average pyrolysis temperature is 500°C, the energy for heating to that temperature is about 250 calories per g of rock, and the calorific value of kerogen is 10,000 calories per g. If the kerogen content of the rock is 2.5%, the total calorific value is used for heating the rock. This converts to three US gallons per short ton. The literature has used 10 US gallons per short ton as a lower limit of economic attractiveness for oil shales.

It can be said that the equivalent of an oil shale, sufficiently buried, can be a petroleum source rock. The converse may not be true due to the richness requirement.

At different times, countries have developed an oil shale industry: France (1838), Scotland (1850), Australia (1865), Brazil (1881), New Zealand (1900), Switzerland (1915), Sweden (1921), Estonia (now USSR) (1921), Spain (1922), China (1929), and South Africa (1935). The highest point of development was reached during or immediately after World War II.

Relatively inexpensive access to liquid hydrocarbons by most countries has limited production activity in the oil shales.

MIGRATION AND ACCUMULATION OF OIL AND GAS

Most oil and gas is found in relatively coarse-grained porous, permeable rocks that contain little or no insoluble organic matter. It is unlikely that the petroleum found in these rocks could have originated in them as no trace of the solid organic matter remains. Therefore, most oil and gas reservoirs are traps for migrating hydrocarbons.

The release of petroleum compounds from kerogen in source beds and their transport within the narrow pores of a fine-grained source bed are called primary migration. The oil and/or gas expelled from a source bed passes through the wider pores of more permeable rock units. This is called secondary migration. Figure 1-4 illustrates primary and secondary migration of hydrocarbons and their ultimate trapping.

Since most permeable rocks in the subsurface are water saturated, movement of hydrocarbons has to be due to active water flow or occur independently of the aqueous phase, either by displacement or by diffusion. Since the gas and oil densities are usually lower than that of water, accumulation in traps is usually at a structural high. Relatively impermeable cap rocks limit further migration.

The processes of primary and secondary migration are still poorly understood. Data on pore geometry, porosity and permeability relationships and the distribution of water in buried source rocks are rare. Likewise, information on petroleum migration is largely theoretical and needs further quantification. Much has been said in the geological literature on the topic, and more remains to be said. These arguments are very important in oil and gas exploration but are less critical to reservoir engineering as practiced after the resource has been located.

It is interesting to speculate on distances of secondary migration. We would expect short range secondary migration in isolated sand lenses imbedded in organic shales in Tertiary or Cretaceous rocks. The same could be expected of hydrocarbon accumulations in pinnacle reefs in the Devonian

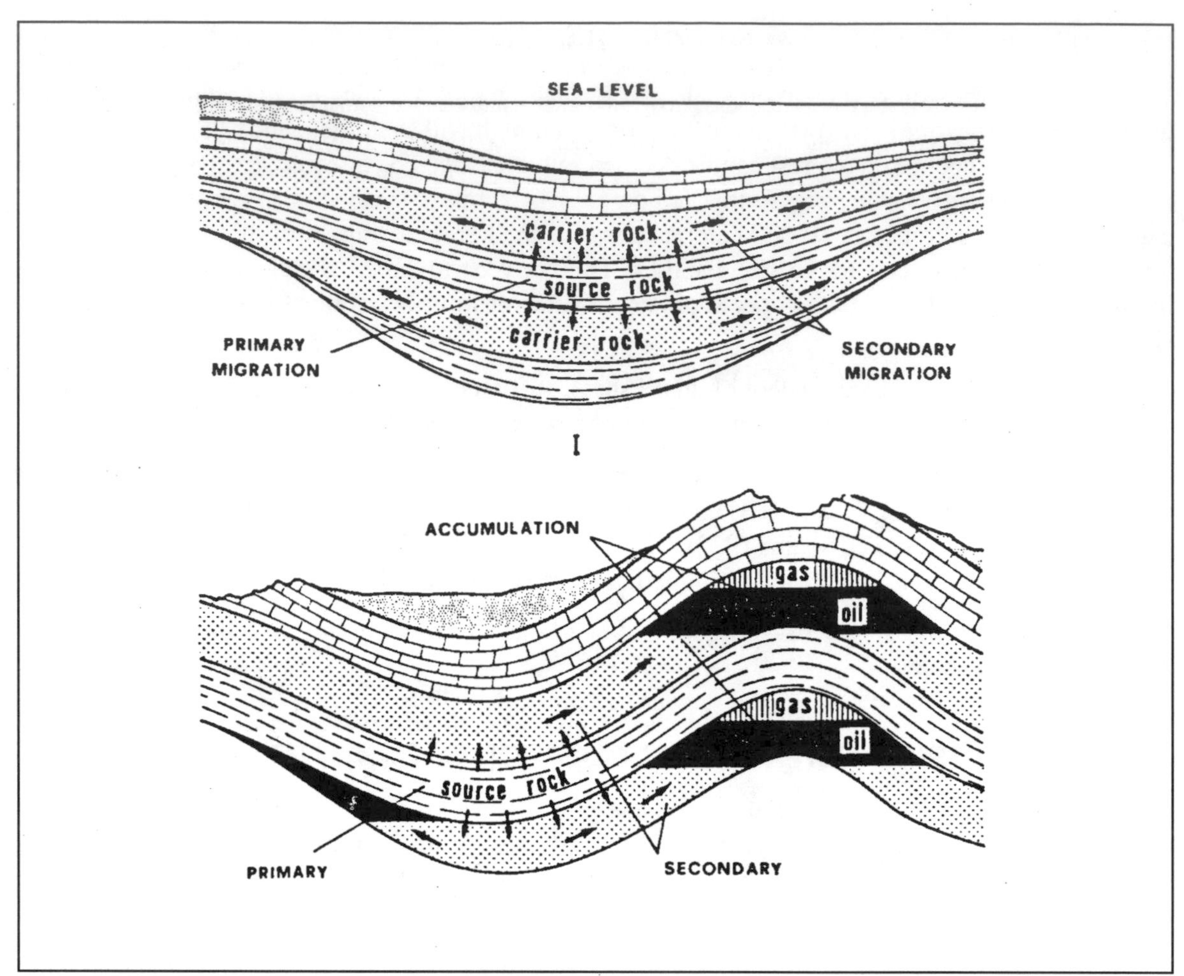

Fig. 1-4. Accumulation of oil and gas in traps showing primary and secondary migration (from Tissot and Welte[1]). Permission to bublish by Springer-Verlag.

of Western Canada. Long distance secondary migration would be expected for the very large hydrocarbon deposits since an approximate material balance calculation suggests that there should be a relationship between the size of the accumulation and the drainage area. Good examples should be the Athabasca heavy oil and tar sands in Canada and the large Middle East oil fields.

Secondary migration may be over substantial distances. It has been suggested that the Athabasca hydrocarbon accumulations require secondary migration of 200 miles or more. In Western Canada, this could have been along the major Cretaceous-Paleozoic unconformity.

Secondary migration over vertical distances should only be possible through faults, fracture systems and other preferred avenues such as dikes, thrust planes and mud volcanoes. Weber and Daukoru[3] have reported that growth faults are probably pathways for vertical migration of hydrocarbons from shale source rocks. Figure 1-5 illustrates the possible kinds of oil accumulations and migration avenues due to growth faulting in the Niger delta.

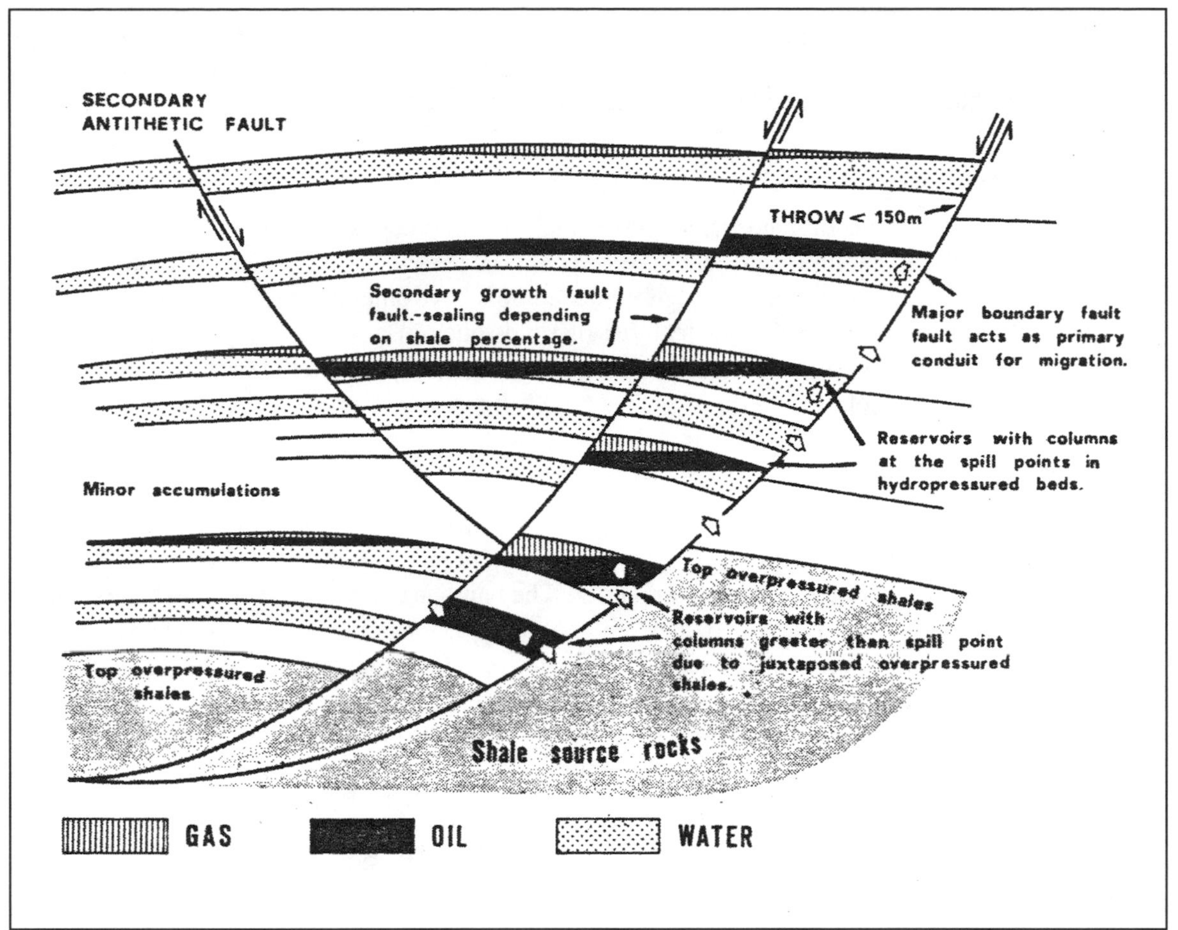

Fig. 1-5. Section through a growth fault in the Niger delta illustrating various kinds of oil accumulations and migration paths between reservoirs (from Weber and Daukoru[3]). Permission to publish by Applied Science Publishers.

SANDSTONES AS RESERVOIR ROCKS

About 60% of the oil and gas reserves in the world's giant fields occur in sandstone reservoirs. If the carbonate fields of the Middle East are excluded, the proportion is 80%.[4]

I. DEPOSITIONAL FACTORS

Before embarking on a treatment of the influence of depositional factors on sandstones as reservoirs, it is important to remember what properties are important for profitable oil and gas extraction. Obviously, large reservoir size, high porosity and permeability, and low minimum water saturation are desirable—these yield high producing rates and long producing life. But poor permeability can be offset by a sufficiently thick productive interval. Low producing rate can be tolerated if product price is adequate. Of course, the continuity and regularity of the pore system needs to be considered. An isotropic reservoir would seem the best arrangement. But, the isotropic reservoir is a rarity since the sedimentary processes tend to segregate different-sized particles into separate laminae. The reservoir engineering literature is replete with treatments of the effects of reservoir inhomogeneities on fluid production. It is the geometric arrangements of these inhomogeneities that have a profound effect on the rate of production and completeness with which a reservoir can be drained. This does not touch on the recoveries that can be achieved from secondary and tertiary oil (and even gas) exploitation practices.

II. INFLUENCE OF ENVIRONMENT

Pettijohn *et al.*[5] have divided sand bodies into seven broad sedimentary environments or associations: desert aeolian sands; alluvial sands; fluviatile sands; deltaic sands; beaches, barriers and bars; shelf sands; and deep-sea sands. Other authors have other divisions. Figure 1-6 presents a compilation of some significant features of sandstones in a variety of geological settings. Dipmeter expressions are over-simplified. Instrumentation improvements show that patterns are more complex. A brief discussion of each setting follows.

Desert Aeolian Sands

Desert aeolian sandstones are characterized by even, parallel, continuous laminae that are inclined at angles that range up to 30°, as illustrated in Figure 1-7. Figure 1-8 presents diagrams of aeolian dunes and dominant wind directions. The barchan dune has a crescent shape with the thin ends pointing in the direction of the dominant wind. The barchanoid ridge is a large structure and is comprised of a series of connected crescents when the supply of sand is larger than would be the case for single barchan dunes. The transverse ridge forms when the sand supply is abundant.

Reservoir sandstones of aeolian origin are comprised of many dunes and can form thick and extensive systems. These can offer the most favorable conditions of all clastic reservoirs provided there are adequate seals, trapping geometry, and source rocks. Sometimes these conditions are met. Thicknesses can be enormous. An interrupted sequence of Permian Rotliegendes dune sands up to 200 m thick in the Lehman gas field of the southern North Sea has been described by van Veen[9].

Figure 1-9 presents a cross section through the Rotliegendes depositional basin of the North Sea. The sections are in the southern North Sea and eastern Netherlands south of the mid-North Sea high and Fyn Grinstad high, respectively. A wadi is a channel that is dry except during periods of rainfall. Sabkhas are associated with the margin or edge of ephemeral desert lakes and may consist of horizontal bedded clays with nodular anhydride and little sand. Sabkhas can also exist between dunes and can be considered to be arid tidal flats.

Other sandstones interpreted to be aeolian in origin have been described in many locations in the world. The Tensleep sandstone of Wyoming (Permo-Pennsylvanian), the equivalent Weber sandstone of northwestern Colorado, and the Casper sandstone of southeastern Wyoming, are now considered to be aeolian in origin. Most of these sandstones were formed during periods of widespread aridity as indicated by associated anhydrides or red beds.

In the Gulf Coast basin, the Jurassic Norphlet sandstone is an aeolian reservoir that was deposited as coastal dunes following the evaporites of the Louann Salt. Figure 1-10 provides a structure map on top of the Smackover Formation on the Hatters Pond and Chunchula fields, Mobile County, Alabama. The underlaying Norphlet sandstone reportedly can reach a maximum thickness of more than 700 ft in the Louisiana and Mississippi salt basin provinces.

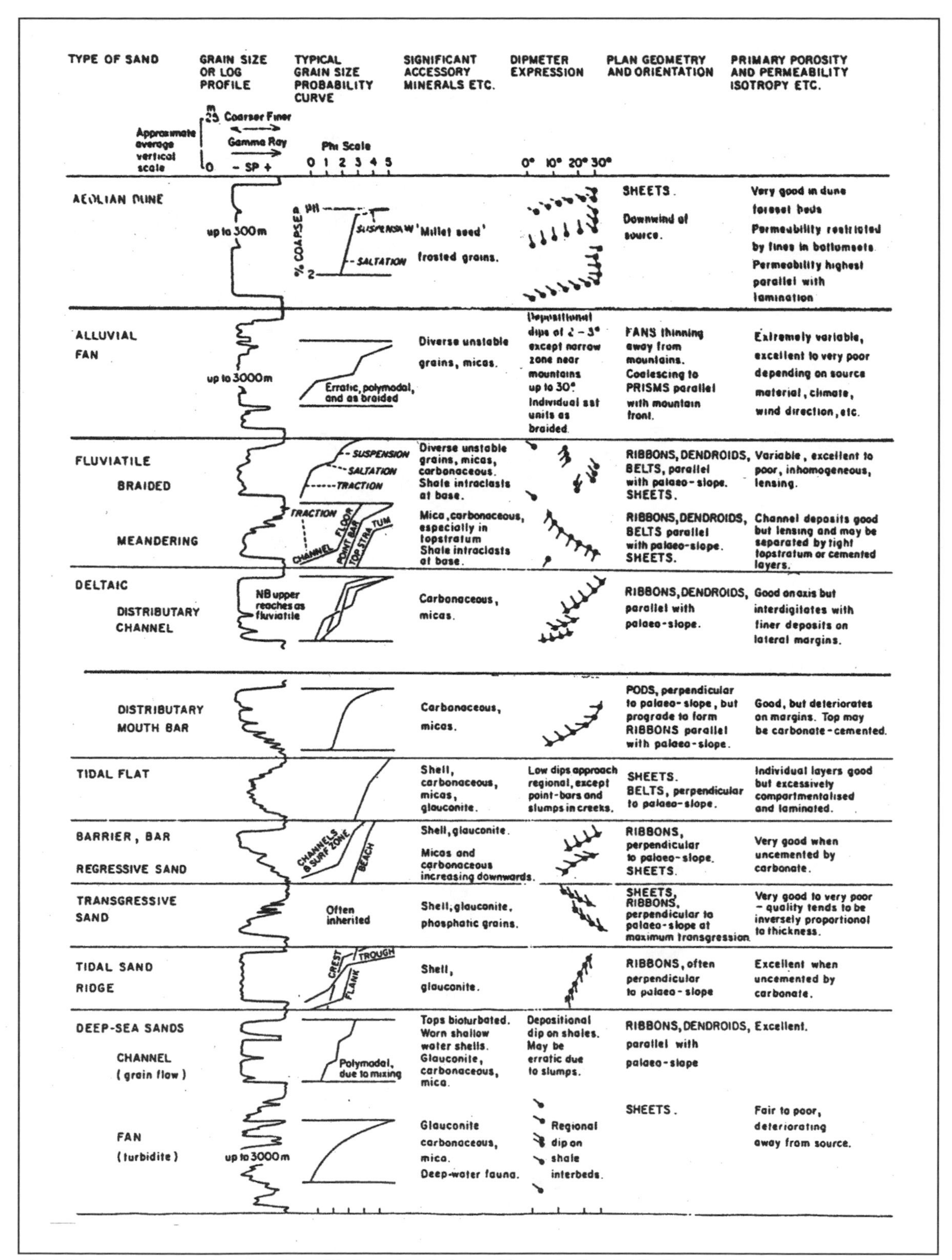

Fig. 1-6. Common reservoir sand types, with characteristic log profiles, grain size distribution, significant accessory minerals, dipmeter expression, plan geometry and orientation, and reservoir characteristics (from Taylor as edited by Hobson[6]). Permission to publish by Applied Science Publishers.

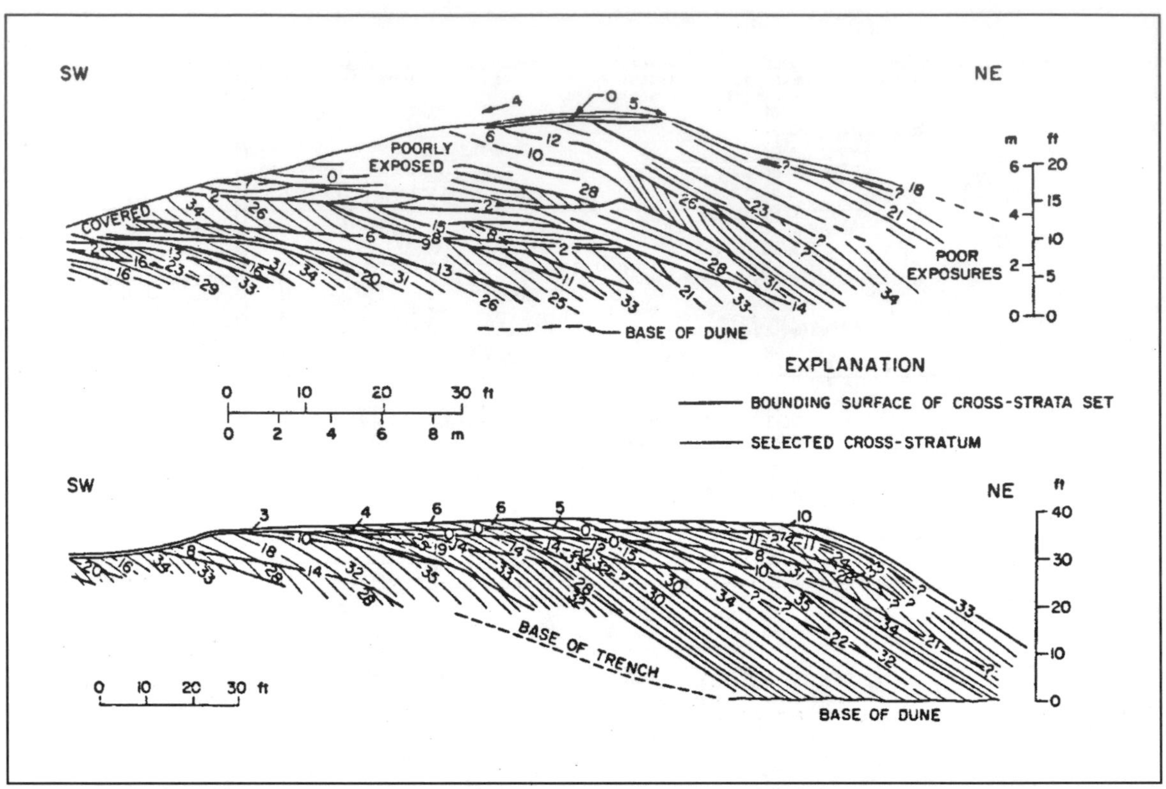

Fig. 1-7. Cross sections of dunes exposed in trenches: Upper diagram is a barchanoid ridge dune with section parallel to dominant wind direction; Lower diagram is a transverse dune with section parallel to dominant wind direction. Small numbers are apparent dip in degrees (from McKee[7]). Courtesy U. S. Geological Survey.

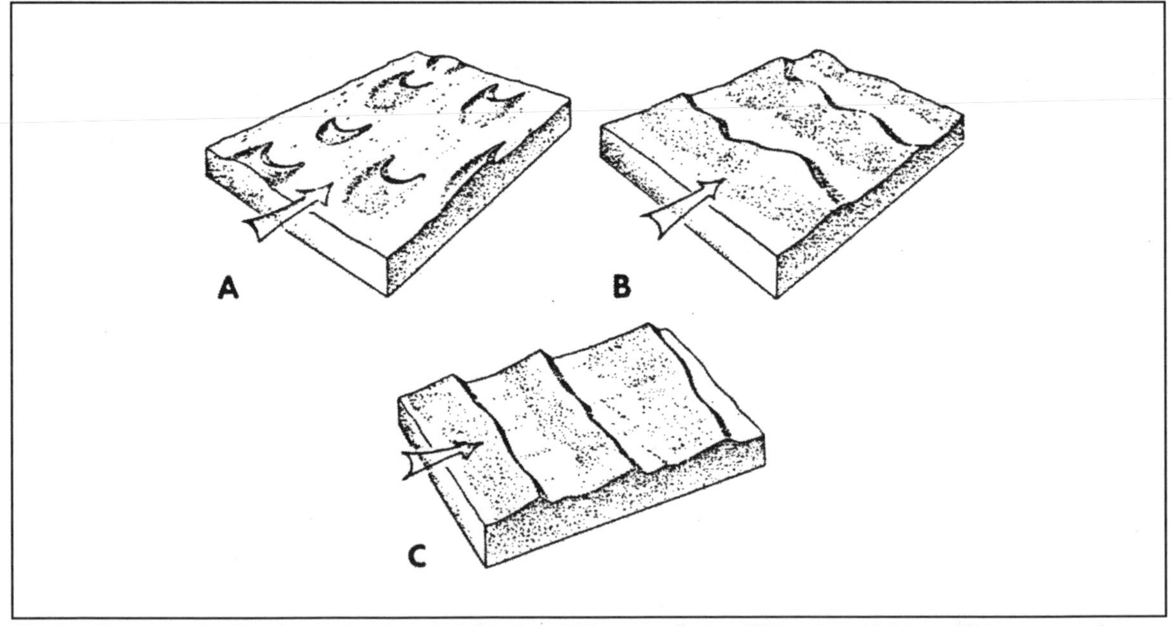

Fig. 1-8. Diagrams of eolian dunes showing wind direction: (A) barchan dunes; (B) barchanoid ridges; (C) transverse dunes (from McKee[8]). Courtesy U. S. Geological Survey.

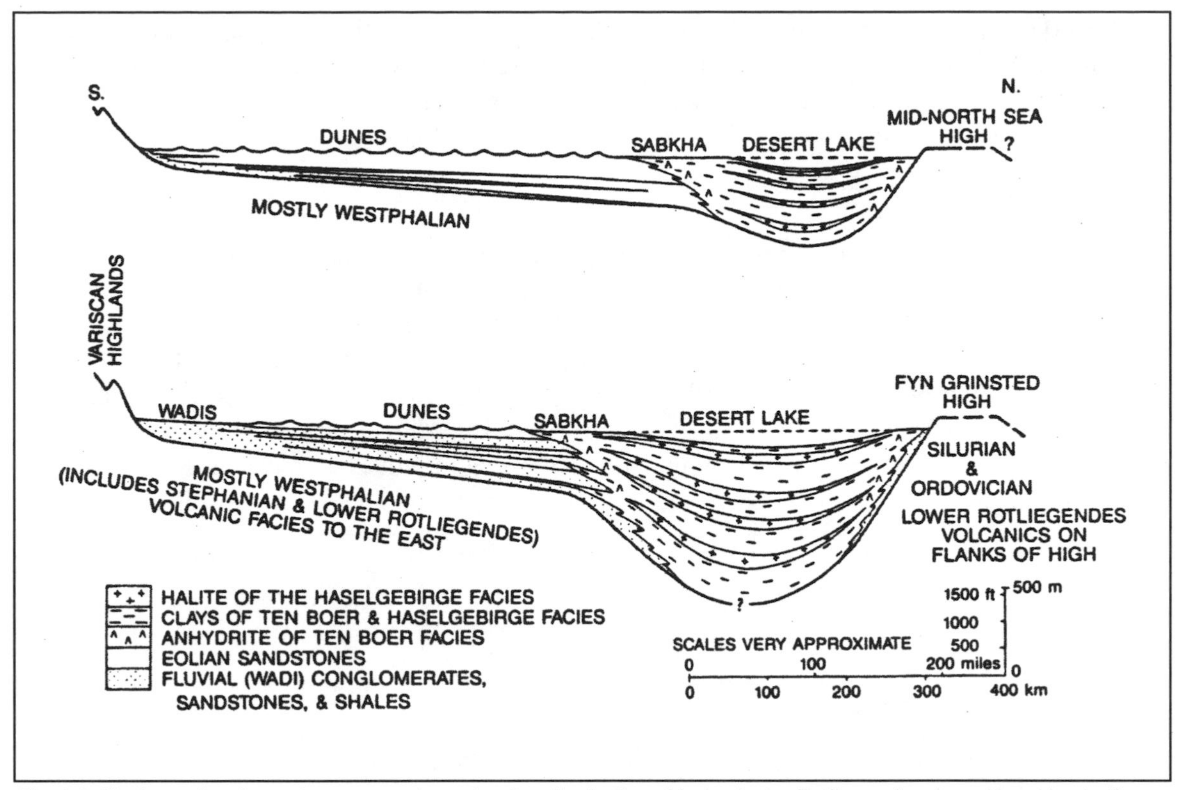

Fig. 1-9. North-south schematic cross sections showing distribution of facies in the Rotliegendes depositional basin (from Glennie[10]). Permission to publish by AAPG.

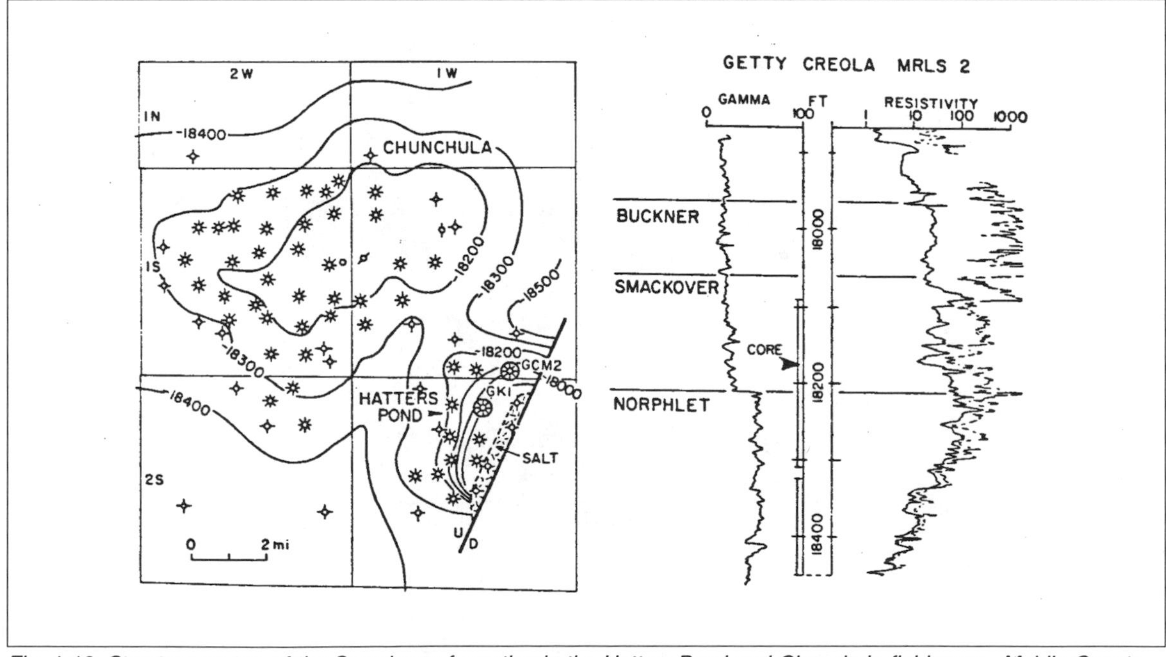

Fig. 1-10. Structure on top of the Smackover formation in the Hatters Pond and Chunchula field areas, Mobile County, Alabama (from Berg[11]). Permission to publish by Prentice-Hall, Inc.

Desert aeolian sands produce reservoirs that have highly variable porosity and permeability, and lamination produces strong anisotropy. Dune sands can attain wide areal extent and are properly called blanket deposits. They make traps for oil and gas by structural, rather than by stratigraphic, closure. But, there are always exceptions.

Alluvial Sands

Reference to Figure 1-6 summarizes the characteristics of alluvial sands. Running water and debris flow provide the elements necessary for the formation of an alluvial fan or on a repeating basis, an alluvial sheet. The result is a linear wedge of coarse detritus running along the foot of a mountain chain. When the contact between the mountains and the plain is an active fault zone, as would often be the case, the wedge can reach phenomenal thicknesses, 10,000 ft or more. Figure 1-11 presents the schematic of a fan surface. Deposits are extremely variable, varying from boulders to clay. Grain size diminishes and sorting improves with distance of transport. If porosity and permeability are preserved, reservoir traps for oil and gas can result.

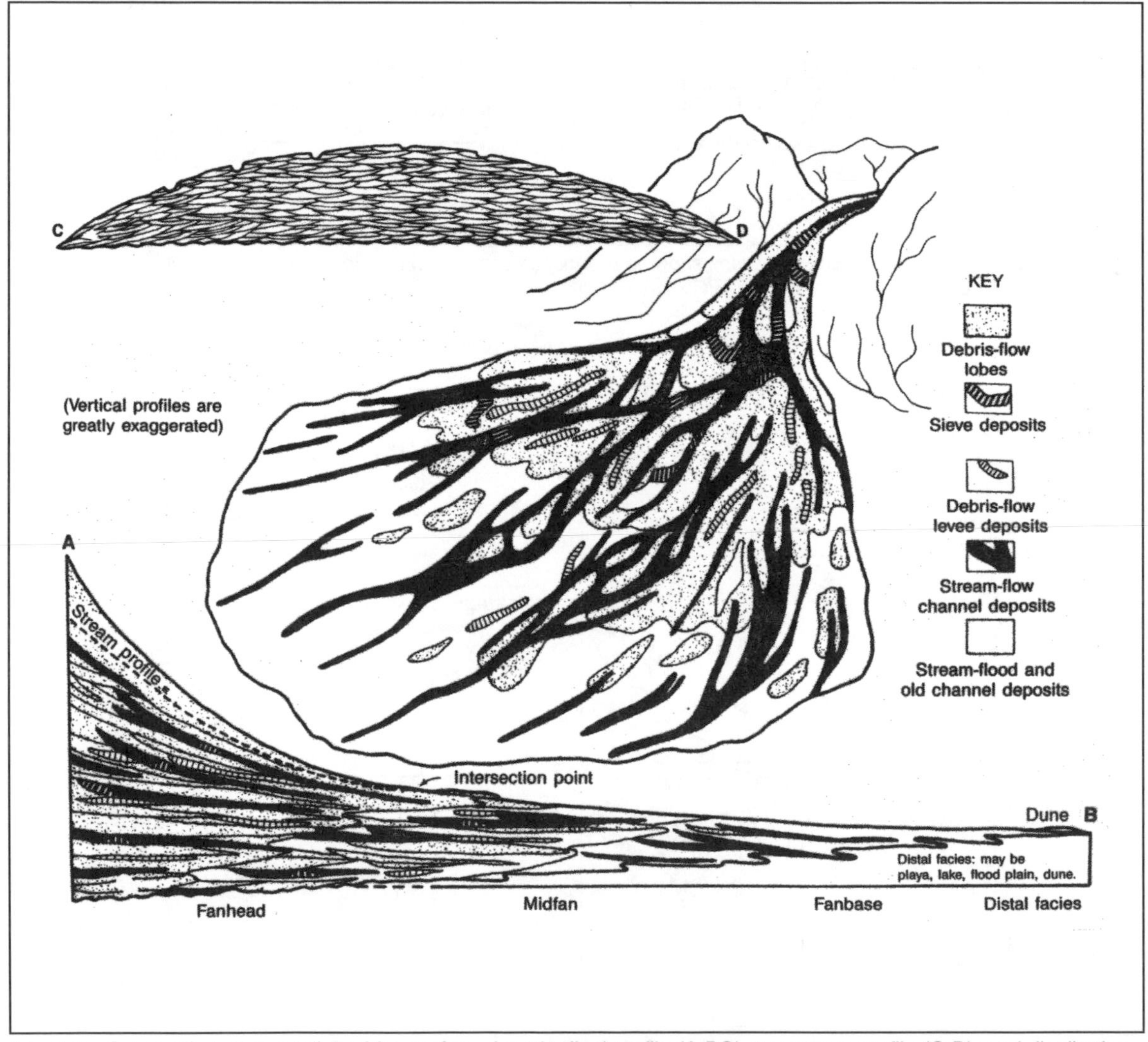

Fig. 1-11. Schematic representation of fan surface. Longitudinal profile (A-BO), transverse profile (C-D), and distribution of sedimentary facies (from Spearing[12]). Courtesy Geological Society of America.

The Quiriquire field of northeastern Venezuela produces from the Quiriquire formation, a Plio-Pleistocene sequence of coarse and fine sediments that unconformably overlies Miocene and Cretaceous rocks, as shown in Figure 1-12. Cores show the reservoir to consist of alternating layers of clay and conglomerate, with lignite and boulder beds.

The fan sediments were derived from the highlands to the northwest. As would be expected, correlations in the field are difficult. Beds are lenticular and show lateral variability. Hydrocarbons (all less than 20°API) are trapped by permeability barriers and by the underlaying unconformity. The field is a stratigraphic trap.

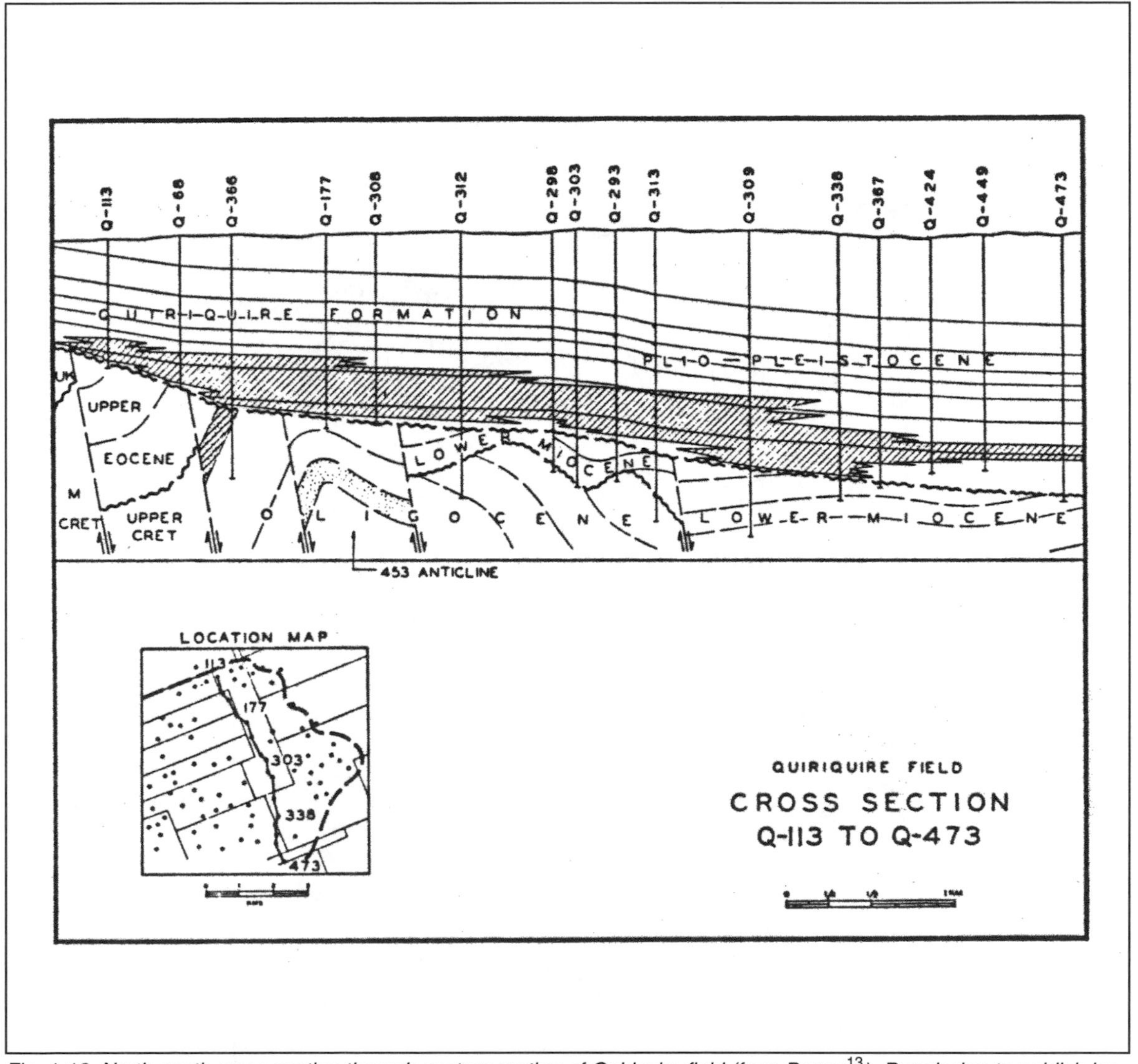

Fig. 1-12. North-south cross section through eastern portion of Quiriquire field (from Borger[13]). Permission to publish by AAPG

Fluviatile Sands

Sands may be deposited from two distinct fluvial regimes—braided and meandering streams or rivers. Braided streams tend to have a higher gradient and a more intermittent discharge, while meandering streams usually deposit finer sediments. While writers have taken great pains to distinguish between braided and meandering streams, the two types may be difficult to identify in wellbores. Figure 1-13 provides a plan view of sediment morphology in a meander system (A) and a braided system (B). In a meandering system, the channel bank is eroded at the outside of the meander loops, and the sands are deposited on the inside as point bars. Channel migration is diagonal and lateral in a downstream direction. Abandonment of the channel may take place when the river cuts through the narrow neck between meander loops. The abandoned channel remains as an oxbow lake. The sequence can repeat many times in the meander belt.

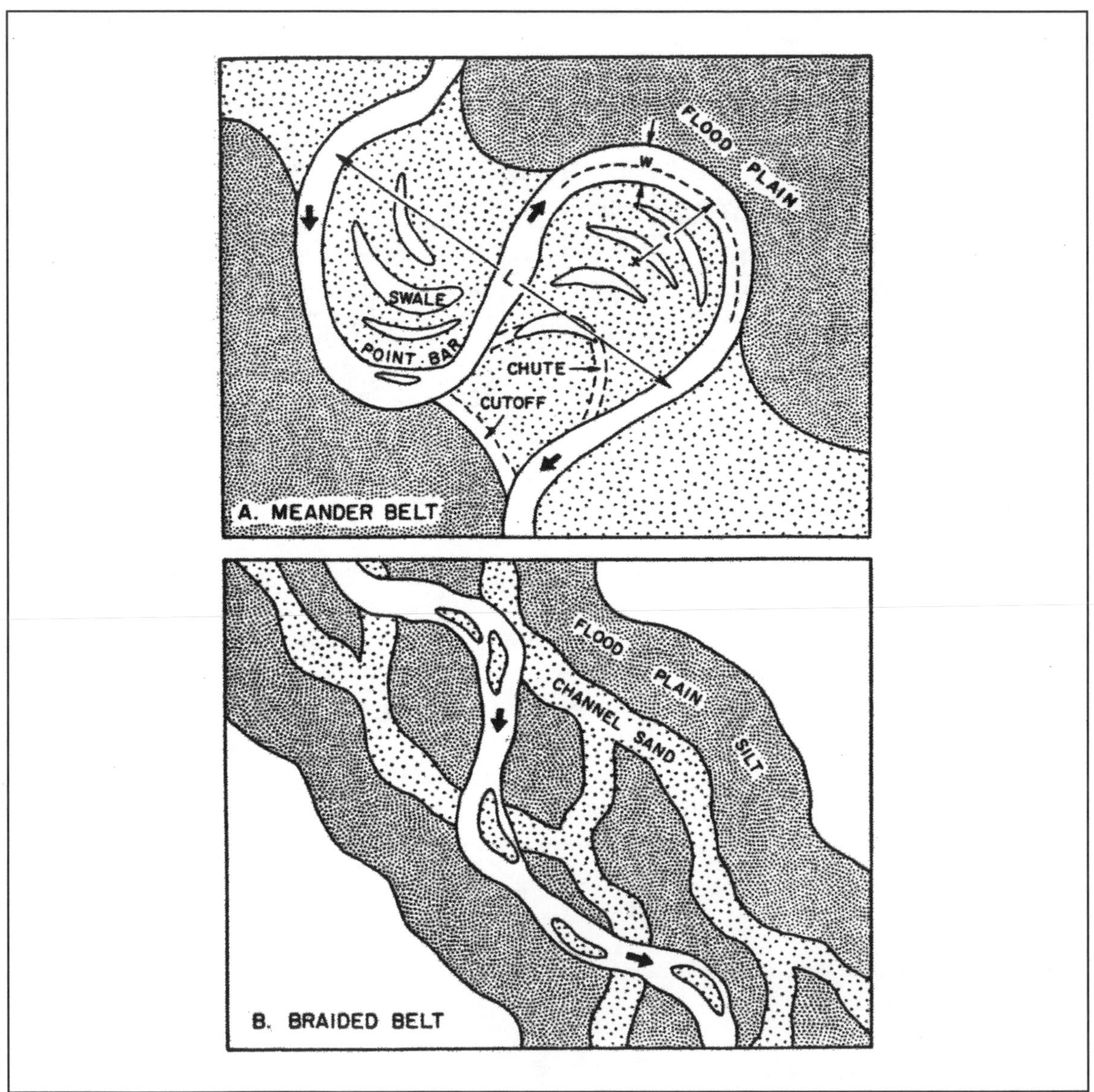

Fig. 1-13. Plan view of sediment morphology in a meandering system (A) and a braided system (B) (from Berg[14]). Permission to publish by Prentice-Hall, Inc.

In a braided system, Figure 1-13B, the sinuosity of the channel is much less. During times of flooding, the channel may scour into previously deposited sediments in a downward rather then a lateral direction. When flow decreases, sand is deposited vertically. Abandonment of the braided channel takes place during decreased flow when the channelway becomes filled with sediment. Figure 1-14 provides a diagrammatic cross section of sand distributions in a meander belt and a braided belt. An isolated well log(s) and core description and analysis will not allow a determination to be made as to whether the reservoir has a braided or meandering channel origin. Both systems produce a fining-upward sequence with a thickness comparable to the river depth during flooding.

The Robinson sandstone of the Fry area, Illinois, is probably the deposit of a single fluvial channel, typical of many thin sandstones of Pennsylvanian age in the Illinois basin. It probably has a nonmarine fluvatile depositional origin as a result of an extensive river that meandered across an alluvial valley or plain.[15] The sandstone is of Pennsylvanian age. Figure 1-15 provides information on Robinson sandstone at the Ohio—L. B. Wampler No. 14. In general, porosity

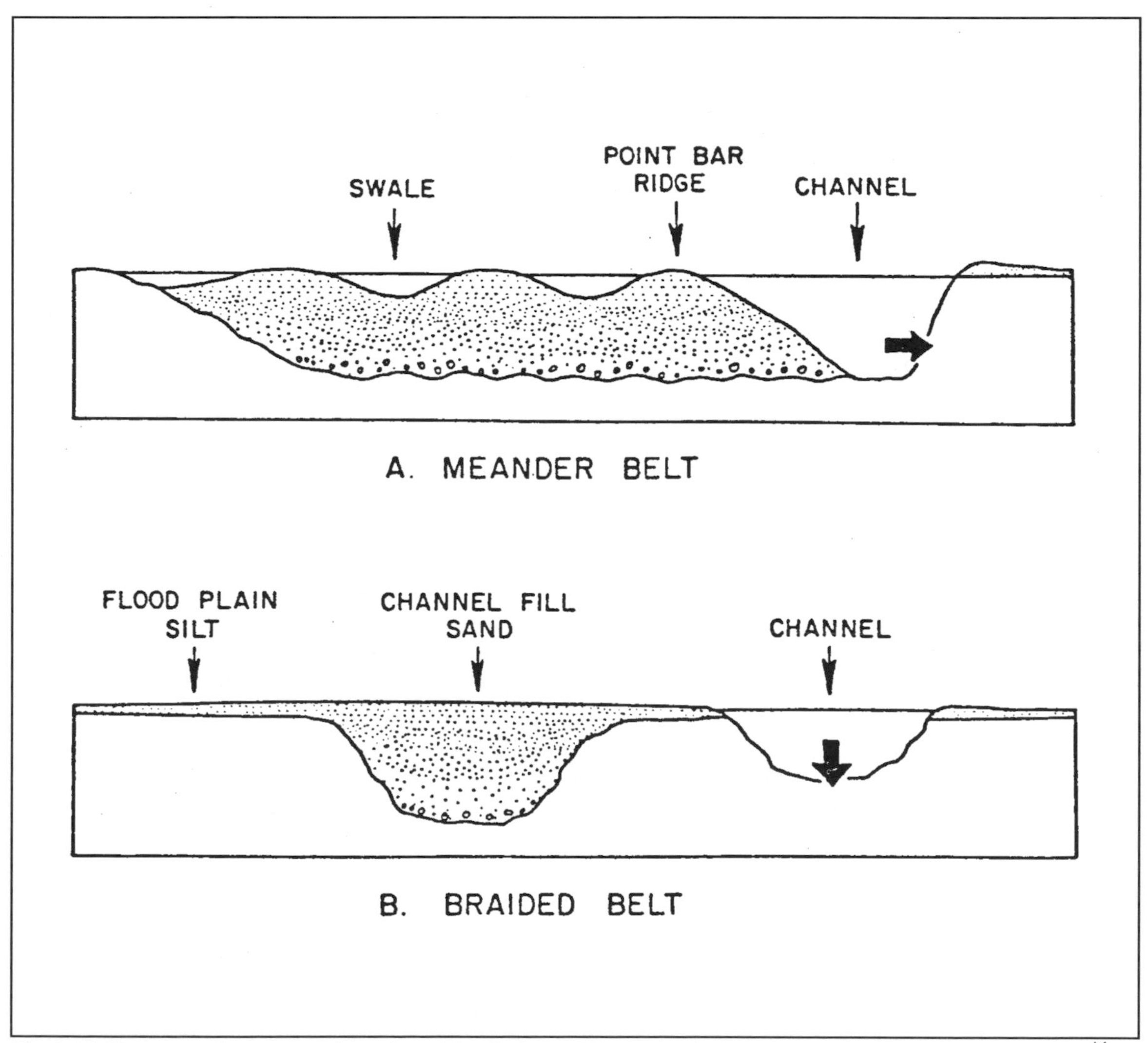

Fig. 1-14. Diagrammatic cross sections of sand distributions within a meander belt (A) and a braided belt B) (from Berg[14]). Permission to publish by Prentice-Hall, Inc.

and permeability decrease from bottom to top for each of the lithologic zones with decreasing depth. An in-situ combustion project, in addition to waterflooding, has been carried out in the Fry area. The reservoir at the Fry combustion site is 12,000 ft long, 3,500 ft wide, zero to 55 ft thick, and trends southwest. The layered nature of the sandstone and its directional properties have controlled performance of the injection and production operations. The depth of the formation for the in-situ combustion test was between 880 and 936 ft, the oil was 28.6°API, and had a viscosity of 40 cp at the reservoir temperature of 65°F. Figure 1-16 provides an isopach map of the Robinson sandstone, Fry area.

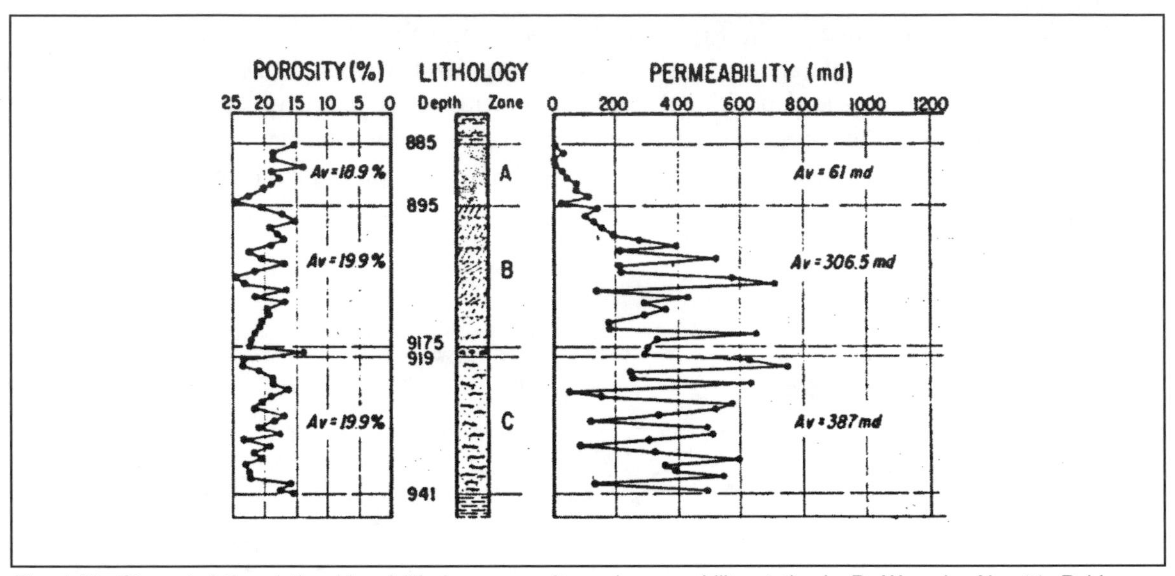

Fig. 1-15. Characteristic relationship of lithology, porosity and permeability at the L. B. Wampler No. 14, Robinson sandstone, Fry Unit, Illinois (from Hewitt and Morgan[15]). Permission to publish by JPT.

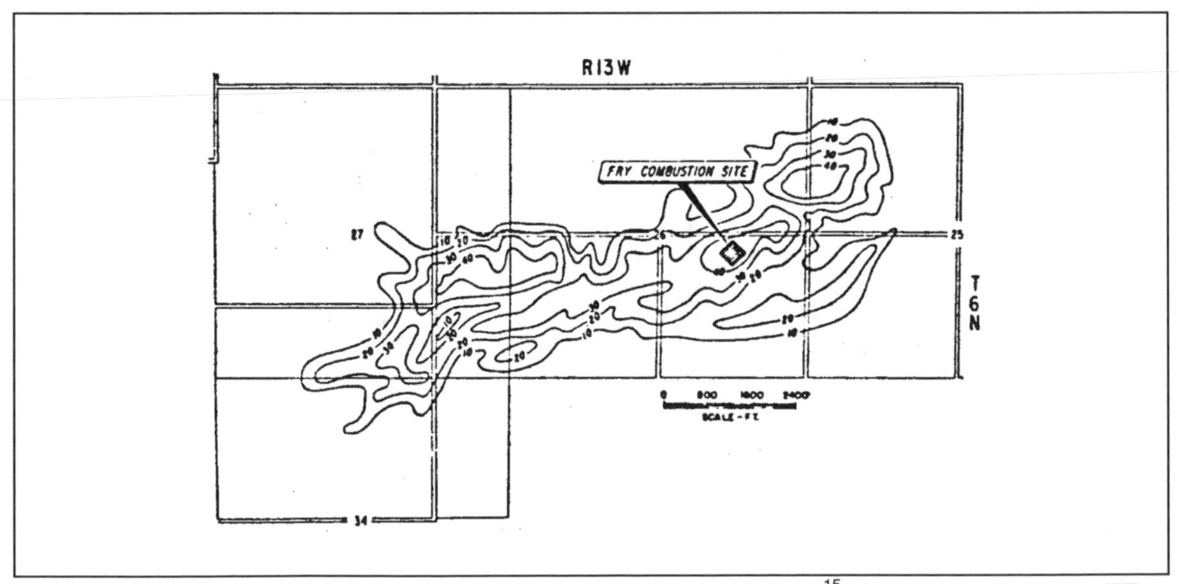

Fig. 1-16. Isopach map of the Robinson sandstone, Fry area (from Hewitt and Morgan[15]). Permission to publish by JPT.

Many other reservoir sandstones have a fluviatile origin—the Bartlesville sandstone (Middle Pennsylvanian) of southeastern Kansas and northeastern Oklahoma; and the Lower Tuscaloosa sandstones (Upper Cretaceous) of southern Mississippi. Prudhoe Bay field on the north slope of Alaska is the most productive oil field in the United States. Production is mainly from braided stream sediments of the Ivishak sandstone in the Sadlerochit group (Permo-Triassic).

Deltaic Sands

Deltas often contain large quantities of oil and gas since the delta building process injects porous reservoir sands far out into marine basins having abundant source beds. Figure 1-17 provides the components of a delta, while Figure 1-6 provides characteristic log profiles, grain size distributions, and the like. A prograding delta-front section is recognized by the vertical sequence that is gradational upward from shallow marine at the base to distributary mouth bars to fluvial-transported sands and thence to nonmarine beds above. Grain size increases upward, with distributary mouth bars and other delta front sands. The sequence may be modified because of the complex shifting depositional conditions and progradation of deltaic plain sediments over the previous delta front. Figure 1-18 presents diagrams showing the development of deltaic facies from studies of the modern Mississippi delta.

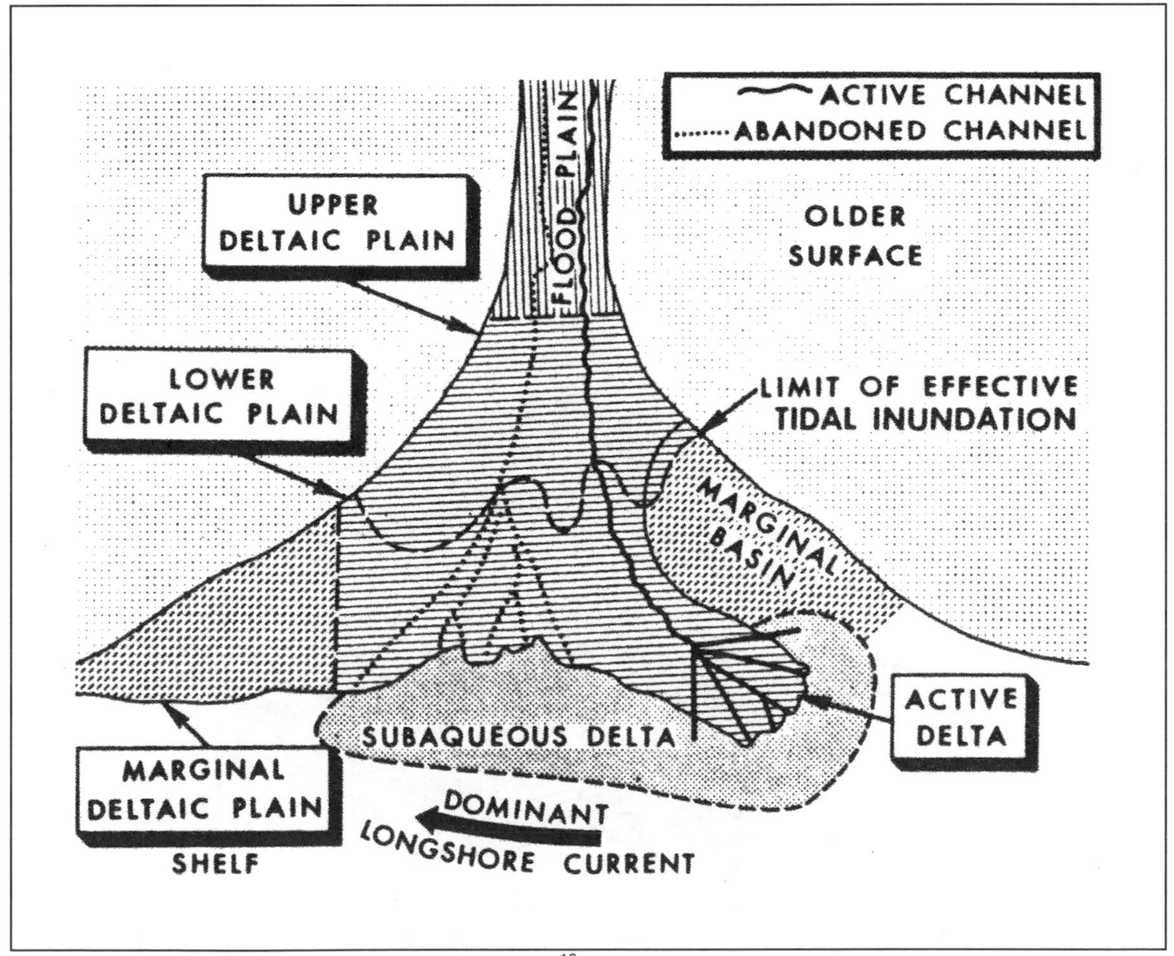

Fig. 1-17. Components of a delta (from Coleman and Prior[16]) Permission to publish by AAPG.

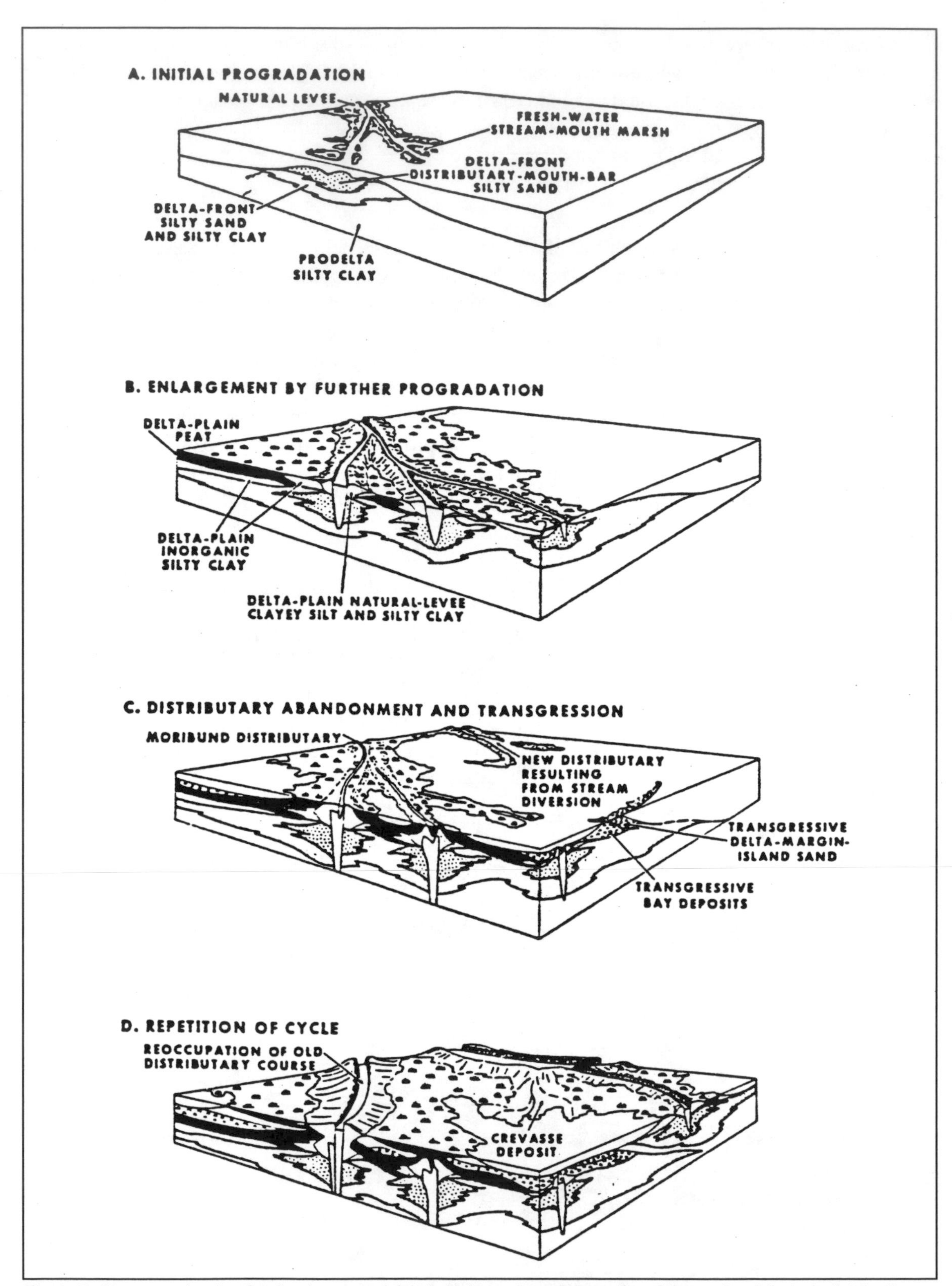

Fig. 1-18. Diagrams showing the development of deltaic facies as based on studies of the modern Mississippi delta (from Frazier[17]). Permission to publish by Gulf Coast Association of Geological Societies.

Ancient delta systems exist in many parts of the world. Berg[18] reports that the Morrowan sections of the Pennsylvanian at the South Empire Field of southeastern New Mexico has at least four different prograding deltaic sequences, with three of these shown in Figure 1-19, as intervals A, B, and C.

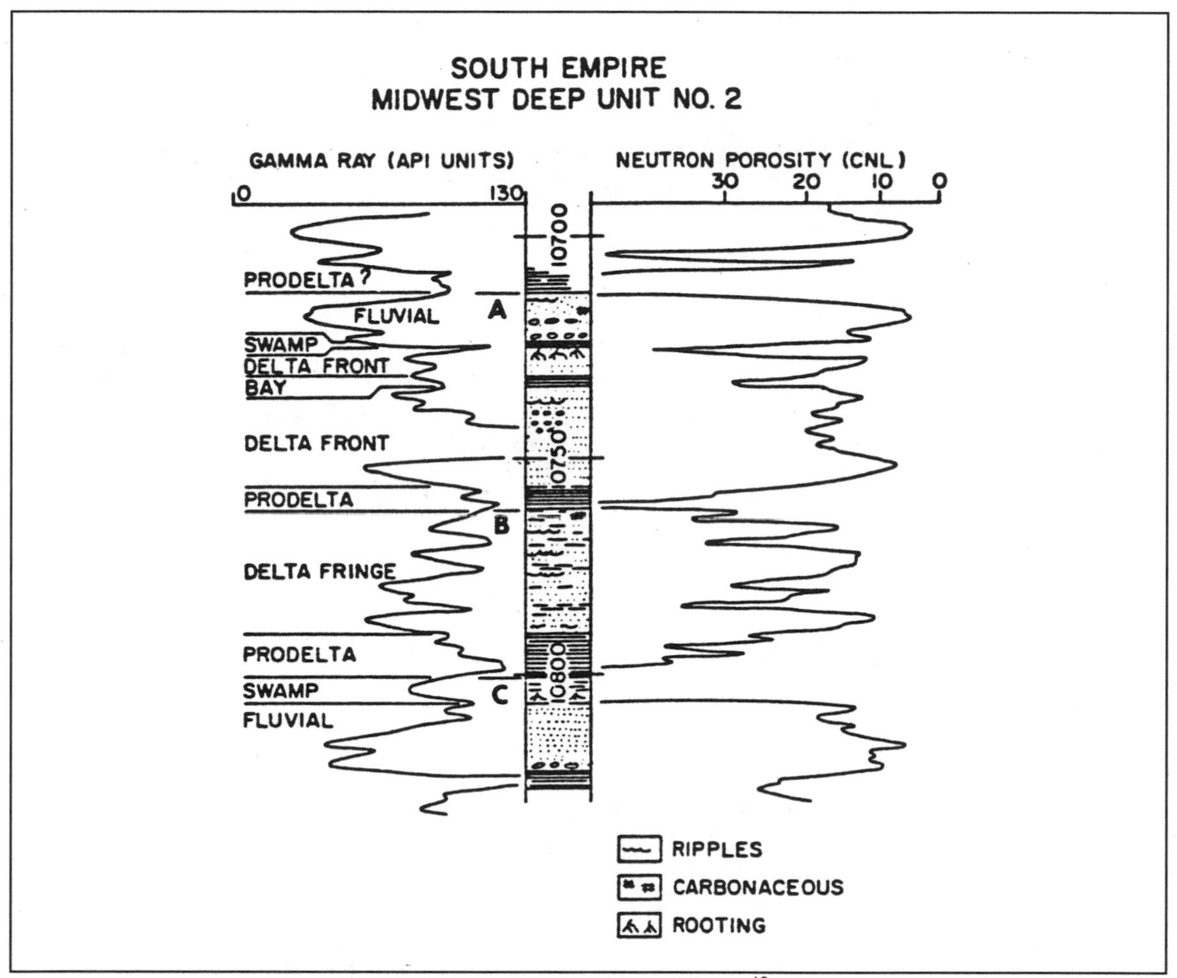

Fig. 1-19. Gamma ray and neutron log South Empire Deep Unit 2 well (from Berg[18]). Permission to publish by Prentice-Hall, Inc.

Figure 1-20 provides an interpretation of the lateral relationships. On the left is a hypothetical cross section and on the right a diagrammatic view of delta lobes.

The Red Fork delta of the McAlester basin of Noble County, Oklahoma, is productive of hydrocarbons from upper deltaic plain distributary channel sands. Scott[19] has provided such an interpretation at the South Ceres field (*see Fig. 1-21*). The productive sandstone interval is Middle Pennsylvanian in age.

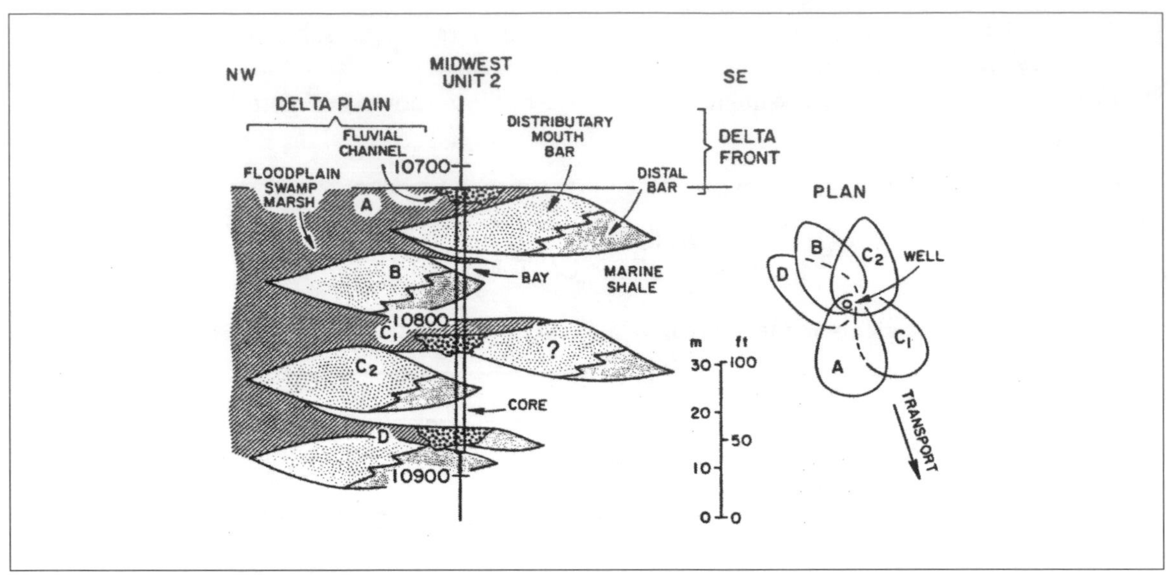

Fig. 1-20. Interpretation of lateral relationships in deltaic facies of the Morrowan sandstone, Midwest Deep Unit 2, South Empire Field (from Berg[18]). Permission to publish by Prentice-Hall, Inc.

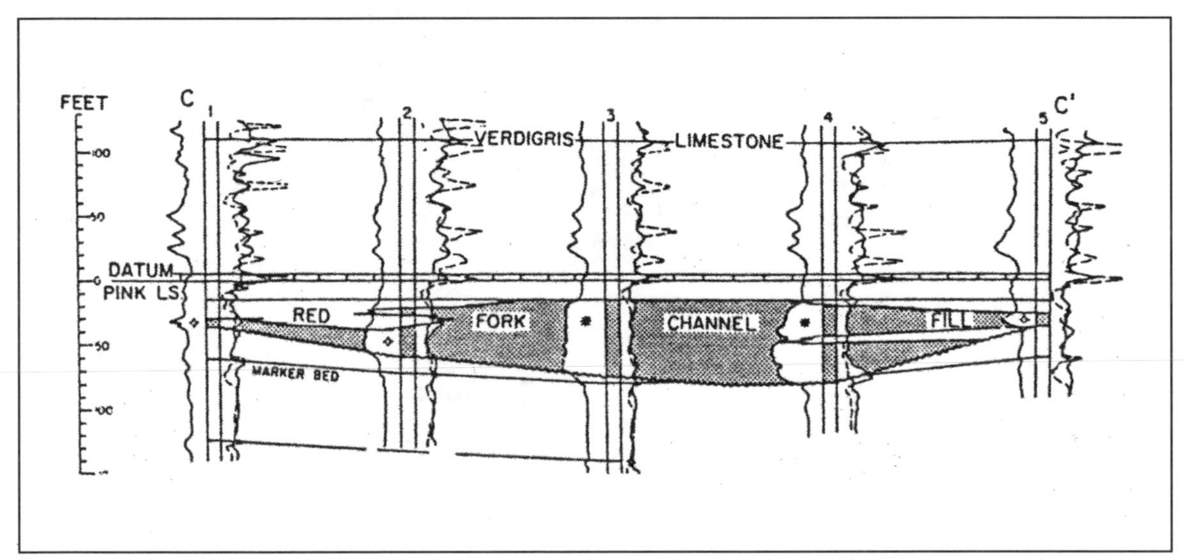

Fig. 1-21. A cross section at South Ceres Pool, Oklahoma, showing the Red Fork channel fill that is predominantly sandstone (from Scott[19]). Permission to publish by University of Oklahoma.

Figure 1-22 provides an isopach map of the Red Fork sandstone. Notice that maximum thickness is 60 ft, maximum width of the channel is one mile, and that length is more than 20 miles. Of interest are the downdip limb to the west which produces 42° API oil and an up-dip limb that yields mainly gas. The drive mechanism is at least partly water drive.

The Brent formation (Middle Jurassic) in the Viking graben area of the North Sea contains sandstone members resulting from a deltaic environment. Oil and gas production has been established at Statfjord, Ninian, Brent, and Heather.

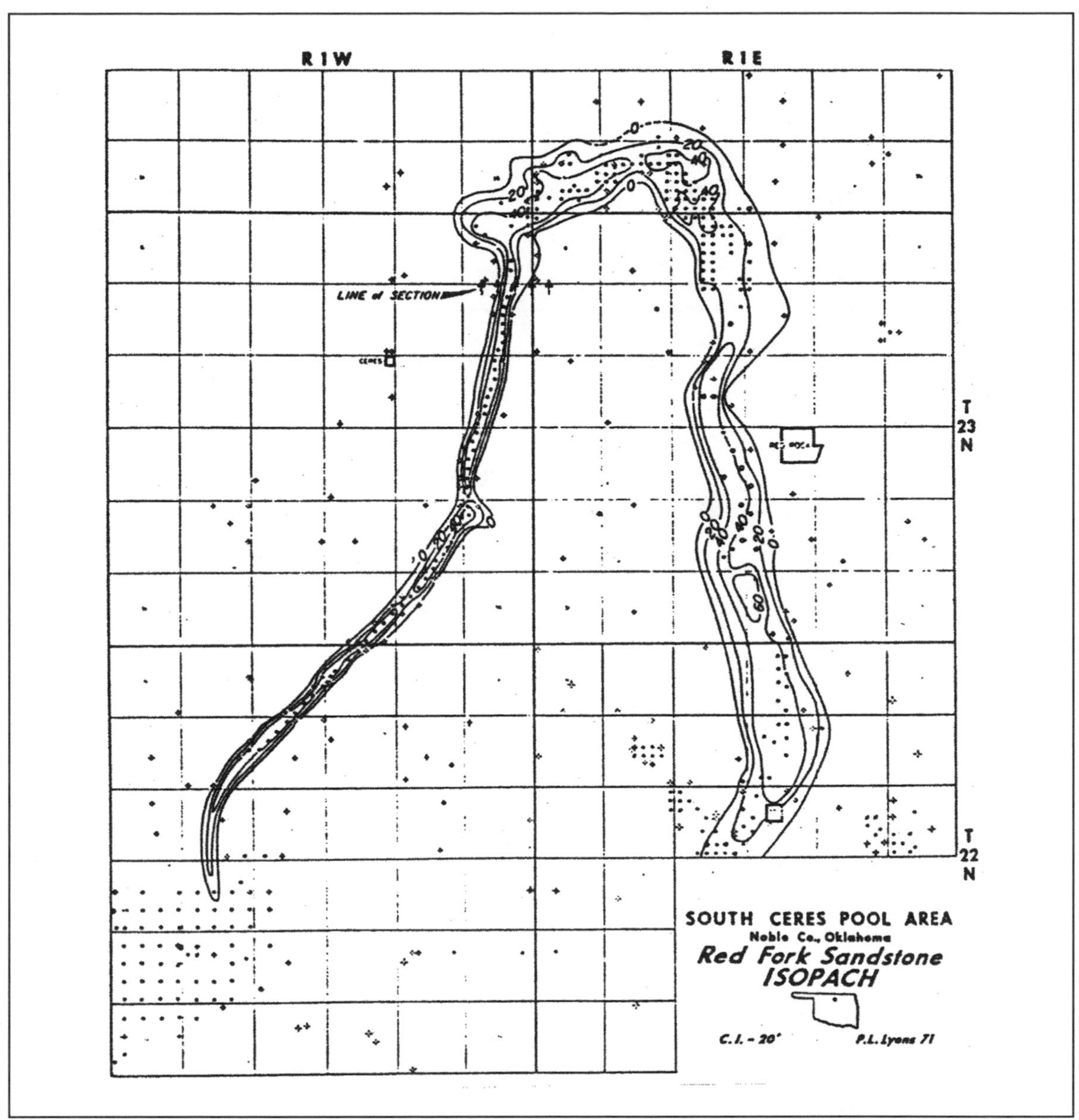

Fig. 1-22. Isopach map of the Middle Pennsylvanian Red Fork sandstone, South Ceres Pool, Oklahoma (from Lyons and Dobrin[20]). Permission to publish by AAPG.

The ancient delta sequence underlaying the modern Niger Delta provides much of Nigeria's current oil and gas production from Cretaceous-age rocks.

Beaches, Barriers And Bars

Waves approaching a shoreline buildup, spill, then break toward the shoreline, the energy of which can transport large amounts of sand to the shoreface. The wave action also sets up a littoral current which can transport large amounts of sand long distances parallel to the coastline. Figure 1-23 provides the characteristics of waves for steeply sloping and more gently sloping beach faces.

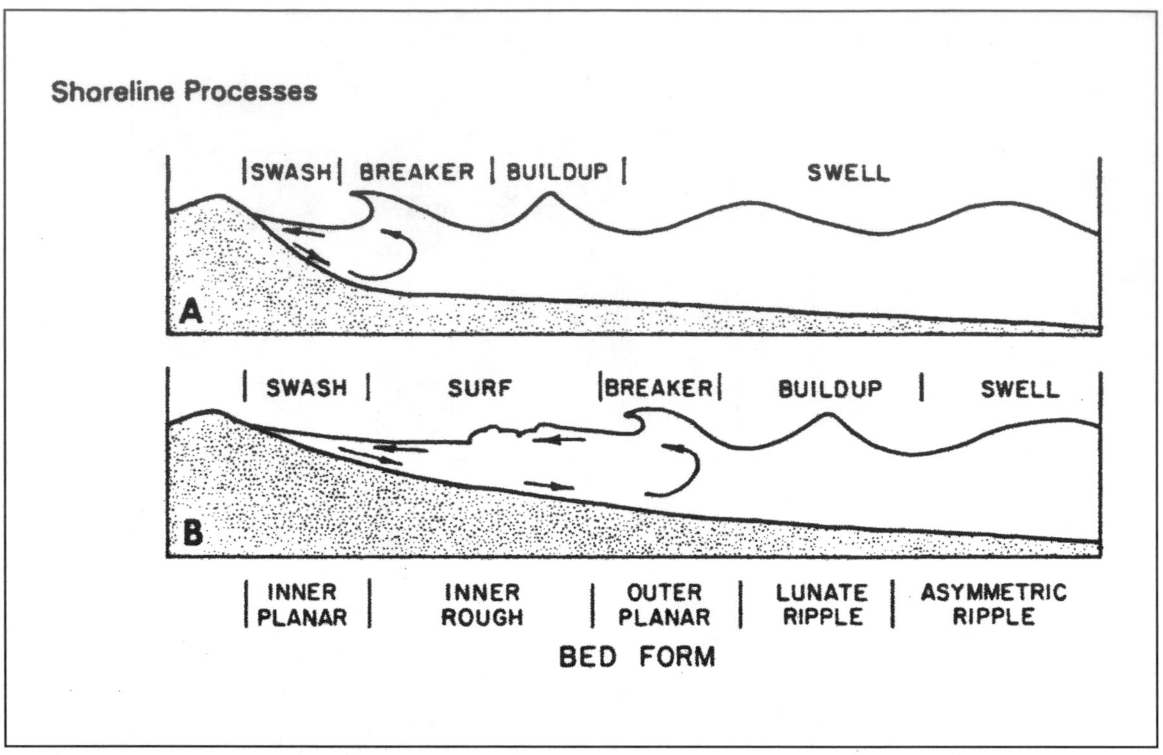

Fig. 1-23. Characteristics of waves and bed forms in the near shore area: (A) steeply sloping beach face; and (B) gently sloping beach face (from Berg[21]). Permission to publish by Prentice-Hall, Inc.

Two types of sandy coastlines can be found: those with barrier islands separated from the shore by a lagoon or bay; and those with beach ridges attached to the shoreline. In the barrier island coast, sands are concentrated in narrow trends that can parallel the coast for many miles. Beach ridge types can have a succession of beach sands forming a broad sandy flat.

Figure 1-24 presents cross sections through prominent barrier-bar sands of the Texas gulf coast—Galveston Island and Padre Island. Figure 1-6 provides the reservoir sand types and characteristic log profiles and the like which would be expected in an ancient barrier bar that could be hydrocarbon productive. The SP curve tends to be funnel-shaped, representing coarser sands to the top. Sandstone bodies are quartzose and can have good porosity and permeability if not extensively altered by diagenesis and cementation.

Sabins[22] has interpreted the Bisti oil field in north-central New Mexico as a barrier-bar complex having three overlapping sandstone bodies—this in the Upper Cretaceous Gallup formation. An isopach map is provided in Figure 1-25. The barrier-bar sands are funnel shaped on the electric logs.

Other examples of barrier bars containing oil and gas are: the Sallyards Trend oil fields of Butler and Greenwood Counties, Kansas (middle Pennsylvanian sandstones, 40 - 100 ft thick, 60 mi long and 1.5 mi wide); Watika Trend of the Anadarko Basin, Oklahoma, oil and gas fields (Pennsylvanian Red Fork sandstones); and the Bell Creek Oil Field, Powder River Basin, southeast Montana and northwest Wyoming (composite bar complex of linear sandstone lenses, Muddy sandstone of the Lower Cretaceous).

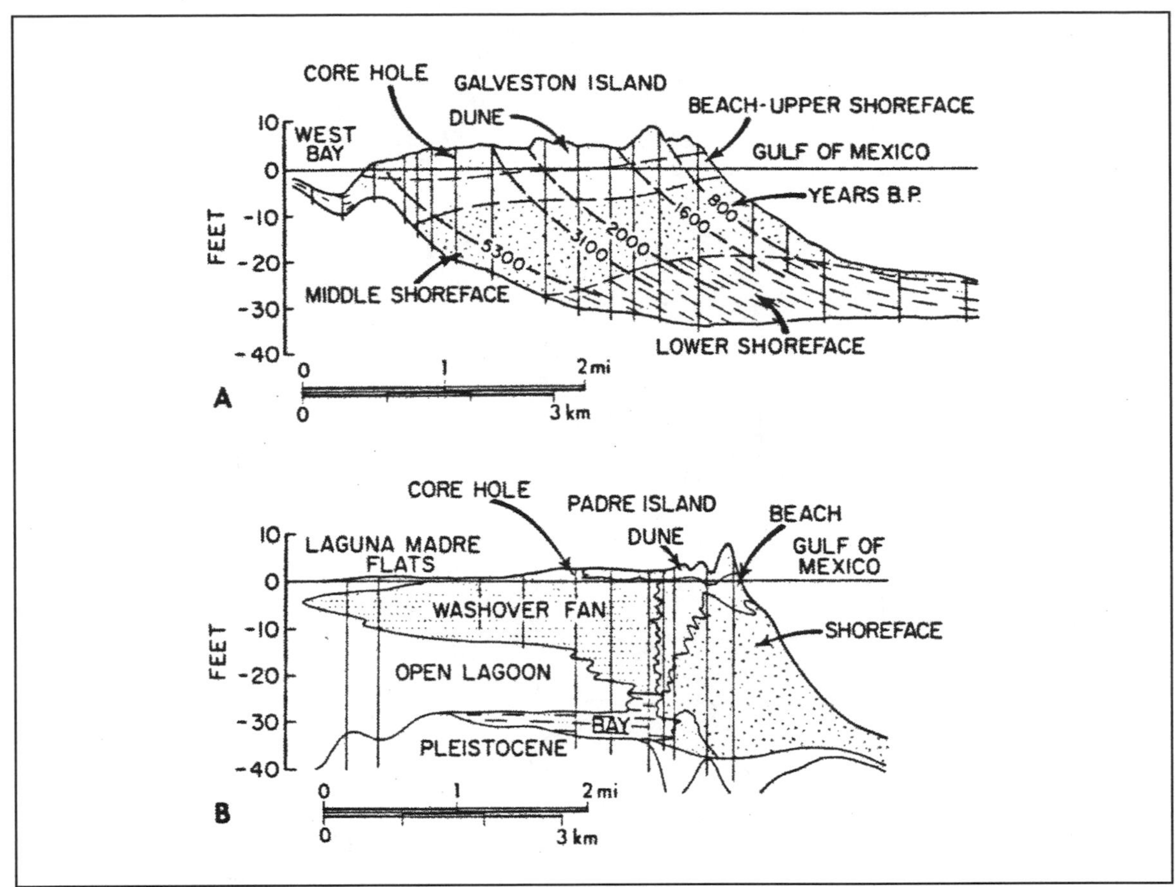

Fig. 1-24. Cross sections through barrier islands of the Texas gulf coast: (A) Galveston Island; and (B) Padre Island (from Berg[21]). Permission to publish by Prentice-Hall, Inc.

Figure 1-26 presents an interpretation of depositional environments of reservoir sandstones in the Muddy formation of the northeast Powder River basin.

The Piper oil field, with recoverable oil reserves of 650 to 900 million STB is an example of a northern North Sea Jurassic shoreline reservoir. The gross sandstone thickness varies from 35 to 100 m and is made up of transgressive and regressive units ranging from 2 m to 30 m thick. Porosities average 23% and permeabilities average 1.5 darcies.

Shelf Sands

In many parts of the world, the shelf is a mud environment, but sand can accumulate as sand waves or ridges. The process of wave motion, tidal currents, storm currents, and turbidity currents can result in important reservoirs for oil and gas in ancient shelf sands. Berg[24] reports that shelf sandstones are of three general types: the shoal mound; the shelf-terrace sand; and the deeper-water current sands. Figure 1-27 provides the relative locations and morphologies of such systems.

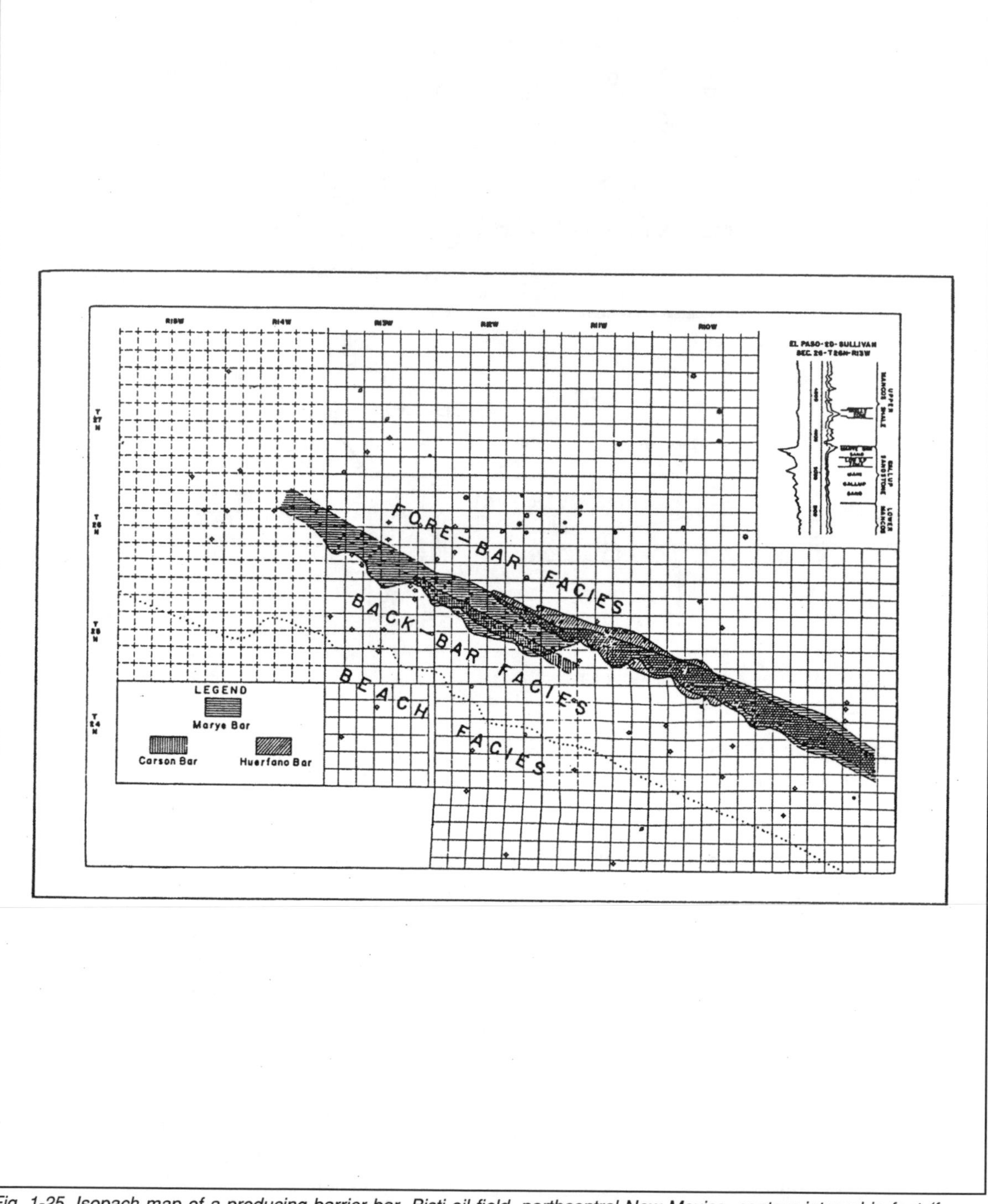

Fig. 1-25. Isopach map of a producing barrier bar, Bisti oil field, northcentral New Mexico, contour interval in feet (from Conybeare[23]). Permission to publish by Elsevier Scientific Publishing Company.

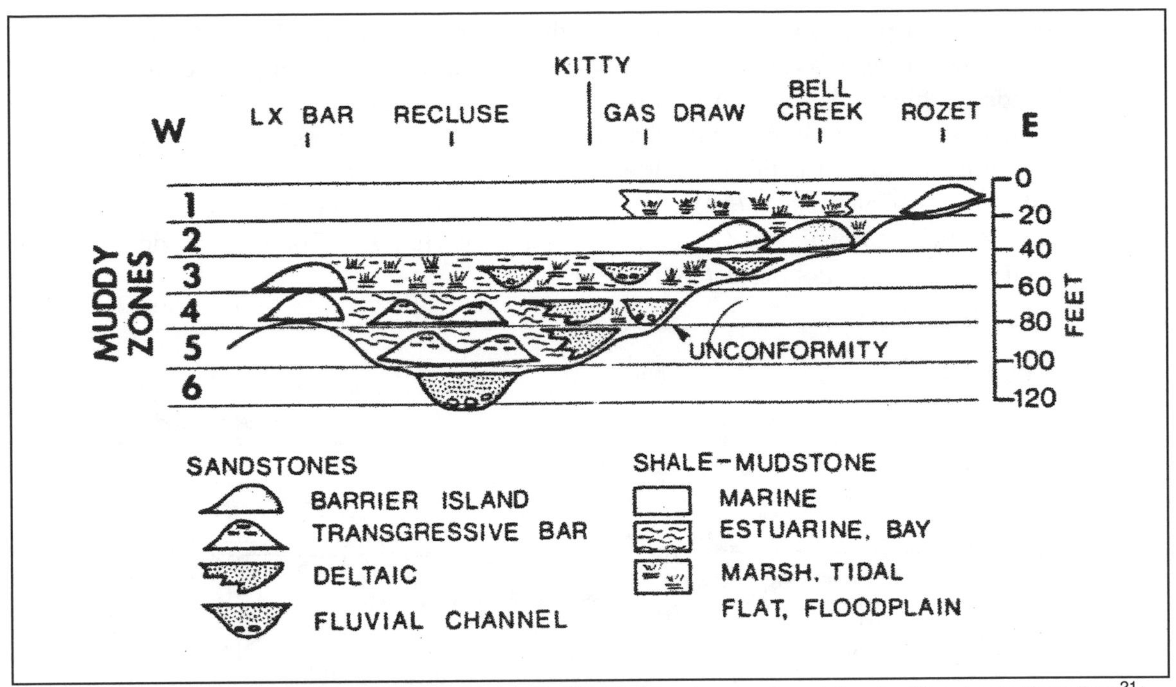

Fig. 1-26. Interpretation of depositional environments in the Muddy formation, northeast Powder River basin (from Berg[21]). Permission to publish by Prentice-Hall, Inc.

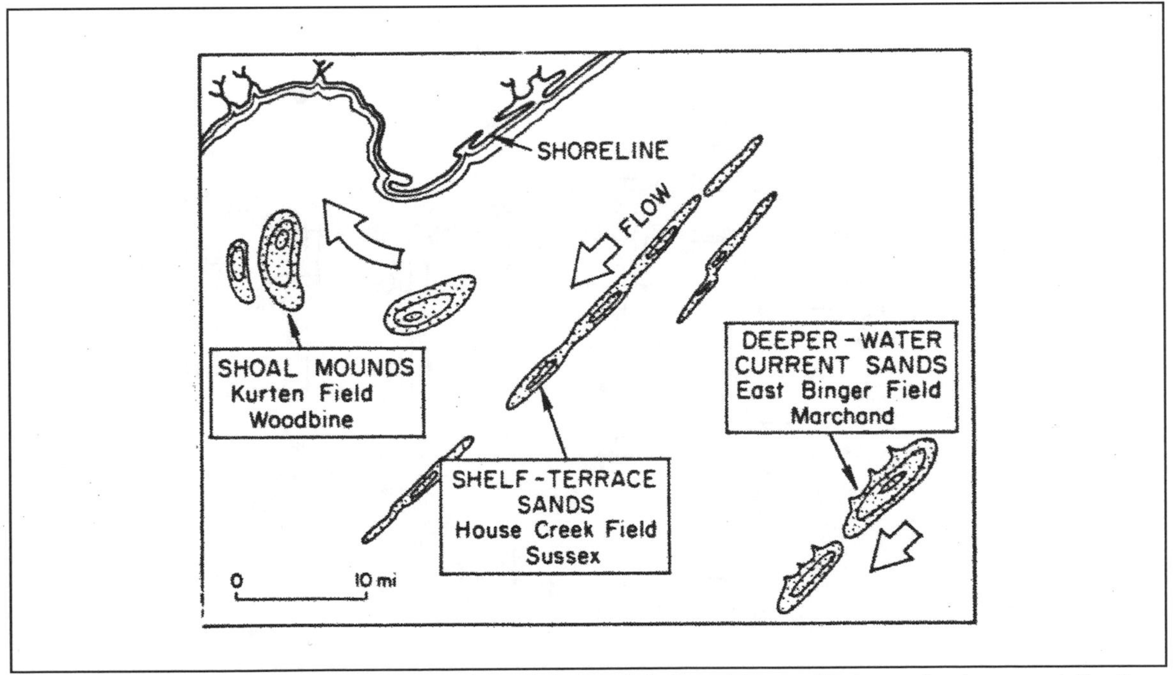

Fig. 1-27. Diagram showing relative locations and morphologies of shelf sandstones with arrows showing current direction (from Berg[24]). Permission to publish by Prentice-Hall, Inc.

The Kurten Field produces oil from the Woodbine sandstone of the Upper Cretaceous. The sandstones were formed as shoal sands on a shallow marine shelf. Somewhat characteristic of shoal sands are thin ripple lenses of sand and shale with evidence of bioturbation giving a churned appearance to the reservoir rock. Porosity and permeability can be poor to moderately good in such systems. In the case of the Kurten Field, northeast of College Station, Texas, low permeability is offset by size (19,000 acres in area). Figure 1-28 presents north-south and east-west sections through the oil-productive interval. Gulf Oil Company is doing a CO_2 improved oil recovery pilot project in the field.

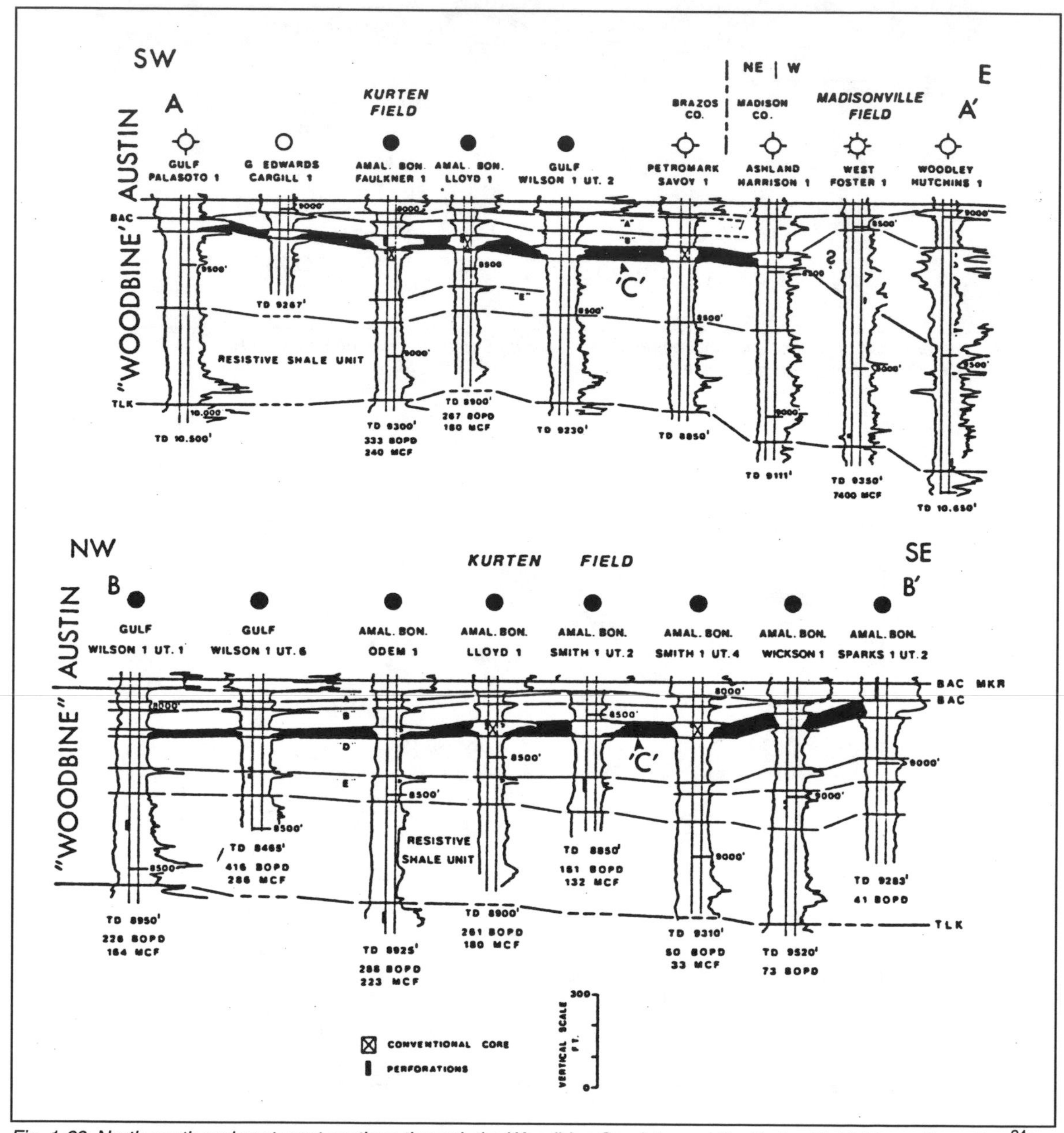

Fig. 1-28. North-south and east-west sections through the Woodbine Sandstone at the Kurten Field, Texas (from Berg[24]). Permission to publish by Prentice-Hall, Inc.

The Sussex Sandstone (Upper Cretaceous) oil production of the Triangle "U" (Anderman[25]) and the House Creek fields of the central Powder River basin, Wyoming, were probably deposited in middle- to outer-shelf locations by currents increasing in velocity. The sandstones are found as isolated lenses within thick shales of the Cody and Lewis formations, and could be thought of as shelf-terrace sands.

Many shelf sandstones can be thought of as shaly sand reservoirs because of intermittent current flow. However, the wide areal extent of some bodies and enclosure in thick organic shale source rocks allow conditions for substantial accumulations of oil and gas.

Deep-Sea Sands

Deep-sea or deep-basin sands were deposits of turbidity flows in which sand is supplied through channels and canyons to prograding deltaic slope by a mechanism of slumping, sliding, and turbidity currents. Figure 1-29 shows the distribution of sediment facies at different locations in a submarine fan. Figure 1-6 illustrates the common reservoir sand types, with characteristic log profiles, and the like. Many sands in basinal areas are the result of turbidity flows, and commonly form reservoirs for oil and gas.

The Val Verde basin on the southeastern edge of the larger Permian basin of West Texas is host to gas-productive sandstones in the Canyon section of the late Pennsylvanian and Early Permian.

Figure 1-30 presents a diagrammatic cross section in the vicinity of the Sonora field. Sediment was carried across the shelf and thick sections of shale with interbedded shales were deposited as prograding slopes.

Canyon sands have produced gas for many years in West Texas, but at moderate rates. Increased demand for gas in 1974 stimulated exploration resulting in production along a trend between the already existing Ozona and Sonora fields. The productivity of the wells is determined by the sand/shale ratios, and this determined by the position of individual wells in the submarine fan. Figure 1-31 shows the thickness of sandstone present in the Ozona and Sonora fields. Reserves calculations can be subject to large error due to the discontinuous nature of the sand lenses and the variable sand/shale ratios in the productive intervals. Also, wells can have low productivity.

Parker[29] has identified the reservoir sands in the Forties and Montrose fields of the North Sea as turbidite sandstones. Sand thicknesses up to 60 m have been found. The productive interval is in the upper part of the Paleocene. In the Frigg field, gas is produced from Frigg sandstones of the Lower Eocene. Producing deep marine sands are found in many of the Tertiary California oil and gas fields. Turbidites are the most important California reservoirs.

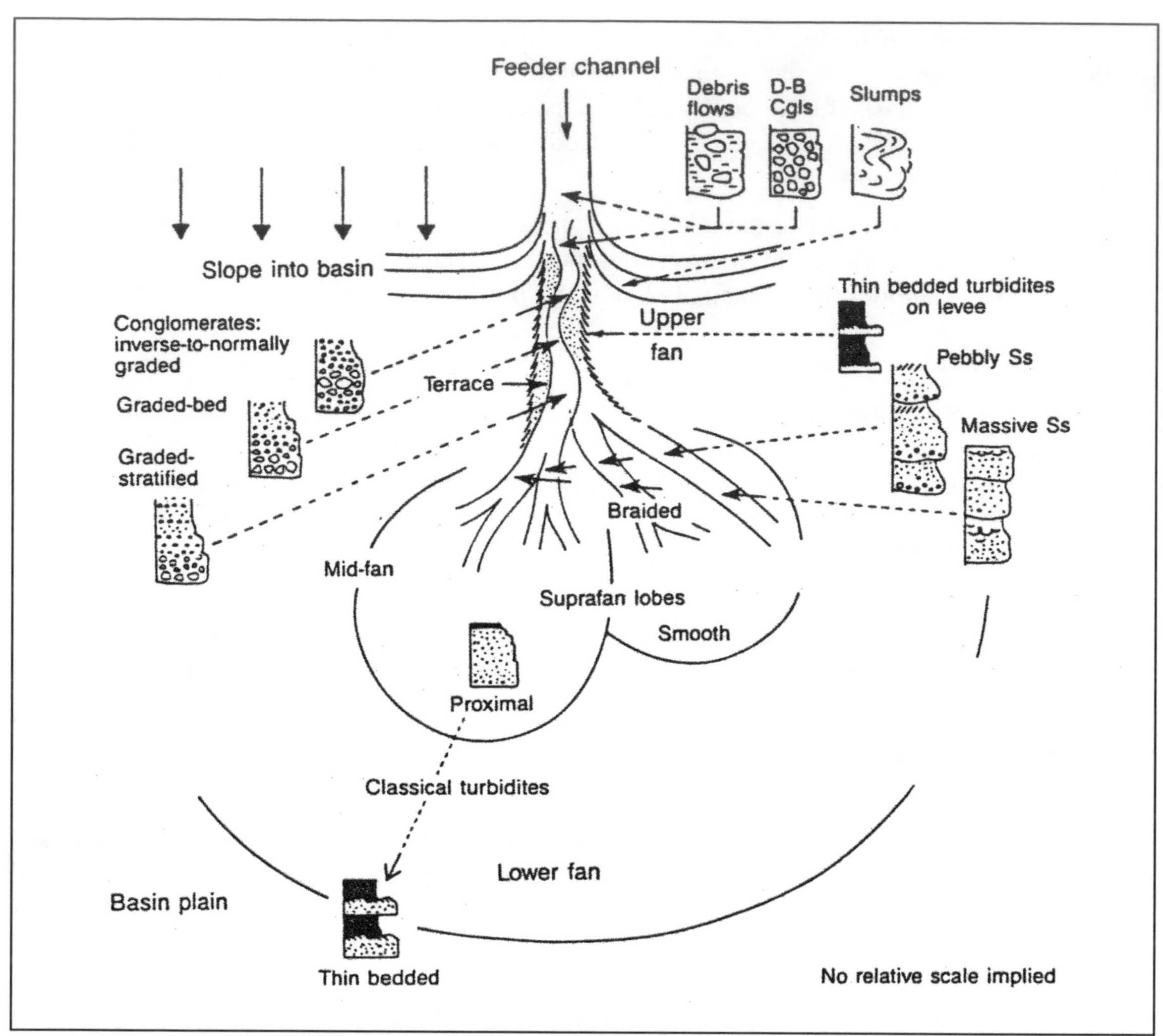

Fig. 1-29. Distribution of sediment facies in a submarine fan (from Walker[26]). Permission to publish by AAPG.

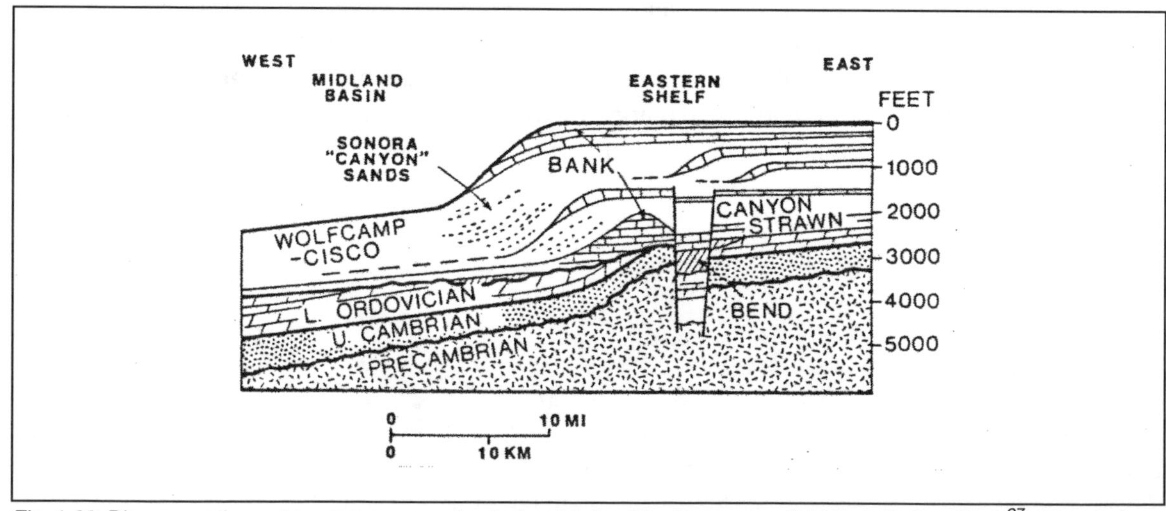

Fig. 1-30. Diagrammatic west-to-east cross section in the vicinity of the Sonora gas field (from Rall and Rall[27]). Permission to publish by AAPG.

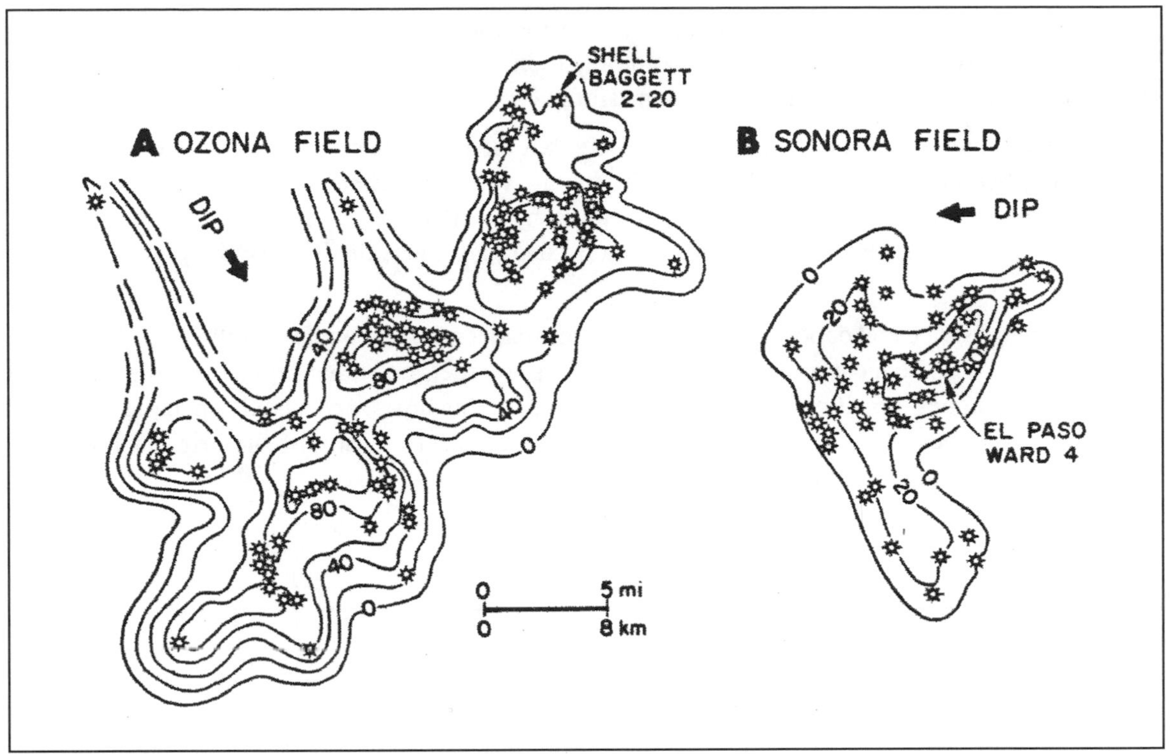

Fig. 1-31. Net thickness of Canyon reservoir sandstones of the Ozona and Sonora fields, Crockett and Sutton counties, respectively, West Texas (from Berg[28]). Permission to publish by Prentice-Hall, Inc.

CARBONATES AS RESERVOIR ROCKS

About one-third of the oil and gas reserves in the world's giant fields occur in carbonate reservoirs. Although averaging only about one-half of the effective porosity of sandstones, carbonate reservoirs often offset this large disadvantage by thicker productive intervals and, on occasion, higher permeabilities.

I. DEPOSITIONAL FACTORS

What is wanted in oil and gas reservoirs is common in both sandstone and carbonate reservoirs—high porosity and permeability, lower water saturations, and large thicknesses. Sufficient productivity and project life must exist to make exploration, development and operational expenses worthwhile. Depending on resource location, taxes, and product price, less than ideal reservoirs can be exploited.

II. INFLUENCE OF ENVIRONMENT

During quiescent periods of orogeny, shallow continental shelves protected from the influx of terrigenous sediments formed vast platform areas that provided an opportunity for the development of abundant marine life. Many of the organisms living in that environment have hard parts (skeletons) of calcium carbonate, most frequently of the unstable variety, aragonite. When the organisms die, the accumulation of their skeletons forms the original limestone deposits. Later on, diagenetic processes during burial change the aragonite into the more stable calcite. Also, some of these limestones are affected by the process of dolomitization that results from the replacement of calcite by the double carbonate of calcium and magnesium.

There are many types of depositional porosity, but the main ones are intergranular and growth framework. Secondary chemical leaching producing vugs and channels, coupled with vertical fractures caused by bending stresses from tectonic movements, often enhance the original rock properties. On the other hand, circulating solutions also may cause the infill of the pores by sparry calcite.

The general shape and depositional settings of the main carbonate bodies are illustrated in Figure 1-32. Among them, the most promising reservoir rocks are: reefs and other organic buildups; bedded platform grainstones resulting from the lithification of the lime sand shoals; pelagic (basinal) limestones; and secondary dolomites, which may be derived from any of the original limestone bodies. A brief discussion of each reservoir type follows.

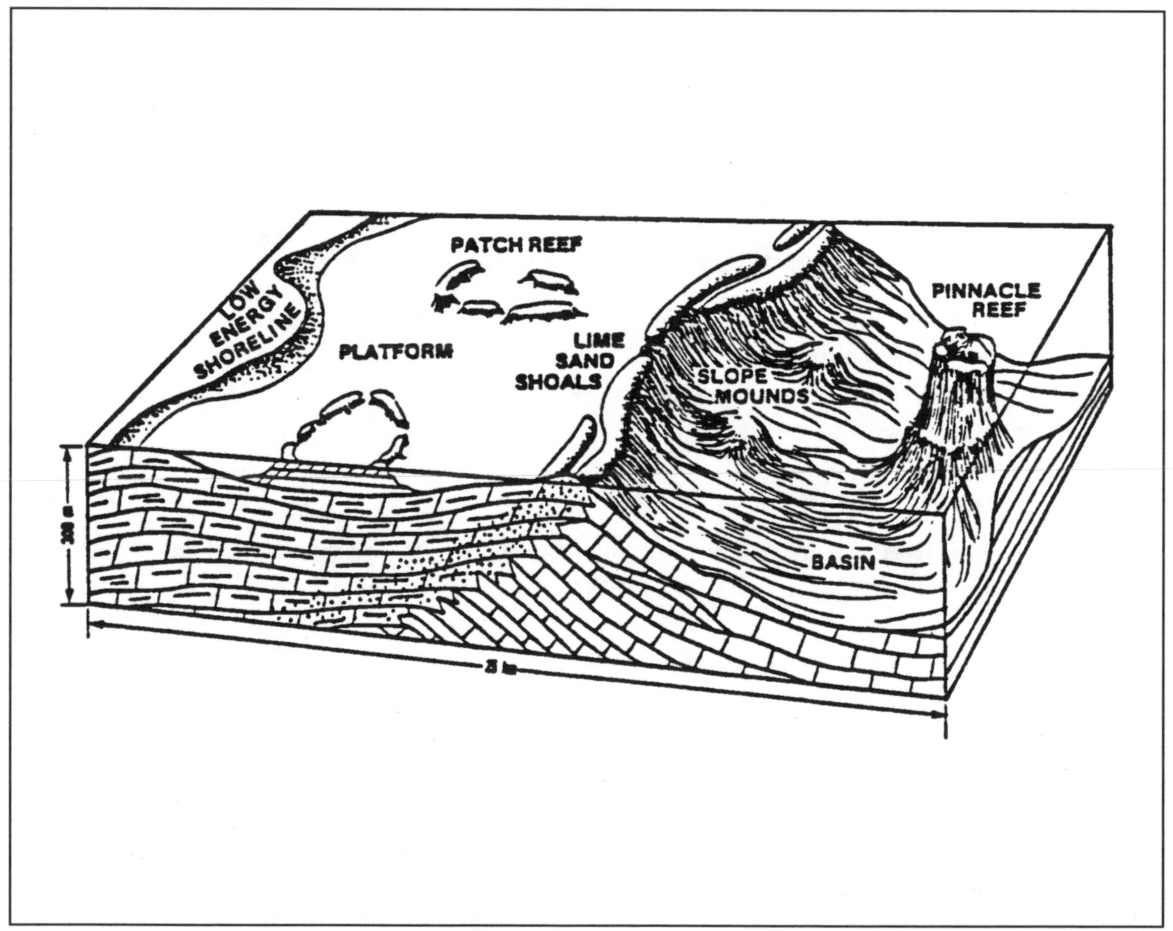

Fig. 1-32. Block diagram illustrating the configuration of carbonate bodies in various depositional settings. Sheet-like deposits of lime mudstone occur on the platform and in the basin. Linear or ribbon-like bodies such as barrier reefs or lime sand shoals are found at the platform margin. Patch reefs, mounds and pinnacle reefs with a circular outline may occur on the platform top or slope ofr in the basin.

III. REEFS AND OTHER ORGANIC BUILDUPS

A true reef consists of a calcareous framework secreted by sea dwelling organisms. If the deposit is an accumulation of dead organisms without a supporting framework, it is a reef-mound. Terms like bioherm and biostrome are not very precise; they only imply shape and can be applied to both types of reefs. Another frequently used term is "carbonate bank," which describes reef-mounds. For example, the Aneth field in Utah is formed by an extensive phylloid algal mound or bank. On the other hand, the D-3 chain of carbonate buildups of Alberta, Canada, consist mainly of reefs.

A well-studied example of bioherm is the Wizard Lake D-3A pool in Alberta, Canada. It is a dolomitized reef of Devonian age and has a maximum, original oil zone of 648 ft (197.5 m) above the Cooking Lake aquifer. Figure 1-33 shows that it is a member of a chain of Leduc D-3 reef pools.

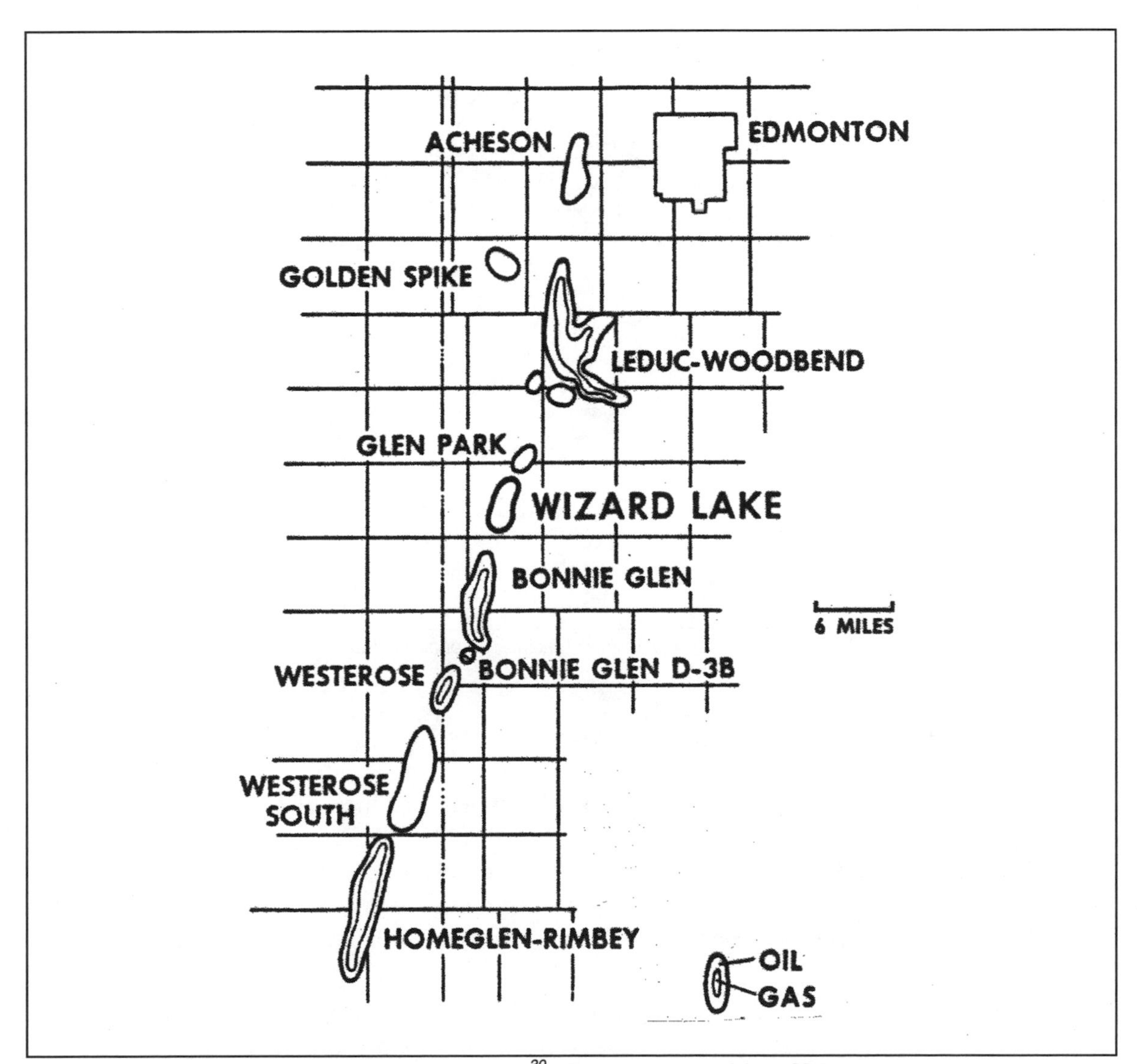

Fig. 1-33. Leduc D-3 reef pools (from Backmeyer, et al., [30]). Permission to publish by SPE.

Figure 1-34 provides a profile of the Rimbey-Acheson reef trend. The ocean side was to the west, and the source of the hydrocarbons was the rich Ireton shale which encases the reefs. Wizard Lake had an original oil-in-place of 390.3 MMSTB, an average porosity of 9.55%, and an arithmetic average horizontal permeability of 1.375 darcies, with a 36°API oil gravity. The pool was produced under primary depletion from 1951 to 1969 (a combination gas expansion, waterdrive and gravity segregation) and evidenced a 66% displacement efficiency. The miscible hydrocarbon process should raise recovery to 88.6%. The operator[30] plans to miscibly displace the residual oil bypassed by early water encroachment, which when completed is projected to increase oil recovery to 95.9% of the OOIP.

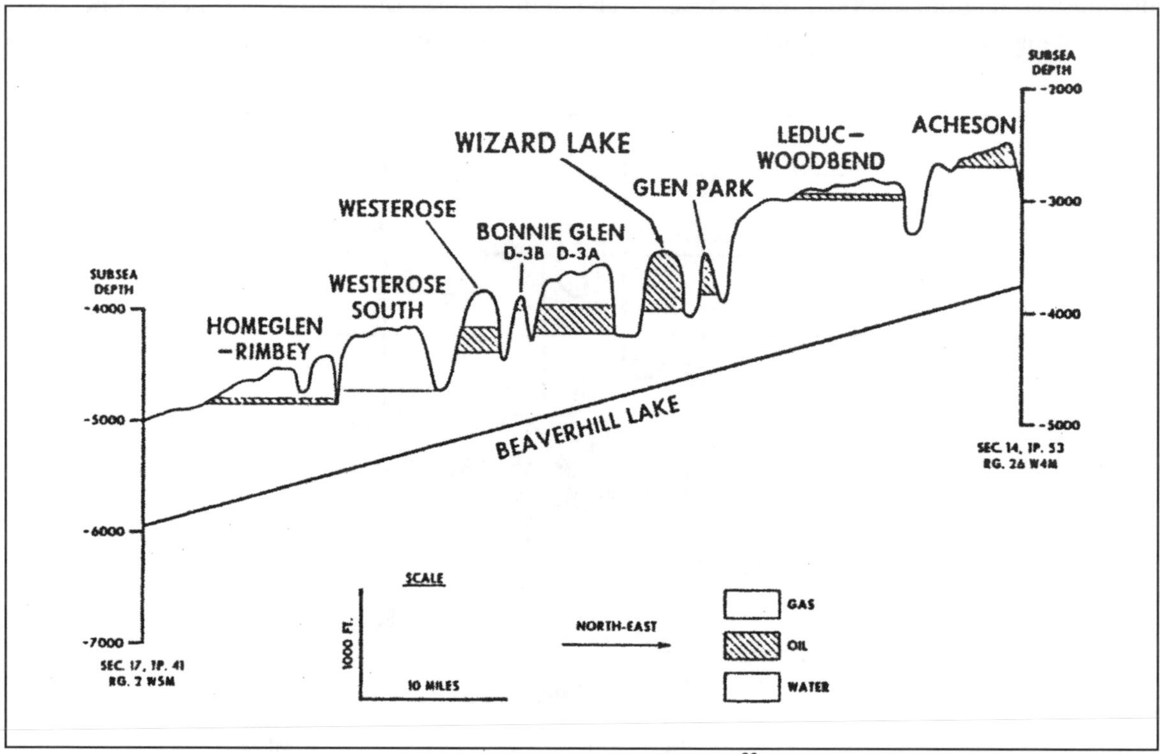

Fig. 1-34. Profile of Rimbey-Acheson Leduc reef trend (from Backmeyer, et al.[30]). Permission to publish by SPE.

IV. BEDDED PLATFORM GRAINSTONES

These are bedded regressive deposits formed by an accumulation of carbonate grains, either detrital (from the erosion of older carbonates), skeletal or oolitic. These carbonate sands, sometimes called calcarenites, are formed in shallow areas near the platform margin where the waves and currents are strong and winnow the finer material. The oolites are nearly spherical, and consist of concentric layers of carbonate. Depositional pore space in these granular carbonates is of the intergranular type.

The Magnolia field, Columbia County, Arkansas, produces from the Reynolds Oolite Member of the Jurassic Smackover formation at a depth approximating 7,350 feet. Figure 1-35 presents structure contours drawn on top of the Smackover Limestone. The field has been an important oil producer in Arkansas. Most of the giant fields of Saudi Arabia produce from granular carbonates.

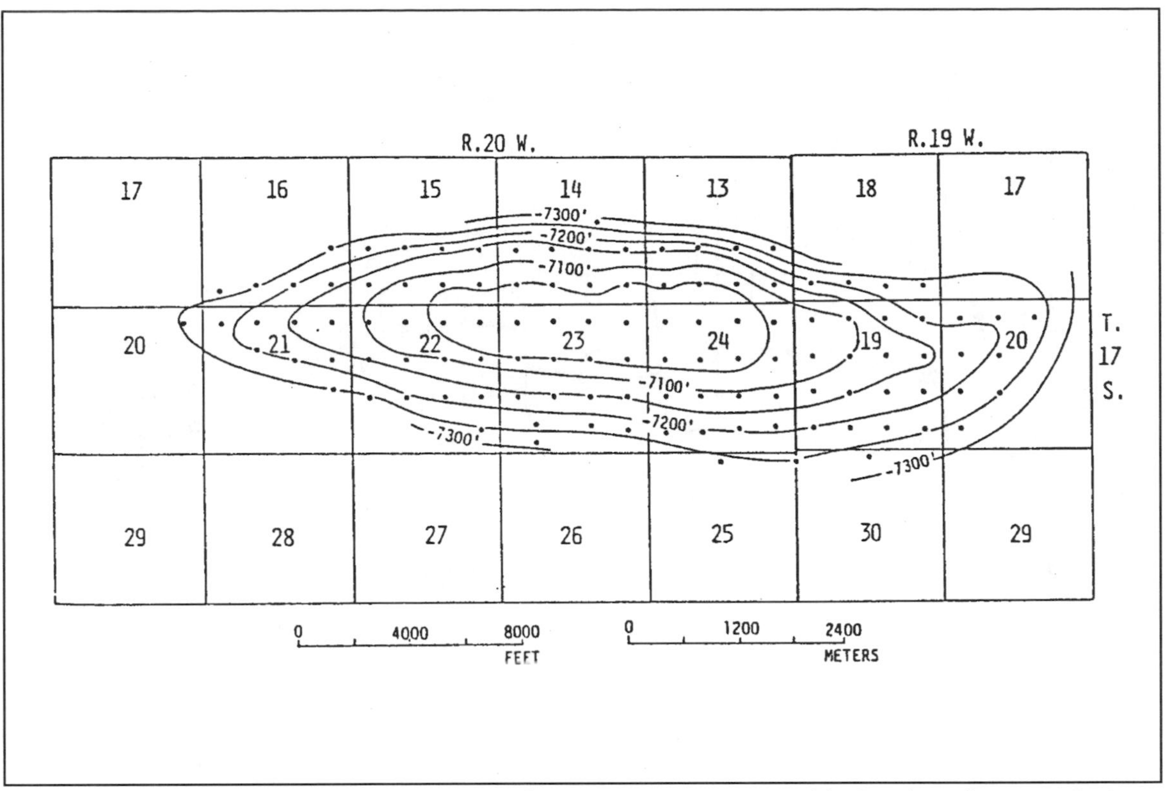

Fig. 1-35. Magnolia Field, Columbia County, Arkansas. Structure contour map on top of the Smackover limestone. Contour interval, 50 ft (from Landes[32]). Permission to publish by Wiley-Interscience.

V. PELAGIC LIMESTONES

Pelagic limestones are formed in the open sea, generally in fairly deep water, by the slow settling of skeletal remains of pelagic plankton. Typical of these deposits are the chalks. The main contributors are microscopic wheel-shaped calcite remains of coccolithospores, a unicellular algae. Some chalks are important producers of oil and gas.

The Ekofisk Field in the North Sea produces oil from two, low permeability, fractured coccolithospores chalk formations, the Ekofisk formation of Danian age and the deeper Tor formation of Maastrichtian age. Porosity ranges from 20 to 40% with block permeability ranging from 0.1 to 8 md. Some intervals are highly fractured, allowing high well productivities.

Although not a conventional chalk deposit, the Shuaiba chalk of the Lower Cretaceous in the Yibal Field of Oman is oil productive. The chalk is a soft, earthy, fine-textured buff/tan limestone formed by shallow-water accumulation of skeletal debris, principally of the phytoplankton, Nannoconus. Figure 1-36 provides a structure cross section through the reservoir.

The porosity ranges from 25 to 42% in the upper oil-bearing portion. Oil column thickness is 370 ft. Log analysis on 70 wells indicates original oil in place to be 2.85 billion STB. Wells have been able to produce 3,500 b/d of 40°API oil.

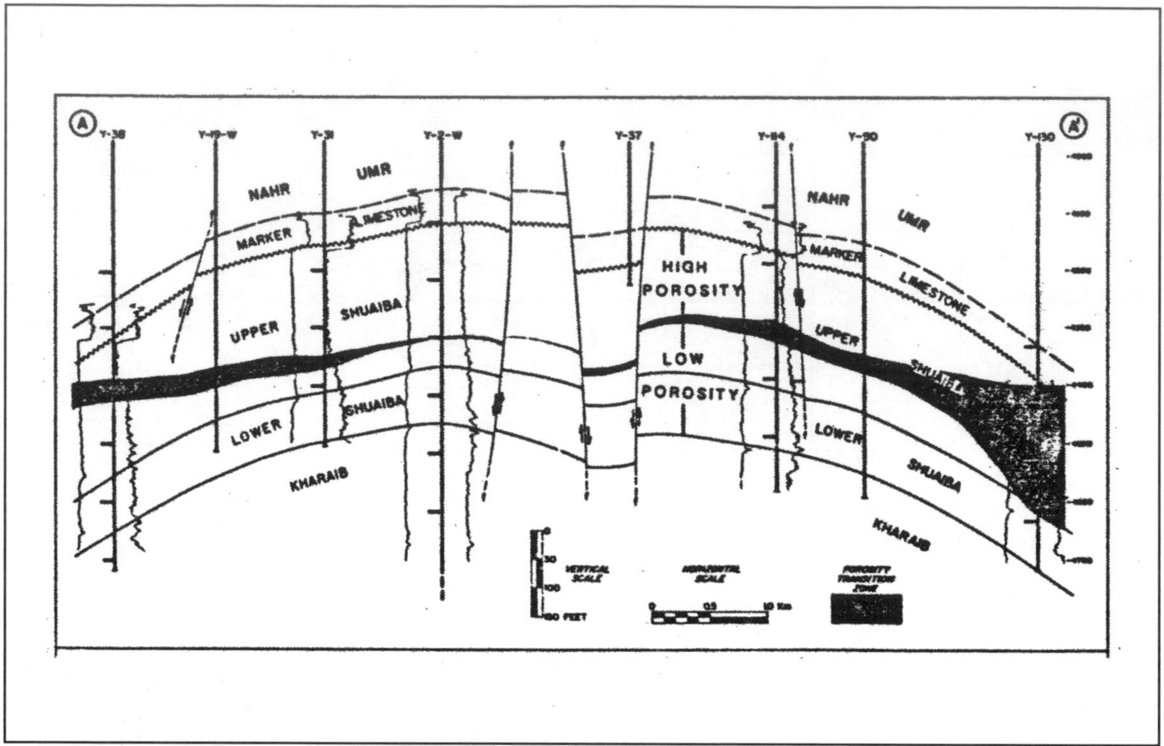

Fig. 1-36. Structural cross section Shuaiba reservoir, Yibal Field, Oman (from Litsey, et al.[33]). Permission to publish by JPT.

VI. DOLOMITES

As mentioned previously, dolomites are secondary carbonates resulting from the replacement of pre-existing limestones. The process of dolomitization is important for reservoir development because it generally increases porosity by forming intercrystalline pores, and also it tends to homogenize the pre-existing rock. The prolific Asmari formation of Iran is a fractured dolomite, with fair matrix porosity and permeability. Initial production rates of early wells reached tens of thousands of barrels per day.

In West Texas, the uplift of the Central Basin platform provided a shallow platform where prolific biological activity could occur, thus allowing the accumulation of abundant carbonate sediments. Figure 1-37 shows the depositional environment present during Leonardian time. Such reservoirs are usually very stratified and lenticular and requiring close well spacing for relatively complete exploitation.

The Robertson field is productive of oil and gas from the Clearfork dolomite of Permian age. Most of the trapping is stratigraphic, being controlled by lateral and vertical limits of porosity and permeability. Figure 1-38 provides a location map of the field and other Clearfork reservoirs of the region. The gross vertical interval at Robertson is 1,200 to 1,400 ft thick with the top at 5,800 ft. Net pay varies from 200 to 400 ft. Porosity types are dominantly moldic and intergranular, although some vugular porosity is present. Porosity approximates 6.0 percent and the permeability 0.1 and 2.0 md. Infill drilling has increased oil reserves, both by primary and waterflooding operations.

SHALE RESERVOIRS

Oil and gas production has been found in fractured shales, but has not had substantial economic impact. Examples include production at Florence, Colorado, and gas production in Kentucky.

EVAPORITES

Evaporites consist of rock salt, anhydrite, and gypsum. Anhydrite can be a reservoir rock (although not common) where circulating water has developed porosity and permeability. Gas is produced from one horizon at the Cotton Valley field, and oil is produced from the Upper Comanche at the Pine Island Field of Louisiana.

IGNEOUS AND METAMORPHIC ROCK RESERVOIRS

These rocks do not normally contain oil and gas reservoirs due to their high temperature origin. Where hydrocarbons are present, it is presumed that effective porosity and permeability were developed after cooling. The serpentine plugs of south central Texas have produced oil and gas in commercial quantities.

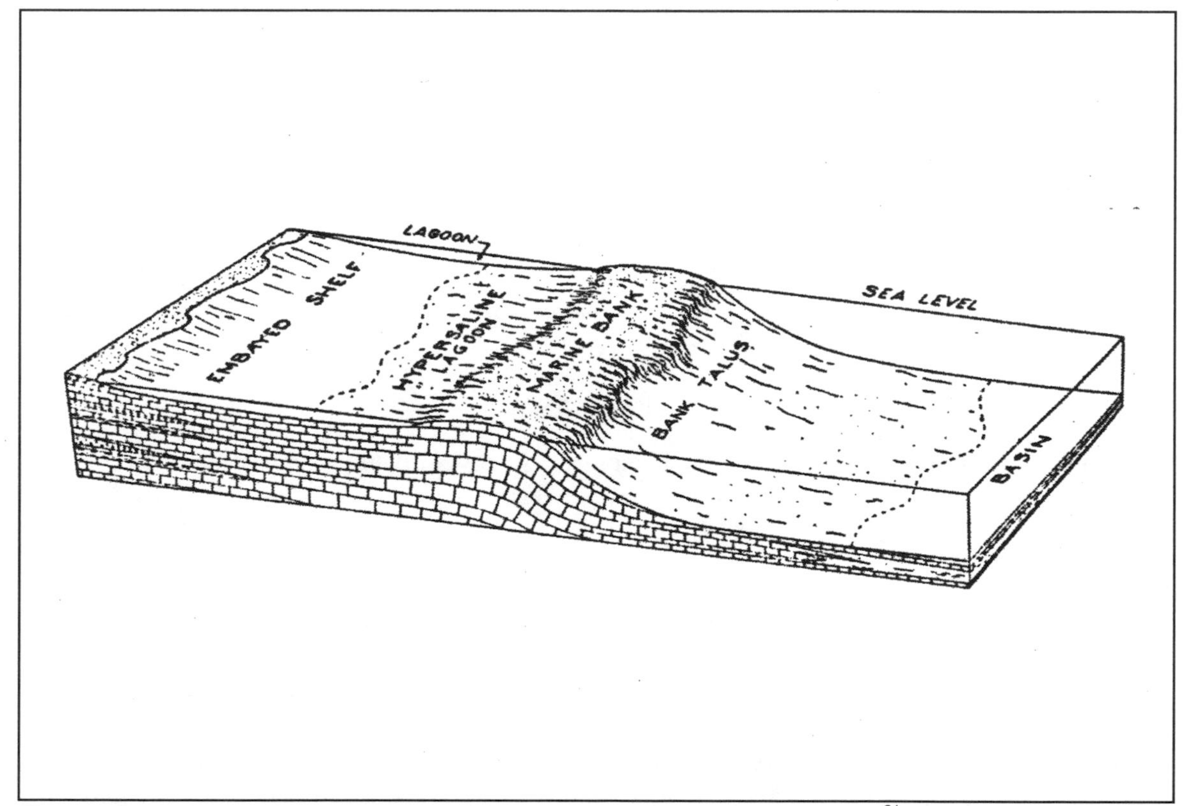

Fig. 1-37. Depositional environment during Leonardian time (from Barbe and Schnoebelen[31]). Permission to publish by SPE.

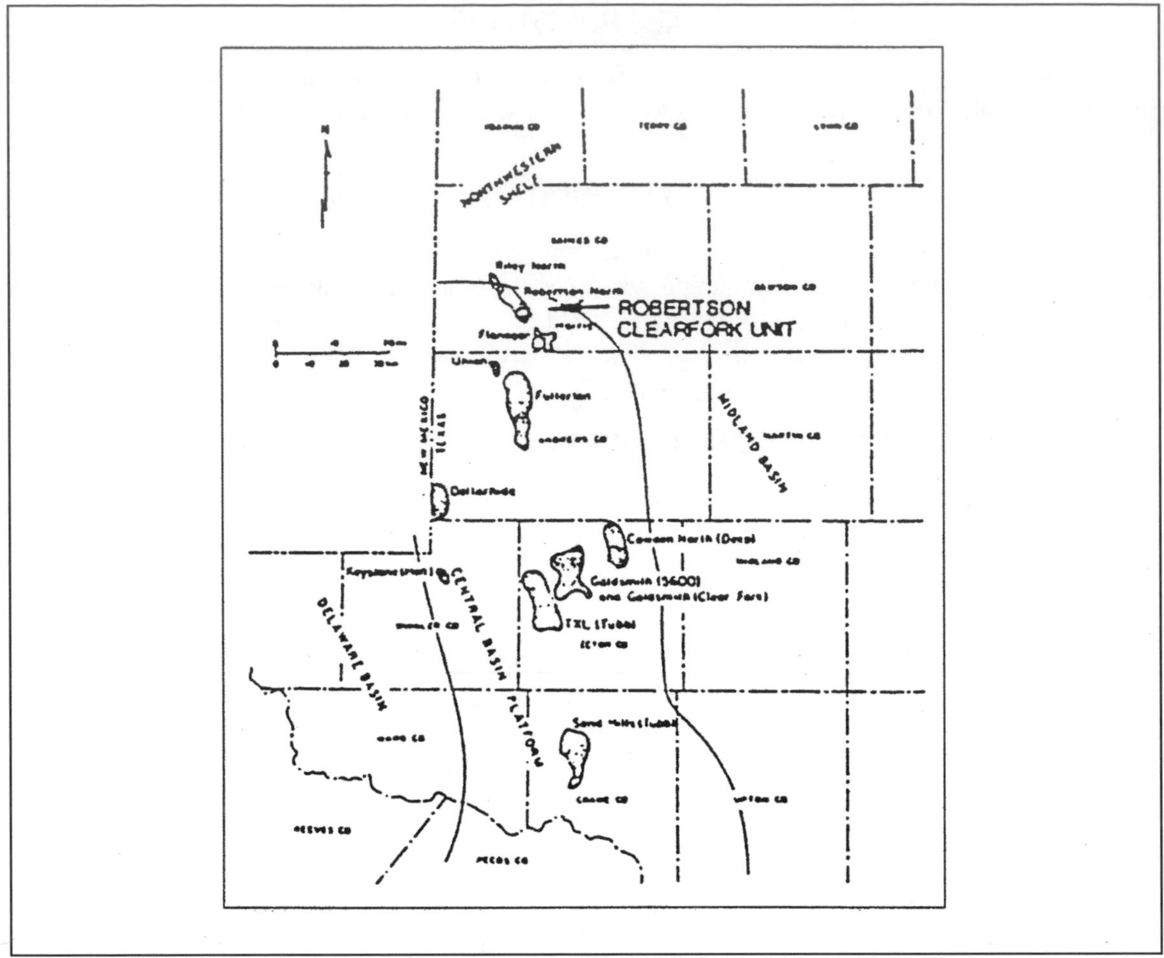

Fig. 1-38. Location map of Robertson Field and other Clear Fork reservoirs of the Central Basin platform area (from Barbe and Schnoebelen[31]). Permission to publish by SPE.

SUMMARY

Much more could be written about the geological aspects of oil and gas reservoirs. No detail has been provided on the stratigraphic and structural aspects of hydrocarbon traps. Salt dome traps and structure caused by shale diapiric structures have not been mentioned.

The main point is that meaningful reservoir engineering studies should begin with careful consideration of the nature of the hydrocarbon trap, the internal structure and composition of the reservoir rock, and with physical aspects that influence hydrocarbon quantities present. All-important to reservoir engineering is the permeability of the rock and its continuity, horizontally and vertically, between wells. With this in mind, meaningful quantitative calculations can be made regarding the original hydrocarbons in place, the expected daily producing rates as time passes, what the producing mechanism(s) are, and what improved oil and gas recovery techniques should be used to maximize recoveries, all within the confines of reasonable profitability.

Every reservoir engineer should own and/or make use of the many good geology books available. Studies on reservoirs are invariably improved when allied disciplines are brought to bear on the problems that invariably occur and require solution.

REFERENCES

1. Tissot, B. P. and Welte, D. H.: *Petroleum Formation and Occurrence*, Springer-Verlag, Berlin, Heidelberg, New York (1978).
2. Deuser, W. G.: Organic-Carbon Budget of the Black Sea, Deep-Sea Res., (1971) Vol. 18, 995-1004.
3. Weber, K. J. and Daukoru, E.: "Petroleum Geology of the Niger Delta," *Proc.* Ninth World Pet. Cong., Tokyo, Applied Science Publishers, London (1975) Vol. II, 209-221.
4. van der Knapp, W. and van der Vlis: *Proc.* Seventh World Pet. Cong., Mexico City (1967) Vol. 3, 85-95.
5. Pettijohn, F. J., Potter, P. E., and Siever, R.: *Sand and Sandstones*, Springer-Verlag, Berlin, Heidelberg, New York (1973).
6. Hobson, G. D.: *Developments in Petroleum Geology - 1*, Applied Science Publishers, London (1977) 147-196.
7. "Ancient Sandstones Considered to be Eolian," E. D. McKee (ed.), U. S. Geol. Sur. Prof. Paper 1052 (1979) 187-251.
8. "A Study of Global Sand Seas," E. D. McKee (ed.), U. S. Geol. Sur. Prof. Paper 1052, 429.
9. van Veen, F. R.: "Geology of the Leman Gas Field in Petroleum and the Continental Shelf of Northwest Europe," A. W. Woodland (ed.), Applied Science Publishers (1975) 223-232.
10. Glennie, K. W.: "Permian Rotleingendes of Northwest Europe Interpreted in Light of Modern Desert Sedimentation Studies," *AAPG Bulletin*, (1972) Vol. 56, No. 6, 1048-1071.
11. Berg, Robert R.: *Reservoir Sandstones* Prentice-Hall, Inc., Englewood Cliffs, N. J. (1986) 116.
12. Spearing, D. R.: Alluvial Fan Deposits: Summary Sheets of Sedimentary Deposits, Sheet 1, Boulder, Colorado, GSA (1974).
13. Borger, H. D.: "Case History of the Quiriquire Field, Venezuela," *AAPG Bulletin*, (1952) Vol. 36, No. 12, 2291-2330.
14. Reference 11, 133-200.
15. Hewitt, C. H. and Morgan, J. T.: "The Fry In-Situ Combustion Test—Reservoir Characteristics," *JPT* (March 1965) 337-342.
16. Coleman, J. M. and Prior, D. B.: Deltaic Sand Bodies, AAPG Education Course Note Series No. 15 (1980) 171 pages.
17. Frazier, D. E.: Recent Deltaic Deposits of the Mississippi River—Their Development and Chronology, Gulf Coast Assn. of Geological Societies, *Trans.*, Vol. 17, 287-311.
18. Reference 11, 201-271.
19. Scott, J. D.: "Cherokee Sandstones of a Part of Noble County, Oklahoma," Master's Thesis, Univ. of Oklahoma (1970).
20. Lyons, P. L. and Dobrin, M. B.: Seismic Exploration for Stratigraphic Traps, part of AAPG Memoir 16 (1972) 225-243.
21. Reference 11, 272-333.
22. Sabins, F. F.: "Anatomy of Stratigraphic Trap, Bisti Field, New Mexico," *AAPG Bulletin*, (1963) Vol. 47, 193-228.
23. Conybeare, C. E. B.: "Geomorphology of Oil and Gas Fields in Sandstone Bodies," Dev. in Petr. Sci., Elsevier Science Publishers, Amsterdam (1876) 174.
24. Reference 11, 334-399.
25. Anderman, G. G.: "Sussex Sandstone Production, Triangle-U Field, Campbell County, Wyoming," Wyoming Geological Assn. Guidebook (1976) 107-113.
26. Walker, R. G.: "Deep-Water Sandstone Facies and Ancient Submarine Models for Stratigraphic Traps," *AAPG Bulletin*, (1978) Vol. 62, No. 6, 932-966.
27. Rall, R. W. and Rall, E. P.: "Pennsylvanian Subsurface Geology of Sutton and Schleicher Counties, Texas," *AAPG Bulletin*, (1958) Vol. 42, 839-869.
28. Reference 11, 400-472.
29. Parker, J. R.: *Lower Tertiary Sand Development in the Central North Sea*, A. W. Woodland (ed.), *Petroleum and the Continental Shelf of North-West Europe*, Applied Science Publishers, London (1975) Vol. 1, 447-452.
30. Backmeyer, L. A., Guise, D. R., and MacDonell, P. E.: "The Tertiary Extension of the Wizard Lake D-3A Pool Miscible Flood," paper SPE 13271 presented at the 1984 Annual Technical Conference and Exhibition, Houston, Texas, Sept. 16 - 19, 12 pages.

31. Barbe, J. A. and Schnoebelen, D. J.: "Quantitative Analysis of Infill Performance: Robertson Clearfork Unit," paper SPE 15568 presented at the 1986 SPE Annual Technical Conference and Exhibition, New Orleans, Oct. 5-8, 13 pages.
32. Landes, K. K.: *Petroleum Geology of the United States*, Wiley-Interscience, New York (1970) 176-187.
33. Litsey, L. R., MacBride, W. L. Jr., Al-Hinai, K. M., and Dismukes, J. B.: "Shauiba Reservoir Geological Study, Yibal Field, Oman," *JPT* (June, 1986) 651-661.

2 PROPERTIES OF RESERVOIR ROCKS

POROSITY

DESCRIPTION AND DEFINITION

All reservoir rocks containing commercial quantities of hydrocarbons have porosity and permeability. Porosity is a measure of the void space within a rock expressed as a fraction (or percentage) of the bulk volume of the rock. This can be written as:

$$\phi = \frac{V_b - V_g}{V_b} = \frac{V_p}{V_b} \tag{2-1}$$

where:

ϕ = porosity
V_b = bulk volume of the rock
V_g = net volume occupied by solids, or the grain volume
V_p = pore volume

Figure 2-1 illustrates the various arrangements for packings of spheres having equal diameters. Simple considerations of the geometries involved will show that porosity is 47.6%, 39.5%, and 25.9%, respectively, for cubic, hexagonal, and rhombohedral packings of spheres. For each of these packing possibilities, grain size has no influence on porosity where the grains are uniform. Of course, sands are not uniform and these porosity values would represent the upper limit of possible porosities in actual reservoir rocks. Sandstones also have cementing material between the grains which further reduces the volume available for the storage of hydrocarbons.

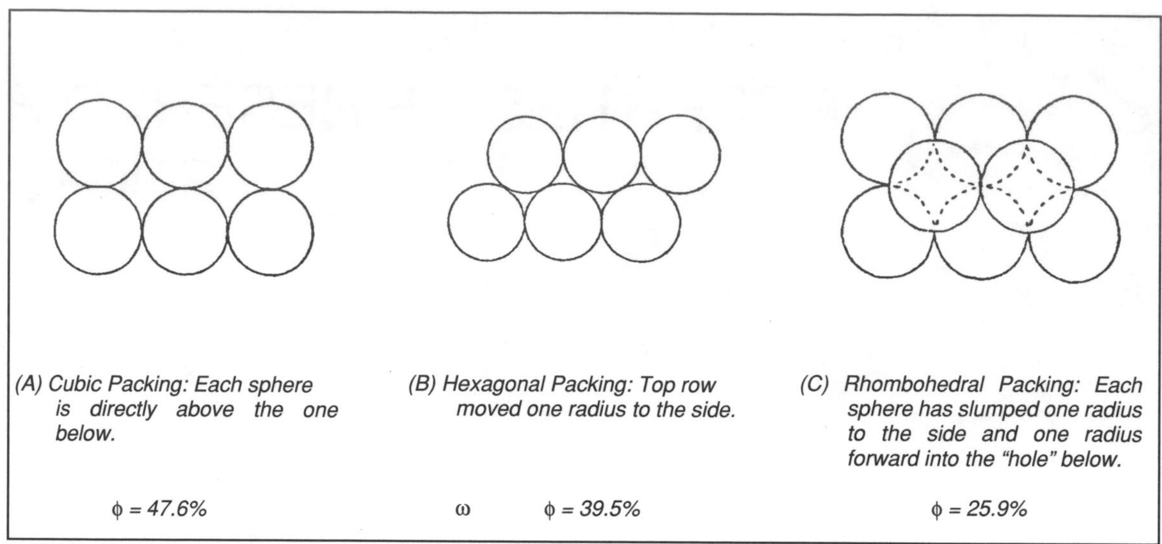

Fig. 2-1. Cubic, hexagonal, and rhombohedral packing of spheres. (Adapted from Gatlin[1]). Permission to publish by Prentice-Hall, Inc.

Porosity can be of two classes, absolute and effective. Absolute porosity is the total porosity of a rock without regard to connections between the voids. Effective porosity is that where the voids are interconnected. While absolute and effective porosity will be essentially the same in most reservoir rocks, only effective porosity is of interest to the reservoir engineer.

Porosity can be classified into two types according to the time of deposition. Primary porosity refers to that present at the time of deposition; e.g., sandstone porosity. Secondary porosity refers to that formed after deposition and is typified by vugular limestones and any of the reservoir rocks having fractures, fissures and joints. Dolomitization is also a process by which relatively dense limestone can acquire additional effective porosity.

Typical sand and sandstone porosities range from 8 to 39% with a worldwide average being near 18%. Sandstone hydrocarbon reservoirs with porosities less than 8% normally have very low permeabilities and are usually not economic to develop. Factors affecting porosity in sandstones may be listed as:

1. Sorting or grain size distribution
2. Cementation
3. Packing
4. Shape
5. Chemical action
6. Fracturing
7. Deformation by the overburden

Carbonates (limestones and dolomites) normally have porosities between 3 to 15% with an average of about 8%. Carbonate storage systems are much more difficult to characterize. Porosity is made up of the following types of voids: (1) intergranular, (2) vugular, (3) fractures and fissures, and (4) intercrystalline. All of these may be present or not, and they may be put together in series, parallel, or combination. Therefore, the performance of carbonates is difficult to predict.

Shales have porosities approximating 40%. This is not an effective porosity, however, and those few hydrocarbon shale reservoirs that have been developed produce from fracture systems where porosity is about 1 to 4%.

MEASUREMENT OF POROSITY WITH CORES

The literature is extensive on the topic of porosity measurement. The reader is referred to particularly complete treatments by Gatlin,[1] Clark,[2] Pirson,[3] Amyx *et al.*,[4] Guerrero,[5] and Calhoun.[6] Reference to Equation 2-1 will show that two quantities need to be measured to calculate porosity: (a) bulk volume and (b) either pore volume or grain volume. Note that:

$$Pore\ Volume = (Bulk\ Volume) - (GrainVolume).$$

Pore Volume or Grain Volume Determination

1. Boyle's Law Porosimeter: Grain volume may be measured via gas displacement in a Boyle's Law Porosimeter (Fig. 2-2). The core sample is placed in a steel chamber and then pressurized with gas (usually helium, nitrogen, or air) to a known pressure about 4 or 5 atmospheres. The gas is then allowed to expand into a fairly large calibrated volume. This is done with and without the sample inside the core chamber, each time noting the resulting system pressure. Using Boyle's law ([p][V] = constant, over modest pressure ranges), the grain volume of the core may be calculated.

2. Liquid Saturation Method: The core fluids are extracted; and then the core is dried and weighed. Through proper laboratory techniques, the core is completely saturated with a suitable liquid such as kerosene or toluene. The saturating fluid should have low viscosity and low surface tension, and it should wet and penetrate the sample freely. The saturated core is weighed. Then, with the density of the saturation fluid known, one can calculate pore volume.

3. Vacuum Techniques: Other porosimeters, such as the Washburn-Bunting (Fig. 2-2) and Coberly-Stevens types, attempt to evacuate the air inside a clean, dry core. A vacuum is imposed on the core by manipulation of a mercury leveling bulb. Several expansions are necessary, and, of course, all the air cannot be removed from the core. Therefore, indicated porosity usually is low.

Bulk Volume Determination: Displacement Methods

The bulk volume of core samples is usually obtained by measurement of core dimensions or by fluid displacement. Measurement by displacement is accurate so long as none of the displaced fluid enters the core and/or no air bubbles adhere to the surface of the core.

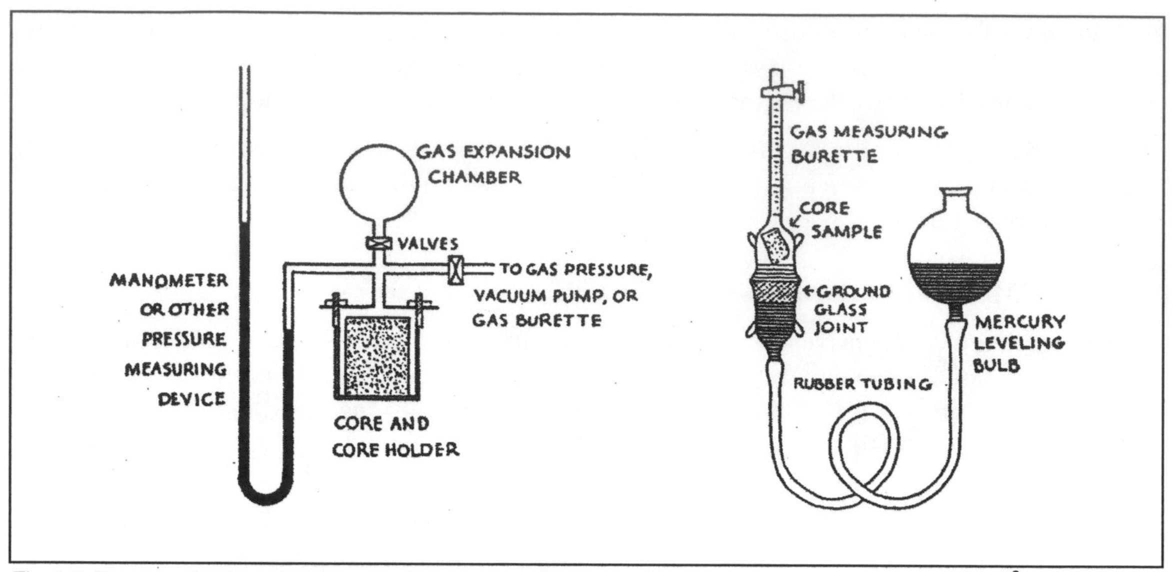

Fig. 2-2. Two porosimeters: At left is a Boyle's law type. At right is a Washburn-Bunting type (from Calhoun[6]). Permission to publish by Oklahoma University.

1. Pycnometer: When using a pycnometer (Fig. 2-3), first fill with mercury. Then measure the volume of mercury that overflows when the core is inserted (or measure the loss of weight of the mercury in the pycnometer).

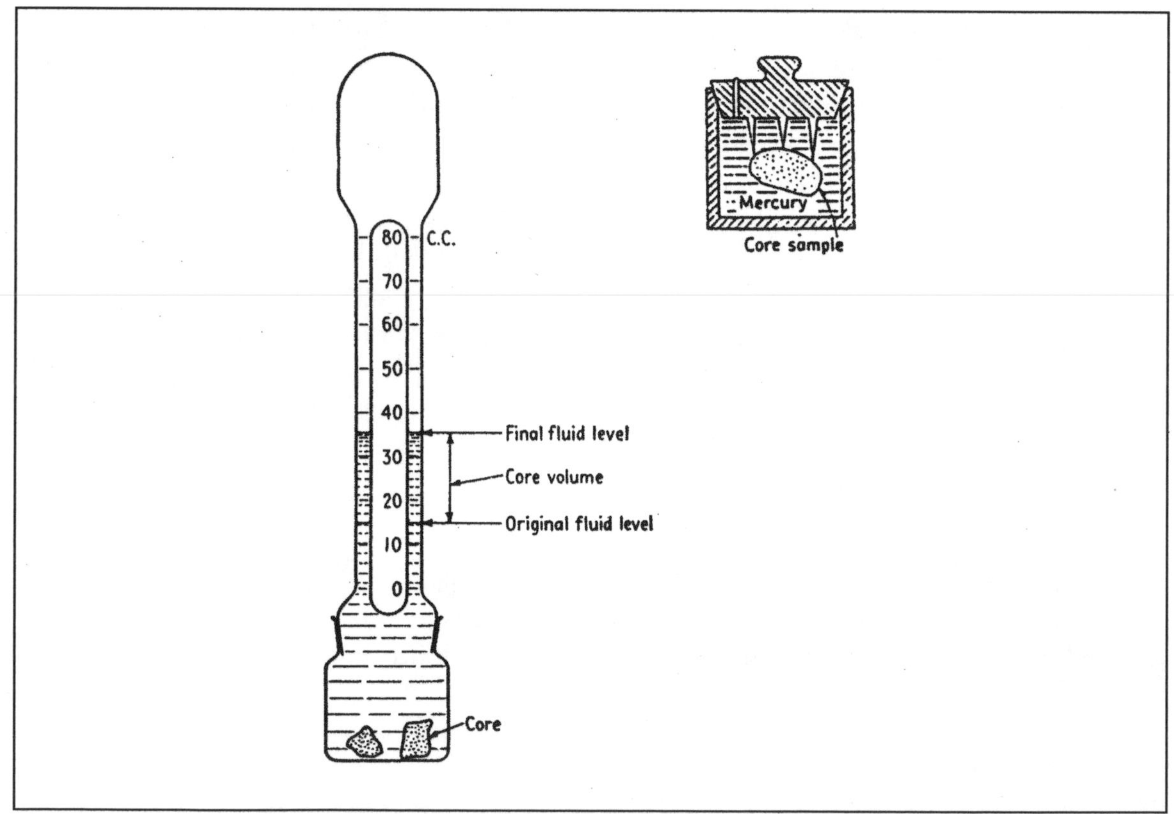

Fig. 2-3. At left: Russell volumeter; At right: Steel pycnometer. (from Pirson[3]). Permission to publish by McGraw-Hill.

2. Westman Balance: This method utilizes Archimede's Principle (Fig. 2-4), and the net force is measured that is required to submerge the rock sample in mercury. Since the density of mercury is known, the volume of mercury displaced may be calculated.

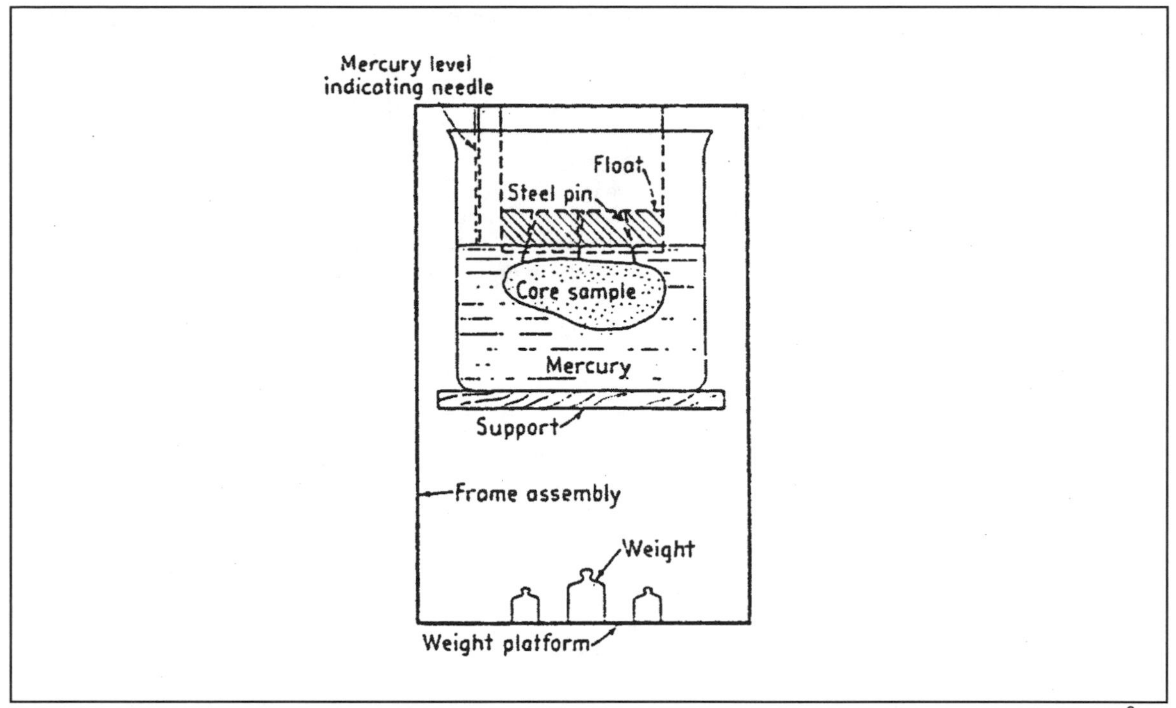

Fig. 2-4. Apparatus for the determination of bulk-volume by mercury displacement (Westman balance). (from Pirson[3]). Permission to publish by McGraw-Hill.

3. Submergence: A saturated core sample is submerged in a suitable liquid in a suitable container. This technique is ideally suited for cases where pore volume has been determined by the liquid saturation methods, and the liquid should have properties already mentioned under this method. The displaced volume of liquid is determined in a wide-mouthed pycnometer or Russell volumeter (Fig. 2-3).

SUMMARY

Table 2-1 provides a partial summary of the available methods for determining porosity and notes the errors which can result. Accuracy in measurement of porosity is important since it is a direct indication of the storage capacity of a reservoir rock for hydrocarbons. A group of major oil companies made a series of tests comparing the gas expansion and saturation techniques. Table 2-2 summarizes the results obtained.

It is useful to note that the gas expansion method results in slightly higher porosity values due to gas adsorption on the rock surface, while incomplete saturation causes the saturation methods to yield slightly lower values for porosity. These errors are more pronounced for lower porosity cores. Normally, however, these methods yield results of acceptable accuracy if carefully performed.

Table 2-1
Methods of Determining Porosity
(from Amyx *et al*[4]).

	Effective porosity	Effective porosity	Effective porosity	Effective porosity	Effective porosity	Effective porosity	Effective porosity	Total porosity
Method	Washburn-Bunting	Stevens	Kobe porosimeter	Boyle's law porosimeter	Saturation	Core laboratories Wet Sample	Core laboratories Dry Sample	Sand density
Type sampling	One to several pieces per increment (usually one)	One to several pieces per increment (usually one)	One to several pieces per increment (usually one)	One to several pieces per increment (usually one)	One to several pieces per increment (usually one)	Several pieces for retort, one for mercury pump	One to several pieces per increment (usually one)	Several pieces per increment
Preparation	Solvent extraction and oven-drying. Occasionally use retort samples.	Solvent extraction and oven-drying. Occasionally use retort samples.	Solvent extraction and oven-drying. Occasionally use retort samples.	Solvent extraction and oven-drying. Occasionally use retort samples.	Solvent extraction and oven-drying. Occasionally use retort samples.	None	Solvent extraction and oven-drying. Occasionally use retort samples.	Extraction, then in step 2, crush sample to grain size
Functions measured	Pore-volume and bulk volume	Sand-grain volume and unconnected pore volume and bulk volume	Sand-grain volume and unconnected pore volume and bulk volume	Sand-grain volume and unconnected pore volume and bulk volume	Pore-volume and bulk volume	Volumes of gas space, oil and water, and bulk volume	Sand-grain volume and unconnected pore volume and bulk volume	Bulk volume of sample and volume of sand grains
Manner of measurement	Reduction of pressure on a confined sample and measurement of air evolved. Bulk volume from mercury pycnometer	Difference in volume of air evolved from a constant-volume chamber when empty and when occupied by sample. Bulk volume by Russell tube	Difference in volume of air evolved from a constant-volume chamber when empty and when occupied by sample. Bulk volume by Russell tube	Difference in volume of air evolved from a constant-volume chamber when empty and when occupied by sample. Bulk volume by Russell tube	Weight of dry sample, weight of saturated sample in air, weight of saturated sample immersed in saturant	Weight of retort sample, volume of oil and water from retort sample, gas volume and bulk volume of M.P.S.	Difference in volume of air evolved from a constant-volume chamber when empty and when occupied by sample.	Weight of dry sample, weight of saturated sample immersed weight, and volume of sand grains
Errors	Air from dirty mercury, possible leaks in system, incomplete evacuation due to rapid operation or tight sample	Mercury does not become dirty. Possible leaks in system, incomplete evacuation due to rapid operation or tight sample	Mercury does not become dirty. Possible leaks in system, incomplete evacuation due to rapid operation or tight sample	Mercury does not become dirty. Possible leaks in system, incomplete evacuation due to rapid operation or tight sample	Possible incomplete saturation	Obtain excess water from shales. Loss of vapors through condensers	Possible leaks in system, incomplete evacuation due to rapid operation or tight sample	Possible loss of sand grains in crushing. Can be reproduced most accurately

Permission to publish by McGraw-Hill.

Table 2-2
Characteristics of Samples used in Porosity-Measurement Comparisons
(from Dotson *et al.*[7]).

Type of material	Sample No.	Approximate gas permeability, millidarcys	Porosity, %				
			Average	Average from gas methods	Average from saturation methods	Value from high observation	Value from low observation
Limestone	1	1.	17.47	17.81	16.96	18.50	16.72
Fritted glass	2	2.	28.40	28.68	27.97	29.30	27.56
Sandstone	3	20.	14.00	14.21	13.70	15.15	13.50
Stonestone	4	1,000.	30.29	31.06	29.13	31.8	26.8
Semiquartzitic sandstone	BZE	0.2	3.95	4.15	3.66	4.60	3.50
Semiquartzitic sandstone	BZG	0.8	3.94	4.10	3.71	4.55	3.48
Alundum	61-A	1,000.	28.47	28.78	28.00	29.4	27.8
Alundum	722	2.	16.47	16.73	16.08	17.80	16.00
Chalk	1123	1.6	32.67	33.10	32.03	33.8	31.7
Sandstone	1141-A	45.	19.46	19.68	19.12	20.2	18.8

Permission to publish by SPE.

PROBLEMS IN MEASUREMENT OF POROSITY ON CORES

In relatively well-consolidated, high-porosity sandstones, the normal measurement techniques provide accurate determinations of the storage capacity (porosity) of the rock involved. Low porosity sandstones (with low permeability) may contain clay minerals that swell in laboratory fluids or which take longer times for equilibrium to be reached (for instance in gas expansion methods). Porosity of siltstones and shales also pose special problems due to the very low permeabilities that exist. Here, the permeability may be so low that no effective porosity exists.

Carbonates, which are vugular, fractured, fissured, and/or dolomitized, create a special sampling problem since porosity measured may not be typical of the reservoir. For these rocks, whole core analysis can usually be used for a much better evaluation. However, for extensively fractured systems, no satisfactory analysis technique is available since the samples cannot be put back together to their natural state. Measurement of carbonate bulk volume may require a special coating to avoid penetration of the displacing fluid (usually mercury).

Other potential problems are removal of the water of crystallization or fine particles during cleaning and alteration of the sample during coring in the subsurface or while being transported to the laboratory. A good example would be the essentially unconsolidated rocks recovered from a gas-producing reservoir in the permafrost at Umiat, Alaska. Here, the problem was compounded due to the unconsolidated nature of the core and the frozen interstitial water.

OTHER POROSITY MEASUREMENT METHODS

While measurement of porosity in rock samples is a direct method, many wells are drilled and completed for hydrocarbon production without cores being taken. In this event, reliance is placed on indirect measurement techniques. Some well logs can provide good measurement of porosity where conditions are favorable. Typical logs from which porosity can be derived are: the density log, the neutron log, the sonic log, several of the resistivity logs (such as the microlog), and, of course, the sidewall coring equipment.

Well logs have the advantage of measuring a larger volume of reservoir rock than can be done with cores. In some instances, logs can provide superior measurements of porosity (Helander[8]). A knowledge of well log analysis is important to the practicing reservoir engineer, development geologist, and anyone involved in the discovery, development, or depletion of a hydrocarbon reservoir.

ALTERATION OF POROSITY

The usual laboratory porosity measurement methods yield a core porosity at the surface. However, in-situ porosity (that down in the reservoir) will be less than the laboratory porosity unless special techniques are used which incorporate overburden and internal pressure effects.

Porosity is altered by overburden pressure. Instances have been recorded of surface subsidence due to the removal of water and/or hydrocarbons. The subsidence is a visual record of the decrease in subsurface porosity in the reservoir rock undergoing depletion. In these instances, the reservoirs involved are usually thick and poorly consolidated.

In the more general case (no visible subsidence), porosity still tends to decrease as depth of burial increases as shown in Figure 2-5 (Timmerman[9]). The more shaley the rock, the greater will be the reduction. At depth, the formation is compressed, which causes a corresponding decrease in porosity, due to the weight of the overlying sediments. Thus, laboratory-measured porosity will likely be optimistic due to the lack of overburden forces on the core.

Not only can the pore volume of reservoir rock be altered by external elements such as overburden and compressive forces within the earth's crust, but porosity is also a function of internal or pore pressure. Reservoir pressure decreases with production of fluids. As pore pressure goes down, a greater percentage of the overburden weight is transmitted to the rock matrix. This in turn causes a compression of the formation bulk volume. The net result is a decrease in porosity as internal pressure decreases. This provides part of the driving mechanism by which fluids are expelled. In Figure 2-6, Hall[10] has provided a plot of consolidated formation effective pore volume compressibilities, this being the change in pore volume/unit pore volume/psi. For both sandstones and limestones, compressibility increases as porosity decreases.

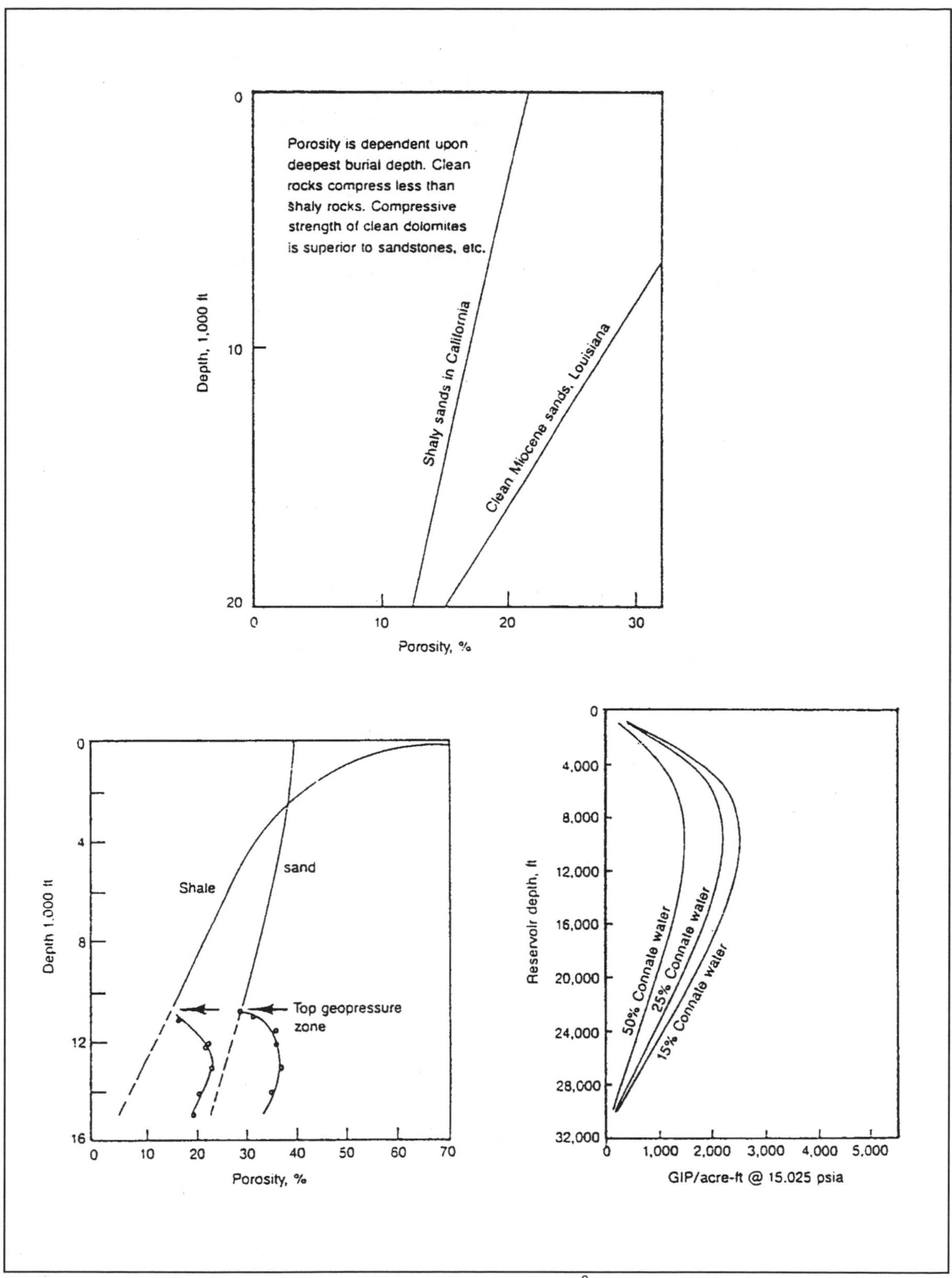

Fig. 2-5. Curves showing decrease in porosity with depth. (from Timmerman[9]). Permission to publish PennWell Publishing Company.

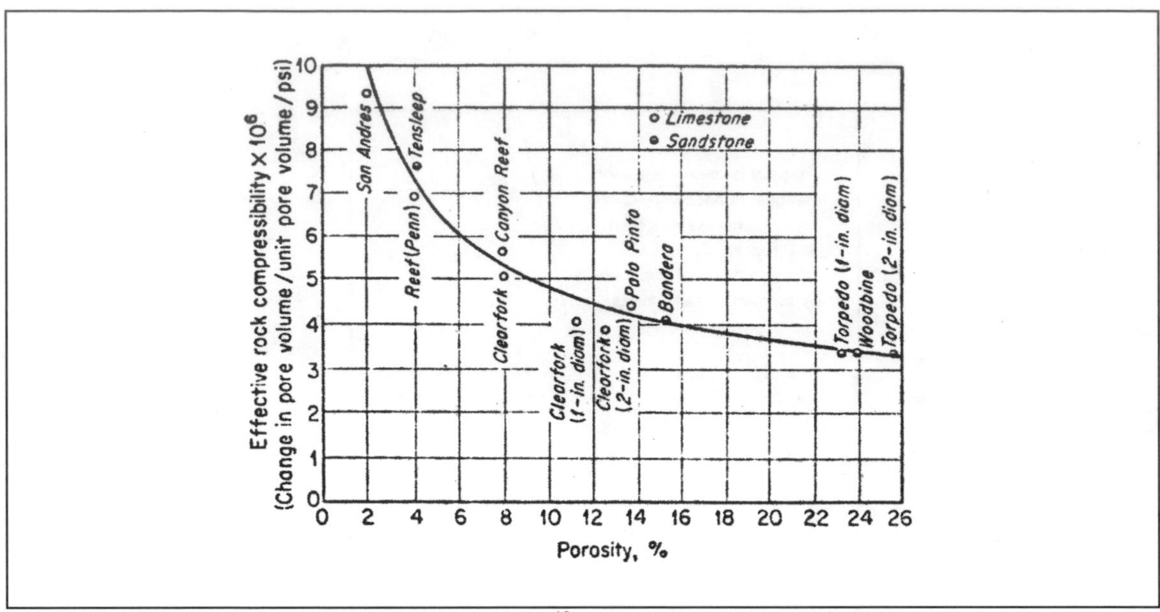

Fig. 2-6. Effective pore volume compressibilities (from Hall[10]). Permission to publish by SPE.

To consider the effects of internal pressure and overburden on porosity, the usual approach is to define an effective stress or effective pressure.

$$p_{eff} = OB - p \tag{2-2}$$

where:

p_{eff} = effective pressure (or stress),
OB = overburden pressure, and
p = internal or reservoir pressure.

As effective stress increases, the in-situ porosity decreases. In most areas, the overburden pressure increases approximately 1 psi per foot of depth. In normally pressured regions, the internal pressure gradient ranges from 0.433 to 0.465 psi/foot depending on the salinity of the formation waters. Therefore, with depth, effective stress is increasing, and porosity is decreasing.

Work by Van der Knaap[11] led to the following empirical equations for adjusting porosity in porous media with elastic behavior.

Sandstone equation:

$$\phi = \phi_s - 0.000337\,(p_{eff})^{0.30} \tag{2-3}$$

Limestone equation:

$$\phi = \phi_s - 0.0000432\,(p_{eff})^{0.42} \tag{2-4}$$

where:

ϕ = in-situ porosity or porosity at reservoir conditions, fraction,
ϕ_s = core porosity measured at the surface with no overburden or internal pressure, fraction,
p_{eff} = effective pressure from (2-2), psi.

It should be emphasized that these equations are not to be used for formations that are incompetent, unconsolidated, or that exhibit inelastic behavior.

Problem 2-1

Problem Statement:

(A) A sandstone formation in the Rocky Mountain region of the United States is at a depth of 10,000 ft from the surface. The hydrostatic gradient is 0.4335 psi/ft (fresh water). What is the initial effective stress on the formation rock at reservoir depth?

(B) The average Boyle's law, laboratory-measured, core porosity is 10.1%. What is the approximate porosity in the reservoir at discovery conditions?

(C) What is the approximate porosity at an abandonment of 1,100 psig?

Solution:

(A) OB = overburden pressure = (10,000 ft) x (1.0 psi/ft) = 10,000 psig
p = internal pressure = (10,000 feet) x (0.4335 psi/ft) = 4,335 psig
p_{eff} = 10,000 - 4335 = 5665 psi

(B) Using the sandstone Eq. 2-3:

$$\phi = 0.101 - (0.000337)(5665)^{0.3}$$
$$= 0.0965$$
$$= 9.65\ \%$$

(C)
$$\phi = 0.101 - (0.000337)(10{,}000 - 1{,}100)^{0.3}$$
$$= 0.0958$$
$$= 9.58\ \%$$

In 1973, G. H. Newman[12] made some interesting observations about porosity and pore volume compressibility. He used a qualitative rock-typing system: consolidated, friable, and unconsolidated. Newman's conclusions included the following: (1) Laboratory compressibility measurements are needed in evaluating pore volume compressibility for a given reservoir because laboratory-determined values (his and those in the literature) are in poor agreement with published correlations. In other words, the Hall correlation shown in Figure 2-6 should be used only when laboratory values are not available. (2) Pore volume compressibilities for a given porosity can vary greatly depending on rock type. (3) Consolidated sandstones differ greatly from limestones and even from friable and unconsolidated sands. Well defined trends were found only in the consolidated sandstones and limestones. Friable and unconsolidated samples showed little, if any, correlation. Reasonable correlations (compressibility vs. porosity) may be obtained from well consolidated formations, sandstone or limestone, if the lithologies are similar (e.g., for samples from the same reservoir where the lithologic variations are small).

If the proper laboratory equipment is available, Jones[35] has presented empirical equations that fit pore volume and pore volume compressibility versus effective stress, p_{eff} (Eq. 2-2).

$$Ln \left\{ V_p \left[1 + (p_{eff})(3 \times 10^{-6}) \right] \right\} = Ln\, c_1 - c_2 \left[1 - e^{-(p_{eff}/3000)} \right] \quad (2\text{-}5)$$

where

V_p = pore volume, cm^3
p_{eff} = effective stress, psi
c_1 = equation constant
c_2 = equation constant

To use Eq. 2-5, pore volume is measured on a core at two or more different stress states. Then, using least squares, the constants c_1 and c_2 may be evaluated. With c_1 and c_2 in hand, Eq. 2-5 may be used to calculate pore volume versus effective stress. According to Jones[35], if a two point calibration of Eq. 2-5 is done using measured pore volumes taken at 1500 psi and 5000 psi confining or "overburden" stress, then this equation should give excellent results up to an effective stress of 10,000 psi. However, for a new reservoir, multiple samples should be analyzed.

If the experimental and mathematical work has been performed to evaluate the constants in Eq. 2-5, then pore volume compressibility may be predicted with:

$$c_f = \frac{c_2 e^{-(p_{eff}/3000)}}{3000} + \frac{3 \times 10^{-6}}{1 + (p_{eff})(3 \times 10^{-6})} \; psi^{-1} \quad (2\text{-}6)$$

where:

p_{eff} = effective pressure, psi, and
c_f = pore volume compressibility, psi^{-1}

PERMEABILITY CONCEPTS

A. Single-phase Systems

By definition, absolute permeability (denoted "k") of a given porous medium is the ability to pass a fluid through its interconnected pore and/or fracture network, if the medium is 100% saturated with the flowing fluid. Permeability has been found to be related to the size of the entrance passageways into the pore spaces. Other factors affecting absolute permeability are grain packing, petrofabric of the rock, grain size distribution, grain angularity, and degree of cementation and consolidation.

In the mid-1800's, working in France with the Dijon water filtration system, Henri Darcy discovered that the water rate through a sand pack was proportional to the head or pressure drop across the pack. It is instructive to write the general form of the Darcy equation for flow in a linear horizontal system:

$$q = \frac{-Ak(\Delta p)}{\mu L} \tag{2-7}$$

where:

k = permeability, darcies,
q = outlet flow rate, cm^3/sec,
μ = fluid viscosity at temperature of the system, cp,
L = system length, cm,
A = system cross-sectional area, cm^2,and
Δp = pressure differential across system, atm.

Darcy's law states that the velocity (q/A) of a homogeneous fluid is proportional to the fluid's mobility, k/μ, and the pressure gradient, $\Delta p/L$. The assumptions behind this equation are:

1. Homogeneous rock
2. Non-reactive medium or rock
3. 100% saturated with single phase homogeneous fluid
4. Newtonian fluid
5. Incompressible flow
6. Laminar flow
7. Steady state
8. Constant temperature

The flow rate involved in Eq. 2-7 is called the apparent flow rate (q/A) because the entire cross-sectional area, "A" is not available for flow (most of "A" is occupied by grain volume). The actual rate inside the porous medium is equal to the apparent flow rate divided by the porosity. Because the pores are all different inside a rock, the Darcy law flow rate is more of a statistical quantity than an actual one.

The constant of proportionality of the porous medium (absolute permeability), in Eq. 2-7, has units of darcies. To obtain a physical picture of this unit, a dimensional analysis can be made:

$$k = \left[\frac{L^3}{T} \cdot \frac{M}{LT} \cdot L \cdot \frac{1}{L^2} \cdot \frac{LT^2}{M} \right] = \left[L^2 \right]$$

where

$q = [L^3/T]$

$\mu = \dfrac{shearing\ stress}{rate\ of\ shearing\ strain} = \dfrac{F/A}{dv/dl} = \left[\dfrac{ML/T^2)/L^2}{(L/T)/L}\right] = \left[\dfrac{M}{LT}\right]$

$L = [L]$

$A = [L^2]$

$\Delta p = \dfrac{force}{unit\ area} = \dfrac{[ML/T^2]}{[L^2]} = [M/LT^2]$

Here M, L, F, and T refer to the units of mass, length, force and time, respectively. From this analysis, it is seen that permeability has the units of $[L^2]$. In fact, 1 Darcy = 0.987×10^{-12} m^2. This can be visualized as an area presented at right angles to the flow direction.

Recently, the Society of Petroleum Engineers[36] has suggested the use of the "SI" units (metric system) instead of "practical field" units. With SPE SI units, the unit of permeability is μm^2.

B. Multiphase Systems

Unfortunately, the authors are unaware of any hydrocarbon reservoirs that have been discovered with absolutely no water contained in the pore spaces. The water saturation, which is considered to be indigenous to the formation, is termed connate or interstitial water saturation. Not only does this connate water reduce the pore space available for hydrocarbons, but it causes at least two fluid phases to be present within the porous medium: the hydrocarbon and the connate water. Therefore, we must define *effective permeability* which is the permeability to a particular fluid, i.e., oil, gas, or water: k_o, k_g, or k_w. The units are the same for effective permeability and absolute permeability. Further, it has been found that:

$$O \leq k_o, k_g, k_w \leq k\ (absolute)$$

The definition for *relative permeability* follows immediately:

$$k_{ri} = k_i / k$$

where i = oil, water, or gas

$$O \leq k_{ro}, k_{rg}, k_{rw} \leq 1$$

Considering an oil/water system, a formation is said to be water-wet when the capillary forces are such that the water resides within the pore spaces next to the walls while the oil stays in the center of the pores. It is also possible to have a formation that is oil-wet or of intermediate wettability. We usually use the terminology "preferentially" water-wet or oil-wet because interaction with various chemicals can change formation wettability.

Consider an oil/water, intergranular (probably sandstone), preferentially water-wet system. For illustrative purposes, a very simple model which depicts the reservoir flow channels as a bundle of capillaries, each of which has a non-constant, smoothly-varying diameter, is shown in Figure 2-7:

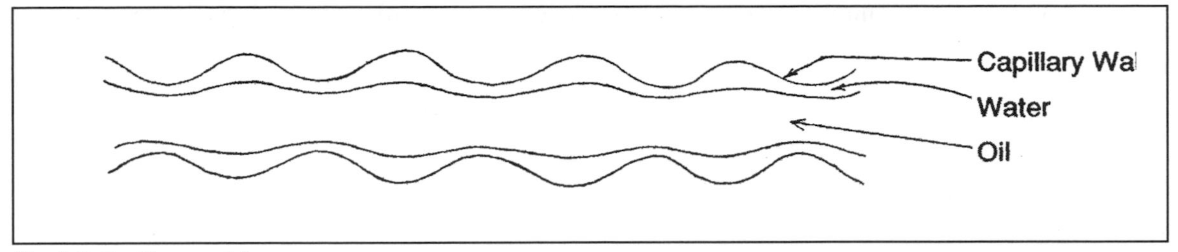

Fig. 2-7. Capillary model.

At all locations, $S_w + S_o = 1.0$. Note that the water is situated next to the pore walls, while the oil is in the center of the pore channels. Thus, for the preferentially water-wet case, it is easier for the oil to flow than for the water.

Relative permeability is usually represented as a function of saturation. A typical graph of water-wet relative permeabilities vs. fluid saturation is shown in Figure 2-8. It should be clear that a graph of effective permeabilities would have the same shape as relative permeabilities; the only difference is that the vertical axis scale would range from zero to "k" (absolute permeability) instead of zero to one.

The relative permeability relationships depicted in Figure 2-8 may be divided into three regions. Region A represents water saturations from zero to S_{wi}. The significance of "S_{wi}," the irreducible water saturation, is quite important. This is the minimum water saturation that can be obtained in the reservoir under normal operations. So, Region A would only be of interest in the laboratory or possibly in some exotic recovery process. Region A exhibits "funicular" (continuous, cord-like) saturation in the oil phase and "pendular" (not continuous, but still in grain contact) saturation in the water phase.

Region B ($S_{wi} \leq S_w \leq 1 - S_{or}$), Figure 2-8, shows the relationships existing when both the water and oil phases have funicular saturations. "S_{or}," the residual oil saturation, is the oil saturation where the oil can no longer flow in a normal immiscible water/oil system. So, during primary and secondary recovery, the oil saturation cannot be reduced below this value. For non-exotic reservoir recovery processes, actual reservoir behavior will be contained within Region B. Notice that the sum of k_{ro} and k_{rw} is less than one; so with two immiscible phases in porous media, each phase hinders the flow of the other phase.

Region C of Figure 2-8 is usually only of academic interest. When the oil saturation is less than S_{or}, the oil phase becomes insular (discontinuous droplets or "islands" in the flow channels) and the water saturation is funicular. These conditions cannot be reached throughout the reservoir under normal primary or secondary recovery operations.

Laboratory Determination of End Point Saturations

Normal steps in the laboratory are as follows:

1. Clean core; evacuate all fluids; and vacuum-dry, in a low temperature oven.

2. Saturate with water. Then, $S_w = 1.0$, therefore $k_{rw} = 1.0$. At this point, absolute permeability can be measured.
3. Inject oil until no further water is produced. Water saturation now equals S_{wi}.
4. Now inject water until no further oil is produced. Oil saturation is equal to S_{or}.

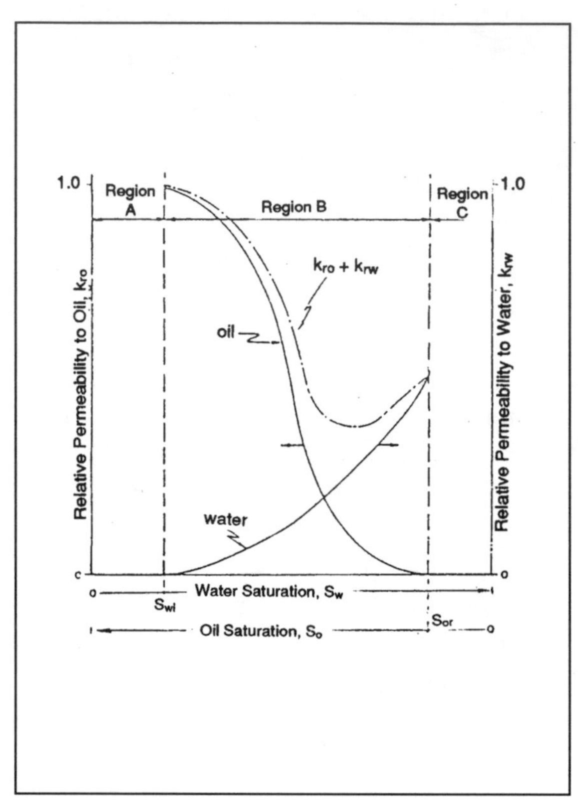

Fig. 2-8. Typical graphical relationships of the relative permeabilities to oil and water in water-wet porous media.

Fig. 2-9. Typical graphical relationships of the relative permeabilities to oil and water in oil-wet porous media.

If the oil and water densities are different enough, then if the core is weighed at the end of each step, then porosity, S_{wi}, and S_{or} may be calculated. Otherwise, material balance considerations may be used.

For preferentially water-wet sandstones, irreducible water saturation normally will range from 10 to 50%, with an average value being near 25%. Residual oil saturation typically ranges from 5 to 30%, with an average of 15%. Most sandstone oil reservoirs are preferentially water-wet.

This discussion concerns immiscible displacement, named because the two phases (oil and water) cannot mix with each other and become a homogeneous fluid due to the interfacial tension between the two phases at the oil/water interface. This type of displacement, which is the most natural, only occurs in the range of saturations where both phases can flow.

Our capillary model, as shown in Figure 2-7, implies that coaxial flow (or two phases flowing in a single capillary flow channel) would occur in the movable saturations region.

However, this is not normally seen in the laboratory. In fact, it has been shown[13] that in intergranular porous media, each small capillary-like channel is either flowing 100% water or 100% oil, not both phases together.
Now, let's look at a piece of the rock:

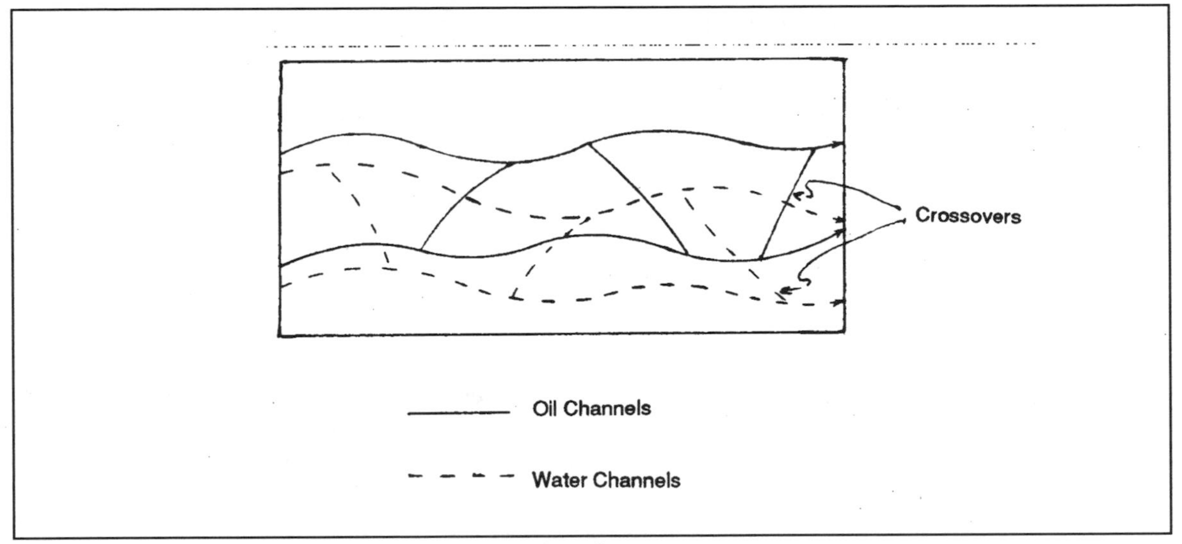

Fig. 2-10. Intergranular porous medium: oil-water multiphase flow.

In Figure 2-10, both phases are in the movable saturation range. For preferentially water-wet systems, all the oil "pipelines" are also thought to have a water layer along the channel walls.

Steady State Permeability Measurement in the Laboratory

First, oil and water with a particular constant ratio, e.g., 1 to 1, or $q_o = q_w$, are injected into the core. The injection at this constant rate ratio of oil to water, is maintained until flow of each phase out of the core is equal to flow into the core. At this point, water saturation is determined (by weighing the core, material balance, or through some other method). Having already determined needed core properties (absolute permeability, cross-sectional area, and length) and fluid properties (water viscosity and oil viscosity), we can use Darcy's Law in its relative form:

$$q_i = \frac{k A \Delta p}{\mu_i L} k_{ri} \tag{2-8}$$

to calculate both k_{rw} and k_{ro}. Indeed, for this particular ratio of oil to water flow rates (1 to 1), $k_{rw} / \mu_w = k_{ro} / \mu_o$. See Figure 2-11.

Fig. 2-11. Steady state relative permeability measurement with example (1 to 1) oil to water flow rate ratio.

Now, using other rates and Darcy's Law, the relative permeability curves may be generated.

Oil-Wet Systems

Figure 2-9 shows the relative permeability relationships for the same intergranular rock, water/oil system as before, but now, it is preferentially oil-wet. Therefore, water is the non-wetting phase and is situated in the middle of the pore spaces.

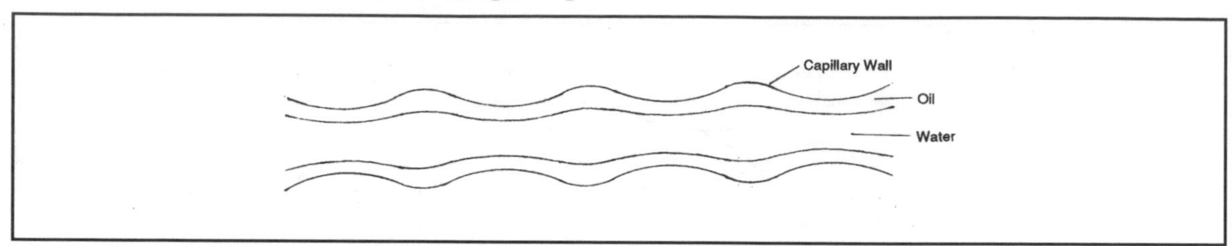

Again, S_{wi} and S_{or} are the endpoints of the movable saturations region. Note that the oil-wet curves are basically mirror images of the water-wet relationships. This is to be expected since the water and oil have merely exchanged places in the pores. In the oil-wet case, water is out in the center of the flow channels, making it easier for the water to flow. This is not a good situation because at the economic limit water cut, higher oil saturations will remain in the reservoir compared to the water-wet case. Thus, oil-wet reservoirs generally have lower percentage recoveries.

In the preferentially oil-wet case, the irreducible water saturation range is lower than for the water-wet case and is usually between 5 and 30%, with the average being about 15%. Residual oil saturation is in the range of 10 to 50%, with a 25% average. End point ranges and averages are just reversed from the water-wet case. So, although water moves with greater ease in an oil-wet system, the initial oil saturation is generally higher than in a water-wet system.

Current theories usually have water in the formation first. Then, at a later time, oil migrates to the formation. Therefore, to make the formation oil-wet, the oil must somehow get the last layer of water off of the rock surface. How could this be accomplished? Another theory targets the composition of the oil. Higher API gravity (light) oils will usually tend to be water-wet; whereas, very heavy (low gravity) crudes will more often be contained in oil-wet reservoirs. Rock wettability can be altered by adsorbable (surface-active) crude oil components. When these components adsorb onto the rock, the surface is made more oil-wet. Although these surface-active components are present in many different types of oil, they are more prevalent in the heavy crudes. Exceptions to this rule are: (1) Massive Wilcox formation, Oklahoma City Field, 41° API, $S_{wi} = 3\%$ (which is strongly indicative of oil-wetness); and (2) Brooks Sand, Cat Canyon Field, California, 4 to 7° API, unconsolidated sand, which is known to be water-wet.

A higher percentage of carbonates are oil-wet than are sandstones. Perhaps the reason for this stems from the fact that both oil and carbonate rock have organic backgrounds.

To determine wettability from a set of relative permeability curves (say Fig. 2-8 or Fig. 2-9), go to the 50% saturation point on the horizontal axis. From this point, move up vertically to the lowest curve. This curve is for the fluid which flows with the greater difficulty and represents the permeability to the wetting phase. In Figure 2-8, the lower curve at the 50% saturation point is for the water, suggesting that the formation is preferentially water-wet. Similarly, in Figure 2-9, the curves indicate the formation to be oil-wet. For intermediate wettability, the two curves will be fairly close to each other at the 50% saturation point.

Gas/Oil Systems

Figure 2-12 shows the typical graphical relationships for effective permeabilities of gas and oil in porous media. As in the oil/water case, a relative permeability plot will have the same shape as Figure 2-12, but the vertical axis ranges from zero to one instead of zero to "k." With the same reasoning applied to the oil/water case, the shapes of the curves indicate oil to be the wetting phase in this system, and gas is the non-wetting phase. It is highly unlikely that the gas could ever exist as the wetting phase. In most reservoirs of practical interest, all three phases are actually present. Where the flow of gas and oil is of primary importance, the assumption that the water present does not exceed its irreducible value is necessary.

With this assumption, the water can be considered as being immobile, serving merely to reduce the pore space, and simplify the pore configuration. Note that Figure 2-12 has an abscissa which represents the hydrocarbon pore space. The gas or oil saturation on a total pore space basis can be readily determined by multiplying each saturation by $(1 - S_{wi})$, which is the fraction of the total pore volume that the hydrocarbons occupy. Note that in the case represented by Figure 2-12, in Region A the oil exists in a discontinuous state, while gas exists in a funicular or continuous filament state. In Region B, the gas and oil both exist in a funicular state. In Region C, the gas exists in an insular state, while the oil exists in the funicular state.

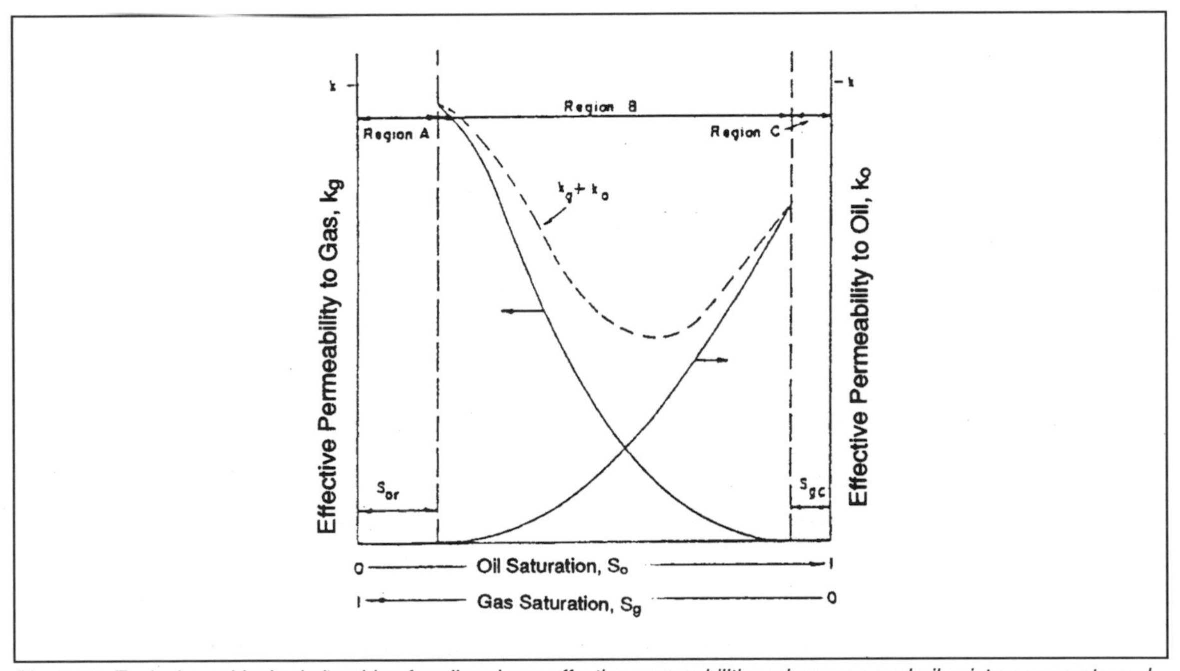

Fig. 2-12. Typical graphical relationships for oil and gas effective permeabilities where gas and oil exist as seperate and distinct phases.

An oil zone is discovered with a free gas saturation equal to zero. At the original conditions, any free gas in the reservoir is above the oil zone in the "gas cap." With production, the reservoir pressure within the oil zone will decline. If the pressure drops low enough (to the "bubblepoint" pressure), then free gas begins to be liberated from the oil. Thus, with decreasing pressure below the bubblepoint, S_g is increasing in the oil zone. The equilibrium gas saturation, S_{gc} (also called the critical gas saturation), represents the saturation at which the first permeability to gas is attained. Lesser gas saturations exist in isolated pockets which are not contiguous; but at the critical gas saturation, a continuous filament is formed which then allows flow of gas within the porous medium. Similarly, the loss of permeability to the oil phase occurs when the oil saturation is reduced to the residual value, S_{or}.

Frequently, what is needed in reservoir engineering calculations (especially with solution gas drive reservoirs) is the ratio, k_g/k_o. (Note that $k_g/k_o = k_{rg}/k_{ro}$). A typical gas/oil, relative permeability curve is illustrated in Figure 2-13. As shown, semilog paper is normally used for this plot. Here, k_g/k_o is plotted vs. liquid saturation. However, it is also commonly plotted versus gas saturation.

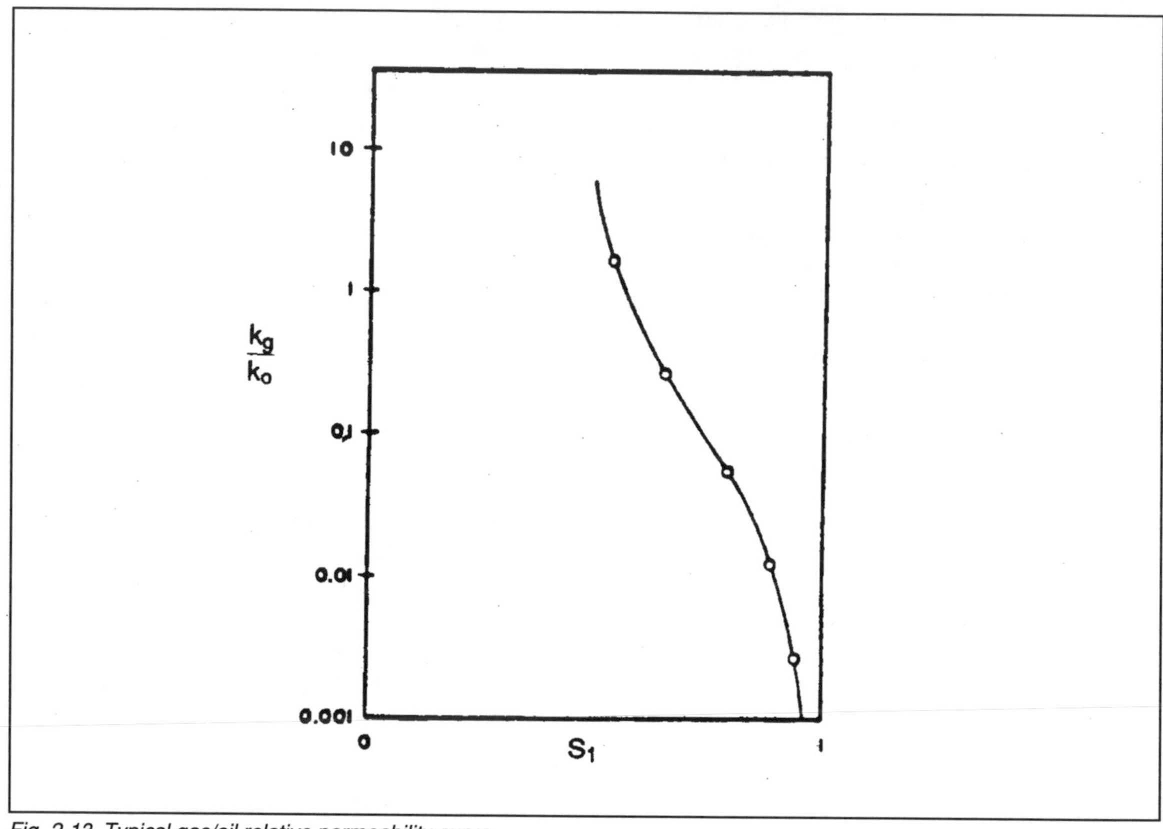

Fig. 2-13. Typical gas/oil relative permeability curve.

The reader may misinterpret previous comments on multiphase flow as indicating that each individual pore contains oil, gas and water saturations in the percentages that the figures would indicate. This is not the case. As was discussed in the section on oil/water systems, laboratory experiments[13] indicate that for two phase flow of oil and gas in intergranular porous media, each capillary-size channel will be transmitting 100% oil or 100% gas, not both phases together in the same channel. For a carbonate rock, it is probably possible to have both phases flowing in a large vug, but this does not usually happen in a sandstone. The gas, being the non-wetting phase, will generally be found in the center of the larger pores, while the smallest pores very likely have no gas saturation at all. The only possible way that each pore could have exactly the same saturation percentages of each phase would be if each pore had exactly the same size, shape, and wettability. Such a formation does not exist in nature. The saturations considered are therefore a statistical average representation of the entire system.

In the case of gas/oil flow, visual experiments[13] have shown that as the pressure is reduced to develop a gas phase, gas bubbles form at certain nucleation sites, and that the effective permeability to gas occurs when a dendritic pattern, or interconnection of pores containing primarily gas saturation, extends to the outlet of the porous medium. In like fashion, a dendritic system is conducting oil through the system. While it is useful to consider the position of the various phases on a single pore basis, it must be recognized that the flow patterns and saturations which exist in the reservoir are considerably more complex.

C. Determining Relative Permeability Curves

Recall that when dealing with more than one fluid saturation within porous media, "effective permeability" was defined as the permeability to a given fluid (such as oil: k_o). Effective permeabilities vary from zero to absolute formation permeability. Then relative permeability was defined as effective permeability divided by absolute permeability: $k_{ri} = k_i/k$; where i = oil, water, or gas. And $0 \leq k_{ro}, k_{rg}, k_{rw} \leq 1.0$. Relative permeabilities for a particular core or formation are usually represented as functions of the fluid saturations.

To make a prediction of future reservoir performance, we need the relative permeability relationships for the active fluids involved in the reservoir. Three principal ways of determining these curves are: (1) from cores through laboratory determination, (2) from published correlations, and (3) from field data.

If the formation in question has been cored, then the relative permeability relationships may be experimentally determined. As this is usually considered to be a special analysis, not all laboratories can perform such tests. Meads *et al.*[14] have suggested that it is very difficult to obtain representative average data from experimentally determined relative permeabilities. This is because most experimental work involves a limited number of small samples of core or plugs that were likely altered by the drilling fluids. Even so, when carefully performed (including the coring operation itself), this method often offers the best means of determining these data.

When no laboratory data are at hand, then generalized correlations are often used to represent the much-needed relative permeability information. However, if the reservoir characteristics deviate much from those of the reservoirs used to prepare the correlation, then the results can be in considerable error. Fortunately, many correlations are available which represent a wide variety of reservoirs and conditions. When used in conjunction with field data, such correlations can be quite accurate.

For some reservoirs, relative permeability data, over a limited range, can be derived from production data. This is a statistical treatment of the entire reservoir. It reflects actual rock and fluid properties and includes heterogeneities peculiar to the actual system. The data are only valid, however, when simultaneous production of the different fluids occurs from the formation of interest. Coning or intrusion from other formations of water or gas will invalidate the data. A requirement here is that sufficient production from the reservoir has occurred so that meaningful data on the reservoir-wide pressures and quantities of fluids removed are available. It should also be noted that this technique provides ratios of effective or relative permeabilities instead of the individual relative permeability values, for example, k_g/k_o, not k_g and k_o.

We will consider in this discussion two two-phase systems: water/oil and gas/oil. Another possible two-phase system is gas-water; but it will not be considered here because the treatment is quite similar to the analysis for the gas/oil system. For those interested in gas-water systems, reference 14 will be useful.

Water/Oil Systems

Again, for this system, using care, the curves may be determined in the laboratory.

In the absence of laboratory data, relative permeabilities for water drive reservoirs are usually estimated from correlations rather than from production data. Some reasons for this follow: (1) The equations used to predict water-displacement performance need not only k_o/k_w data (note that $k_{ro}/k_{rw} = k_o/k_w$), but also k_{ro} vs. saturation data as well. (2) Many water drive reservoirs do not produce water immediately, therefore too much production data are required to generate sufficient relative permeabilities to be of much assistance in predicting performance. (By the time you get the relative permeability relationship, not much performance is left to predict). (3) Many mechanisms exist by which water can be transported to the well for production with the oil other than coming simultaneously with the oil from the formation of interest. These other mechanisms, such as coning, water production from another formation, faulty cement job, etc., are so common that the chance of using field data to generate a valid k_o/k_w curve is small.

However, for those interested in the field data approach, the pertinent equations are:

$$k_{ro} / k_{rw} = k_o / k_w = \frac{q_o B_o \mu_o}{q_w B_w \mu_w} \tag{2-9}$$

$$S_o = \frac{(N - N_p)(B_o)(1 - S_{wc})}{N B_{oi}} \tag{2-10}$$

where:

k_{ro}/k_{rw} = the oil/water relative permeability ratio,
μ_o = oil viscosity, cp,
μ_w = water viscosity, cp,
q_o = oil rate, STB/D,
q_w = water rate, STB/D,
B_o = oil formation volume factor, Bbl/STB,
B_w = water formation volume factor, Bbl/STB,
S_o = oil saturation, fraction,
N = original oil in place, STB,
N_p = cumulative oil produced, STB,
S_{wc} = connate or initial water saturation, fraction, and
B_{oi} = initial oil formation volume factor, Bbl/STB.

Notice that two equations are needed: (1) k_o/k_w calculation, and (2) a liquid saturation equation.

Another big problem with this approach for a water/oil system is the saturation equation. Because water is not generated at each point in the reservoir, but comes in from the side or from the bottom, no one equation will adequately provide information on the saturation state.

The method of derivation and use of these equations will be illustrated with the gas/oil system. Two definitions are necessary before discussing correlations. Imbibition is the process in which the wetting phase (usually water for water/oil systems) saturation is increasing with time. Drainage implies that the wetting phase saturation is decreasing.

From petrophysical considerations (Pirson[15]) and laboratory work (Pirson *et al.*[16]), the following equations were developed for water-wet intergranular porous media, imbibition conditions, and water/oil systems:

$$k_{rw} = S_w^{\ 4} \left[\frac{S_w - S_{wi}}{1 - S_{wi}} \right]^{1/2} \qquad (2\text{-}11)$$

$$k_{ro} = \left[1 - \frac{S_w - S_{wi}}{1 - S_{wi} - S_{nwt}} \right]^{2} \qquad (2\text{-}12)$$

where:

S_{wi} = irreducible water saturation, fraction, and

S_{nwt} = irreducible non-wetting phase saturation, fraction (in most cases, this is the residual oil saturation).

for an oil-wet rock, essentially the fluid phases have changed places compared to the water-wet case. Therefore, the same equations may be used if the "w" subscripts are changed to "o", and the "o" subscript is changed to a "w". However, Pirson's equations in the oil-wet case should be used with considerable caution (as should all oil-wet correlation equations) because many rocks have a "dalmatian" or "speckled" wettability. This means that only a fraction of the total rock surface is truly oil-wet. In this case, laboratory relative permeability relationships are necessary.

Taking an entirely empirical approach, Honarpour *et al.*[17] developed a number of different relative permeability equations using laboratory measurements made on reservoir rocks from reservoirs around the world. Although not extensively tested, these statistically determined, multivariate regression correlations cover many different situations: water/oil, gas/oil, water-wet, oil-wet, intermediate-wet, sandstone, conglomerate, and carbonates. The equations presented in Table 2-3 are for consolidated rocks, and are described in Tables 2-4, 2-5, and 2-6.

Table 2-3
Empirical Relative Permeability Relationships for Consolidated Formations
(from Honarpour *et al.*[17])

$$k_{rw}^{wo} \equiv 0.035388\,\frac{(S_w - S_{wi})}{(1 - S_{wi} - S_{orw})} - 0.010874\left[\frac{(S_w - S_{orw})}{(1 - S_{wi} - S_{or})}\right]^{2.9} + 0.56556\,(s_w)^{3.6}\,(S_w - S_{wi}) \quad \text{(A-1)}$$

$$k_{rw}^{wo} = 1.5814\left(\frac{S_w - S_{wi}}{1 - S_{wi}}\right)^{1.91} - 0.58617\left(\frac{S_w - S_{orw}}{1 - S_{wi} - S_{orw}}\right)(S_w - S_{wi}) - 1.2484\phi\,(1 - S_{wi})(S_w - S_{wi}) \quad \text{(A-2)}$$

$$k_{ro}^{wo} = 0.76067\left[\frac{\left(\frac{S_o}{1 - S_{wi}}\right) - S_{orw}}{1 - S_{orw}}\right]^{1.8}\left(\frac{S_o - S_{orw}}{1 - S_{wi} - S_{orw}}\right)^{2.0} + 2.6318\phi\,(1 - S_{orw})(S_o - S_{orw}) \quad \text{(A-3)}$$

$$k_{ro}^{og} = 0.98372\left(\frac{S_o}{1 - S_{wi}}\right)^{4}\left(\frac{S_o - S_{org}}{1 - S_{wi} - S_{org}}\right)^{2} \quad \text{(A-4)}$$

$$k_{rg}^{og} = 1.1072\left(\frac{S_g - S_{gc}}{1 - S_{wi}}\right)^{2} k_{rg(S_{org})} + 2.7794\,\frac{S_{org}(S_g - S_{gc})}{(1 - S_{wi})}\,k_{rg(S_{org})} \quad \text{(A-5)}$$

$$k_{rw}^{wo} = 0.0020525\,\frac{(S_w - S_{wi})}{\phi^{2.15}} - 0.051371\,(S_w - S_{wi})\left(\frac{1}{k_a}\right)^{0.43} \quad \text{(A-6)}$$

$$k_{rw}^{wo} = 0.29986\left(\frac{S_w - S_{wi}}{1 - S_{wi}}\right) - 0.32797\left(\frac{S_w - S_{orw}}{1 - S_{wi} - S_{orw}}\right)^{2}\left(S_w - S_{wi}\right) + 0.413259\left(\frac{S_w - S_{wi}}{1 - S_{wi} - S_{orw}}\right)^{4} \quad \text{(A-7)}$$

$$k_{ro}^{wo} = 1.2624\left(\frac{S_o - S_{orw}}{1 - S_{orw}}\right)\left(\frac{S_o - S_{orw}}{1 - S_{wi} - S_{orw}}\right)^{2} \quad \text{(A-8)}$$

$$k_{ro}^{og} = 0.93752\left(\frac{S_o}{1 - S_{wi}}\right)^{4}\left(\frac{S_o - S_{org}}{1 - S_{wi} - S_{org}}\right)^{2} \quad \text{(A-9)}$$

$$k_{rg}^{og} = 1.8655\,\frac{(S_g - S_{gc})(S_g)}{1 - S_{wi}}\,k_{rg(S_{org})} + 8.0053\,\frac{(S_g - S_{gc})(S_{org})^2}{(1 - S_{wi})} - 0.025890\,(S_g - S_{gc})\left(\frac{1 - S_{wi} - S_{org} - S_{gc}}{1 - S_{wi}}\right)^{2}$$

$$\cdot\left(1 - \frac{1 - S_{wi} - S_{org} - S_{gc}}{1 - S_{wi}}\right)^{2}\left(\frac{k_a}{\phi}\right)^{0.5} \quad \text{(A-10)}$$

where:

k_a = air permeability, md
k_o = oil permeability, md
$k_{o(Swi)}$ = oil permeability at irreducible water saturation, md
k_{rg} = gas relative permeability, fraction
$k_{rg(Sor)}$ = gas relative permeability at residual oil saturation, fraction
k_{ro} = oil relative permeability, fraction
k_{rw} = water relative permeability, fraction
S_g = gas saturation, fraction
S_{gc} = critical gas saturation, fraction
S_o = oil saturation, fraction
S_{org} = residual oil saturation to gas, fraction
S_{orw} = residual oil saturation to water, fraction
S_w = water saturation, fraction
S_{wi} = irreducible water saturation, fraction
ϕ = porosity, fraction

Superscripts

og = oil and gas system
wo = water and oil system

Table 2-4
Summary of Data Used for Relative Permeability Equations
(from Honarpour et al.[17])

Equation	Relative Permeability Predicted	Fluids in System	Number of Sets	Number of Data Points	Lithology	Wettability
A-1	k_{rw}	water and oil	84	361	sandstone and conglomerate	water
A-2	k_{rw}	water and oil	101	478	sandstone and conglomerate	oil and intermediate
A-3	k_{ro}	water and oil	185	1,000	sandstone and conglomerate	any
A-4	k_{ro}	oil and gas	133	822	sandstone and conglomerate	any
A-5	k_{rg}	oil and gas	133	766	sandstone and conglomerate	any
A-6	k_{rw}	water and oil	8	57	limestone and dolomite	water
A-7	k_{rw}	water and oil	26	197	limestone and dolomite	oil and intermediate
A-8	k_{ro}	water and oil	54	593	limestone and dolomite	any
A-9	k_{ro}	oil and gas	30	273	limestone and dolomite	any
A-10	k_{rg}	oil and gas	30	227	limestone and dolomite	any

Table 2-5
Ranges of Rock Properties and Fluid Saturations Used in Developing Equations for Water/Oil Systems
(from Honarpour et al.[17])

Equation	Porosity Range (%)	Air Permeability Range (md)	Water Saturation Range (%)	Residual Oil Saturation Range (%)
A-1	9.9 to 30.3	4.1 to 2,640	6.7 to 70.0	16.4 to 51.4
A-2	9.1 to 37.1	0.2 to 4,000	3.6 to 64.0	7.3 to 50.0
A-3	9.1 to 37.1	0.2 to 4,000	3.6 to 70.0	7.3 to 56.0
A-6	10.1 to 15.7	0.05 to 800	18.5 to 43.2	10.0 to 36.4
A-7	8.0 to 29.1	0.04 to 490	6.5 to 40.5	12.7 to 46.0
A-8	6.5 to 31.1	0.05 to 713.4	8.6 to 43.2.	10.0 to 53.7

Table 2-6
Ranges of Rock Properties and Fluid Saturations Used in Developing Equations for Oil-Gas Systems
(from Honarpour et al.[17])

Equation	Porosity Range (%)	Air Permeability Range (md)	Water Saturation Range (%)	Gas Saturation Range (%)	Residual Oil Saturation Range (%)
A-4	9.0 to 45.0	0.2 to 4,000	0 to 60.0	0.0001 to 34	4.0 to 66
A-5	9.0 to 45.0	0.52 to 4,000	0 to 47.7	0.0001 to 34	4.0 to 66
A-9	6.7 to 29.1	0.2 to 2,000	0 to 40.0	1.0 to 40	5.0 to 56
A-10	6.7 to 29.1	0.2 to 2,000	0 to 40.0	1.0 to 40	5.0 to 56

Gas/Oil Systems

Gas/oil relative permeabilities may be determined from a special core analysis in the laboratory. However, this is often not done,and other approaches are needed. Gas/oil relative permeability data are needed to predict the performance of solution-gas drive and gas-cap drive reservoirs and those reservoirs where gas injection is being considered as a possible secondary recovery method.

One of the equations necessary for determination of a "field" gas/oil relative permeability curve follows naturally from the definition of the instantaneous producing gas/oil ratio, or GOR:

$$GOR = R = Gas \ produced \ per \ day \, / \, Oil \ produced \ per \ day$$

Gas produced per day = (Gas liberated from the produced oil per day) + (Free gas produced per day)

Hence,

$$R = R_s + \frac{C\, k_g\, h\, (\Delta p)}{B_g\, \mu_g \ln (r_e / r_w)} \Big/ \frac{C\, k_o\, h\, (\Delta p)}{B_o\, \mu_o \ln (r_e / r_w)}$$

Simplifying:

$$R = R_s + \frac{B_o\, k_g\, \mu_o}{B_g\, k_o\, \mu_g} \qquad (2\text{-}13)$$

where:

R = instantaneous or producing gas/oil ratio, SCF/STB,
R_s = solution gas/oil ratio, SCF/STB,
B_o = oil formation volume factor, Bbl/STB,
B_g = gas formation volume factor, (volume at reservoir conditions divided by volume at standard conditions), Bbl/SCF,
k_g/k_o = effective (or relative) gas/oil permeability ratio,
μ_o = oil viscosity, cp,
R_s = gas viscosity, cp, and
C = units constant.

The "instantaneous gas/oil ratio equation," Eq. (2-13), contains the instantaneous producing gas/oil ratio, R, which can be obtained from production data. The factors B_o, B_g, μ_o, μ_g and R_s are mainly functions of pressure and can be determined from laboratory PVT data or correlations.

Rearranging Eq. 2-13:

$$k_g / k_o = (R - R_s) \cdot \frac{B_g}{B_o} \frac{\mu_g}{\mu_o} \qquad (2\text{-}14)$$

The gas/oil permeability ratio is a function of the total liquid saturations existing in the reservoir. This can be expressed as:

$$S_l = S_w + S_o = S_w + \left[\frac{N - N_p}{N} \cdot \frac{B_o}{B_{ob}} \cdot (1 - S_w) \right] \qquad (2\text{-}15)$$

where:

the initial conditions for this equation are the bubblepoint,
S_l = total liquid saturation, fraction,
S_w = water saturation (assumed to be at the irreducible level), fraction,
S_o = oil saturation, fraction,
N = oil in place, STB,
N_p = cumulative oil production (from the bubblepoint), STB,
B_o = oil formation volume factor, Bbl/STB,
B_{ob} = B_o at the bubblepoint, Bbl/STB.

Actually, for Eq. 2-15, N should be the oil in place at the bubblepoint.

This equation may be written directly by noting the following:

1. The initial condition is the bubblepoint, because this is the pressure at which free gas begins to leave the oil. So, only two fluids are present: oil and connate water. Therefore, the initial oil saturation is equal to (1 - S_w).

2. The original volume of the oil at the bubblepoint is NB_{ob}.

3. The reservoir volume of oil remaining in the reservoir after production of N_p barrels of stock-tank oil is $(N - N_p)B_o$.

Therefore, the oil saturation after N_p barrels of oil production is just the ratio of the oil volume in (3) divided by the original oil volume (2), and then multiplied by the original oil saturation in (1). Of course, the total liquid saturation, S_l, is equal to the sum of the water saturation and the oil saturation, and the gas saturation is equal to the quantity $(1 - S_l)$. In the equation for S_l, the water saturation would be obtained from log and/or core analysis; and the original oil in place, N, is usually approximated from volumetric considerations and/or material balance calculations. The cumulative stock-tank oil produced, N_p, would be available from production data. The oil pressure dependent factors (B_o, R_s and μ_o) needed to solve Equation 2-14 and Equation 2-15 are best determined from laboratory tests on a representative fluid sample. As will be discussed in the Fluid Properties chapter, the gas pressure dependent functions of B_g and μ_g are often obtained from correlations. Table 2-7 suggests a form which might be convenient for calculation of k_g / k_o data vs. saturation.

Figure 2-13 is a typical plot of gas/oil relative permeability data where considerable production was available to define the curve down to low liquid saturation values. Such will not normally be the case. Most often, only sufficient production and fieldwide pressure data will be available to define the shape of the curve over the high liquid saturation portion of the curve. However, this is of considerable value within the limitations imposed by the quality of the production data, since it permits the calibration of data from some other method such as published correlations or laboratory data.

Table 2-7
CALCULATION OF GAS/OIL RELATIVE PERMEABILITY VALUES FROM PRODUCTION DATA

(1) p	(2) N_p	(3) R	(4) $\frac{N - N_p}{N}$	(5) B_o / B_{ob}	(6) $S_l = S_w + (1 - S_w) \cdot (4) \cdot (5)$	(7) $F = \frac{B_o \mu_o}{B_g \mu_g}$	(8) R_s	(9) $\frac{k_g}{k_o} = \frac{R - R_s}{F}$
p_1	N_{p1}	R_1	$\frac{N - N_{p1}}{N}$	B_{o_1} / B_{ob}	S_{l_1}	F_1	R_{s_1}	$(k_g / k_o)_1$
p_2	N_{p2}	R_2	$\frac{N - N_{p2}}{N}$	B_{o_2} / B_{ob}	S_{l_2}	F_2	R_{s_2}	$(k_g / k_o)_2$
p_3	N_{p3}	R_3	$\frac{N - N_{p3}}{N}$	B_{o_3} / B_{ob}	S_{l_3}	F_3	R_{s_3}	$(k_g / k_o)_3$
↓	↓	↓	↓	↓	↓	↓	↓	↓

Using petrophysics theory, the efforts of several different investigators[18,19,20,21] led to the following relative permeability correlations for a gas/oil system:

$$k_{ro} = \left[\frac{S_o}{1 - S_{wi}} \right]^4 \tag{2-16}$$

$$k_{rg} = \left[1 - \frac{S_o}{S_m - S_{wi}} \right]^2 \left[1 - \left(\frac{S_o}{1 - S_{wi}} \right)^2 \right] \tag{2-17}$$

where:

$S_m = 1 - S_{gc}$
S_{gc} = " critical gas saturation" or that saturation of gas above which flow is possible, and below which gas cannot flow.

Equation 2-16 and 2-17 are generally attributed to Corey.[21] These relationships have been found to be valid for reservoir materials not having extensive stratification, large solution channels, or large amounts of cementing materials. Further, these equations have application for a drainage mechanism in which the wetting phase (oil) is decreasing in saturation. This is normally the case in a solution-gas-drive reservoir.

The gas relative permeability equation contains the quantity, S_m, which is equal to 1 - S_{gc}. S_{gc} is the critical gas saturation which is that saturation where the free gas becomes mobile. Normally, for sandstones, $4\% < S_{gc} < 7\%$. So any time that $S_g > S_{gc}$, then, there will be free gas flow in the reservoir.

Of course, the ratio, k_g/k_o or k_{rg}/k_{ro}, can be determined by dividing Equation 2-17 by Equation 2-16.

Another correlation developed by Wahl *et al.*[23] is based entirely upon relative permeability data from field measurements in sandstone reservoirs:

$$k_{rg} / k_{ro} = \xi \, [\, 0.0435 + 0.4556 \, (\xi) \,] \tag{2-18}$$

where:

$$\xi = \frac{1 - S_{gc} - S_w - S_o}{S_o - C}$$

S_{gc} is the equilibrium or critical gas saturation as a fraction of the total pore space, while C is a constant normally taken to be 0.25. Closer examination of the constant, C, indicates that is is likely that the residual oil saturation, S_{or}, is being represented.

Torcaso and Wyllie[22] have compared the equations for gas/oil relative permeability as presented by Wahl *et al.* and Corey, with the results illustrated in Figure 2-14. The agreement between the two sets of curves is excellent except at high values of irreducible water saturation ($S_{wi} > 0.40$). The close agreement between the equations justifies the mathematical model used by Corey, Burdine, Wyllie, and Spangler, etc., since the Wahl *et al.* equation was based upon actural field measurements.

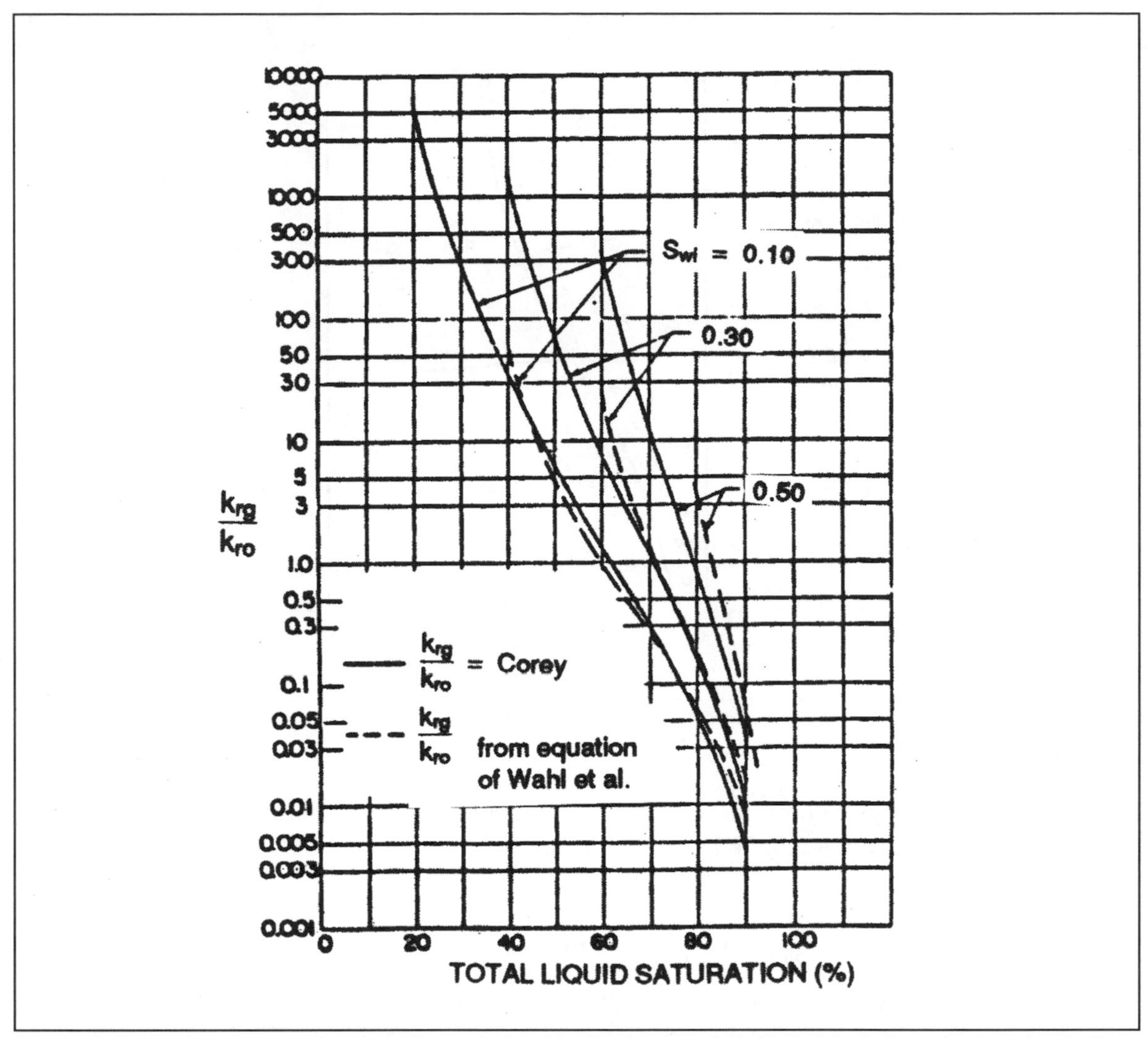

Fig. 2-14. Comparison between k_{rg}/k_{ro} values calculated by the Corey and Wahl et al. techniques for three irreducible water saturation values (Torcaso and Wyllie[22]). Permission to publish by SPE.

Equations 2-16 through 2-18 have application where reservoir rocks contain an irreducible water saturation and where the gas saturation increases at the expense of the oil saturation during the production process.

Other gas/oil relative permeability empirical equations are presented in Tables 2-3 through 2-6. Strictly speaking, these Honarpour *et al.*[17] relationships are valid for consolidated rocks only.

Correlations are particularly powerful when used in conjunction with field data. First, the field-data-calculated relative-permeability ratio points are plotted as shown in Figure 2-15, and then an appropriate correlation is plotted on the same graph.

The correlation curve and the field data points hopefully will be roughly parallel. As seen in Figure 2-15, the field data points only extend over a small range of gas saturation values, which is the usual case. Now, the field data points are extrapolated parallel to the published correlation line.

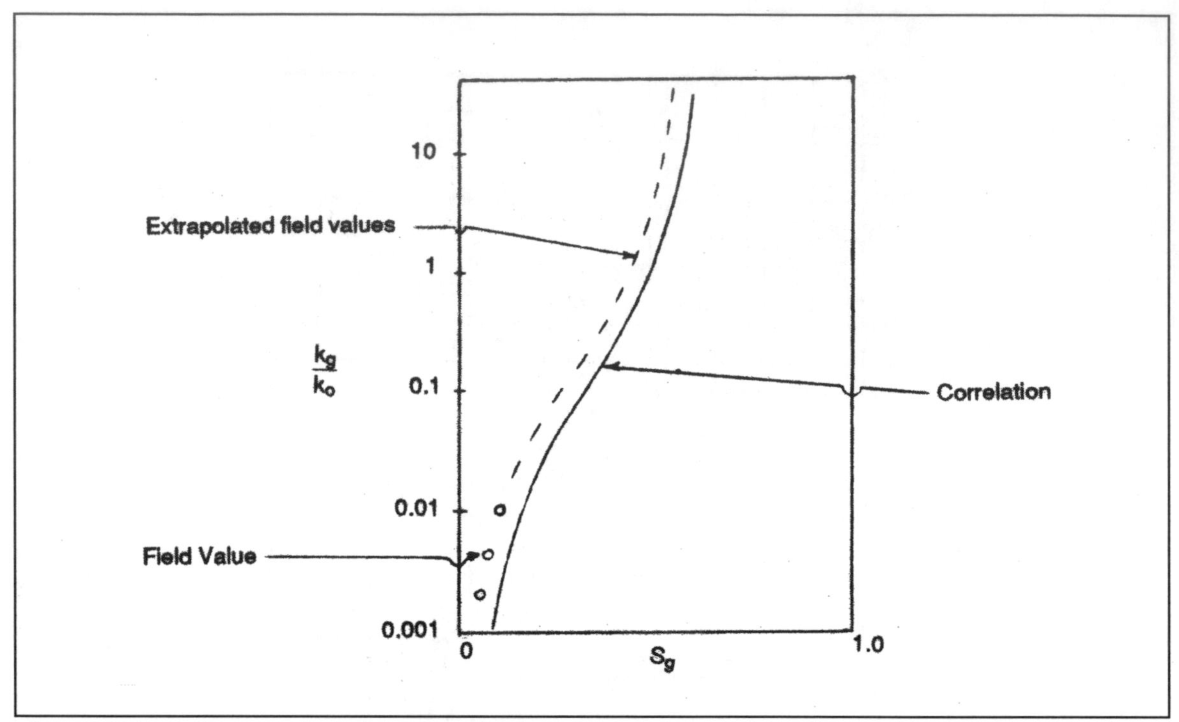

Fig. 2-15. Example illustrating the extrapolation of field k_g/k_o values (calculated from production data) parallel to a correlation k_g/k_o curve.

This is what is meant by "calibration" of a general correlation to a particular system. Of course, the relative permeability values could have been graphed versus liquid saturation instead of gas saturation.

THREE-PHASE RELATIVE PERMEABILITY

It is the authors' belief that three mobile phases (gas, oil, and water) seldom exist at the same point in a reservoir. For this reason, two-phase relative permeability relationships usually suffice for understanding or predicting fluid flow behavior. However, Wyllie and Gardner[20] have presented three-phase relative permeability relationships for water-wet systems, operating on the drainage cycle with respect to water and oil. In this instance, the gas saturation is increasing at the expense of the water and oil saturations. Where the formation can be represented as an unconsolidated sand with well-sorted grains, the following relationships may apply:

$$k_{rg} = \frac{S_g^3}{(1 - S_{wi})^3} \tag{2-19}$$

$$k_{ro} = \frac{S_o^3}{(1 - S_{wi})^3} \tag{2-20}$$

$$k_{rw} = \left[\frac{S_w - S_{wi}}{1 - S_{wi}} \right]^3 \tag{2-21}$$

Where the reservoir rock is a cemented sandstone, oolitic limestone, or vugular rock, the equation forms are:

$$k_{rg} = \frac{S_g^2 \, [\, 1 - S_{wi})^2 - (S_w + S_o - S_{wi})^2]}{(1 - S_{wi})^4} \tag{2-22}$$

$$k_{ro} = \frac{S_o^3 \, (2S_w + S_o - 2S_{wi})}{(1 - S_{wi})^4} \tag{2-23}$$

$$k_{rw} = \left[\frac{S_w - S_{wi}}{1 - S_{wi}} \right]^4 \tag{2-24}$$

The six preceding equations may be modified for totally oil-wet systems where the oil is the wetting phase, water the non-wetting phase and the gas non-wetting, with respect to both oil and water. In this case, the formulas apply if S_o is substituted for S_w, and vice versa. It should be noted that the above equations give reasonable values for three-phase flow case in porous media, the equations should be used with caution where a mixed or "speckled" wettability might exist, or for a cycle of changing saturations which is not parallel in concept to the drainage cycle as described. Figures 2-16 and 2-17 present a method for three-phase relative permeability mapping and specifically relate oil isoperms and water isoperms, respectively, for a water-wet Berea sandstone core containing oil, gas, and water.

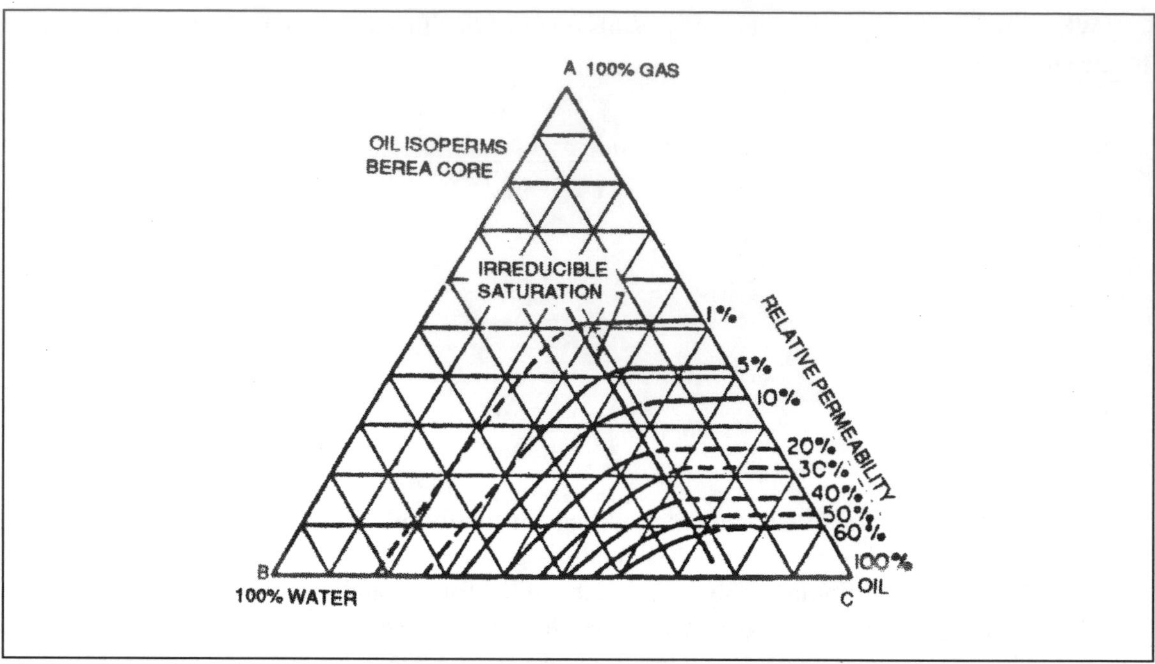

Fig. 2-16. Lines of constant k_{ro} for a water-wet Berea sandstone core containing gas, oil and water-drainage cycle with respect to water and oil (Wyllie and Gardner[20]). Permission to publish by World Oil.

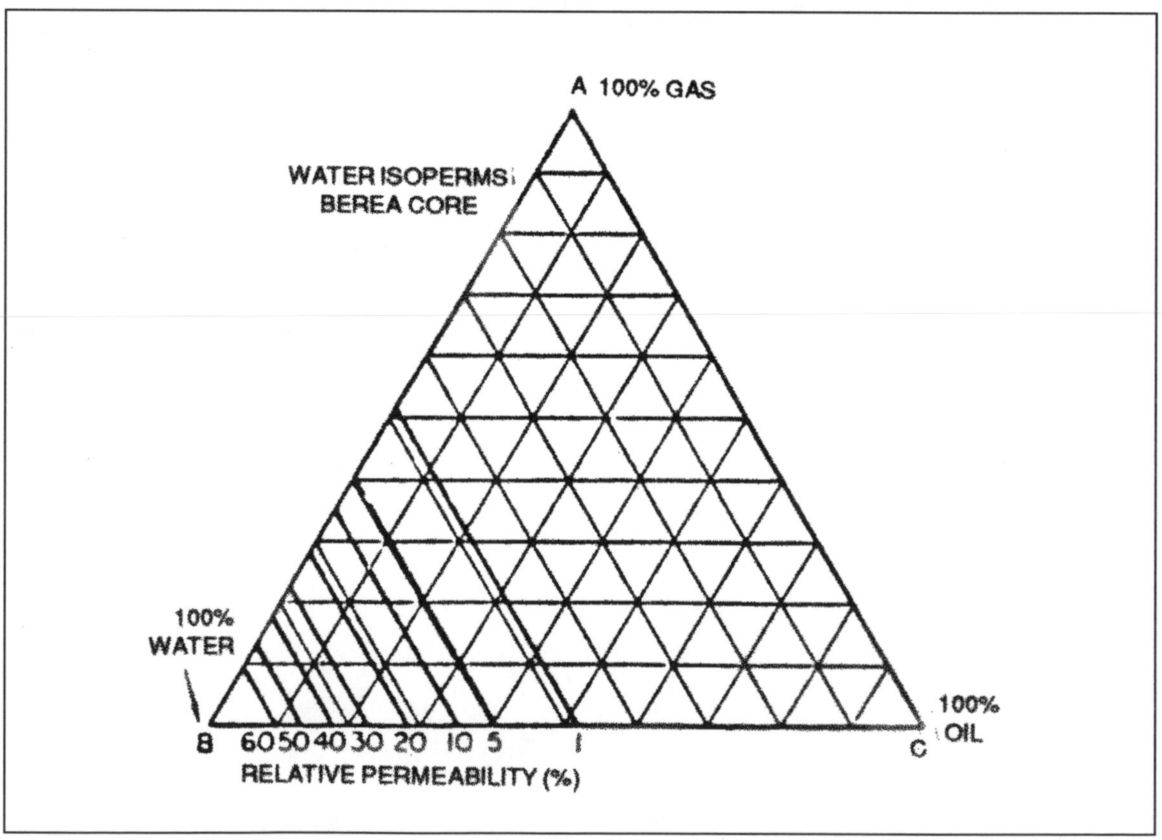

Fig. 2-17. Lines of constant k_{rw} for a water-wet Berea sandstone core containing gas, oil and water-drainage cycle with respect to water and oil (Wyllie and Gardner[20]). Permission to publish by World Oil.

RELATIVE PERMEABILITY AND ENHANCED OIL RECOVERY

Literally billions of barrels of discovered oil remain in the ground today. Most of this is held by capillary forces, a major factor of which is interfacial tension. This will be discussed in greater detail under the section on capillary pressure. Basically, it is a force acting at the interface between two immiscible fluids due to the lack of similarity between the two molecular species. The higher the interfacial tension is, the less affinity the two fluids have for each other, and the higher the resistance to mixing. Capillary forces are largely responsible for the residual saturations that are experienced in two-phase systems, e.g., in an oil/water system, a residual oil saturation and an irreducible water saturation. The residual oil saturation is normally unproducible by conventional primary and/or secondary recovery mechanisms. Many enhanced oil recovery (EOR) methods use as the basic mechanism some means of reducing interfacial tension which then reduces the capillary forces that are involved in "holding" the residual oil saturation. A typical preferentially water-wet oil/water set of relative permeability curves may be seen in Figure 2-18a.

It has been observed that residual saturations and relative permeabilities are dependent on the ratio of viscous forces to capillary forces.[24,25] This ratio, which may be written in different ways, is defined as the capillary number:

$$N_c = \frac{\mu v}{\sigma \phi}$$

where:

N_c = capillary number, fraction,
μ = fluid viscosity, cp,
v = flow velocity, cm/sec,
σ = fluid interfacial tension, dynes/cm, and
ϕ = fractional porosity.

The viscous forces are represented as the fluid viscosity, flow velocity, and flow path length (related to $1/\phi$); while the capillary force is represented by the interfacial tension. Laboratory studies indicate that as the capillary number increases (or as the interfacial tension decreases), both the oil and water relative permeability curves were found to shift upwards, indicating less resistance to flow for each phase, or that the two phases are interfering less with each other. Decreasing interfacial tension straightens out the relative permeability curves and decreases residual oil and irreducible water saturations. It has been reported[25] that at zero interfacial tension, the theoretical relative permeability curves attain an "X" shape as shown in Figure 2-18b.

It should be noted that the reduction in irreducible water saturation with decreasing interfacial tension is not always a simple matter. In instances where the interfacial tension is being decreased, the irreducible water saturation first increases[22] and then later decreases. In any event, irreducible water saturation does not decrease as quickly as does residual oil saturation in preferentially water-wet systems. This is due to the smaller pores being occupied by water. Oil, as the nonwetting phase, resides in the larger pores and is easier to displace. Conversely, preferentially oil-wet rocks have larger and more difficult residual oil saturations to influence by those EOR processes which function (in part) by interfacial tension reduction. The main objective is to lower the residual oil saturation to as close to zero as possible.

The typical range of oil/water interfacial tension is 5 to 40 dynes/cm, and this value normally must be decreased to a value below one dyne/cm before meaningful reductions in residual oil saturation occur. Total recovery of oil requires interfacial tensions less than 0.01 dyne/cm. This can be achieved with surface active agents (surfactants).

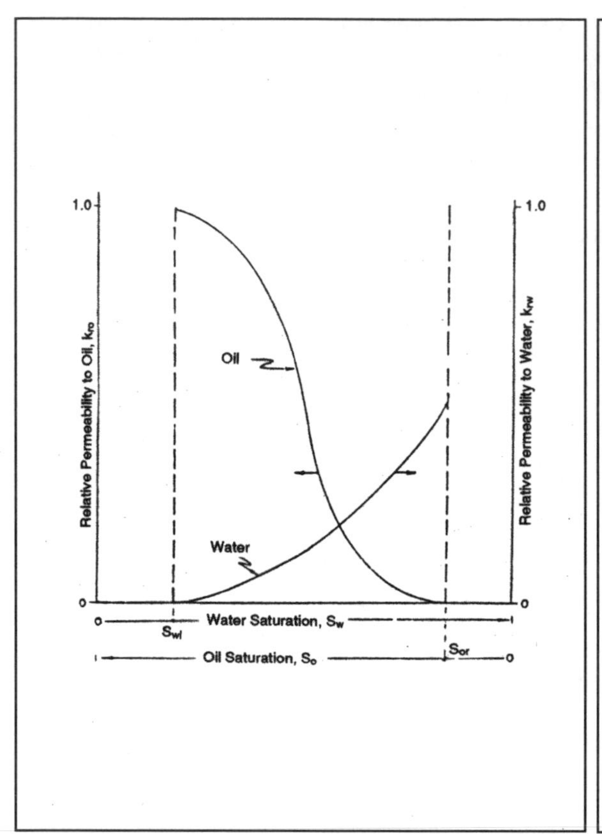

Fig. 2-18a. Typical graphical relationships of the relative permeabilities to oil and water in water-wet porous media.

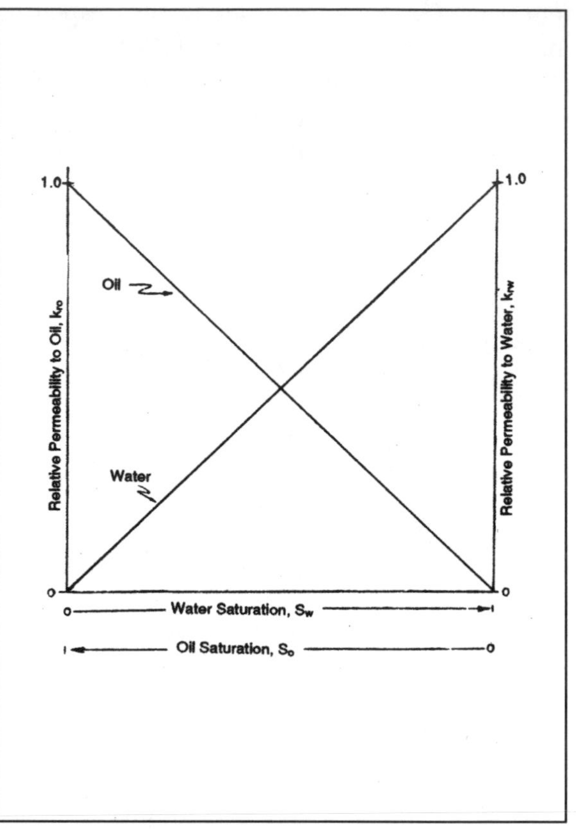

Fig. 2-18b. Typical graphical relationships of the relative permeabilities to oil and water in porous media for the theoretical case of zero interfacial tension.

CORING AND CORE ANALYSIS

In order to characterize a reservoir system, the properties of the rock need to be known. Using logging tools, reservoir properties (such as porosity, water saturation, gross formation thickness, amount of shale, etc.) often can be inferred with good accuracy. Nothing, though, can totally replace measurements made directly on reservoir rock samples.

Samples of reservoir rock can be obtained in four ways: rotary coring, sidewall coring, cable-tool coring, and drilling cuttings. Gatlin[26] and Anderson[27] may be consulted for details relative to coring practices, procedures, and tools used.

1. Conventional Coring and Resulting Fluid Saturations

Conventional coring takes place after normal (rotary) drilling down to a point just above the desired coring interval. Before running in the hole with the core barrel, care is taken that the hole is clean and the mud well conditioned. As in normal drilling, when coring, the mud pressure is greater than formation pressure. However, best results will be obtained when this differential is as small as safely possible. Conventional coring is normally done with either a (1) water-base mud or (2) an oil-base mud.

It is instructive to trace the fluid saturations that result when conventional coring a preferentially water-wet, oil-bearing sandstone that is above the bubblepoint pressure (no gas saturation) and at the irreducible water saturation. For discussion purposes, assume that the reservoir oil has an API gravity of 35° (specific gravity of 0.85) and has good mobility (k_o / μ_o). The formation is an "average" sandstone that could have the following undisturbed (before coring) saturations:

S_o	S_g	S_w	, %
75	0	25	(S_{wi})

During the coring operation, pressure within the core will be raised from formation pressure to mud column pressure, while the temperature will remain close to formation temperature (it may be lowered slightly due to mud circulation). As the coring operation is accomplished, the formation (or core) face acts as a filter with resulting water-base mud filtrate (mainly water) invading the core. Thus, a waterflood is performed on the core which operates on the movable oil saturation: $1 - S_{wi}$ to S_{or}. The oil in the core will generally be left at the residual saturation. So, saturations after coring, while still at formation depth, should approximate:

S_o	S_g	S_w	, %
15 (S_{or})	0	85	

The core is recovered with pressure and temperature becoming those existing at the surface. Gas (dissolved in the oil) comes out of solution causing mobile liquids to be ejected from the core. Since the oil is at residual saturation, the water is the only mobile liquid in the core. The temperature drop imposed on the core fluids causes a small amount of thermal shrinkage. The total effect of the gas drive and the thermal shrinkage is to reduce the water saturation by roughly one-half. The oil shrinks for two reasons: (1) the loss of gas in solution, and (2) the drop in temperature. The amount of oil shrinkage is usually in the range of 10 to 50% of the saturation before recovery. A good approximation of surface oil saturation is residual oil saturation (that saturation just before core recovery) divided by the oil formation volume factor. (Oil formation volume factor is equal to the reservoir volume of oil and solution gas divided by the corresponding stock tank volume of the oil). So, surface fluid saturations could be the following:

S_o	S_g	S_w, %
10	50–55	35–40

Consider the same water-wet, "average" sandstone formation again, but this time the coring is done with an oil-base drilling fluid. The fluid saturations in the core before coring will be the same as in the last example: irreducible water saturation, no gas, and the balance being oil. During the coring operations, the mud filtrate will be oil (probably diesel fuel). This time movable saturations are subjected to an "oilflood." Because water was at the irreducible saturation before coring began, none is displaced by the oilflood. The pressure in the core has been raised to that of the mud system; hence, no gas has come out of solution. So, gas saturation in the core remains zero. Therefore, the oil, gas, and water saturations at depth are the same after the coring operation as before. However, it is likely that oil composition in the core has been substantially altered.

After core recovery, the oil saturation will be reduced to roughly one-half of its value at depth. The gas associated with the remaining reservoir oil expands, according to Boyle's Law, as it comes out of solution during core recovery and provides a gas drive. Some oil shrinkage due to the temperature drop also occurs.

The water saturation usually changes only in a minor way during core recovery. This change is due to thermal contraction and the evolution of gas. The reservoir water saturation divided by the water formation volume factor (typically 1.01 to 1.10 reservoir volume/surface volume) should approximate surface water saturation.

Thus, when coring with an oil-base mud, a summary of the saturation history for an average sandstone core might be:

	S_o	S_g	S_w, %
Before Coring (@ depth)	75	0	25 (S_{wi})
After Coring (@ depth)	75	0	25 (S_{wi})
Surface	40-45	Rest	23-24

Core Saturations Review

The presence of gas in the core analyses does not necessarily indicate that free gas is in the reservoir. When coring with a water-base fluid, the oil saturations will tend to be at or below residual values even though the in-situ oil saturation may be (1-S_{wi}). With a conventional core analysis of cores recovered using water-base muds, any adjusted residual oil saturation usually suggests the presence of an oil reservoir.

2. Rubber Sleeve Coring

If conventional coring is attempted in a poorly consolidated formation, the operation normally will not be successful since the matrix will wash away. For such unconsolidated formations or for soft, friable, or semi-consolidated reservoir zones, rubber sleeve coring can be attempted. Here, a rubber sleeve is drawn over the core upon entry into the inner core barrel.

This type of coring may also be done with a polyvinyl chloride solid sleeve (or fiberglass sleeve for even greater strength), which may be better because it cannot be twisted. When a rubber sleeve core is twisted, the core properties are altered.

Unfortunately, the traditional rubber sleeve core barrel is unsatisfactory for coring hard, fractured formations as sharp edges cut the rubber sleeve. In addition, the rubber sleeve cannot be used at high temperatures. A new coring tool is now available that reportedly[28] overcomes these limitations. The most outstanding feature of the new "anti-jamming" core barrel is a specially designed wire-mesh sleeve. This metal-braided sleeve, similar in principle to a "Chinese finger trap," has been incorporated into the inner core barrel. When compressed, the wire mesh increases in diameter, making loading easy. When placed in tension, the wire-mesh sleeve decreases in diameter and grips the core, keeping the core to original diameter and preventing jamming. The tool is also reported as being able to withstand high temperatures.

3. Pressure Core Barrel

Pressure coring has become an accepted reservoir engineering tool.[29] As we have seen, when gases are present (even in the form of solution gas), conventional, nonpressurized, coring techniques are unable to recover meaningful in-situ fluid saturations. Pressure coring solves this problem by maintaining pressure on the core specimen until the core can be brought to the surface, at which time the core is frozen to immobilize core fluids.

Pressure coring basically involves four steps. (1) The core is cut with essentially the same technology as in conventional coring. (2) Trapping of pressure is accomplished by mechanical actions which create a seal at the top and bottom of the tool. This makes a pressure vessel out of the core barrel while it is at the bottom of the wellbore. The system also maintains pressure as the core is withdrawn from the wellbore. (3) Freezable coring fluid within the core barrel annulus is displaced by a non-freezing medium. This is necessary to allow the inner barrel, which contains the core specimen, to be removed for transport to the laboratory after freezing. (4) Freezing the core is necessary to immobilize the fluids within the core. At this point, the core barrel may be depressurized and the core removed. The core is then transported to the laboratory in the frozen state. The freezing is accomplished by placing the entire core barrel in a dry ice (or liquid nitrogen) trough for eight to twelve hours.

Unfortunately, pressure coring is expensive: up to 10 times the cost of a conventional core. Also, fluids that are movable at bottomhole conditions are displaced as in conventional coring. The special barrel simply retains coring pressures during core recovery.

4. "Sponge" Coring

The core barrel needed for this type of coring has an annulus surrounding the core that is filled with a porous and permeable sponge material. As the core barrel is raised to the surface from formation depth, the evolving gas (originally in solution in the oil) forces liquids out of the core that are absorbed in the sponge material surrounding the core. So, in the laboratory, not only is the core processed but also the sponge, to extract the liquids that were pushed out of the core. By analyzing both the core and the sponge, the saturation state existing in the core (@ depth) just after the coring operation can be determined. Of course, this is not the original saturation state of the reservoir, but at least it should be closer to the original state than that obtained with a conventional core barrel. The cost associated with sponge coring is approximately three times that of conventional coring.

5. Sidewall Coring

The sidewall coring tool is used to take small core samples from formations in holes already drilled to a predetermined depth. The tool (gun) is lowered to the desired depth and then "bullets" (tiny core collection chambers) are fired electrically. The samples are caught in the "bullets", and the tool moved to other desired depths to shoot additional sidewall cores. Each gun can recover a number of different cores (approximate range of 30 to 50), and sometimes two guns can be run in tandem. The tool is usually run after the well logs have been run and reviewed on site. Much can be said supporting sidewall coring as a way to get a "piece of the rock." Conventional core properties usually can be measured.

6. Standard Core Analysis

After a core has been sent to the laboratory with a request for a standard core analysis, then the returned report will normally have a format similar to:

Depth	Δh	ϕ	k_h	k_{90°	k_v	S_o	S_g	S_w	Comments

This information includes surface porosity, permeability, and fluid saturations of the core versus depth. Comments give brief results of a visual examination. In the standard analysis report, Δh is the analyzed interval and often is one foot.

Porosity is usually measured on a plug cut from the core (except with formations such as carbonates where whole core analysis may be needed). Porosity measurement methods were discussed near the beginning of this chapter.

For permeability measurement, core plugs (in unfractured formations) are also used. Anderson[27] reports the following core plug procedure where the operating company desires two different horizontal permeability measurements: If the core has dipping beds, the first plug (for k_h) is cut along the dip, and the second (for $k_{90°}$) is cut along the strike. Vertical permeability (k_v) is measured perpendicular to the strike of the beds and parallel to the side of the core.

Permeability determination using core plugs is normally done with an instrument called a permeameter. In the permeameter a pressure drop is imposed across the core, and then the flow rate through the core is measured. With knowledge of the core cross-sectional area, length, and fluid viscosity, permeability may be calculated using Darcy's Law ($q = (kA\,\Delta p)/\mu L$). Gas is the fluid generally used because (1) flow rates stabilize quickly due to the low gas viscosity, (2) some liquids interact with the rock matrix thereby negating results, and (3) 100% saturation of a dry core with gas is readily obtained. The gas normally used is helium or nitrogen. Use of air can cause unstable materials within the core to oxidize.

Recall that the definition for absolute permeability required the rock to be 100% saturated with a non-reacting, incompressible liquid. Gas does not exactly fit this definition. Even in a core where the absolute permeability is totally invariant with pressure, the measured permeability using gas is not independent of pressure. At low pressures, the mean free path of a gas molecule is relatively large compared to the pore opening sizes. But as pressure increases, the mean free paths of the molecules approach those of a liquid. It turns out that at low pressures, gas has a higher permeability than at high pressures. This phenomenon is referred to as the "Klinkenberg effect" or "gas slippage."

Often this is corrected by plotting k_g vs. $1/\bar{p}$ (where $\bar{p}$ is the mean pressure in the core during a test; i.e., $\bar{p} = [P_{upstream} + P_{downstream}]/2$) and extrapolating to $1/\bar{p} = 0$. This is, of course, actually where the mean pressure goes to infinity (Fig. 2-19). This extrapolated k_g is referred to as k_∞, and is taken to be the formation liquid permeability or absolute permeability.

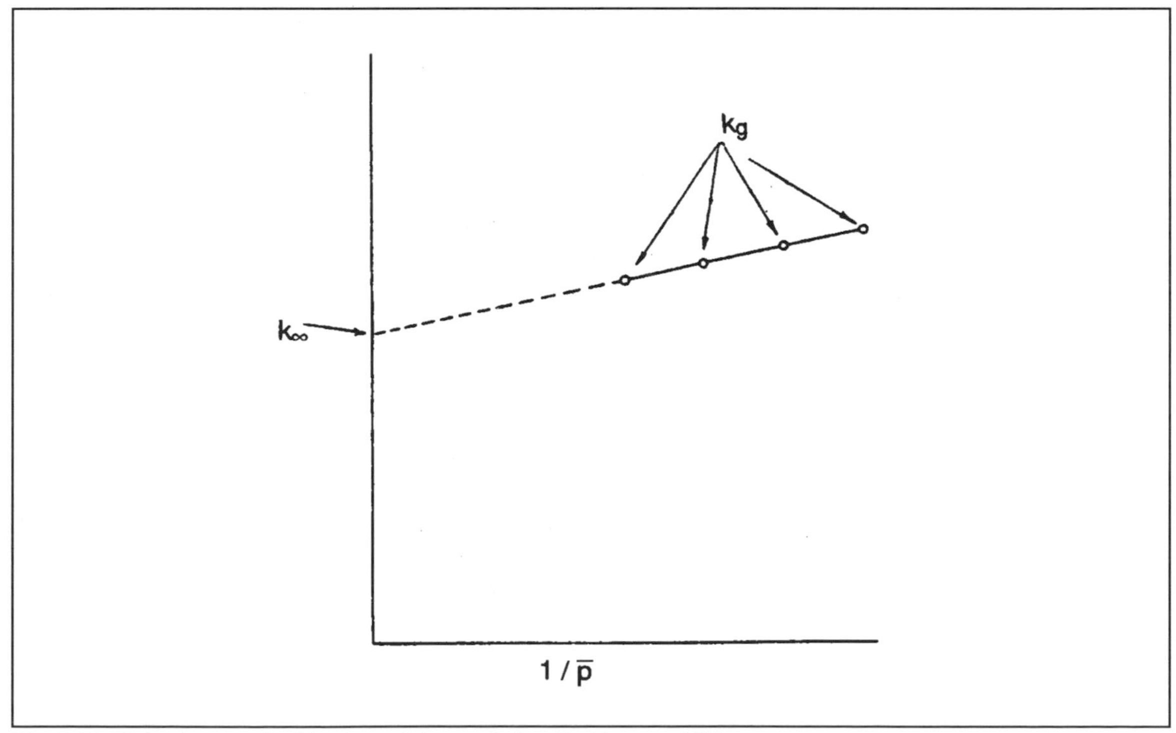

Fig. 2-19. Klinkenberg method for determining equivalent liquid permeabilities.

Klinkenberg[38] developed an equation for determination of k:

$$k_{\infty} = \frac{k_g}{1 + b/\overline{p}} \tag{2-26}$$

where:

k_{∞} = Klinkenberg permeability (absolute)
k_g = measured gas permeability at $\overline{p}$,
b = parameter showing influence of pore size and mean free path,
$\overline{p}$ = mean pressure

It should be noted that the correction to measured gas permeability can be substantial with particularly tight reservoir rocks at low pressure. For high permeability cores, testing laboratories will often make their tests with one high mean pressure, and then quite validly neglect the Klinkenberg correction.

The previous discussion of the Klinkenberg correction to gas-measured permeability, involving Figure 2-19 and Equation 2-26, assumed the absolute permeability to be invariant with pressure. Jones[35,37] has discussed the procedures and equipment needed to relate gas-measured permeabilities where the formation permeability varies significantly with pressure.

Laboratory determinations of fluid saturations are probably the least reliable of the basic measurements made on reservoir rock samples. In addition, the values obtained are those at laboratory (not reservoir) conditions.

Ruska saturation stills are commonly used for determining oil and water saturations in cores. In this distillation method, liquid saturations are determined using a solvent, such as xylene, toluene, or a gasoline fraction, which boils at about 150°F. Figure 2-20a illustrates a distillation apparatus consisting of a flask, water trap, and reflux condenser. Gas saturation is then obtained by difference: $S_g = 1 - S_o - S_w$.

Another saturation determination method is the retort procedure in which the core sample is placed in a retort that, using heat alone, drives off the water and oil (Fig. 2-20b). The vaporized fluids are condensed and collected in a graduated cylinder. The main objection to this technique is the cracking of oil with the resulting production of gases and the possibility of driving off water of crystallization from the rock itself.

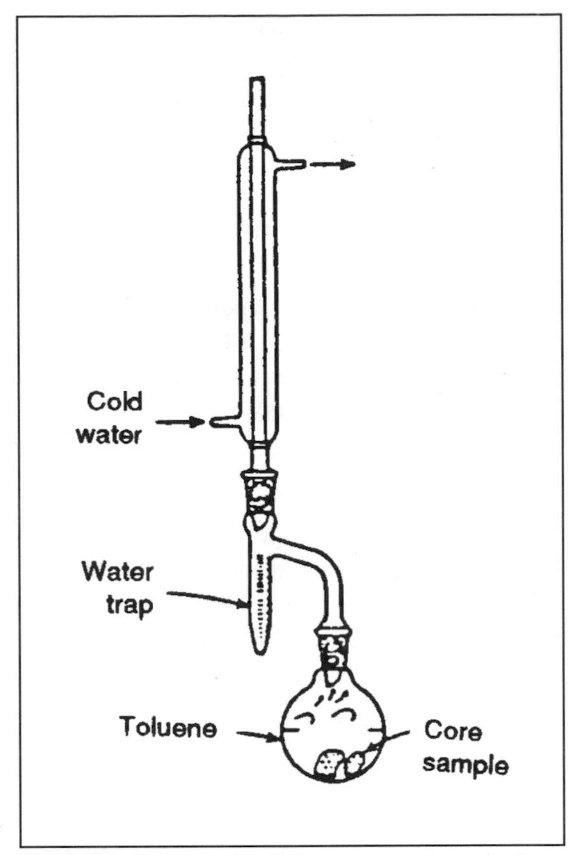

Fig. 2-20a. Solvent extraction-type water-determination apparatus (from Gatlin[26]).

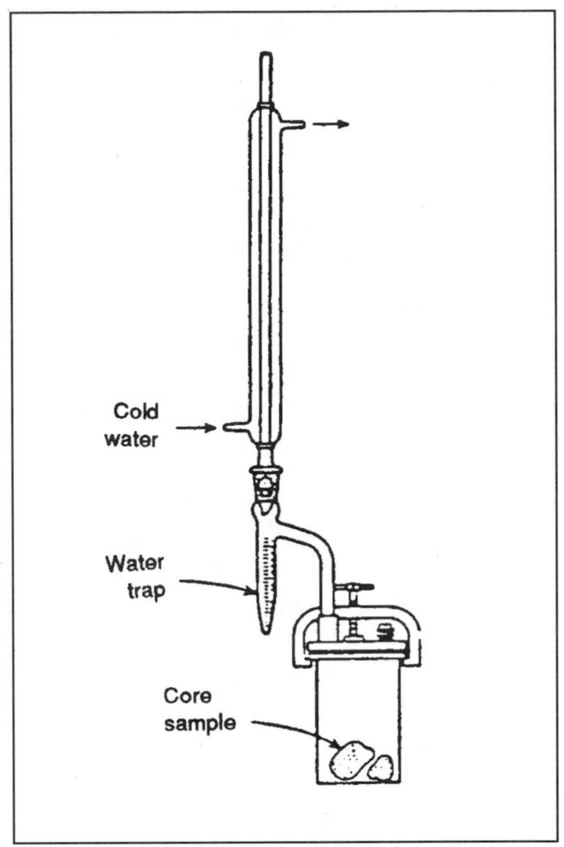

Fig. 2-20b. Retort-type water-determination apparatus (from Gatlin[26]).

7. Rig Core Handling, Core Preservation, and Laboratory Handling

Especially in whole core analysis, the core should be removed from the barrel in segments as long as possible, and care should be taken to prevent excessive breaking up of the core. Jarring and hammering on the core barrel is often necessary, but this should be done as carefully as possible to avoid crushing the core or opening fractures. Each piece should be wiped clean with dry rags as soon as it is removed from the barrel and laid out on the pipe rack. Do not wash off the drilling fluid. The core should then be marked as to top and bottom. After all of the core is removed from the barrel, the core is measured with a tape and marked off into units of length.

For conventional "plug-type" core analysis, the procedure is the same with the exception that extra precautions to recover long pieces are not necessary. In sidewall core analysis, care should be taken in removal from the hollow bullet. Samples should be securely sealed in small containers immediately upon removal from the coring instrument.

Different methods are used for core preservation depending on what is intended, and on condition of the core. Often air-tight cans or tubes are used where the core's pieces very nearly fill the container. For transporting, the cores are normally wrapped in foil or placed in plastic bags before being inserted in the cans.

Dry ice may be used to freeze and consolidate the core and fluids in place. This is quite useful when the laboratory is not a great distance away. Care must be taken when thawing the core to avoid atmospheric condensation on the core and because thawing done slowly can cause fluid redistribution in the core.

Plastic or paraffin coatings may be used on the outside of the core to preserve it when being shipped over long distances. It also helps maintain fluid saturations. Sidewall cores are usually kept in bottles supplied by the coring service.

Permanent preservation of the cores is not routinely done by many oil-producing companies or agencies. This is unfortunate since, occasionally, information is needed at a later time that can best be obtained by resort to the cored material. Some state or governmental agencies have developed core depositories and require that at least part of every core taken be submitted for storage.

After the samples arrive in the laboratory, they are placed in order of depth and sample number. If frozen, they are allowed to thaw until they can be handled. They are wiped clean again and an ultraviolet examination and a visual (microscopic) description made and recorded. A detailed notation of fractures and vugs is made at this time. Sometimes, whole core sections are photographed to permit later detailed study of fractures and vugs. A radioactivity log can be made at the customer's request.

Possible problems to avoid during laboratory processing include: (1) removing water of crystallization (retorting), (2) causing fine particles to migrate within the core during cleaning and handling, and (3) wettability alteration (caused by the core coming into contact with certain fluids and by weathering).

8. Coring Review

The philosophy should be to get as many cores as the budget will permit. Why? Reservoir studies and meaningful performance predictions require data. Cores are not only used for the information already discussed, but for special analyses as well. More specialized determinations include: relative permeability curves, capillary pressure relationships, wettability preference, and other specific tests as needed. Cores should be catalogued and stored in a clean and dry place for future reference; i.e., for geological and reservoir engineering needs not currently envisioned.

9. Absolute Permeability Alteration with Pressure

Quite often, absolute permeability is represented as being invariant with pressure or effective stress (Eq. 2-2). Actually, this is never true. As the pressure in a reservoir declines, absolute permeability also decreases. For many formations, the change in permeability with pressure is relatively insignificant; however, in some gas sands where permeability is basically due to micro-fractures, the permeability at abandonment is only a small fraction of the initial permeability.

Given the proper laboratory apparatus, i.e., a permeameter that allows permeability measurements on a core at different effective stress states, Jones[35] has developed an empirical equation that describes absolute permeability as a function of effective stress.

$$\ln \left\{ k \left[1 + (p_{eff})(3 \times 10^{-6}) \right] \right\} = \ln a_1 - a_2 \left[1 - e^{-(p_{eff}/3000)} \right] \quad (2\text{-}27)$$

where:

k = absolute permeability, md, at effective stress, p_{eff},
p_{eff} = effective stress = overburden pressure minus pore pressure, psi,
a_1 = equation constant, and
a_2 = equation constant.

To use Eq. 2-27, absolute permeability is measured on a core at two or more different stress states. Then, using least squares, the constants a_1 and a_2 may be evaluated. According to Jones[35], if a_1 and a_2 are determined using the core permeabilities measured at 1500 psi and 5000 psi confining or "overburden" stress, then Eq. 2-27 should yield results that are likely within the experimental accuracy of the permeameter itself up to an effective stress of 10,000 psi.

CAPILLARY PRESSURE

Typically, reservoir fluids are not miscible. For instance, oil and water in physical contact exhibit an interface with a pressure differential across it. This difference in pressure between the two immiscible phases (in this case, oil and water) is referred to as capillary pressure. At normal reservoir conditions, free hydrocarbon gas and oil are also immiscible. Therefore, there is a pressure difference (capillary pressure) across the interface between the gas and oil. The discussion here will consider capillary pressure in the reservoir between oil and water. The reader should note that all of the concepts and equations presented have an analogous counterpart for the gas/oil case.

Capillary pressure normally is defined as the pressure in the non-wetting phase minus the pressure in the wetting phase at the same location. (However, in immiscible displacement processes, it is sometimes defined as the displacing phase pressure minus the displaced phase pressure.) Therefore, in a water-wet formation, capillary pressure is usually taken to be the pressure in the oil phase minus the pressure in the water phase.

Capillary pressure has been shown to have a large influence on (1) the initial fluid distribution within a reservoir and (2) the fraction of each fluid flowing in an immiscible displacement such as a waterflood. Some typical initial reservoir saturation distributions are shown in Figure 2-21 as a function of permeability for a particular consolidated, water-wet formation. Here, the saturations grade from 100% water in the water zone to an irreducible water saturation some vertical distance in the reservoir above the oil/water contact. If a gas cap exists, a similar "transition zone" would exist between the oil and the gas zones.

Figure 2-21 illustrates the gradual change in oil and water saturations with height above the oil/water contact ranging from inches to tens of feet. In the transition zone where water saturation is varying between S_{wi} and $(1-S_{or})$, both oil and water will flow. Usually the transition zone is not perforated for oil production.

Surface Forces

Within a fluid substance, there is an attraction between molecules that is inversely proportional to the distance between the molecules:

$$F \alpha 1/d$$

In the absence of other forces, this force of cohesion will cause a fluid to contract to minimum surface area. Within the body of a liquid, a molecule has other molecules completely surrounding it resulting in a balance (net force of zero) of cohesive forces. Because the molecules on the surface of a liquid do not have other such molecules above exerting an attractive force, an imbalance of forces exists. Therefore, these surface molecules exhibit a "free energy" referred to as "surface tension." Actually, some molecules of the liquid are above its surface due to the equilibrium vapor volume above the liquid. Surface tension should be measured between a liquid and its vapor. Nevertheless, it is usually measured between the liquid and air. Surface tension is measured parallel to the surface as force per unit length, usually in dynes/cm. Another approach is that surface tension is the contractile tendency of a liquid's surface when in the presence of its vapor.

The normal designation for surface tension is σ, e.g., for water at 60 F°, σ_w = 72 dynes/cm. Surface tensions generally decrease with increasing temperature. If the interface is between two liquids, then we use the term "interfacial tension," not surface tension. For oil and water at 60 F°, the typical range is 15 to 40 dynes/cm.

In reservoir systems, capillary pressure is affected by the forces at the interfaces: oil/water, oil/rock, and water/rock. Hence, reservoir rock wettability has an important effect.

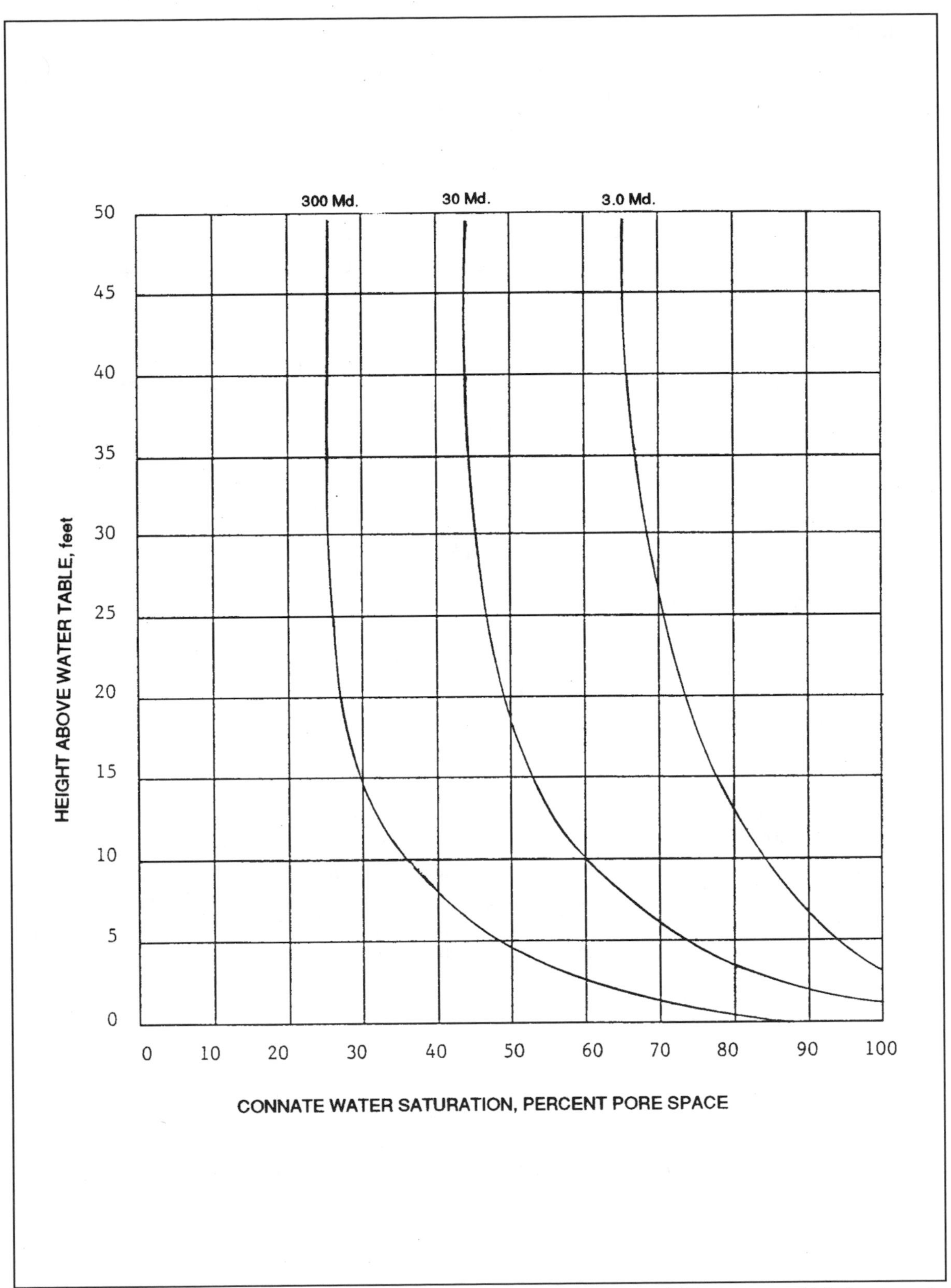

Fig. 2-21. Connate water saturation versus height above water table for various permeabilities in a consolidated, water-wet reservoir.

Wettability

Wettability is a measure of the capacity for a fluid to coat a solid surface. A drop of water will spread on glass indicating that it will "wet" a glass surface, and it will wet most reservoir rock surfaces as well. For a wetting fluid, the contact angle is less than 90°, as illustrated in Figure 2-22.

Mercury does not wet glass (Fig. 2-22) since the forces of cohesion are stronger than the forces of adhesion (the attractive forces of the glass); therefore, the contact angle is greater than 90°.

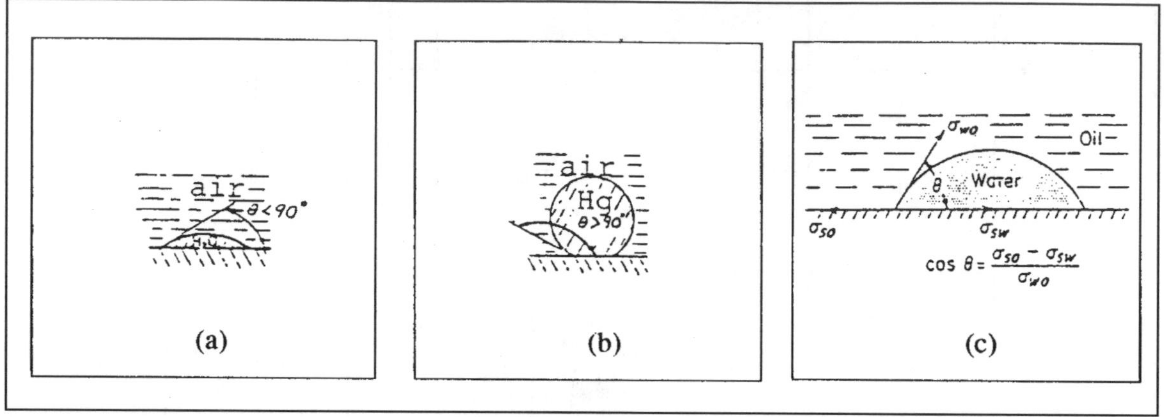

Fig. 2-22. Interfacial contact angles on a glass surface. (a) Water-air; (b) Mercury-air; (c) Equilibrium of forces at a water-oil-glass interface. (From Amyx, Bass and Whiting[4])

For oil and water on glass, water is found to displace oil indicating water to be the wetting phase. As can be seen in Figure 2-22, the situation is much the same as water on glass in the presence of air, except the air is replaced by oil. By convention, the contact angle, θ, is measured through the denser phase (water), and of course, $\theta < 90°$. The contact angle is related to the three different physical interactions present at this point (Fig. 2-22): water/oil, solid-oil, and solid-water. If the contact angle were less than 90° in a reservoir, which it usually is, then we would say that the formation is preferentially water-wet. As mentioned earlier, it is possible to alter wettability through the addition of various chemicals (such as surfactants) to the system. Care must be taken in the laboratory handling of cores not to change wettability. In the reservoir the wetting fluid will tend to occupy the smaller interstices, whereas the non-wetting phase will usually exist in the larger pores.

Adhesion tension is defined to be:

$$A_T = \sigma_{wo} \cos \theta \tag{2-28}$$

(force per unit length)

The magnitude of the adhesion tension determines the ability of the wetting phase to adhere to the solid and to spread over the surface of the solid.

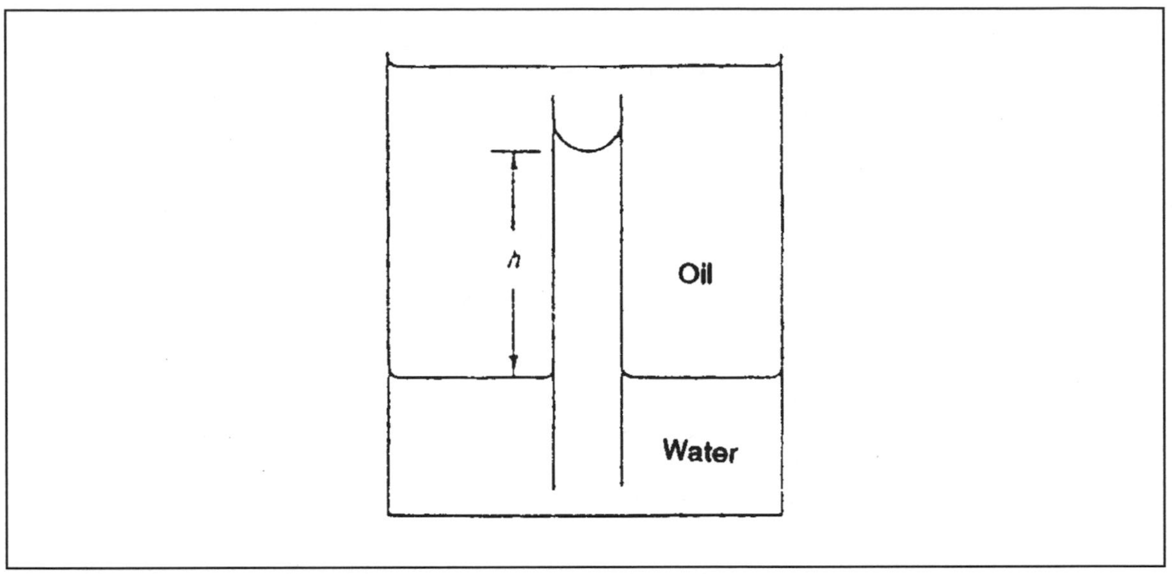

Fig. 2-23. Fluid Rise in a Capillary Tube.

Consider a capillary tube inserted into a beaker containing oil overlying water, Figure 2-23. Recalling that normally:

$$p_c = p_o - p_w$$

then we can see that the difference in oil and water pressure across the interface within the capillary is related to the density difference and height of water rise:

$$p_c = (\rho_w - \rho_o)\, gh \tag{2-29}$$

where:

ρ_w = water density,
ρ_o = oil density,
g = acceleration due to gravity, and
h = height of the rise of water in the capillary above the level in the container.

In practical field units, we have:

$$p_c = 0.433\ (\Delta\gamma)\, h \tag{2-30}$$

where:

$\Delta\gamma$ = specific gravity (or relative density) difference, $\gamma_w - \gamma_o$,
h = height of water rise in capillary, feet, and
p_c = capillary pressure, psi.

Looking at a force balance within the capillary, the force of adhesion tension is equal to the potential energy represented by the water rise inside the capillary or:

$$2\,\pi\,r\,\sigma_{wo}\cos\theta \;=\; \pi\,r^2\,\Delta\rho\,h\,g\,, \tag{2-31}$$

and

$$p_c \;=\; \Delta\rho\,h\,g\,,$$

so

$$p_c \;=\; 2\,\sigma_{wo}\cos\theta\,/\,r \tag{2-32}$$

where:

r = radius of capillary tube,
θ = contact angle (water) in the capillary tube,
σ_{wo} = water/oil interfacial tension,
$\Delta\rho$ = water density minus oil density.

In the reservoir, considering two, equal-sized spherical grains, we would have a volume of water held between the sand grains due to the force of adhesion tension as shown in Figure 2-24.

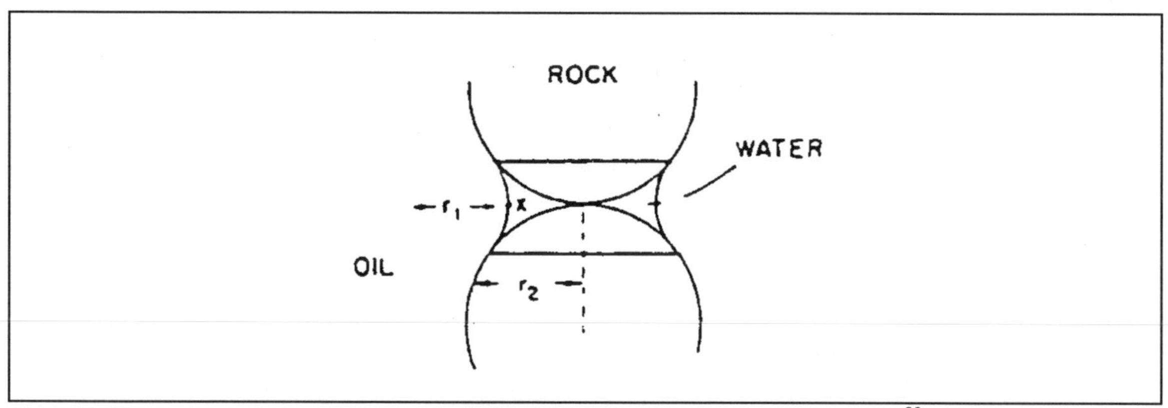

Fig. 2-24. Water entrapment between two spherical grains in a water-wet reservoir (from Dake[30]). Permission to publish by Elsevier Scientific Publishing Company.

Considering this figure, a general expression from Plateau[31] involving interfacial tension and principle radii of curvature of the interface is:

$$p_c \;=\; \sigma_{wo}\left[\frac{1}{r_1} + \frac{1}{r_2}\right] \tag{2-33}$$

Note that r_1 and r_2, the principal radii of curvature of the interface, are dependent on water saturation and grain size. Actually, Figure 2-24, is a representation of the pendular ring formed between two rock grains by the wetting phase at a saturation state such that the wetting phase is not continuous. In this case, the non-wetting phase is in contact with some of the solid surface. The wetting phase occupies the smaller interstices. The Plateau equation allows us to understand experimental and field observations:

(1) Capillary pressure goes up with decreasing water saturations (because r_1 is becoming smaller), and

(2) Lower permeability rocks have higher capillary pressure (r_2 will tend to be smaller with lower permeability rocks), and higher permeability rocks have lower capillary pressures.

If we inspect the force balance Eq. 2-31, we can see that:

$$h \ \alpha \left[\frac{(\sigma_{wo})(\cos\theta)}{(r)(\Delta\rho)} \right] \tag{2-34}$$

where

h = the height the liquid will rise in the capillary tube of radius "r".

From this relation the effect of wettability, interfacial tension, pore size, and specific gravity difference on the height that water will rise in a reservoir above the free water table can be observed. The length of the transition zone is a direct measure of this "h". This relationship also indicates that smaller grain size reservoirs (usually indicative of lower permeability) will have longer transition zones.

Laboratory Measurement

Capillary pressure relationships normally are obtained in the laboratory by first saturating the core with the wetting phase. Then the core is placed in a chamber, subjected to pressure, and invaded by the nonwetting phase (Fig. 2-25). This is done in steps with the pressure and volume of wetting fluid displaced noted at each step. The pressure required to first cause any displacement from the core (or invasion of non-wetting fluid) is called the "threshold pressure." A typical graph of such experimental results is called a "capillary pressure curve" and is illustrated in Figure 2-25.

The most common laboratory fluid combinations are (1) water/air, (2) air/mercury, and (3) water/oil.

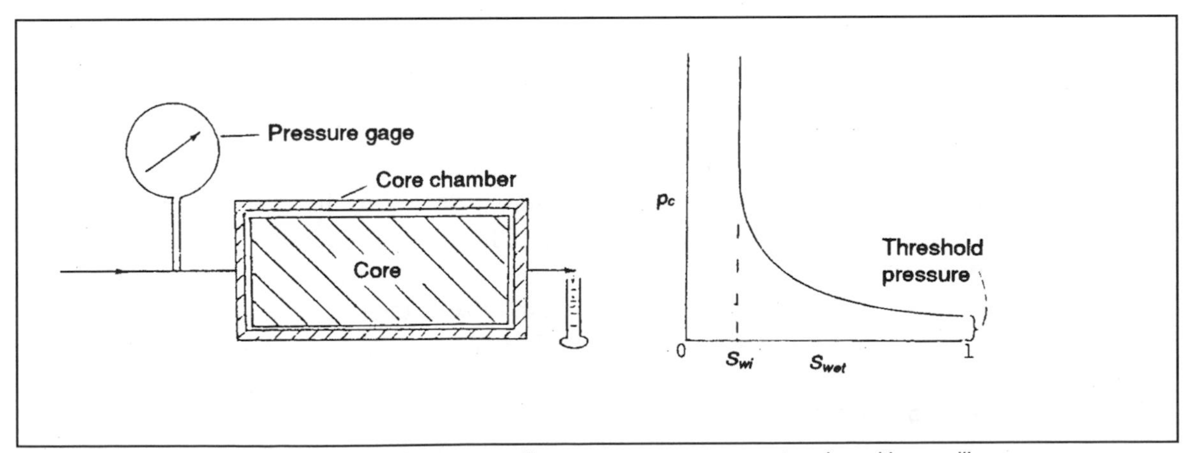

Fig. 2-25. Schematic representation of laboratory capillary pressure measurement and resulting capillary pressure curve.

Calculating Reservoir Capillary Pressure Data from Laboratory Data

Laboratory capillary pressure measurements must be corrected before use in reservoir calculations since different fluids normally are employed in the laboratory. If we know the interfacial tension and wetting angle for the fluids in the laboratory, then we can write:

$$p_{cL} = \frac{2\,(\sigma \cos \theta)_L}{r} \tag{2-35}$$

And the corresponding expression for the reservoir:

$$p_{cR} = \frac{2\,(\sigma \cos \theta)_R}{r} \tag{2-36}$$

So, we divide the equations and solve for reservoir capillary pressure:

$$p_{cR} = p_{cL}\,\frac{(\sigma \cos \theta)_R}{(\sigma \cos \theta)_L} \tag{2-37}$$

We can usually get the interfacial tension and contact angle for the laboratory, but these quantities are difficult to obtain for reservoir conditions. A small amount of interfacial tension data, which can be used on an analogy basis, may be found in Katz[32].

The reservoir contact angle is usually not known and it will suffice to write.

$$p_{cR} = p_{cL}\left[\frac{\sigma_R}{\sigma_L}\right] \tag{2-38}$$

Initial Saturation Distribution from Capillary Pressure Data

With a reservoir capillary pressure curve in hand, the initial saturation distribution in the reservoir can be calculated by rearranging Eq. 2-30:

$$h = p_c \,/\, \{(0.433)(\Delta\gamma)\} \tag{2-39}$$

Looking at this equation, it can be seen that h = 0 when $p_c = 0$. The “free water level” is defined to be that level or depth in the reservoir where $p_c = 0$. Our “h” is measured from this depth.

From logs, DST’s, or production testing, the highest level of 100% water saturation may be determined. This is idealized in the reservoir in Figure 2-26 where the corresponding capillary pressure curve is also shown. Note that the highest 100% water saturation level has a capillary pressure associated with it, equal to the threshold pressure. The free water level is lower than the 100% saturation level by a distance equal to the capillary rise in the largest pore in the reservoir. If the reservoir has very large pores, then the free water level and the 100% water saturation level will be essentially the same. Conversely, for a low permeability, small grain size formation, the difference can be considerable. So, we may write:

$$Free\ water\ level = D_{\underset{water}{100\%}} + p_t \,/\, \{\,(0.433)\,(\Delta\gamma)\,\} \tag{2-40}$$

where

$D_{\underset{water}{100\%}}$ = highest depth of 100% water saturation, feet,

p_t = threshold pressure, psi.

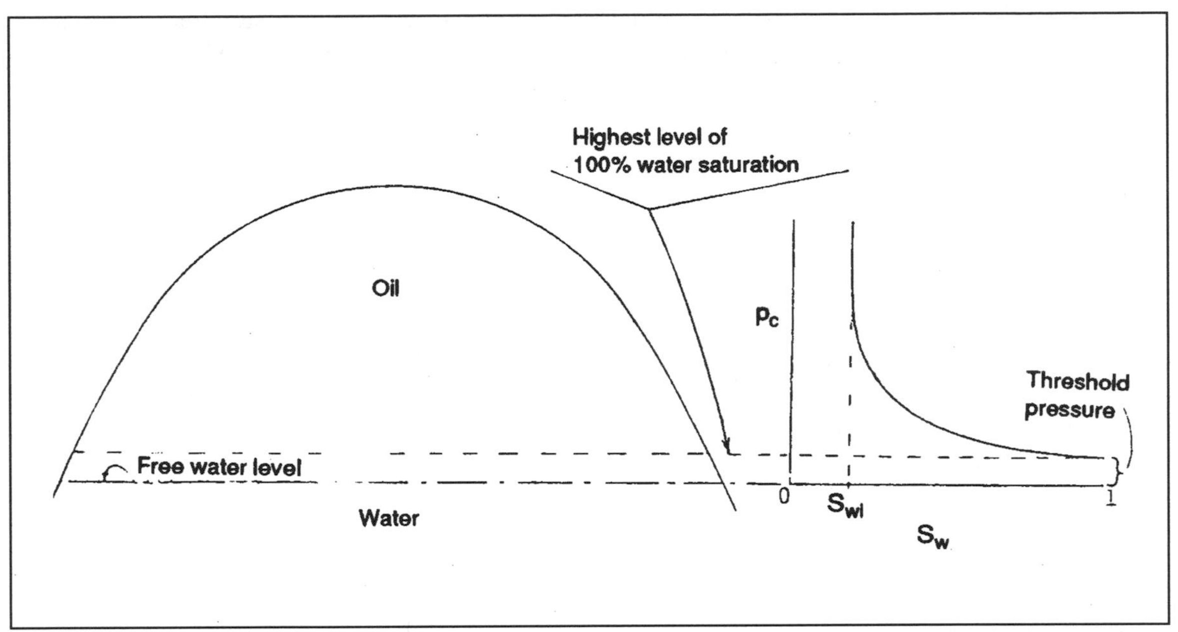

Figure 2-26. Hypothetical reservoir and corresponding capillary pressure curve illustrating the relationship between free water level and highest level of 100% water saturation.

All capillary pressure heights in the h vs. S_w relationship are referenced to the free water level. So, given p_c vs. S_w data, the height (using Eq. 2-39) above the free water level for each S_w can be calculated.

Averaging Capillary Pressure Data

If a large number of capillary pressure curves are available for one reservoir, then the data may be averaged. However, since each rock sample will, no doubt, have varying permeability and porosity, the capillary pressure curves will also be quite different. Leverett[33] has proposed the "J-function" to be used for this averaging:

$$J = \frac{p_c\ (k/\phi)^{0.5}}{\sigma\ \cos\ \theta} \tag{2-41}$$

where:

p_c = capillary pressure, dynes/sq cm
σ = interfacial tension, dynes/cm
k = permeability, sq cm
ϕ = fractional porosity
θ = contact angle; and this term ($\cos\theta$) is often neglected.

Leverett[33] assumed in deriving this relationship that the reservoir behaves as a bundle of capillaries. This is really a means of normalizing capillary pressure measurements. When the "J-function" is plotted, a similar shape to that of the capillary pressure curve results.

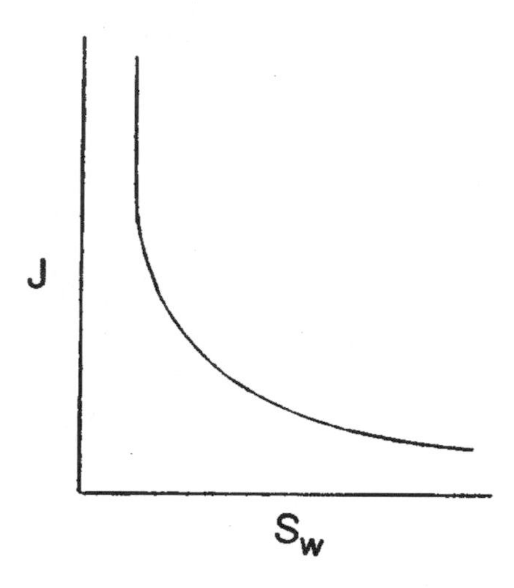

For some reservoirs, this approach does not work well (probably because the reservoir is not behaving as a bundle of capillaries), and a statistical approach is used. This statistical treatment is more widely used than the "J-function" approach. A semilog graph similar to that presented in Figure 2-27 is prepared for constant values of capillary pressure. A straight line is fitted to the data for each constant value of capillary pressure. This "smoothed" data can now be replotted in a form as represented in Figure 2-28, where capillary pressure (or height above a free water surface) is related to permeability and water saturation. It is interesting to note that the capillary pressure curves tend to converge at the higher permeability values. This would be expected due to the larger capillaries being associated with higher permeabilities.

When developed for a particular formation, Figure 2-28 has application in determining average water saturation over an entire oil- or gas-bearing section of interest. Similarly, the graphs shown in Figures 2-27 and 2-28 can be prepared for gas above oil or water zones.

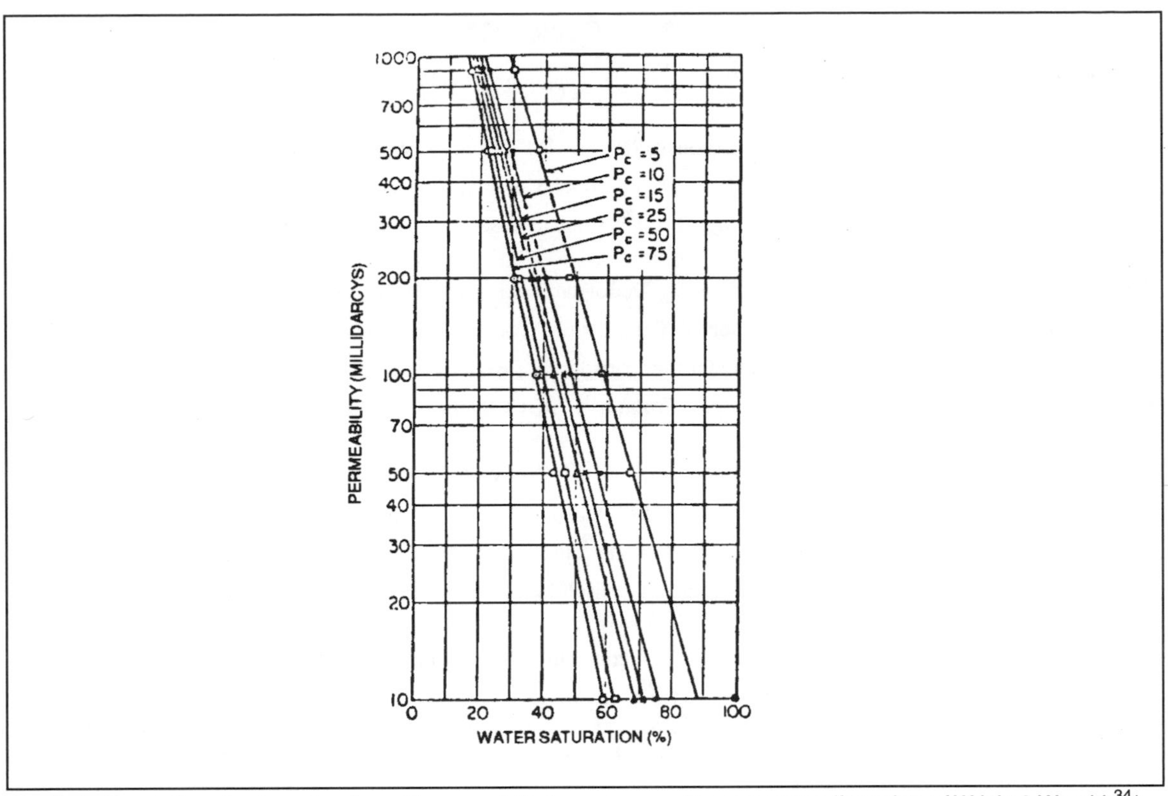

Fig. 2-27. Correlation of water saturation with permeability for various capillary pressures (from data of Wright & Woody[34]). Permission to publish by SPE.

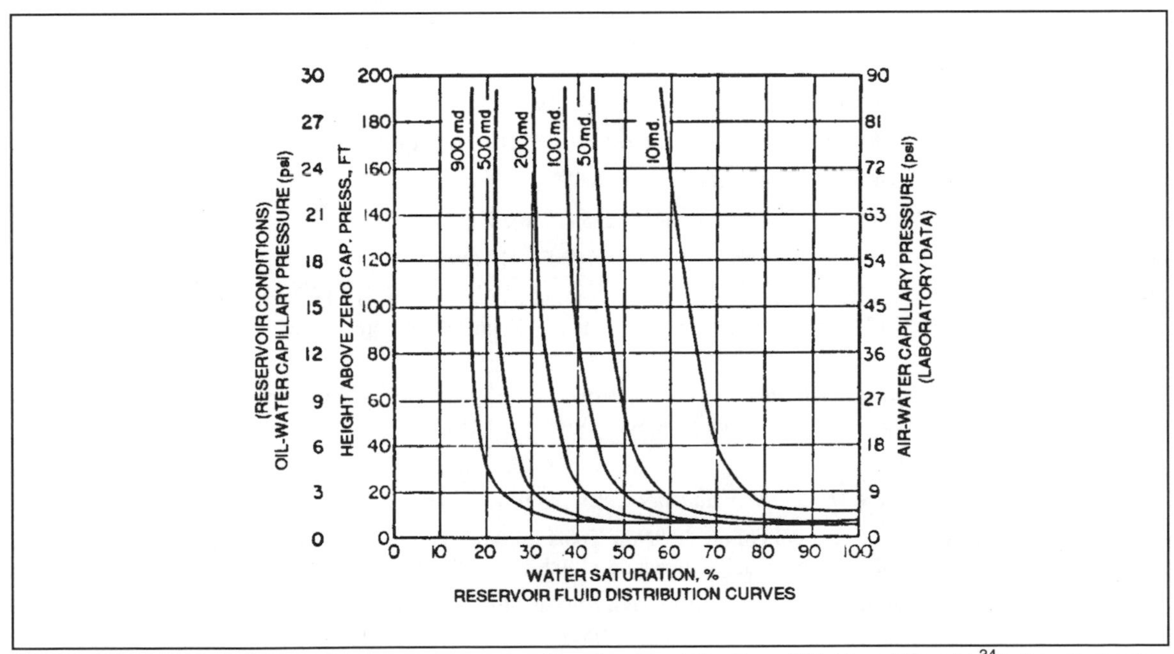

Fig. 2-28. Series of capillary-pressure curves as a function of permeability (from Wright and Wooddy[34]). Permission to publish by SPE.

REFERENCES

1. Gatlin, C.: *Petroleum Engineering - Drilling and Well Completions,* Prentice-Hall, Inc., Englewood Cliffs, N.J. (1960) 21.
2. Clark, N. J.: "Elements of Petroleum Reservoirs," SPE of AIME, Henry Doherty Series (1960) Chapter 2.
3. Pirson, S. J.: *Oil Reservoir Engineering*, McGraw-Hill Book Co, Inc., New York City (1958) Chapter 2.
4. Amyx, J. W., Bass, D. M. Jr., and Whiting, R. L.: *Petroleum Reservoir Engineering*, McGraw-Hill Book Co., Inc., New York City (1960) Chapter 2.
5. Guerrero, E. T,: *Practical Reservoir Engineering*, Petroleum Publishing Co., Tulsa, OK (1968) 2-16.
6. Calhoun, J. C., Jr.: "Fundamentals of Reservoir Engineering," Univ. of Okla. Press (1955) 126.
7. Dotson, B. J., *et al*.: "Porosity - Measurement Comparisons by Five Laboratories," *Trans*., AIME (1951) 341.
8. Helander, D. P.: *Fundamentals of Formation Evaluation*, OGCI Publications, Oil & Gas Consultants International, Inc., Tulsa, OK (1983).
9. Timmerman, E, H.: *Practical Reservoir Engineering*, Vol I, PennWell Publishing Co., Tulsa, OK (1982) 52.
10. Hall, H. N.: "Compressibilities of Reservoir Rocks," *Trans*., AIME (1953) 309.
11. Van der Knaap, W.: "Nonlinear Behavior of Elastic Porous Media," *Trans*., AIME, (1959) 216, 179-187.
12. Newman, G. H,: "Pore-Volume Compressibility of Consolidated, Friable, and Unconsolidated Reservoir Rocks Under Hydrostatic Loading," *JPT* (February 1973) 129-134.
13. Skidmore, F. A.: "Bubble Nucleation and Gas Structure Growth in Synthetic Porous Media," MS Thesis, Petroleum Engineering Dept., Univ. of Texas (August 1962).
14. Meads, R. and Bassiounia, Z.: "Combining Production History and Petrophysical Correlations to Obtain Representative Relative Permeability Data," paper SPE 12113 presented at the 1983 Annual Technical Conference and Exhibition, San Francisco (October).
15. Reference 1, Chapter 3.
16. Pirson, S. J., Boatman, E. M., and Nettle, R. L.: *JPT* (May 1964), 564 .
17. Honarpour, M., Koederitz, L. F., and Harvey, A. H.: "Empirical Equations for Estimating Two-Phase Relative Permeability in Consolidated Rocks," *JPT* (December 1982).
18. Burdine, N. T., *Trans*., AIME 198 (1953), 71.
19. Wyllie, M. R. J. and Spangler, M. R., *AAPG Bulletin*, 36, (1952), 359.
20. Wyllie, M. R. J. and Gardner, G. H. F., *World Oil* (March 1958), 121; (April 1958), 210.
21. Corey, A. T., *Producers Monthly* (November 1954), 38.
22. Torcaso, M. A. and Wyllie, M. R. J., *Trans*., AIME, 213 (1958), 436.
23. Wahl, W. L., Mullins, L. D., and Elfrink, E. B., *Trans*. AIME, 213 (1958), 132.
24. Fulcher, R. A. Jr.: "The Effect of the Capillary Number and its Constituents on Two-Phase Relative Permeability Curves," paper SPE 12170, presented at the 1983 Annual Technical Conference and Exhibition, San Francisco (October 1983).
25. Harbert, L. W.: "Low Interfacial Tension Relative Permeability," paper SPE 12171, presented at the 1983 Annual Technical Conference and Exhibition, San Francisco (October 1983).
26. Reference 1, Chapter 10.
27. Anderson, G.: *Coring and Core Analysis Handbook,* PennWell Books (1975).
28. Filshtinsky, M., Aumann, J. T., and Quinn, J.: "New Tools Improve the Economics of Coring," paper SPE 12092, presented at the 1983 Annual Technical Conference and Exhibition, San Francisco (October 1983).
29. Hyland, C. R.: "Pressure Coring — an Oilfield Tool," paper SPE 12093, presented at the 1983 Annual Technical Conference and Exhibition, San Francisco (October 1983).
30. Dake, L. P.: *Fundamentals of Reservoir Engineering,* Elsevier Scientific Publishing Co. Amsterdam, The Netherlands (1978).

31. Plateau, J. A. F.: "Experimental and Theoretical Research on the Figures of Equilibrium of a Liquid Mass Withdrawn from the Action of Gravity," Smith Inst. Ann. Repts. (1863-1866).

32. Katz, D. L., Cornell, D., Kobayashi, R., Poettman, F. H., Vary, J. A., Elenbaas, J. R., and Weinaug, C. F.: *Handbook of Natural Gas Engineering*, McGraw-Hill Book Co., Inc., New York City(1959).

33. Leverett, M. C.: "Capillary Behavior in Porous Solids," *Trans.*, AIME (1941) Technical Publication 1223.

34. Wright, H. T. Jr. and Wooddy, L. D. Jr.: "Formation Evaluation of the Borregas and Seeligson Fields, Brooks and Jim Wells Counties, Texas," Symp. of Formation Evaluation., AIME (October 1955) 135.

35. Jones, S. C.: "Two-Point Determinations of Permeability and PV vs. Net Confining Stress," paper SPE 15380, presented at the 1986 Annual Technical Conference and Exhibition, New Orleans (October 5-8, 1986).

36. Society of Petroleum Engineers: "The SI Metric System of Units and SPE METRIC STANDARD," Published by the SPE, Richardson, TX., Second Printing (June 1984).

37. Jones, S. C.: "A Rapid Accurate Unsteady-State Klinkenberg Permeameter," *SPEJ* (October 1972) 383-397.

38. Klinkenberg, L. J.: *Drill. & Prod. Prac.*, API (1941) 200.

3 FLUID PROPERTIES

INTRODUCTION

The substances of interest to the reservoir engineer are oil, gas, and water. Normally we would expect these materials to be fluid; i.e., either liquid or vapor. In some instances, though, the oil can be quite viscous or even solid. While we would usually think that the water should be liquid, the interstitial water is solid in some locations. This can occur in permafrost regions.

In reservoir studies, we normally prefer to use data obtained from laboratory analysis of actual fluids recovered from the reservoir early (hopefully) in field life.

Where analyses are not available or the accuracy of the information is in question, the reservoir engineer will need to rely on published correlations, analyses of similar fluids from nearby reservoirs, etc. Rarely does the reservoir engineer have all of the data necessary without some reliance on published correlations.

The oil industry fortunately has extensive literature relative to fluid properties in the form of books, symposia, trade and technical journals.

PROPERTIES OF NATURALLY OCCURRING PETROLEUM DEPOSITS

Petroleum deposits vary widely in properties as to producing horizon, geographical location, and producing depth. The bulk of the chemical compounds present are hydrocarbons and, as the name implies, are comprised of hydrogen and carbon. Since the carbon atom has the ability to combine with itself and form long chains, the number of possible compounds is very large. A typical crude oil contains hundreds of different chemical compounds and normally is separated into crude fractions according to the range of boiling points of the compounds included in each fraction.

Hydrocarbons may be gaseous, liquid, or solid at normal temperature and pressure, depending on the number and arrangement of the carbon atoms in the molecules. Those compounds with up to four carbon atoms are gaseous; those with twenty or more are solids; and those in between are liquid. Liquid mixtures, such as crude oils, may contain either gaseous or solid compounds or both in solution. For instance, some oils are liquid at the wellhead, but are solid upon cooling due to crystallization of the solid compounds.

The simplest hydrocarbon is methane, a gas consisting of one carbon atom and four hydrogen atoms. The methane molecule can be represented as:

```
        H
        |
    H — C — H        or  CH4
        |
        H
```

This is the first of the so-called paraffin series of hydrocarbons having the general formula C_nH_{2n+2}. Paraffin-base crude oils contain predominantly paraffin series hydrocarbons and are often referred to as straight chain hydrocarbons since the carbon atoms are linked together to form a straight chain. Other arrangements are possible by branching, ring structures, double and triple bonds between adjacent carbon atoms, and combinations of these. These result in the isoparaffins, cycloparaffins, olefins, acetylenes, and aromatics.

Crudes containing mainly paraffin-base materials give good yields of paraffin wax and high-grade lubricating oils. Asphaltic base oils are comprised largely of naphthenic (ringed, mostly aromatic) compounds. Asphaltic crudes yield lubricating oils that are more viscosity sensitive to temperature and require special refining methods and additives. Mixed-base crudes are also common.

A number of nonhydrocarbons may occur in crude oils and gases, and though usually small in quantity, these compounds can have a considerable influence on physical properties and product quality. The most important elements in non-hydrocarbons are sulfur (S), nitrogen (N), and oxygen (O). Small quantities of vanadium (V), nickel (Ni), sodium (Na), and potassium (K) are in some crude oils.

Table 3-1 provides a summary of typical components in natural gas. The first part describes gas from a gas reservoir while the second part is for "rich" gas originating in a gas condensate field or gas associated with oil (casinghead gas or gas produced from a gas cap).

Table 3-1
Components of Typical Natural Gases (from McCain[1])

Natural gas	
Hydrocarbon	
Methane	70-98%
Ethane	1-10%
Propane	trace - 5%
Butanes	trace - 2%
Pentanes	trace - 1%
Hexanes	trace - 1/2%
Heptanes +	trace - (usually none)
Nonhydrocarbon	
Nitrogen	trace - 15%
Carbon dioxide*	trace - 1%
Hydrogen sulfide*	trace occasionally
Helium	up to 5%, usually trace or none

*Occasionally natural gases are found which are predominately carbon dioxide or hydrogen sulfide.

Gas from a well which is also producing liquid	
Hydrocarbon	
Methane	50 - 92%
Ethane	5 - 15%
Propane	2 - 14%
Butanes	1 - 10%
Pentanes	trace - 5%
Hexanes	trace - 3%
Heptanes +	none - 1 1/2%
Nonhydrocarbons	
Nitrogen	usually trace - up to 10%
Carbon dioxide	trace - 4%
Hydrogen sulfide	none-trace - 6%
Helium	none

Considering the hydrocarbon components, note that from butane on, more than one kind exists for each component type. These compounds that have the same chemical formula, but a slightly different structure, are called isomers. Of the nonhydrocarbon components, CO_2 and H_2S are referred to as "acid gases."

The physical properties of the normal constituents in natural gases and crude oils are summarized later in this chapter in Table 3-3.

FLUID SYSTEMS

The following definitions are needed to discuss fluid systems:

Phase — Any homogeneous and physically distinct part of a system that is separated from any other part of the system by definite bounding surfaces. Examples: solid, liquid, gas.

**(Note that this is a thermodynamic definition of phase and is intended only to apply to the following discussion of fluid systems. Normally in reservoir engineering and elsewhere in this book, the word "phase" indicates a fluid that will not mix readily with the other fluids present due to interfacial tension. Examples: oil, gas, water.)

Component— A pure substance. The number of components in a thermodynamic system is the smallest number of independently variable constituents by means of which the composition of each phase can be expressed for a system in equilibrium.

Intensive—An intensive physical property is one that is independent of the quantity of material present. Example: density.

Extensive—An extensive physical property is one that is determined by the amount of material present. Example: volume.

Bubblepoint—That point (condition of temperature and pressure) at which the first few molecules leave the liquid and form a small bubble of gas is called the bubblepoint. In an oil reservoir, which is at a constant temperature, with production the pressure is decreasing. Therefore, the first pressure at which gas starts to break out of solution is the bubblepoint.

Dewpoint— That point (condition of temperature and pressure) at which only a small drop of liquid is in the fluid system. With a gas system that is undergoing decreasing pressure and/or temperature (the reservoir is isothermal; the wellbore is not), the first point at which a small drop of liquid forms is the dewpoint.

A. Single-Component Systems (Pure Substances)

For a pure substance, phase behavior depends on three variables: pressure, temperature, and volume. Figure 3-1 presents the pressure-temperature phase diagram for a pure substance (or for a single component, e.g., propane). The line segment TC, which divides the liquid and gas regions, is termed the vapor-pressure line. For this reason, the single component P - T diagram is often called a vapor-pressure curve. The pressure temperature points which fall exactly on TC define the conditions where both gas and liquid can coexist in equilibrium. Point C is the critical point that is defined to be the point of critical pressure and critical temperature. The critical pressure is the pressure above which liquid and gas cannot coexist regardless of the temperature. The critical temperature defines the point above which a gas cannot be liquified regardless of the pressure applied.

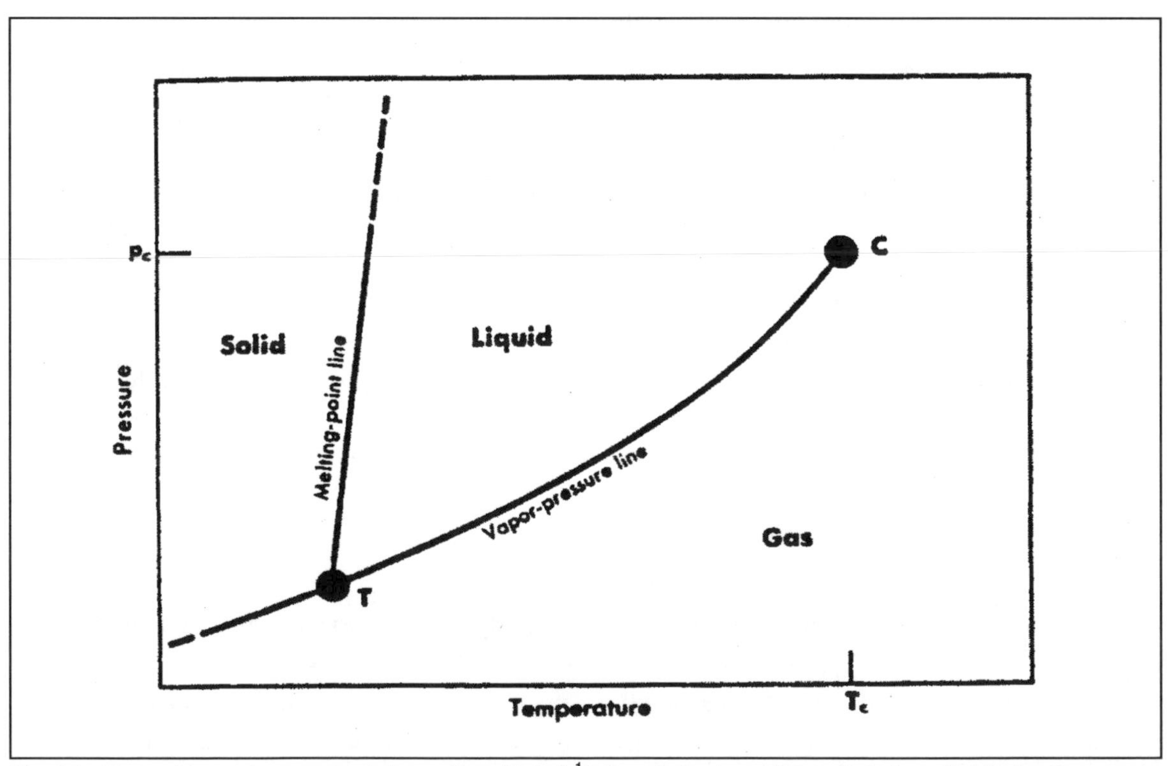

Fig. 3-1. Phase diagram for a pure substance (from McCain[1]). Permission to publish by Petroleum Publishing Company.

The triple-point, T, represents the pressure and temperature at which gas, liquid, and solid coexist under equilibrium conditions. Below the triple-point temperature, the solid region is in direct contact with the gas region. This is the sublimation pressure line. The line separating the solid and the liquid regions is the locus of melting points for the pure component.

In the region above and to the right of the critical point (Fig. 3-1), the pure component can no longer be referred to as a "gas" or as a "liquid." It is usually designated as a "fluid" or "dense phase fluid" and has some properties of a gas (such as the fact that it will fill its container completely) and some properties of a liquid (its density is more like that of a liquid).

To emphasize the interesting nature of this "fluid" region, consider Figure 3-2. If we begin in the gaseous region at point 1 and isothermally increase the pressure to point 5, we will cross the two-phase line and go into the liquid state. During this process, as we go across the vapor pressure line, we will se a meniscus indicating the presence of two phases. On the other hand, we could start at point 1 and isobarically (at a constant pressure) increase the temperature to a point greater than the critical temperature (point 2). At this point, increase the pressure isothermally to a pressure above the critical pressure, point 3. From here, isobarically decrease the temperature until reaching a temperature below the critical, point 4.Now, isothermally reduce the pressure until reaching point 5. In this manner, we went from a gaseous state to a liquid state without ever seeing a meniscus. Instead, we went through the dense phase fluid region.

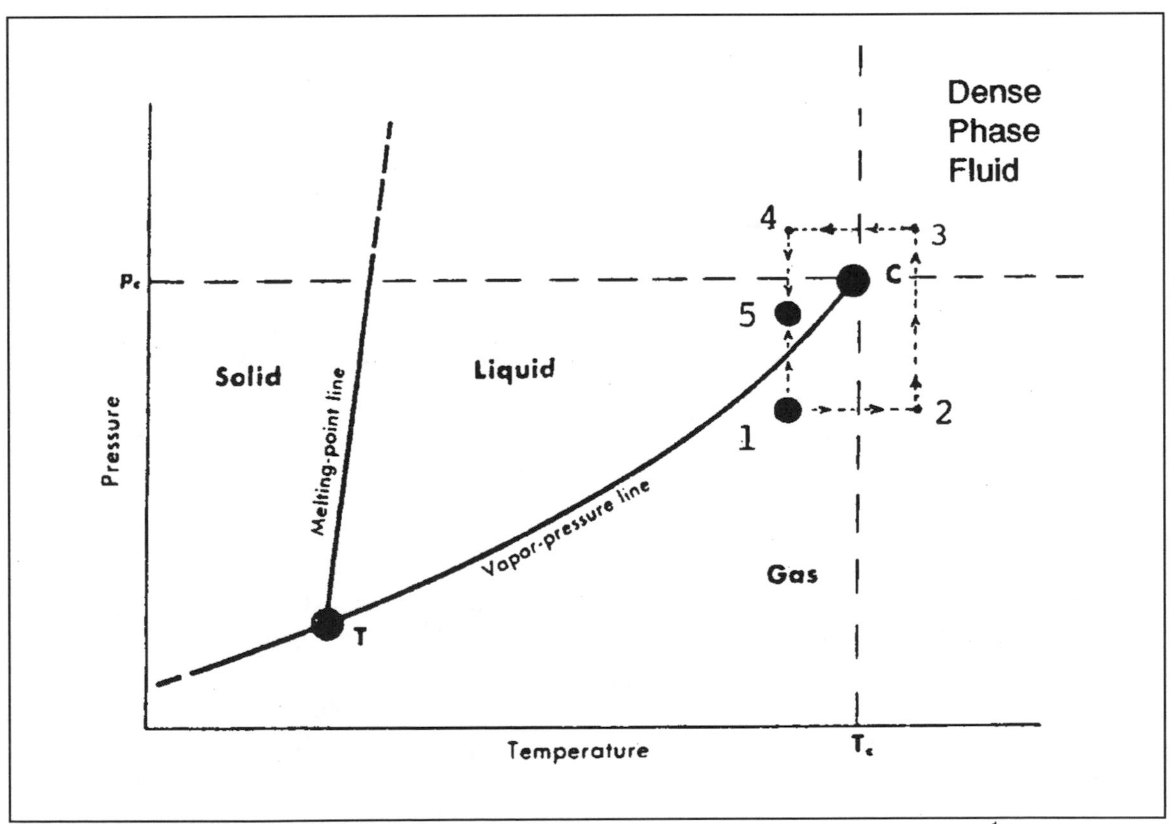

Fig. 3-2. Phase diagram for a pure substance illustrating dense phase fluid region (adapted from McCain[1]). Permission to publish by Petroleum Publishing Company.

Consider Figure 3-3 which is a typical pressure-volume diagram for a pure substance. The solid lines are isotherms. The isotherm corresponding to the critical temperature passes through the critical point (defined by the critical pressure), which is the point at which the dewpoint and bubblepoint curves coincide. The dashed line contains the region in which liquid and gas can coexist. The dashed line to the right of the critical point defines the set of dewpoints, while the one to the left comprises the locus of bubblepoints.

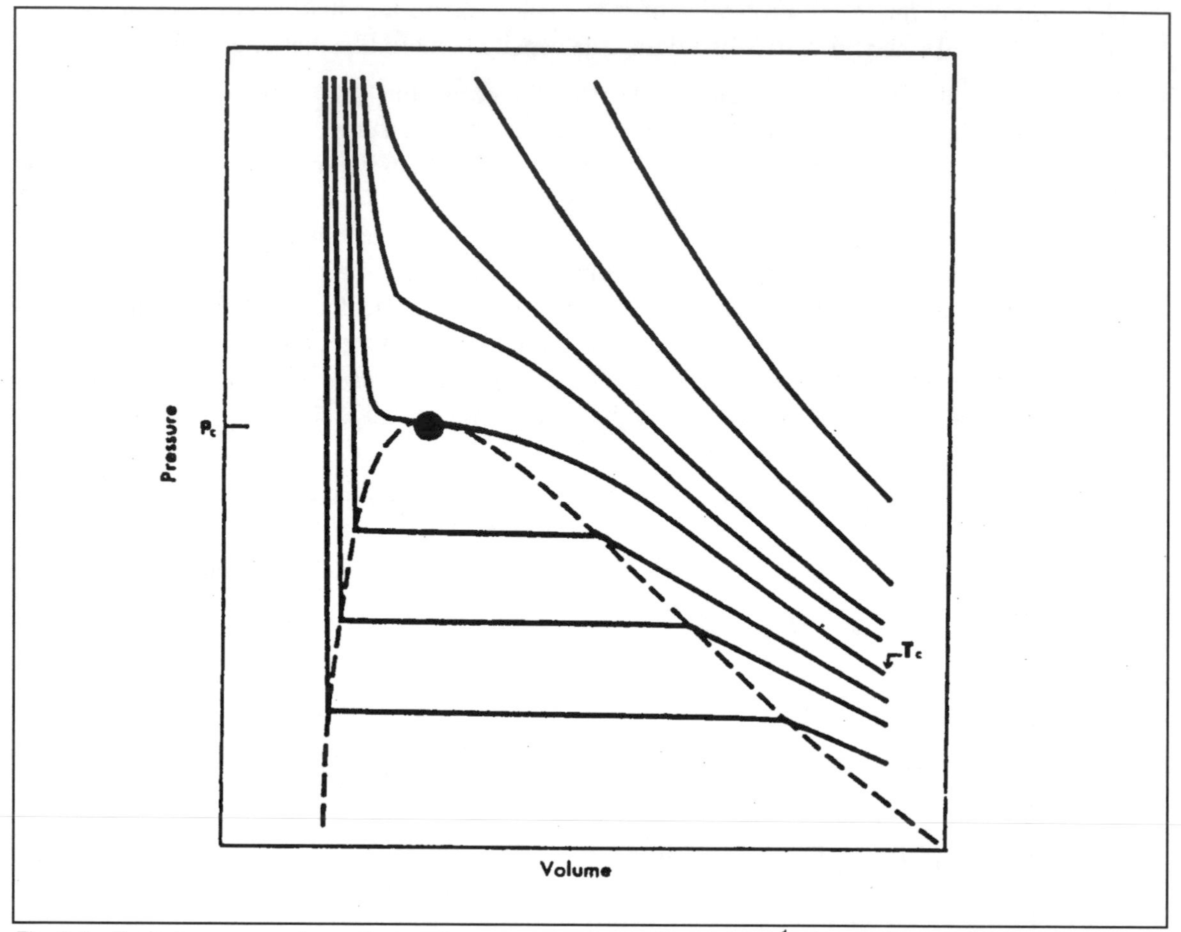

Fig. 3-3. Typical pressure-volume diagram for a pure substance (from McCain[1]).Permission to publish by Petroleum Publishing Company.

B. Two-Component Systems

While the reservoir engineer normally works with multicomponent systems, it is instructive to consider the phase behavior of two components. The differences between two-component systems and pure substances are amplified in the multicomponent systems. Figure 3-4 is the phase (pressure-temperature) diagram for a 50/50 mixture of two components. Notice that no longer is there a mere vapor pressure curve (which describes the points where gas and liquid can coexist in equilibrium), but a "phase envelope" or broad area where the two phases can coexist. The upper boundary of the phase envelope is the bubblepoint line, which represents the locus of bubblepoints. "Quality" lines may be seen within the envelope. Each represents a constant percentage of liquid. The dewpoint line is the lower boundary of the phase envelope.

At point 1 in Figure 3-4, the fluid mixture is 100% liquid. Decreasing the pressure at a constant temperature causes the liquid to expand. This occurs until the upper boundary of the phase envelope (bubblepoint) is contacted. Below the bubblepoint some molecules of the fluid will leave the liquid and form a gas phase. When we reach the lower boundary (dewpoint) of the phase envelope, the fluid has been converted to 100% saturated gas. (This means that even a slight increase in the pressure will cause some liquid to drop out.) Further reduction in pressure, such as down to point 2, will result only in gas expansion.

The definition of critical point changes from that for a pure substance. In the two-component case, the critical point is simply the point at which the dewpoint and the bubblepoint curves join. At this point all properties of the liquid and the gas are identical.

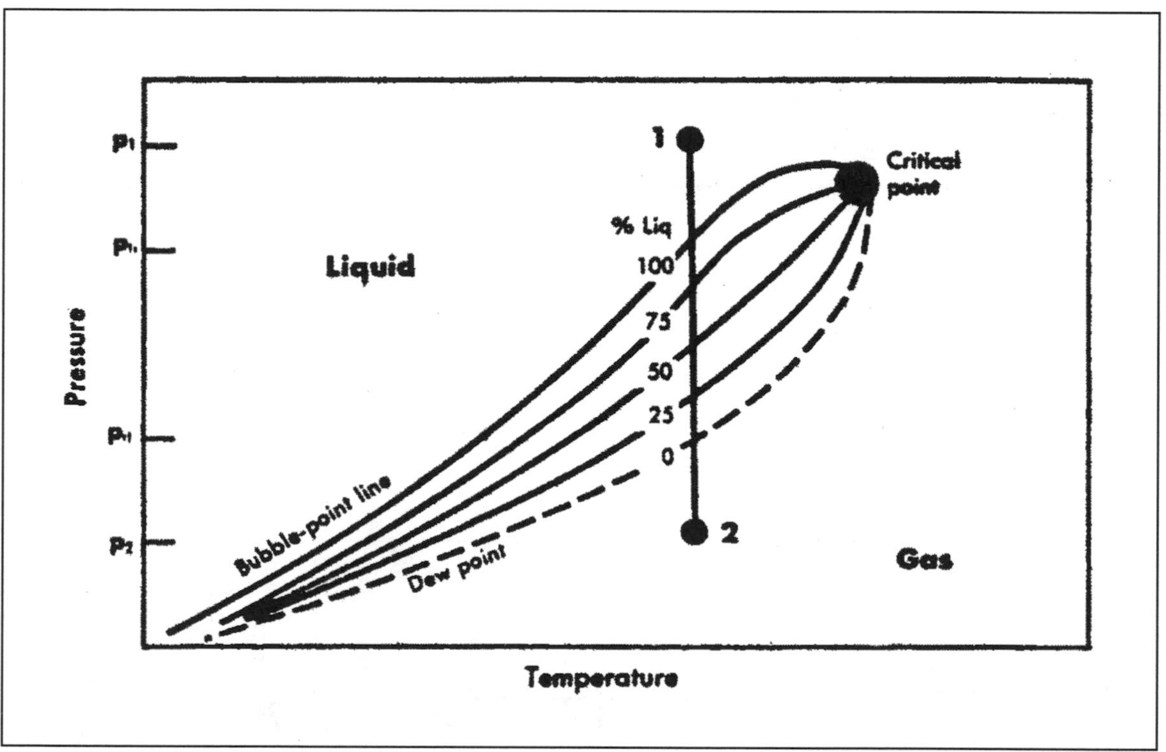

Fig. 3-4. Phase diagram for 50\50 mixture of two components (from McCain[1]). Permission to publish by Petroleum Publishing Company.

Figure 3-5 shows the phase diagram of a two-component system superimposed on that for the individual pure substances. Note that the critical point is higher than that for either pure substance. As might be expected, a change in the proportion of each component present will have a substantial effect on the phase diagram. Figure 3-6 provides a series of phase diagrams for mixtures of methane and ethane. Note that the locus of critical points for all of the different mixtures of methane and ethane is shown as a dashed line.

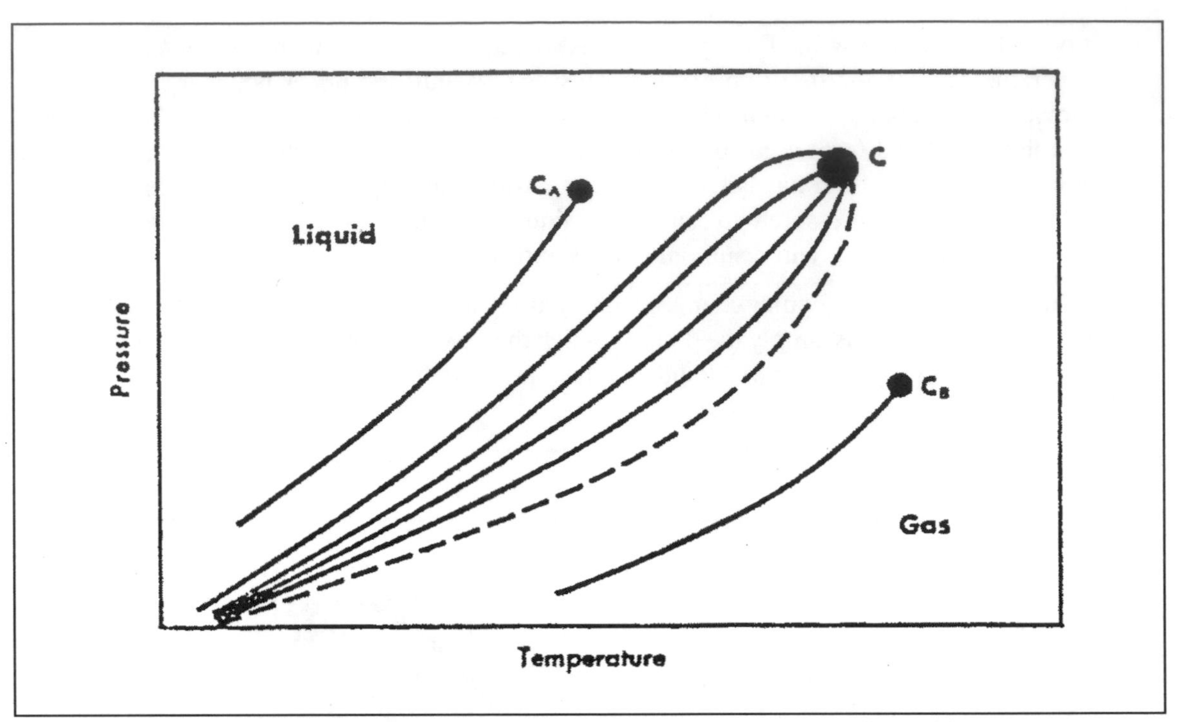

Fig. 3-5. Vapor pressure curves for two pure components and phase diagram for a 50 / 50 mixture of the same components (from McCain[1]). Permission to publish by Petroleum Publishing Company.

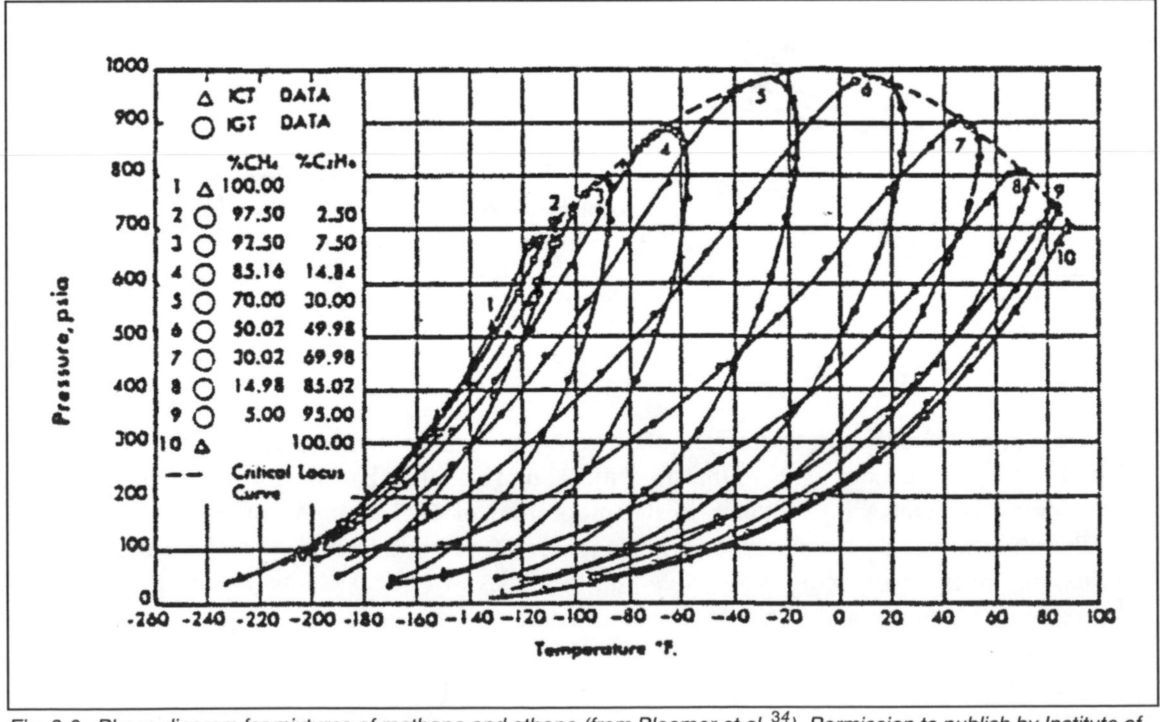

Fig. 3-6. Phase diagram for mixtures of methane and ethane (from Bloomer et al. [34]). Permission to publish by Institute of Gas Technology.

Many different binary mixtures have been studied and published. Figure 3-7 provides a useful plot of the critical loci for two-component systems comprised of normal paraffinic-series hydrocarbons. Notice that large differences in molecular weight of the two components greatly increases the critical pressures of the mixtures. This is often important when miscibility of displacement processes using hydrocarbons is being considered.

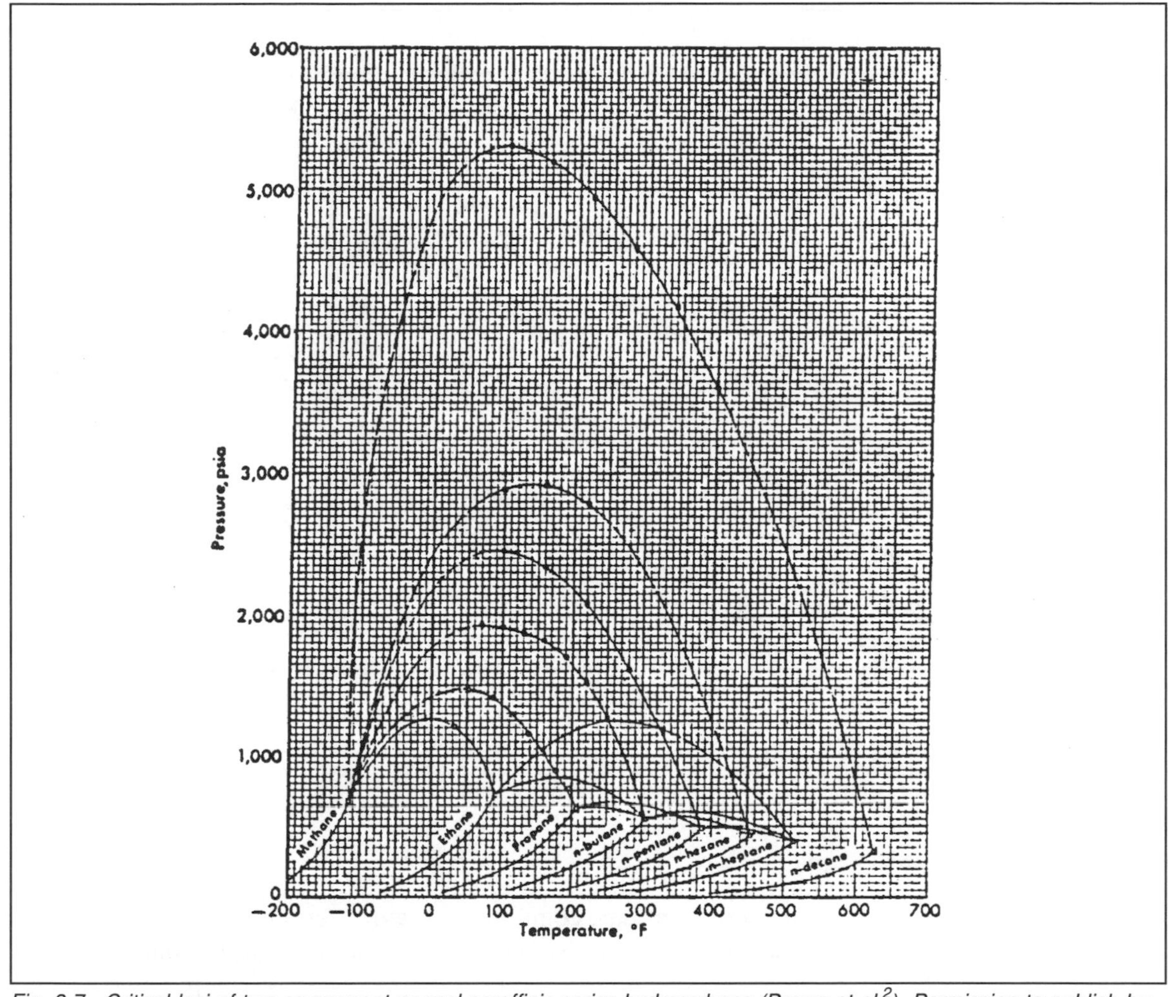

Fig. 3-7. Critical loci of two-component normal-paraffinic series hydrocarbons (Brown et al.[2]). Permission to publish by Natural Gas Association.

Reference to Figure 3-8 will show that the two-phase region can encompass areas where the temperature or pressure is higher than the critical pressure and the critical temperature. The maximum possible temperature of the two-phase region is defined as the cricondentherm, while the maximum possible pressure is called the cricondenbar.

The shaded area of Figure 3-8 to the right of the critical point illustrates the region where retrograde condensation can occur. If the pressure is isothermally decreased from point 1 (all gas) to point 2, then the upper dewpoint is crossed in transit; and a point 2, five percent by volume liquid has been formed. Provided phase equilibrium is maintained, further pressure drop will cause revaporization of the liquid. This phenomenon is observed in the production of some gas condensate reservoirs. Unfortunately, phase equilibrium cannot be conveniently maintained in most actual reservoirs, and the liquids that drop out in the reservoir rock are held by capillary forces. Therefore, while still at point 1, prudent operation would suggest injection of a dry gas and "cycling" this reservoir to keep the pressure above the phase envelope and, consequently, avoid the retrograde region.

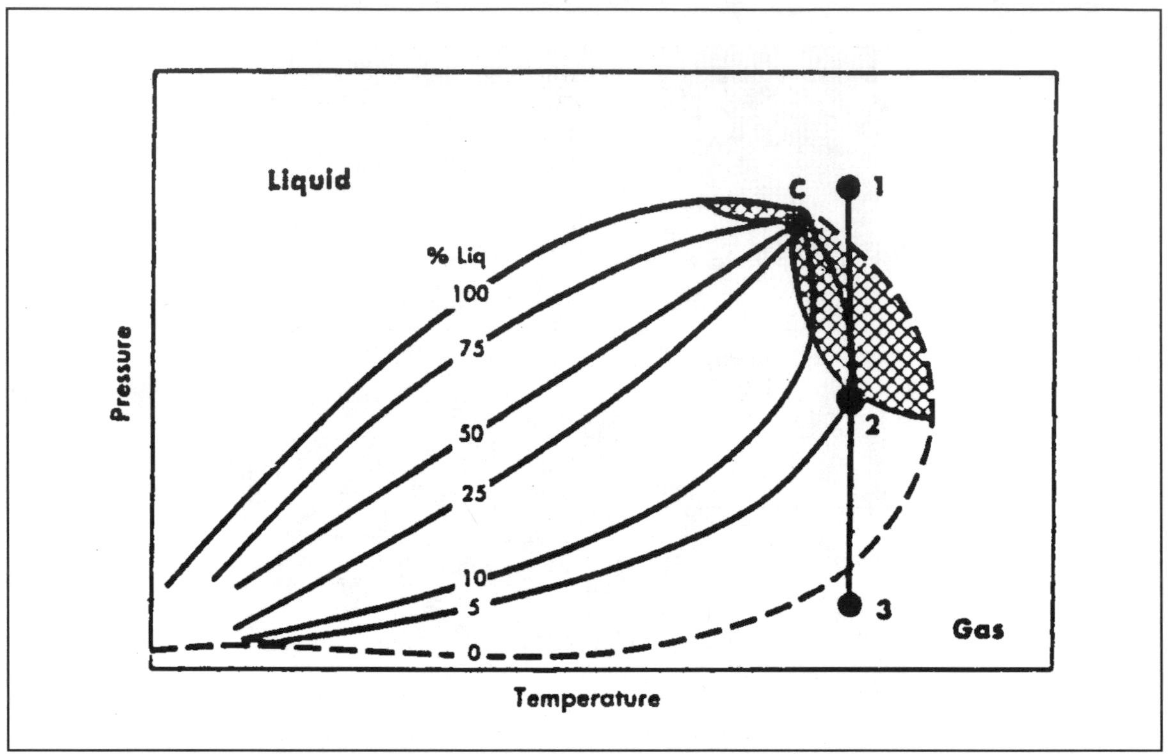

Fig. 3-8. Phase diagram showing regions of retrograde condensation (from McCain[1]). Permission to publish by Petroleum Publishing Company.

C. Three-Component Systems

To study the effects of composition in a three component system, "ternary" diagrams are often used. Figure 3-9 illustrates such a diagram. By convention, the top point of the triangle is used to represent 100% C_1; the lower right point is equal to 100% C_2; and lower left point is 100% C_3. Note that C_1 does not indicate methane, merely component one. Each component has its own linear scale or axis representing mole percent with 100% being at its vertex and 0% being at the opposite side of the triangle. It should be noted that each ternary diagram is valid only for a given constant temperature and pressure. Composition alone is varying on the diagram.

D. Multicomponent Systems

At this point, it is logical to extend the ternary diagram type of analysis to multicomponent hydrocarbon mixtures; however, pressure-temperature diagrams are also used and will be discussed under hydrocarbon classification.

In a normal hydrocarbon mixture (such as a crude oil), hundreds of components probably exist. To be able to use a ternary diagram type analysis, all of these components often are divided by using pseudocomponents: methane (C_1), intermediates (C_2 through C_6), and the heavies (C_{7+}). The resulting plot (Fig. 3-10) is called a pseudoternary diagram because two of the components are really not single components, but are made up of multicomponents themselves. Usually, the light component is situated at the top, the intermediates at the lower right corner, and the heavies are plotted at the lower left corner. It is important to remember that conditions for a given graph are at constant temperature and pressure.

Assume for the system represented by Figure 3-10 that the pressure is 3000 psia, and the temperature is 200°F. At these conditions C_1 is a gas and the C_{7+} pseudocomponent is a liquid. The state of the intermediate pseudocomponent will depend on the proportions of its different fractions. Notice that Figure 3-10 has an area surrounded by a phase envelope within which the mixture exists in two phases: gas and liquid.

The dewpoint curve represents the top part of the two-phase boundary down to the critical point. The bubblepoint curve consists of the lower part of the phase envelop up to the critical point. "Tie lines" extend across the two-phase region terminating on the dewpoint and bubblepoint lines. If the fluid composition falls within the two-phase region, then the corresponding tie line (one on which the composition point falls) will yield the gas composition and the liquid composition. The gas composition is the tie-line intersection with the dewpoint curve, and the liquid composition is the tie line intersection with the bubblepoint curve.

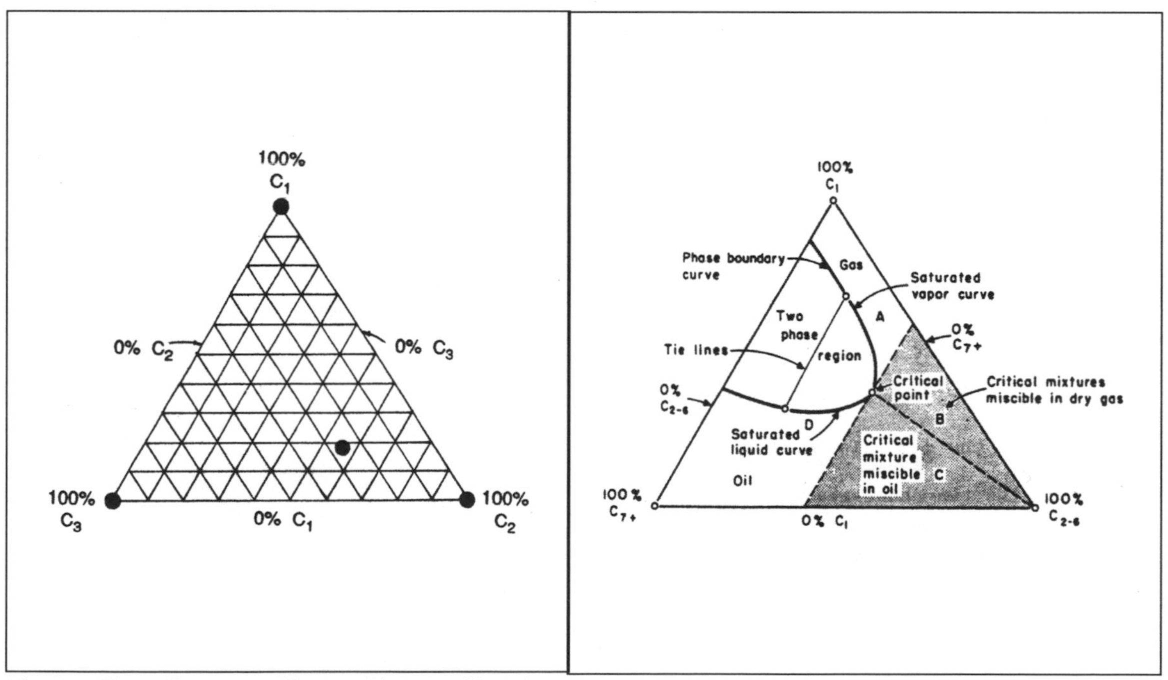

Fig. 3-9. Three Component Ternary Diagram. (Note that the interior point represents a composition of 18% C_1, 56% C_2, and 26% C_3.)

Fig. 3-10. Triangular graph showing physical conditions of hydrocarbon systems at fixed temperature and pressure. (Clark[11]). Permission to publish by SPE.

The tie lines are determined through laboratory experimental procedures, and the limiting tie line determines the location of the critical point. Notice in Figure 3-10 that the limiting tie line has been extended across the pseudoternary diagram for the purpose of showing the different fluid regions. Mixtures in area A are gas; in area D, liquid; in areas B and C, critical mixture; and within the phase envelope, gas and liquid in equilibrium. Inside the critical mixture region, the concentration of intermediates is high enough to cause the fluid to be single phase, either gas or liquid depending on the relative proportions of methane and C_{7+}. If a fluid is found in the critical mixture regions, then a hydrocarbon gas displacement process (at the pressure of the diagram) should be a miscible displacement. Although not thermodynamically rigorous, the qualitative pseudoternary phase diagram can be used to predict the results of some gas-drive processes.

Two fluids are said to be miscible when no interface is between them and will therefore readily mix together to form a uniform solution. If this were the case between a solvent and oil, then no capillary forces would be acting within the reservoir rock to retain the oil in the pore channels through which the solvent flows. Without capillary pressure effects, it is theoretically possible to obtain 100% recovery of oil from those pores contacted by the solvent.

Two basic kinds of miscibility are: (1) single contact miscibility where the displacing and displaced fluids are miscible immediately and in all proportions, and (2) multiple contact miscibility where miscibility is not attained in all proportions. In the multiple contact case, some component transfer must occur between the two fluids before miscibility is attained.

Assume a reservoir oil, the pseudoternary diagram for which is represented in Figure 3-11A. This is a dipping reservoir, and methane gas is planned to be injected updip. The recovery plan is to miscibly displace the oil to downdip producing wells. After all the oil is recovered, then the injected gas will be produced. A simple schematic diagram of this reservoir management scheme is presented in Figure 3-12.

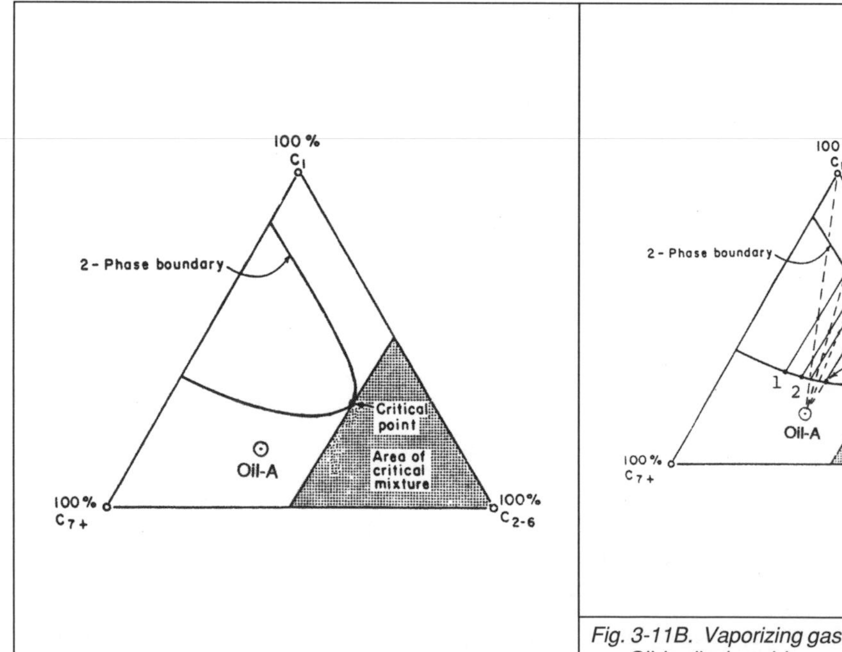

Fig. 3-11A. Pseudoternary diagram for oil-A showing original composition at reservoir conditions.

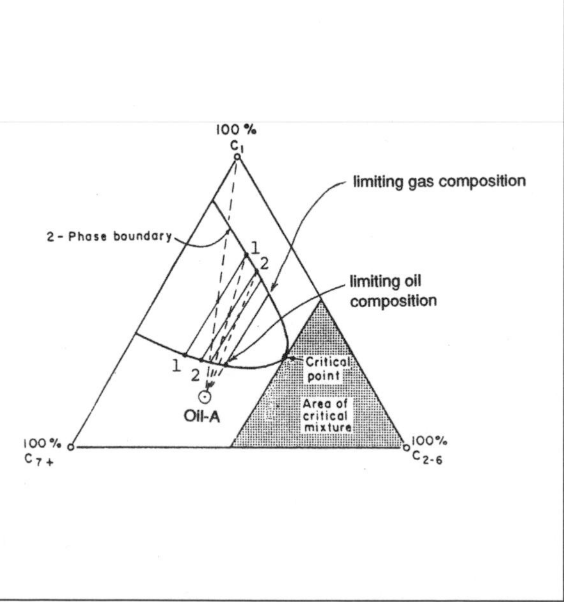

Fig. 3-11B. Vaporizing gas drive, noncritical displacement. Oil is displaced by gas which vaporizes intermediate components from the oil. Here, a critical mixture is not formed; so miscibility is not obtained.

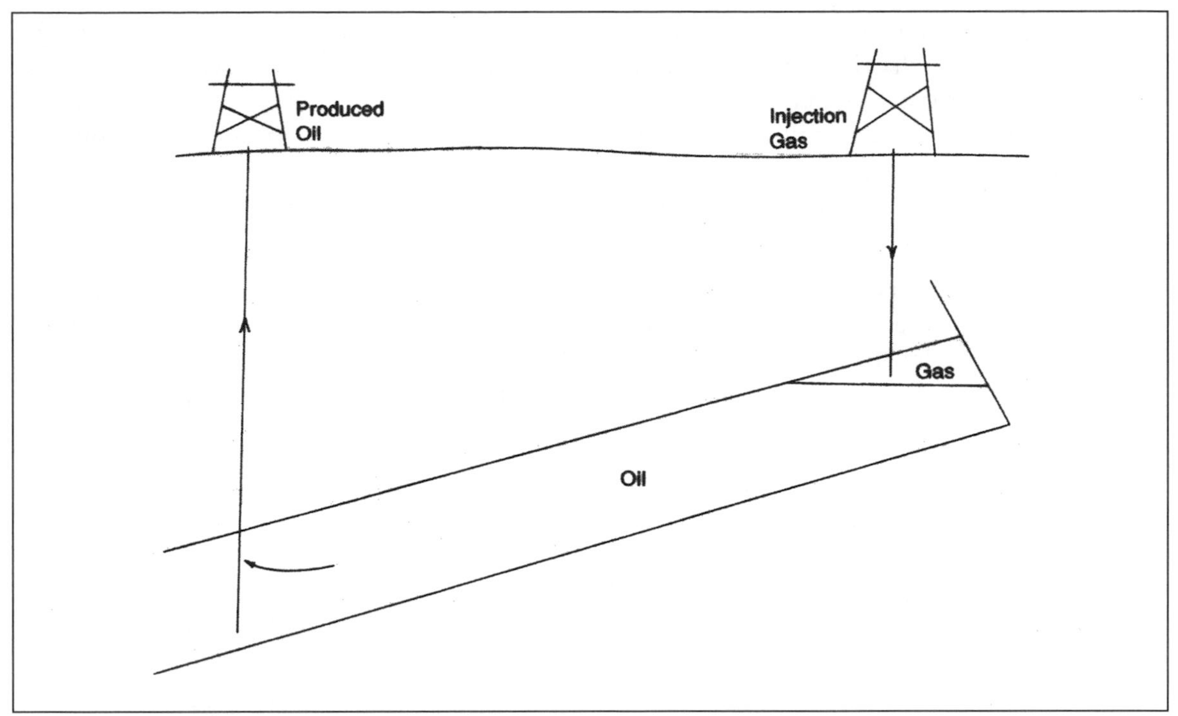

Fig. 3-12. High pressure gas injection oil recovery plan.

To determine whether miscibility will be achieved, we may use a pseudoternary diagram as shown in Figure 3-11. First draw a line between the oil composition point and the injected gas composition point (100% C_1). Because this line goes through the two-phase region (Fig. 3-11B), the diagram is predicting that single contact miscibility will not be achieved. To investigate whether multiple contact miscibility is possible, a few more steps are needed.

What is being hoped for in the multiple contact process is that some interphase mass transfer will occur between the oil and the dry gas, which will change the compositions of both to the extent that miscibility is possible. In this particular case, the methane gas contains no intermediates; so it will vaporize intermediate components from the oil. The process is called a vaporizing gas drive.

As in Figure 3-11B, a line is drawn between the gas composition and the oil composition on the pseudoternary diagram. Shown is the intersection of this line with a typical tie line within the two-phase region. This is the left-most tie line shown on the diagram. When the system composition (in the mixing zone or transition bank at the "front") is represented by the intersection point, then two phases must be present. The gas will have a composition found at the dewpoint end of this tie line (gas 1), and the liquid composition is at the bubblepoint end of the tie line (oil 1). It can be seen that the gas has become richer in intermediates and the oil leaner. With continued contact, the gas in the transition zone, if possible, will continue to vaporize intermediates from the oil. To investigate this, a (dotted) line is drawn from gas 1 composition to the original oil composition. Another tie line is intersected such that the new enriched gas composition is gas 2. The corresponding equilibrium oil moves to the position of oil 2.

Aa this process continues, the tie lines containing the compositions of the transition bank gas and oil move to the right. Note that a tie line, if extended past the two-phase envelope, will intersect the original oil composition. This tie line is as far to the right as the process can go, and it therefore contains the limiting or final gas composition (in the transition bank). Consequently, at the pressure and temperature of the diagram (original reservoir conditions), oil-A being driven by methane cannot escape the two-phase region. Thus, a critical mixture will never be formed, and miscibility cannot be achieved.

A rich injection gas is considered in Figure 3-13. Single contact miscibility is not achieved. However, this time the oil composition moves to the right along the bubblepoint line as the oil is stripping intermediate components out of the injected gas. This process is a condensing gas drive. When the oil composition gets to the critical point, miscibility is achieved.

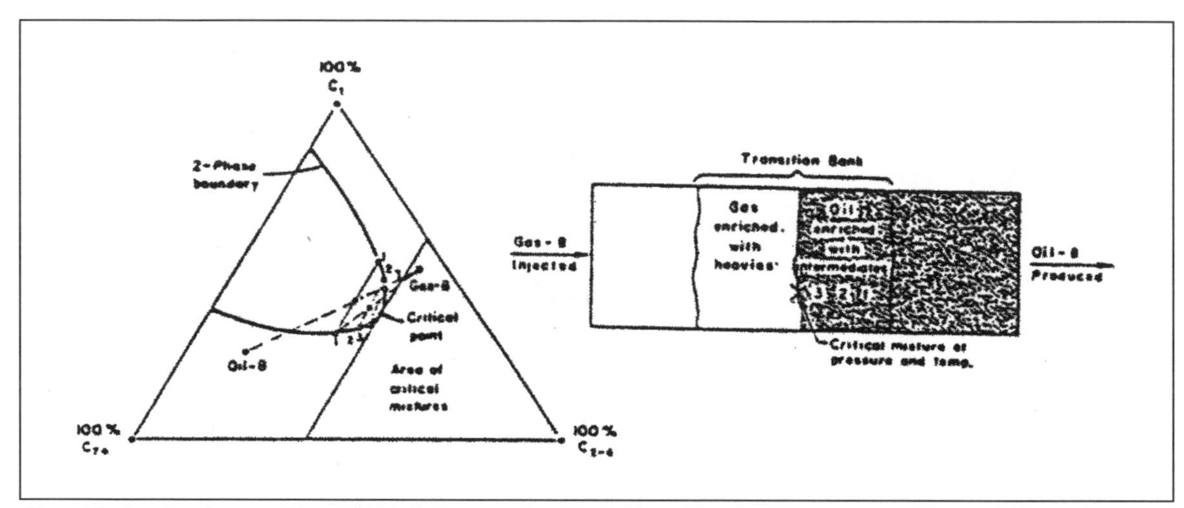

Fig. 3-13. Condensing gas drive, critical displacement—oil is displaced from sand by gas components which condense to form a critical mixture and transition bank at the front. (Clark[11]). Permission to publish by SPE.

According to Novosad and Costain,[32] high temperature rich gas drives are likely not to be condensing processes, but liquid extraction drives. With this mechanism, C_{6+} is extracted from the oil into the gas during miscibility development.

It should be pointed out that not all condensing gas drives achieve miscibility. The gas composition must fall within the region of critical mixtures for a miscible process to be obtained. Should the gas composition be less rich than this, then a condensing gas drive may still occur, but the process will not attain complete miscibility.

Similarly for the vaporizing gas drive, if the oil composition is in the area of critical mixtures, then miscible displacement will develop.

PROPERTIES OF GASES

A gas may be defined as a homogeneous fluid, generally of low density and viscosity, that has no definite volume but fills completely any vessel in which it is placed. To be able to predict the behavior of gases, an equation of state is needed.

A. Gas Laws

1. Perfect Gases

Boyle's law states that for a given mass of gas, at a given temperature, the volume varies inversely as the pressure, or

$$pV = \text{constant} \tag{3-1}$$

Charles' law (also Gay-Lussac's law) states that for a given mass of gas, at a given pressure, the volume varies directly with the absolute temperature:

$$V/T = \text{constant} \tag{3-2}$$

Avogadro's law states that under the same conditions of pressure and temperature, equal volumes of all ideal gases contain the same number of molecules.

Combining Boyle's, Charles', and Avogadro's laws and defining R to be a constant, then the perfect or ideal gas law may be written for "n" moles of gas:

$$pV = nRT \tag{3-3}$$

Table 3-2 presents values for R, the universal gas constant, for different units.

Table 3-2
Values of Gas Constant, R, in Various Units

Units	R
psia, ft^3, lb-mole, °R	10.73
lb/ft^2 abs., ft^3, lb-mole, °R	1545
atm, ft^3, lb-mole, °R	0.730
atm, ft^3, lb-mole, K	1.3145
in. Hg., ft^3, lb-mole, °R	21.85
kPa, m^3, kmole, K	8.314
MPa, m^3, kmole, K	0.00831
kg/cm^2, m^3, kmole, K	0.08478
atm, cm^3, g-mole, K	82.06
atm, liter, g-mole, K	0.08206
mm Hg, liters, g-mole, K	62.37

Now, if w let m = mass and M = molecular weight (relative molecular mass), then n = m/M. And, if we note that density = ρ = m/V, Equation 3-3 can be rewritten in terms of density:

$$\rho_g = \frac{pM}{RT} \tag{3-4}$$

Usually, the reservoir engineer will be interested in the behavior of a mixture of gases and seldom deals with only one component. With a mixture, it is usually better to work with the components in terms of mole fractions or mole percentages.

Dalton's law of partial pressures states that each gas in a mixture of gases exerts a pressure equal to the pressure that it would exert if it alone were present in the volume occupied by the gas mixture. The total pressure exerted by the mixture is the sum of the pressures exerted by its components. The pressure exerted by each component is called its "partial pressure." Dalton's rule is sometimes referred to as the law of additive pressures.

Amagat's law states that the total volume occupied by a gas mixture is equal to the sum of the volumes that the pure components would occupy at the same pressure and temperature. This has also been called the law of additive volumes. It can be shown that:

$$\frac{V_j}{V} = \frac{n_j \dfrac{RT}{p}}{n \dfrac{RT}{p}} = \frac{n_j}{n} = y_j \tag{3-5}$$

This relation indicates that the volume fraction is equal to the mole fraction for an ideal gas.

It is also useful to note that a gas mixture, whether ideal or nonideal, behaves as if it were a pure gas with a definite molecular weight. For instance, air contains approximately 78 mole percent nitrogen, 21 mole percent oxygen, and 1 mole percent argon. The gas mixture molecular weight can then be determined by using a mole fraction weighting procedure:

$$M = (0.78)(28.01) + (0.21)(32.00) + (0.01)(39.94) = 28.97 \cong 29$$

By definition, the specific gravity (relative density) of a gas is the ratio of the density of the gas compared to the density of dry air at standard conditions of temperature and pressure:

$$\gamma_g = \frac{\dfrac{M_g p}{RT}}{\dfrac{M_{air} p}{RT}} = \frac{M_g}{M_{air}} = \frac{M_g}{28.97} \cong \frac{M_g}{29} \tag{3-6}$$

Thus, we may calculate the specific gravity of a gas mixture (such as a natural gas) if we know the mole percentages of the components present.

Problem 3-1:

Component	Mole Fraction y_j	Molecular Weight M_j	y_jM_j
C1	0.70	16.0	11.20
C2	0.10	30.1	3.01
C3	0.10	44.1	4.41
n-C4	0.10	58.1	5.81
			24.43 = M_g

Specific gravity (relative density) of the gas is 24.43/29 or 0.84

2. NonIdeal Gases

The perfect gas law will work reasonably well for modest pressures and temperatures. However, for higher pressures and temperatures, error will result.

For single component gases, a wide variety of equations of state have been presented that take into account the forces of attraction between molecules, the actual volume of the molecules, and behavior at or above the critical point. Well-known equations are those due to van de Waals, Beattie-Bridgeman, Benedict-Webb-Rubin, Redlich-Kwong, and others. Unfortunately, the main use of these equations is for simple gases or very simple mixtures of light gases. However, the equations are of value to identify the factors that are causing nonideal behavior of gases.

As reservoir gases are anything but simple mixtures, another approach was needed. It has become standard practice to use the so-called "real gas law," which modifies the ideal gas law with a compressibility factor or deviation factor, z:

$$pV = znRT \tag{3-7}$$

Density is equal to:

$$\rho_g = \frac{pM}{zRT} \tag{3-8}$$

The z factor is nothing more than a correction factor supported by laboratory experimentation on observed physical behavior of mixtures of gases. For the ideal gas, the compressibility factor is unity. Figure 3-14 provides a plot of the z factor for methane, the simplest of the hydrocarbon gases.

The theorem of corresponding states due to van de Waals, states that different gases are at the same corresponding state if they exist at the same conditions of reduced pressure and reduced temperature. Reduced temperature is defined as absolute temperature divided by absolute critical temperature:

$$T_r = T / T_c \tag{3-9}$$

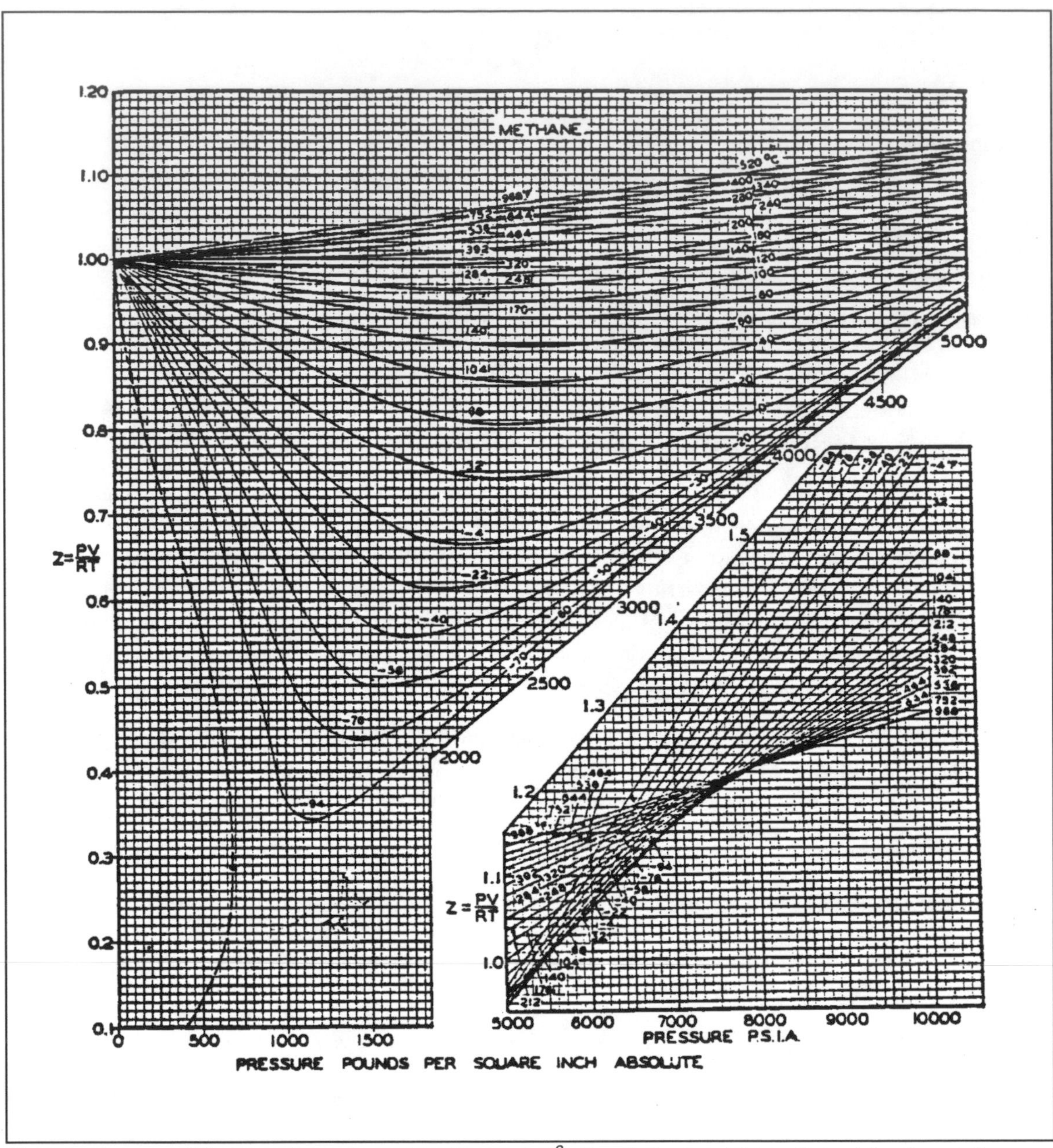

Fig. 3-14. Compressibility factors for methane (from Brown et al.[2]). Permission to publish by Natural Gas Association.

And reduced pressure:

$$p_r = p / p_c \quad (3\text{-}10)$$

where:

T_c = the absolute critical temperature,
p_c = the absolute critical pressure,
T = the absolute temperature at which the gas exists, and
p = the absolute pressure at which the gas exists.

When properly used, corresponding state theory may be applied to gaseous mixtures. For a gas mixture of n components:

$$p_{pc} = \sum_{i=1}^{n} y_i p_{ci} \quad (3\text{-}11)$$

$$T_{pc} = \sum_{i=1}^{n} y_i T_{ci} \quad (3\text{-}12)$$

where:

p_{pc} = pseudocritical pressure, absolute,
T_{pc} = pseudocritical temperature, absolute,
y_i = mole fraction of component i,
p_{ci} = critical pressure, absolute, of component i,
T_{ci} = critical temperature, absolute, of component i, and
n = number of components in the mixture.

The pseudocritical pressure and pseudocritical temperature are mole fraction weighted averages of the critical values for the individual components of a gas mixture.

Then, pseudoreduced pressure and pseudoreduced temperature are defined to be:

$$p_{pr} = p / p_{pc} \quad (3\text{-}13)$$

$$T_{pr} = T / T_{pc} \quad (3\text{-}14)$$

Figure 3-15 provides compressibility factors as a function of pseudoreduced pressures and temperatures. This chart is best for the "sweet" hydrocarbon gases or the paraffin series gases. For these gases, the correlation is quite accurate.

It is sometimes convenient to use gas gravity as a means of determining the approximate pseudocritical temperature and pressure of a hydrocarbon gas mixture. Figure 3-16 provides the necessary graphical correlations of both miscellaneous hydrocarbon gases and condensates. For gases containing substantial amounts of non-hydrocarbon components, Figure 3-16 contains correction graphs that can be used to modify the mixture pseudocritical

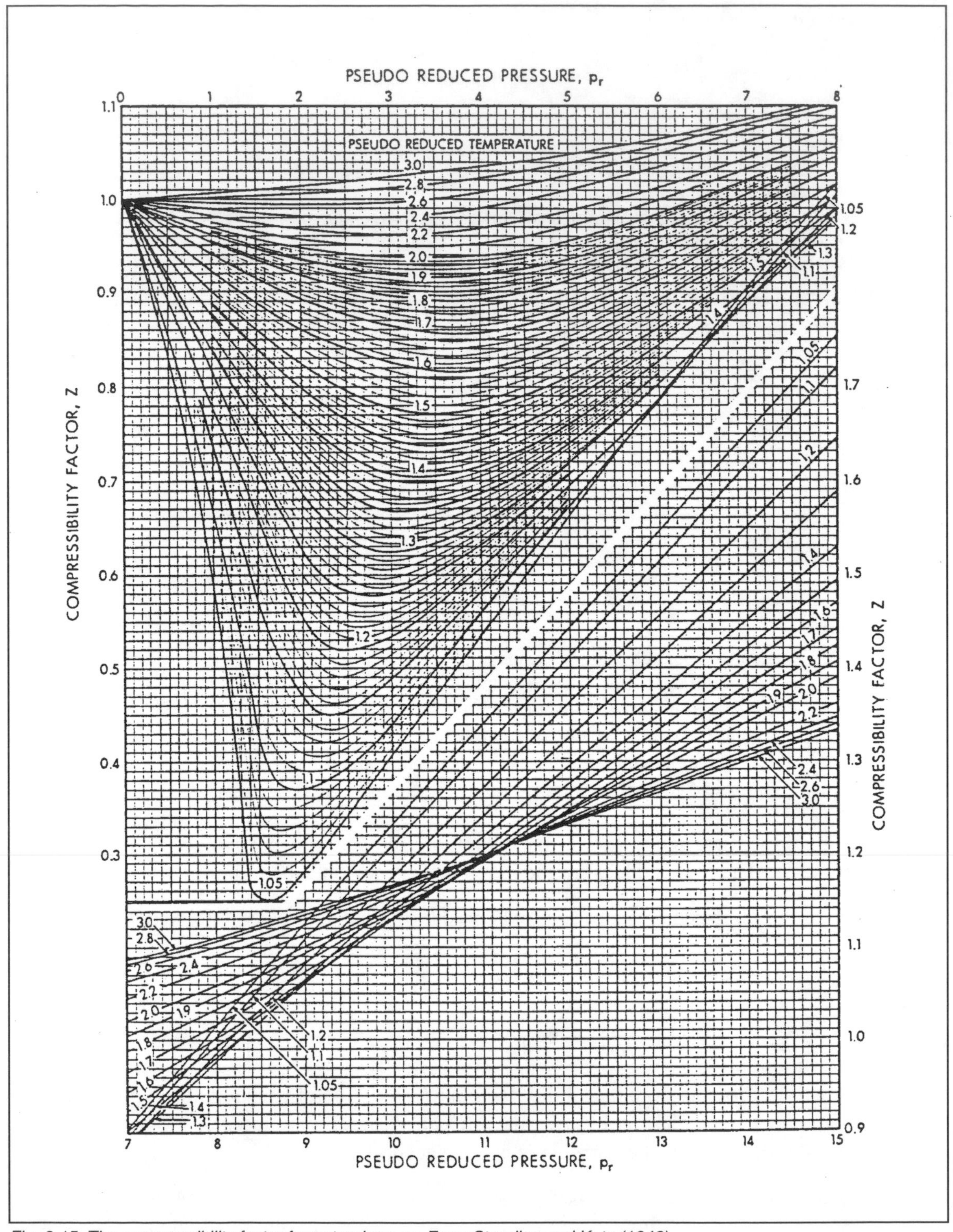

Fig. 3-15. The compressibility factor for natural gases. From Standing and Katz (1942).

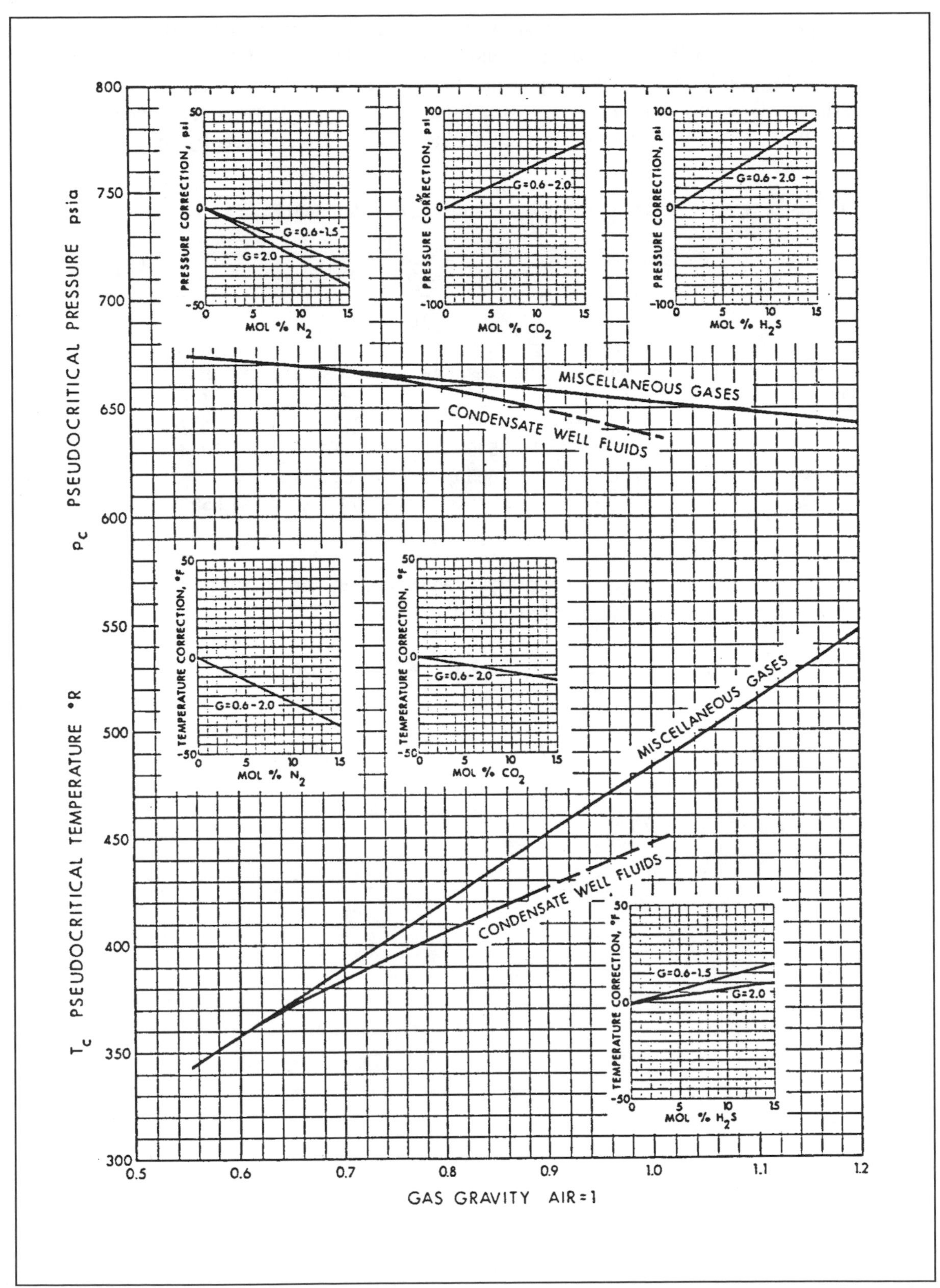

Fig. 3-16. Pseudocritical properties of miscellaneous natural gases. From Brown, et al., (1948); inserts from Carr, et al., (1954). Permission to publish by Natural Gas Association and SPE.

temperature and pressure. Accurate z values can be determined for hydrocarbon gases containing as much as 15 mole percent each of N_2, H_2S, and CO_2. Unless corrections are made for non-hydrocarbon components, Figure 3-15 should not be used to determine the z-factor. However, it is legitimate to use Figure 3-15 without corrections if the gas contains less than 5 percent nonhydrocarbons, with the resulting z-factor usually being less than 1% in error.

The pseudocritical temperature and pressure can be calculated from a gas analysis as follows:

Problem 3-2

Component	Mole fraction y_i	Critical temperature °R T_{cj}	$y_j T_{cj}$	Critical pressure psia p_{cj}	$y_j p_{cj}$
C_1	0.70	343.1	240.2	667.8	467.5
C_2	0.10	549.8	55.0	707.8	70.8
C_3	0.10	665.7	66.6	616.3	61.6
n-C_4	0.10	765.3	76.5	550.7	55.1
	1.00		T_{pc} = 438.3 °R		p_{pc} = 655.0

If this mixture is at a pressure of 2500 psia and a temperature of 168°F, then pseudoreduced properties and z-factor can be calculated as follows:

$$p_{pr} = p / p_{pc} = 2500 / 655 = 3.817$$

$$T_{pr} = T / T_{pc} = \frac{168 + 460}{438.3} = 1.433$$

From Figure 3-15: z = 0.73

Problem 3-3:

Consider a dry natural gas at 168°F and 2500 psia that illustrates the use of Figure 3-16. It is worked two different ways: (1) no correction for nonhydrocarbon components, and (2) correction (reference 22) for nonhydrocarbons.

Component	Mole Fraction x_j	Molecular Weight M_j	x_jM_j
C_1	0.5	16.0	8.0
C_2	0.1	30.0	3.0
C_3	0.1	44.0	4.4
H_2S	0.1	34.0	3.4
CO_2	0.1	44.0	4.4
N_2	0.1	28.0	2.8
			M = 26.0

Specific gravity = γ_g = 26.0 / 29.0 = 0.897

1. No correction for nonhydrocarbon components:

From Figure 3-16, the pseudocritical pressure and temperature are obtained using the "Miscellaneous Gases" curves because this is a dry natural gas:

$$T_{pc} = 450\ °R \text{ and } p_{pc} = 658\ psia$$

Therefore:

$$T_{pr} = (168 + 460) / 450 = 1.396$$

$$p_{pr} = 2500 / 658 = 3.799$$

From Fig. 3-15:

$$z = 0.71$$

2. Correction for nonhydrocarbon components:

From the Fig. 3-16 correction graphs:

	Temperature Correction	Pressure Correction
H_2S:	+13	+61
CO_2:	- 8	+44
N_2:	-24	-20
	-19	+85

In part 1, the uncorrected pseudocritical properties were determined to be:

$$T_{pc} = 450\ °R \text{ and } p_{pc} = 658\ psia.$$

So, the corrected pseudocritical properties are:

$$T_{pc} = 450 - 19 = 431\ °R$$

$$p_{pc} = 658 + 85 = 743\ psia.$$

Then, the pseudoreduced properties are calculated as usual:

$$T_{pr} = \frac{168 + 460}{431} = 1.457$$

$$p_{pr} = 2500 / 743 = 3.365$$

From Fig. 3-15:

$z = 0.745$

In 1972, Wichert and Aziz[28] published a method to calculate the z factor for hydrocarbon gases containing H_2S and CO_2. Particularly for gases containing large amounts of these components, the Wichert -Aziz correlation has become the industry standard. The procedure follows:

(a) Determine the critical pressure and temperature from the gas composition or from the gas gravity.

(b) Adjust the critical properties obtained in (a):

Let: B = mole fraction of H_2S and

A = (mole fraction of CO_2) + B

$$T'_{pc} = T_{pc} - \varepsilon$$

$$p'_{pc} = \frac{p_{pc}\, T'_{pc}}{T_{pc} + \varepsilon\,(B - B^2)}$$

where

$\varepsilon = 120\,(A^{0.9} - A^{1.6}) + 15\,(B^{0.5} - B^4)$

T_{pc}, T'_{pc} = °R

p_{pc}, p'_{pc} = psia

(c) Calculate the corrected reduced pressure and temperature:

$$T_{pr} = T / T'_{pc} \quad ; \quad p_{pr} = p / p'_{pc}$$

(d) Determine the z factor using the corrected reduced properties.

Meehan[33] presents a procedure for using the Wichert-Aziz method when a complete gas analysis is not available. At times, only the gas mixture gas gravity, γ_g, is known plus the mole fractions of N_2, H_2S, and CO_2 (as from an Orsat analysis). In this case:

(a) Calculate the gas gravity of the pure hydrocarbon portion (γ_{ghc}) of the total gas mixture by deleting the effects of the nonhydrocarbon components.

$$\gamma_{ghc} = \frac{\gamma_g - 0.967\, y_{N_2} - 1.52\, y_{CO_2} - 1.18\, y_{H_2S}}{1 - y_{N_2} - y_{CO_2} - y_{H_2S}}$$

where:

γ_{ghc} = gas gravity of the pure hydrocarbon portion of the gas mixture (air = 1.0),
y_{N_2} = mole fraction of N_2 in the gas,
y_{CO_2} = mole fraction of CO_2 in the gas, and
y_{H_2S} = mole fraction of H_2S in the gas.

and: γ_{ghc} must be > 0.55

(b) Determine the pseudocritical pressure and temperature of the hydrocarbon portion using γ_{ghc} and Figure 3-16 or the following equations.

Miscellaneous Gases:

$$p_{pchc} = 709.6 - 58.7\, \gamma_{ghc}$$

$$T_{pchc} = 168 + 325\gamma_{ghc} - 12.5\, (\gamma_{ghc})^2$$

Condensates or Associated Gases:

$$p_{pchc} = 706 - 51.7\, \gamma_{ghc} - 11.1\, (\gamma_{ghc})^2$$

$$T_{pchc} = 187 + 330\, \gamma_{ghc} - 71.5\, (\gamma_{ghc})^2$$

where:

p_{pchc} = pseudocritical pressure for the hydrocarbon portion of the gas mixture, psia, and
T_{pchc} = pseudocritical temperature for the hydrocarbon portion of the gas mixture, deg. R.

(c) Determine the whole gas mixture pseudocritical temperature and pressure:

$$p_{pc} = (1 - y_{N_2} - y_{CO_2} - y_{H_2S})\, p_{pchc} + 493\, y_{N_2} + 1071\, y_{CO_2} + 1306\, y_{H_2S} \quad psia$$

$$T_{pc} = (1 - y_{N_2} - y_{CO_2} - y_{H_2S})\, T_{pchc} + 227.6\, y_{N_2} + 547.9\, y_{CO_2} + 672.7\, y_{H_2S} \quad °R$$

(d) Make the Wichert-Aziz corrections for nonhydrocarbon components.

Problem 3-4:

Using the Meehan procedure, find the z factor for the following miscellaneous sour gas:

$$y_{H_2S} = 0.10 \; ; \; y_{CO_2} = 0.10 \; ; \; y_{N_2} = 0.10 \; ; \; \gamma_g = 0.897$$

Reservoir conditions:

$p = 2500$ psia;
$T = 168$ °F.

SOLUTION:

Hydrocarbon portion:

$$\gamma_{ghc} = \frac{0.897 - (0.967)(0.1) - (1.52)(0.10) - (1.18)(0.10)}{1 - 0.1 - 0.1 - 0.1}$$

$$= 0.7576$$

$$p_{pchc} = 709.6 - (58.7)(0.7576) = 665.1 \text{ psia}$$

$$T_{pchc} = 168 + (325)(0.7576) - (12.5)(0.7576)^2$$

$$= 407.0 \text{ °R}$$

Total mixture:

$$p_{pc} = (1 - 0.1 - 0.1 - 0.1)(665.1) + (493)(0.1) + (1071)(0.1) + (1306)(0.1)$$

$$= 752.6 \text{ psia}$$

$$T_{pc} = (1 - 0.1 - 0.1 - 0.1)(407.0) + (227.6)(0.1) + (547.9)(0.1) + (672.6)(0.1)$$

$$= 429.7 \text{ °R}$$

Wichert-Aziz Corrections:

$$B = 0.10;$$

$$A = 0.10 + 0.10 = 0.20$$

$$\varepsilon = 120\,[\,(0.20)^{0.9} - (0.20)^{1.6}\,] + 15\,[\,(0.10)^{0.5} - (0.10)^4\,]$$

$$\varepsilon = 23.8$$

$$T'_{pc} = T_{pc} - \varepsilon = 429.7 - 23.8 = 405.9 \text{ °R}$$

$$p'_{pc} = \frac{p_{pc} T'_{pc}}{T_{pc} + \varepsilon (B - B^2)} = \frac{(752.6)(405.9)}{429.7 + (23.8)(0.10 - 0.01)}$$

$$p'_{pc} = 707.4 \text{ psia}$$

$$T_{pr} = \frac{168 + 460}{405.9} = 1.55$$

$$p_{pr} = \frac{2500}{707.4} = 3.53$$

From Fig. 3-16:

$$z = 0.79$$

The purpose here was to illustrate the use of the Wichert-Aziz method when a complete gas analysis is not available. Actually, the composition of this gas was given in Example 3-3. Therefore, a more accurate solution is to use Equation 3-11 and 3-12 to calculate the gas mixture pseudocritical properties (p_{pc}=752.4 psia, T_{pc} = 438.1 °R). Then make the Wichert-Aziz adjustments (p'_{pc} = 708.1 psia, T'_{pc} = 414.3 °R). The resulting z factor at 2500 psia and 168 °F is 0.78.

Table 3-3 contains critical temperature and pressure information for many different pure substances.

B. Formation Volume Factor

"B_g" is used to signify gas formation volume factor which is equal to the volume of gas at reservoir temperature and pressure divided by the volume of the same amount of gas at standard conditions of temperature and pressure. With this factor, we can relate gas reservoir volume to its surface volume. Using the real gas law,

$$B_g = \frac{V_{res}}{V_{sc}} = \frac{\frac{znRT}{p}}{\frac{z_{sc}nRT_{sc}}{p_{sc}}} = \frac{zTp_{sc}}{z_{sc}T_{sc}p} \quad (3\text{-}15)$$

Normally, with field units T_{sc} = 520 °R, p_{sc} = 14.7 psia, and z_{sc} = 1. Making these substitutions into Equation 3-15,

$$B_g = 0.0283\ z\ T / p \quad (3\text{-}16)$$

where temperature is °R and pressure is psia. The units of Equation 3-16 are volume per standard volume; e.g., cubic feet per standard cubic foot. At times, especially with oil reservoir material balance calculations, it is convenient to use a gas formation volume factor that has units of barrels/SCF. To arrive at this relation, Equation 3-16 is divided by 5.615.

$$B_g = 0.005034\ z\ T / p \quad \text{barrels / SCF} \quad (3\text{-}17)$$

with T in °R and p in psia.

Table 3-3
Physical Constants of Hydrocarbons and Other Gases and Liquids
(from GPSA[3])
(continued on next two pages)

No.	Compound	Formula	Molecular weight	Boiling point °F., 14.696 psia	Vapor pressure, 100°F., psia	Freezing point, °F., 14.696 psia	Critical constants		
							Pressure, psia	Temperature, °F.	Volume, cu ft/lb
1	Methane	CH_4	16.043	−258.69	(5000)	−296.46[d]	667.8	−116.63	0.0991
2	Ethane	C_2H_6	30.070	−127.48	(800)	−297.89[d]	707.8	90.09	0.0788
3	Propane	C_3H_8	44.097	−43.67	190.	−305.84[d]	616.3	206.01	0.0737
4	n−Butane	C_4H_{10}	58.124	31.10	51.6	−217.05	550.7	305.65	0.0702
5	Isobutane	C_4H_{10}	58.124	10.90	72.2	−255.29	529.1	274.98	0.0724
6	n−Pentane	C_5H_{12}	72.151	96.92	15.570	−201.51	488.6	385.7	0.0675
7	Isopentane	C_5H_{12}	72.151	82.12	20.44	−255.83	490.4	369.10	0.0679
8	Neopentane	C_5H_{12}	72.151	49.10	35.9	2.17	464.0	321.13	0.0674
9	n−Hexane	C_6H_{14}	86.178	155.72	4.956	−139.58	436.9	453.7	0.0688
10	2−Methylpentane	C_6H_{14}	86.178	140.47	6.767	−244.63	436.6	435.83	0.0681
11	3−Methylpentane	C_6H_{14}	86.178	145.89	6.098	—	453.1	448.3	0.0681
12	Neohexane	C_6H_{14}	86.178	121.52	9.856	−147.72	446.8	420.13	0.0667
13	2,3−Dimethylbutane	C_6H_{14}	86.178	136.36	7.404	−199.38	453.5	440.29	0.0665
14	n−Heptane	C_7H_{16}	100.205	209.17	1.620	−131.05	396.8	512.8	0.0691
15	2−Methylhexane	C_7H_{16}	100.205	194.09	2.271	−180.89	396.5	495.00	0.0673
16	3−Methylhexane	C_7H_{16}	100.205	197.32	2.130	—	408.1	503.78	0.0646
17	3−Ethylpentane	C_7H_{16}	100.205	200.25	2.012	−181.48	419.3	513.48	0.0665
18	2,2−Dimethylpentane	C_7H_{16}	100.205	174.54	3.492	−190.86	402.2	477.23	0.0665
19	2,4−Dimethylpentane	C_7H_{16}	100.205	176.89	3.292	−182.63	396.9	475.95	0.0668
20	3,3−Dimethylpentane	C_7H_{16}	100.205	186.91	2.773	−210.01	427.2	505.85	0.0662
21	Triptane	C_7H_{16}	100.205	177.58	3.374	−12.82	428.4	496.44	0.0636
22	n−Octane	C_8H_{18}	114.232	258.22	0.537	−70.18	360.6	564.22	0.0690
23	Diisobutyl	C_8H_{18}	114.232	228.39	1.101	−132.07	360.6	530.44	0.0676
24	Isooctane	C_8H_{18}	114.232	210.63	1.708	−161.27	372.4	519.46	0.0656
25	n−Nonane	C_9H_{20}	128.259	303.47	0.179	−64.28	332.	610.68	0.0684
26	n−Decane	$C_{10}H_{22}$	142.286	345.48	0.0597	−21.36	304.	652.1	0.0679
27	Cyclopentane	C_5H_{10}	70.135	120.65	9.914	−136.91	653.8	461.5	0.059
28	Methylcyclopentane	C_6H_{12}	84.162	161.25	4.503	−224.44	548.9	499.35	0.0607
29	Cyclohexane	C_6H_{12}	84.162	177.29	3.264	43.77	591.	536.7	0.0586
30	Methylcyclohexane	C_7H_{14}	98.189	213.68	1.609	−195.87	503.5	570.27	0.0600
31	Ethylene	C_2H_4	28.054	−154.62	—	−272.45[d]	729.8	48.58	0.0737
32	Propene	C_3H_6	42.081	−53.90	226.4	−301.45[d]	669.	196.9	0.0689
33	1−Butene	C_4H_8	56.108	20.75	63.05	−301.63[d]	583.	295.6	0.0685
34	Cis−2−Butene	C_4H_8	56.108	38.69	45.54	−218.06	610.	324.37	0.0668
35	Trans−2−Butene	C_4H_8	56.108	33.58	49.80	−157.96	595.	311.86	0.0680
36	Isobutene	C_4H_8	56.108	19.59	63.40	−220.61	580.	292.55	0.0682
37	1−Pentene	C_5H_{10}	70.135	85.93	19.115	−265.39	590.	376.93	0.0697
38	1,2−Butadiene	C_4H_6	54.092	51.53	(20.)	−213.16	(653.)	(339.)	(0.0649)
39	1,3−Butadiene	C_4H_6	54.092	24.06	(60.)	−164.02	628.	306.	0.0654
40	Isoprene	C_5H_8	68.119	93.30	16.672	−230.74	(558.4)	(412.)	(0.0650)
41	Acetylene	C_2H_2	26.038	−119[e]	—	−114.[d]	890.4	95.31	0.0695
42	Benzene	C_6H_6	78.114	176.17	3.224	41.96	710.4	552.22	0.0531
43	Toluene	C_7H_8	92.141	231.13	1.032	−138.94	595.9	605.55	0.0549
44	Ethylbenzene	C_8H_{10}	106.168	277.16	0.371	−138.91	523.5	651.24	0.0564
45	o−Xylene	C_8H_{10}	106.168	291.97	0.264	−13.30	541.4	675.0	0.0557
46	m−Xylene	C_8H_{10}	106.168	282.41	0.326	−54.12	513.6	651.02	0.0567
47	p−Xylene	C_8H_{10}	106.168	281.05	0.342	55.86	509.2	649.6	0.0572
48	Styrene	C_8H_8	104.152	293.29	(0.24)	−23.10	580.	706.0	0.0541
49	Isopropylbenzene	C_9H_{12}	120.195	306.34	0.188	−140.82	465.4	676.4	0.0570
50	Methyl Alcohol	CH_4O	32.042	148.1(2)	4.63(22)	−143.82(22)	1174.2(21)	462.97(21)	0.0589(21)
51	Ethyl Alcohol	C_2H_6O	46.069	172.92(22)	2.3(7)	−173.4(22)	925.3(21)	469.58(21)	0.0580(21)
52	Carbon Monoxide	CO	28.010	−313.6(2)	—	−340.6(2)	507.(17)	−220.(17)	0.0532(17)
53	Carbon Dioxide	CO_2	44.010	−109.3(2)	—	—	1071.(17)	87.9(23)	0.0342(23)
54	Hydrogen Sulfide	H_2S	34.076	−76.6(24)	394.0(6)	−117.2(7)	1306.(17)	212.7(17)	0.0459(24)
55	Sulfur Dioxide	SO_2	64.059	14.0(7)	88.(7)	−103.9(7)	1145.(24)	315.5(17)	0.0306(24)
56	Ammonia	NH_3	17.031	−28.2(24)	212.(7)	−107.9(2)	1636.(17)	270.3(24)	0.0681(17)
57	Air	N_2O_2	28.964	−317.6(2)	—	—	547.(2)	−221.3(2)	0.0517(3)
58	Hydrogen	H_2	2.016	−423.0(24)	—	−434.8(24)	188.1(17)	−399.8(17)	0.5167(24)
59	Oxygen	O_2	31.999	−297.4(2)	—	−361.8(24)	736.9(24)	−181.1(17)	0.0382(24)
60	Nitrogen	N_2	28.013	−320.4(2)	—	−346.0(24)	493.0(24)	−232.4(24)	0.0514(17)
61	Chlorine	Cl_2	70.906	−29.3(24)	158.(7)	−149.8(24)	1118.4(24)	291.(17)	0.0281(17)
62	Water	H_2O	18.015	212.0	0.9492(12)	32.0	3208.(17)	705.6(17)	0.0500(17)
63	Helium	He	4.003	—	—	—	—	—	—
64	Hydrogen Chloride	HCl	36.461	−121(16)	925.(7)	−173.6(16)	1198.(17)	124.5(17)	0.0208(17)

Table 3-3 (continued)

14.696 psia							14.696 psia Ideal gas*			60°F., 14.696 psia Cp, Btu/lb/°F.		
Specific gravity 60°F./60°F.a,b	lb/gal*a (Wt in vacuum)	lb/gal*a,c (Wt in air)	Gal/lb Mole*	Temperature Coefficient of density*a	Pitzer acentric factor (18	Compressibility factor of real gas, Z 14.696 psia, 60°F.	Specific gravity Air = 1*	cu ft gas/lb*	cu ft gas/gal liquid*	Ideal gas	Liquid	No.
0.3[i]	2.5[i]	2.5[i]	6.4[i]	—	0.0104	0.9981	0.5539	23.65	59.[i]	0.5266	—	1
0.3564[h]	2.971[h]	2.962[h]	10.12[h]	—	0.0986	0.9916	1.0382	12.62	37.5[h]	0.4097	0.9256	2
0.5077[h]	4.233[h]	4.223[h]	10.42[h]	0.00152[h]	0.1524	0.9820	1.5225	8.606	36.43[h]	0.3881	0.5920	3
0.5844[h]	4.872[h]	4.865[h]	11.93[h]	0.00117[h]	0.2010	0.9667	2.0068	6.529	31.81[h]	0.3867	0.5636	4
0.5631[h]	4.695[h]	4.686[h]	12.38[h]	0.00119[h]	0.1848	0.9696	2.0068	6.529	30.65[h]	0.3872	0.5695	5
0.6310	5.261	5.251	13.71	0.00087	0.2539	0.9549	2.4911	5.260	27.67	0.3883	0.5441	6
0.6247	5.208	5.199	13.85	0.00090	0.2223	0.9544	2.4911	5.260	27.39	0.3827	0.5353	7
0.5967[h]	4.975[h]	4.965[h]	14.50[h]	0.00104[h]	0.1969	0.9510	2.4911	5.260	26.17[h]	(0.3866)	0.554	8
0.6640	5.536	5.526	15.57	0.00075	0.3007	—	2.9753	4.404	24.38	0.3864	0.5332	9
0.6579	5.485	5.475	15.71	0.00078	0.2825	—	2.9753	4.404	24.15	0.3872	0.5264	10
0.6689	5.577	5.568	15.45	0.00075	0.2741	—	2.9753	4.404	24.56	0.3815	0.507	11
0.6540	5.453	5.443	15.81	0.00078	0.2369	—	2.9753	4.404	24.01	0.3809	0.5165	12
0.6664	5.556	5.546	15.51	0.00075	0.2495	—	2.9753	4.404	24.47	0.378	0.5127	13
0.6882	5.738	5.728	17.46	0.00069	0.3498	—	3.4596	3.787	21.73	0.3875	0.5283	14
0.6830	5.694	5.685	17.60	0.00068	0.3336	—	3.4596	3.787	21.57	(0.390)	0.5223	15
0.6917	5.767	5.757	17.38	0.00069	0.3257	—	3.4596	3.787	21.84	(0.390)	0.511	16
0.7028	5.859	5.850	17.10	0.00070	0.3095	—	3.4596	3.787	22.19	(0.390)	0.5145	17
0.6782	5.654	5.645	17.72	0.00072	0.2998	—	3.4596	3.787	21.41	(0.395)	0.5171	18
0.6773	5.647	5.637	17.75	0.00072	0.3048	—	3.4596	3.787	21.39	0.3906	0.5247	19
0.6976	5.816	5.807	17.23	0.00065	0.2840	—	3.4596	3.787	22.03	(0.395)	0.502	20
0.6946	5.791	5.782	17.30	0.00069	0.2568	—	3.4596	3.787	21.93	0.3812	0.4995	21
0.7068	5.893	5.883	19.39	0.00062	0.4018	—	3.9439	3.322	19.58	(0.3876)	0.5239	22
0.6979	5.819	5.810	19.63	0.00065	0.3596	—	3.9439	3.322	19.33	(0.373)	0.5114	23
0.6962	5.804	5.795	19.68	0.00065	0.3041	—	3.9439	3.322	19.28	0.3758	0.4892	24
0.7217	6.017	6.008	21.32	0.00063	0.4455	—	4.4282	2.959	17.80	0.3840	0.5228	25
0.7342	6.121	6.112	23.24	0.00055	0.4885	—	4.9125	2.667	16.33	0.3835	0.5208	26
0.7504	6.256	6.247	11.21	0.00070	0.1955	0.9657	2.4215	5.411	33.85	0.2712	0.4216	27
0.7536	6.283	6.274	13.40	0.00071	0.2306	—	2.9057	4.509	28.33	0.3010	0.4407	28
0.7834	6.531	6.522	12.89	0.00068	0.2133	—	2.9057	4.509	29.45	0.2900	0.4332	29
0.7740	6.453	6.444	15.22	0.00063	0.2567	—	3.3900	3.865	24.94	0.3170	0.4397	30
—	—	—	—	—	0.0868	0.9938	0.9686	13.53	—	0.3622	—	31
0.5220[h]	4.352[h]	4.343[h]	9.67[h]	0.00189[h]	0.1405	0.9844	1.4529	9.018	39.25[h]	0.3541	0.585	32
0.6013[h]	5.013[h]	5.004[h]	11.19[h]	0.00116[h]	0.1906	0.9704	1.9372	6.764	33.91[h]	0.3548	0.535	33
0.6271[h]	5.228[h]	5.219[h]	10.73[h]	0.00098[h]	0.1953	0.9661	1.9372	6.764	35.36[h]	0.3269	0.5271	34
0.6100[h]	5.086[h]	5.076[h]	11.03[h]	0.00107[h]	0.2220	0.9662	1.9372	6.764	34.40[h]	0.3654	0.5351	35
0.6004[h]	5.006[h]	4.996[h]	11.21[h]	0.00120[h]	0.1951	0.9689	1.9372	6.764	33.86[h]	0.3701	0.549	36
0.6457	5.383	5.374	13.03	0.00089	0.2925	0.9550	2.4215	5.411	29.13	0.3635	0.5196	37
0.658[h]	5.486[h]	5.470[h]	9.86[h]	0.00098[h]	0.2485	(0.969)	1.8676	7.016	38.49[h]	0.3458	0.5408	38
0.6272[h]	5.229[h]	5.220[h]	10.34[h]	0.00113[h]	0.1955	(0.965)	1.8676	7.016	36.69[h]	0.3412	0.5079	39
0.6861	5.720	5.711	11.91	0.00086	0.2323	(0.962)	2.3519	5.571	31.87	0.357	0.5192	40
0.615[k]	—	—	—	—	0.1803	0.9925	0.8990	14.57	—	0.3966	—	41
0.8844	7.373	7.365	10.59	0.00066	0.2125	0.929(15)	2.6969	4.858	35.82	0.2429	0.4098	42
0.8718	7.268	7.260	12.68	0.00060	0.2596	0.903(21)	3.1812	4.119	29.94	0.2598	0.4012	43
0.8718	7.268	7.259	14.61	0.00054	0.3169	—	3.6655	3.574	25.98	0.2795	0.4114	44
0.8848	7.377	7.367	14.39	0.00055	0.3023	—	3.6655	3.574	26.37	0.2914	0.4418	45
0.8687	7.243	7.234	14.66	0.00054	0.3278	—	3.6655	3.574	25.89	0.2782	0.4045	46
0.8657	7.218	7.209	14.71	0.00054	0.3138	—	3.6655	3.574	25.80	0.2769	0.4083	47
0.9110	7.595	7.586	13.71	0.00057	—	—	3.5959	3.644	27.67	0.2711	0.4122	48
0.8663	7.223	7.214	16.64	0.00054	0.2862	—	4.1498	3.157	22.80	0.2917	(0.414)	49
0.796(3)	6.64	6.63	4.83	—	—	—	1.1063	11.84	78.6	0.3231[v](24)	0.594(7)	50
0.794(3)	6.62	6.61	6.96	—	—	—	1.5906	8.237	54.5	0.3323[v](24)	0.562(7)	51
0.801[m](8)	6.68[m]	6.67[m]	4.19[m]	—	0.041	0.9995(15)	0.9671	13.55	—	0.2484(13)	—	52
0.827[h](6)	6.89[h]	6.88[h]	6.38[h]	—	0.225	0.9943(15)	1.5195	8.623	59.5[h]	0.1991(13)	—	53
0.79[h](6)	6.59[h]	6.58[h]	5.17[h]	—	0.100	0.9903(15)	1.1765	11.14	73.3[h]	0.238(4)	—	54
1.397[h](14)	11.65[h]	11.64[h]	5.50[h]	—	0.246	—	2.2117	5.924	69.0[h]	0.145(7)	0.325[h](7)	55
0.6173(11)	5.15	5.14	3.31	—	0.255	—	0.5880	22.28	114.7	0.5002(10)	1.114[h](7)	56
0.856[m](8)	7.14[m]	7.13[m]	4.06[m]	—	—	0.9996(15)	1.0000	13.10	—	0.2400(9)	—	57
0.07[m](3)	—	—	—	—	0.000	1.0006(15)	0.0696	188.2	—	3.408(13)	—	58
1.140(25)	9.50[m]	9.49[m]	3.37[m]	—	0.0213	—	1.1048	11.86	—	0.2188(13)	—	59
0.810(26)	6.75[m]	6.74[m]	4.15[m]	—	0.040	0.9997(15)	0.9672	13.55	—	0.2482(13)	—	60
1.414(14)	11.79	11.78	6.01	—	—	—	2.4481	5.352	63.1	0.119(7)	—	61
1.000	8.337	8.328	2.16	—	0.348	—	0.6220	21.06	175.6	0.4446(13)	1.0009(7)	62
—	—	—	—	—	—	—	—	—	—	—	—	63
0.8558(14)	7.135	7.126	5.11	0.00335*	—	—	1.2588	10.41	74.3	0.190(7)	—	64

Table 3-3 (continued)

No.	Compound	Calorific value, 60°F.*				Heat of vaporization, 14.696 psia at boiling point, BTU/lb	Refractive index, nD	Air required for combustion ideal gas*	Flammability limits, vol % in air mixture		ASTM octane number	
		Net		Gross								
		BTU/cu ft, Ideal gas, 14.696 psia (20)*	BTU/cu ft, Ideal gas, 14.696 psia (20)*	BTU/lb liquid (wt in vacuum)	BTU/gal liquid*		68° F.	cu ft/cu ft	Lower	Higher	Motor method D-357	Research method D-908
1	Methane	909.1	1009.7	—	—	219.22	—	9.54	5.0	15.0	—	—
2	Ethane	1617.8	1768.8	—	—	210.41	—	16.70	2.9	13.0	+.05^{f}	+1.6^{j,f}
3	Propane	2316.1	2517.4	21513	91065	183.05	—	23.86	2.1	9.5	97.1	+1.8^{j,f}
4	n-Butane	3010.4	3262.1	21139	102989	165.65	1.3326^{h}	31.02	1.8	8.4	89.6^{j}	93.8^{j}
5	Isobutane	3001.1	3252.7	21091	99022	157.53	—	31.02	1.8	8.4	97.6	+.10^{j,f}
6	n-Pentane	3707.5	4009.5	20928	110102	153.59	1.35748	38.18	1.4	8.3	62.6^{j}	61.7^{j}
7	Isopentane	3698.3	4000.3	20889	108790	147.13	1.35373	38.18	1.4	(8.3)	90.3	92.3
8	Neopentane	3682.6	3984.6	20824	103599	135.58	1.342^{h}	38.18	1.4	(8.3)	80.2	85.5
9	n-Hexane	4403.7	4756.1	20784	115060	143.95	1.37486	45.34	1.2	7.7	26.0	24.8
10	2-Methylpentane	4395.8	4748.1	20757	113852	138.67	1.37145	45.34	1.2	(7.7)	73.5	73.4
11	3-Methylpentane	4398.7	4751.0	20768	115823	140.09	1.37652	45.34	(1.2)	(7.7)	74.3	74.5
12	Neohexane	4382.6	4735.0	20710	112932	131.24	1.36876	45.34	1.2	(7.7)	93.4	91.8
13	2,3-Dimethylbutane	4391.7	4744.0	20742	115243	136.08	1.37495	45.34	(1.2)	(7.7)	94.3	+0.3^{f}
14	n-Heptane	5100.2	5502.9	20681	118668	136.01	1.38764	52.50	1.0	7.0	0.0	0.0
15	2-Methylhexane	5092.1	5494.8	20658	117627	131.59	1.38485	52.50	(1.0)	(7.0)	46.4	42.4
16	3-Methylhexane	5095.2	5497.8	20668	119192	132.11	1.38864	52.50	(1.0)	(7.0)	55.8	52.0
17	3-Ethylpentane	5098.2	5500.9	20679	121158	132.83	1.39339	52.50	(1.0)	(7.0)	69.3	65.0
18	2,2-Dimethylpentane	5079.4	5482.1	20620	116585	125.13	1.38215	52.50	(1.0)	(7.0)	95.6	92.8
19	2,4-Dimethylpentane	5084.3	5487.0	20636	116531	126.58	1.38145	52.50	(1.0)	(7.0)	83.8	83.1
20	3,3-Dimethylpentane	5085.0	5487.6	20638	120031	127.21	1.39092	52.50	(1.0)	(7.0)	86.6	80.8
21	Triptane	5081.0	5483.6	20627	119451	124.21	1.38944	52.50	(1.0)	(7.0)	+0.1^{f}	+1.8^{f}
22	n-Octane	5796.7	6249.7	20604	121419	129.53	1.39743	59.65	0.96	—	—	—
23	Diisobutyl	5781.2	6234.3	20564	119662	122.8	1.39246	59.65	(0.98)	—	55.7	55.2
24	Isooctane	5779.8	6232.8	20570	119388	116.71	1.39145	59.65	1.0	—	100.	100.
25	n-Nonane	6493.3	6996.6	20544	123613	123.76	1.40542	66.81	0.87^{s}	2.9	—	—
26	n-Decane	7188.6	7742.3	20494	125444	118.68	1.41189	73.97	0.78^{s}	2.6	—	—
27	Cyclopentane	3512.0	3763.7	20188	126296	167.34	1.40645	35.79	(1.4)	—	84.9^{j}	+0.1^{f}
28	Methylcyclopentane	4198.4	4500.4	20130	126477	147.83	1.40970	42.95	(1.2)	8.35	80.0	91.3
29	Cyclohexane	4178.8	4480.8	20035	130849	153.0	1.42623	42.95	1.3	7.8	77.2	83.0
30	Methylcyclohexane	4862.8	5215.2	20001	129066	136.3	1.42312	50.11	1.2	—	71.1	74.8
31	Ethylene	1499.0	1599.7	—	—	207.57	—	14.32	2.7	34.0	75.6	+.03^{f}
32	Propene	2182.7	2333.7	—	—	188.18	—	21.48	2.0	10.0	84.9	+0.2^{f}
33	1-Butene	2879.4	3080.7	20678	103659	167.94	—	28.63	1.6	9.3	80.8^{j}	97.4
34	Cis-2-Butene	2871.7	3073.1	20611	107754	178.91	—	28.63	(1.6)	—	83.5	100.
35	Trans-2-Butene	2866.8	3068.2	20584	104690	174.39	—	28.63	(1.6)	—	—	—
36	Isobutene	2860.4	3061.8	20548	102863	169.48	—	28.63	(1.6)	—	—	—
37	1-Pentene	3575.2	3826.9	20548	110610	154.46	1.37148	35.79	1.4	8.7	77.1	90.9
38	1,2-Butadiene	2789.0	2940.0	20447	112172	(181.)	—	26.25	(2.0)	(12.)	—	—
39	1,3-Butadiene	2730.0	2881.0	20047	104826	(174.)	—	26.25	2.0	11.5	—	—
40	Isoprene	3410.8	3612.1	19964	114194	(153.)	1.42194	33.41	(1.5)	—	81.0	99.1
41	Acetylene	1422.4	1472.8	—	—	—	—	11.93	2.5	80.	—	—
42	Benzene	3590.7	3741.7	17992	132655	169.31	1.50112	35.79	1.3^{g}	7.9^{g}	+2.8^{f}	—
43	Toluene	4273.3	4474.7	18252	132656	154.84	1.49693	42.95	1.2^{g}	7.1^{g}	+0.3^{f}	+5.8^{f}
44	Ethylbenzene	4970.0	5221.7	18494	134414	144.0	1.49588	50.11	0.99^{g}	6.7^{g}	97.9	+0.8^{f}
45	o-Xylene	4958.3	5210.0	18445	136069	149.1	1.50545	50.11	1.1^{g}	6.4^{g}	100.	—
46	m-Xylene	4956.8	5208.5	18441	133568	147.2	1.49722	50.11	1.1^{g}	6.4^{g}	+2.8^{f}	+4.0^{f}
47	p-Xylene	4956.9	5208.5	18445	133136	144.52	1.49582	50.11	1.1^{g}	6.6^{g}	+1.2^{f}	+3.4^{f}
48	Styrene	4828.7	5030.0	18150	137849	(151.)	1.54682	47.72	1.1	6.1	+0.2^{f}	>+3.^{f}
49	Isopropylbenzene	5661.4	5963.4	18665	134817	134.3	1.49145	57.27	0.88^{g}	6.5^{g}	99.3	+2.1^{f}
50	Methyl Alcohol	—	—	9760	64771	473.(2)	1.3288(8)	7.16	6.72(5)	36.50	—	—
51	Ethyl Alcohol	—	—	12780	84600	367.(2)	1.3614(8)	14.32	3.28(5)	18.95	—	—
52	Carbon Monoxide	—	321.(13)	—	—	92.7(14)	—	2.39	12.50(5)	74.20	—	—
53	Carbon Dioxide	—	—	—	—	238.2^{n}(14)	—	—	—	—	—	—
54	Hydrogen Sulfide	588.(16)	637.(16)	—	—	235.6(7)	—	7.16	4.30(5)	45.50	—	—
55	Sulfur Dioxide	—	—	—	—	166.7(14)	—	—	—	—	—	—
56	Ammonia	359.(16)	434.(16)	—	—	587.2(14)	—	3.58	15.50(5)	27.00	—	—
57	Air	—	—	—	—	92.(3)	—	—	—	—	—	—
58	Hydrogen	274.(13)	324.(13)	—	—	193.9(14)	—	2.39	4.00(5)	74.20	—	—
59	Oxygen	—	—	—	—	91.6(14)	—	—	—	—	—	—
60	Nitrogen	—	—	—	—	87.8(14)	—	—	—	—	—	—
61	Chlorine	—	—	—	—	123.8(14)	—	—	—	—	—	—
62	Water	—	—	—	—	970.3(12)	1.3330(8)	—	—	—	—	—
63	Helium	—	—	—	—	—	—	—	—	—	—	—
64	Hydrogen Chloride	—	—	—	—	185.5(14)	—	—	—	—	—	—

C. Isothermal Compressibility of Gases

In reservoir engineering, we often need to know how much a gas will compress with an increase in pressure or how much it will expand with a decrease in pressure. This need brings us to compressibility (not compressibility factor, which is the z factor). The general mathematical definition for isothermal compressibility for any material is:

$$c = -\frac{1}{V}\left[\frac{\partial V}{\partial p}\right]_T \qquad (3\text{-}18)$$

The minus sign is a part of the definition by convention due to the fact that the term $(\partial V / \partial p)$ is normally a negative quantity. The mathematical derivation for gas compressibility follows:

$$V = nRT\frac{z}{p}$$

$$\left[\frac{\partial V}{\partial p}\right]_T = nRT\,\frac{p\dfrac{\partial z}{\partial p} - z}{p^2}$$

$$c_g = \left[-\frac{1}{V}\right]\left[\frac{\partial V}{\partial p}\right]_T$$

$$c_g = \left[-\frac{p}{nRTz}\right]\left[\frac{nRT}{p^2}\left(p\frac{\partial z}{\partial p} - z\right)\right]$$

$$c_g = \frac{1}{p} - \frac{1}{z}\frac{\partial z}{\partial p} \qquad (3\text{-}19)$$

It should be evident from Equation 3-19 that we can calculate the compressibility of a real gas if we have data on the gas compressibility factor, z, as a function of pressure. It should also be apparent that for an ideal gas, $c_g = 1/p$.

Because gas compressibility, like so many gas properties, can be related using the theorem of corresponding states, it is convenient to consider reduced states for mixtures of gases. It can be shown that:

$$c_g\,p_{pc} = \frac{1}{p_{pr}} - \frac{1}{z}\left[\frac{\partial z}{\partial p_{pr}}\right] = c_{pr} \qquad (3\text{-}20)$$

The term c_{pr} is the pseudoreduced compressibility. By using Figure 3-15 and doing a little manipulating, it is possible to obtain z and $\partial z / \partial p_{pr}$. Then, c_{pr} (dimensionless number) may be calculated. After this, determine c_g as it is equal to c_{pr} / p_{pc}, psia^{-1}.

Trube[4] has presented graphs for estimating the compressibility of natural gases (Figures 3-17 and 3-18). The pseudoreduced compressibility is read from the graph. This number is divided by pseudocritical pressure to obtain c_g.

A newer method to determine gas compressibility was published in 1975 by Mattar *et al.*[29]. This correlation includes the low pressure range, and is valid for:

$$(1.05 \leq T_{pr} \leq 3.0 \; ; \; 0.2 \leq p_{pr} \leq 15.0)$$

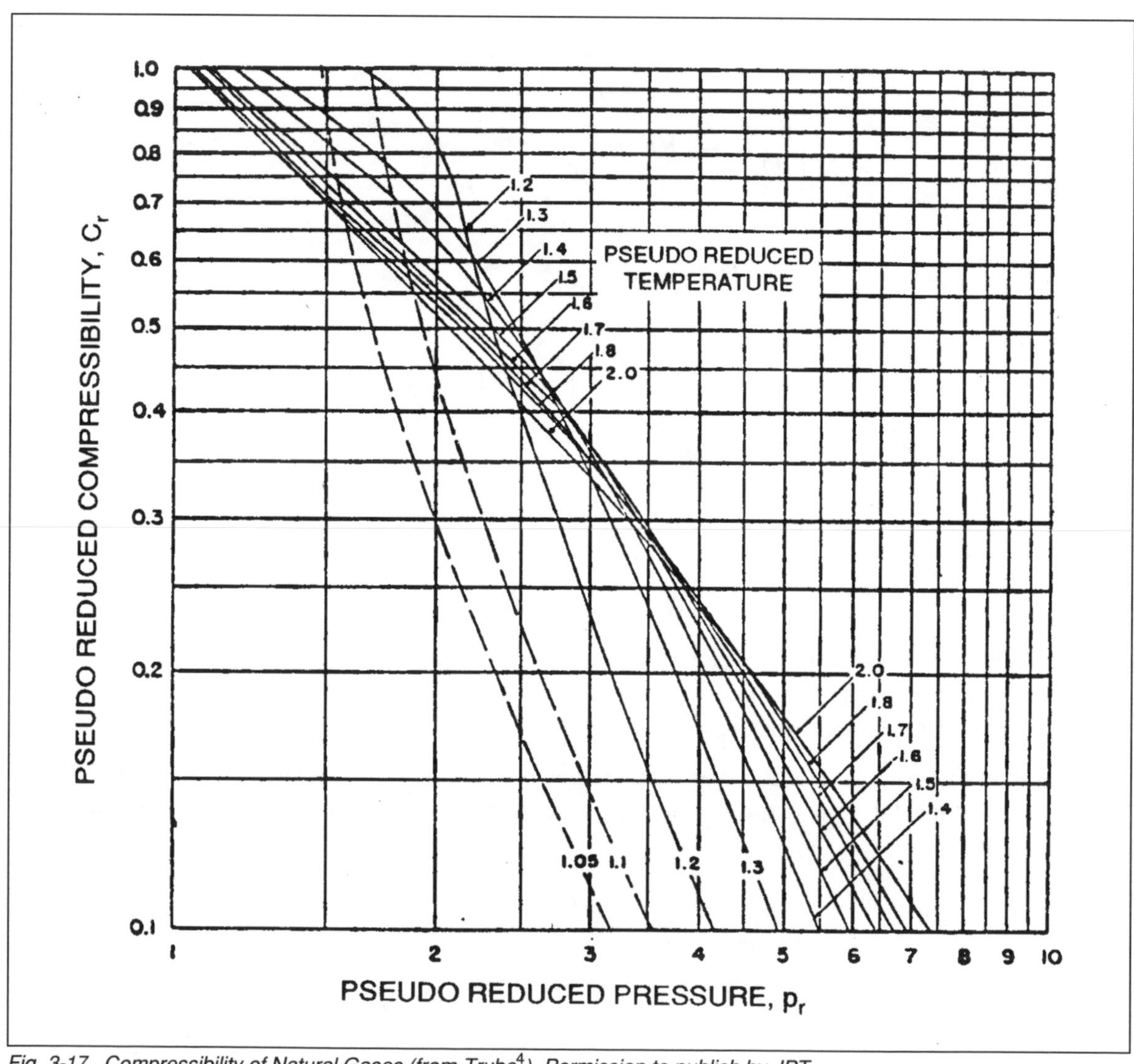

Fig. 3-17. Compressibility of Natural Gases (from Trube[4]). Permission to publish by JPT.

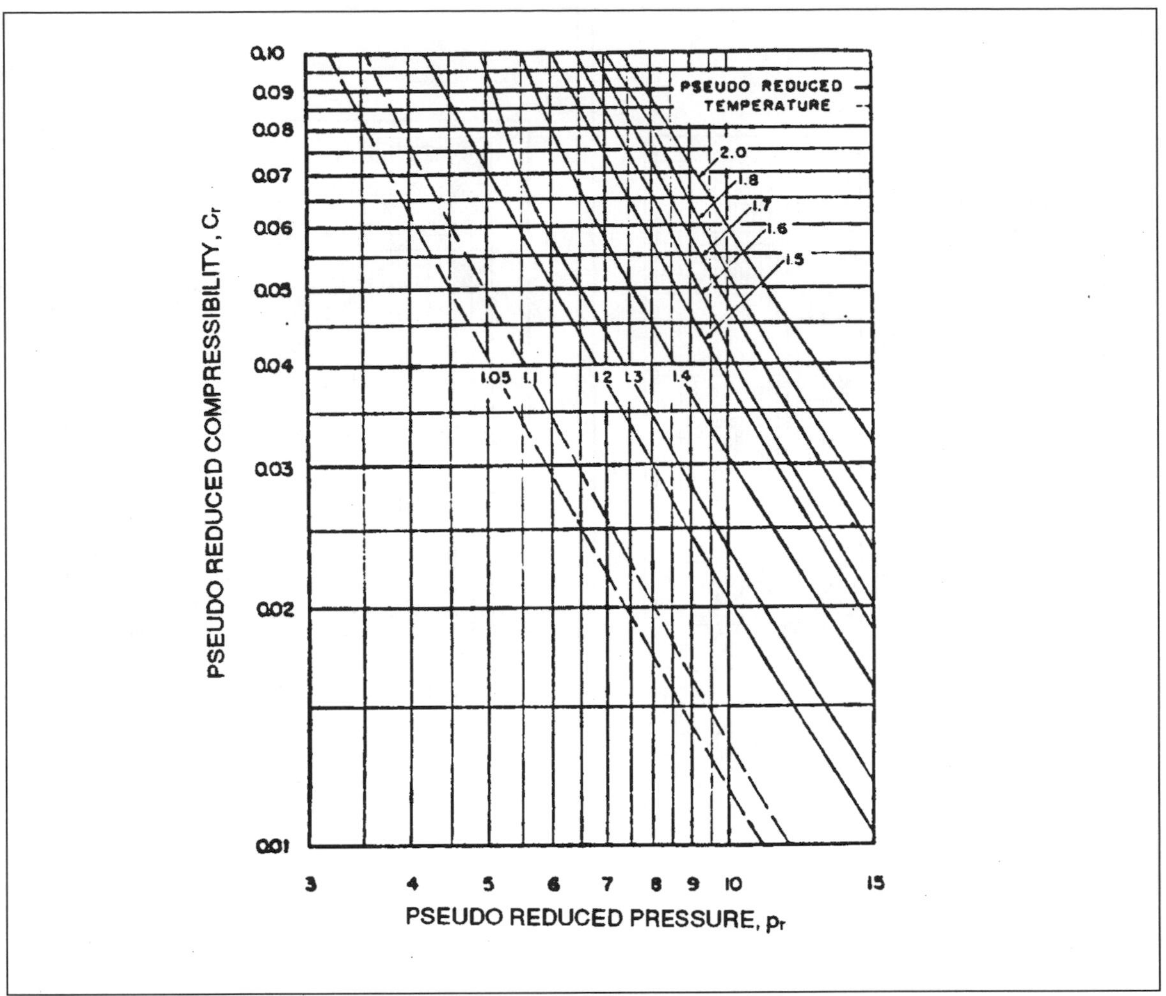

Fig. 3-18. Compressibility of natural gases (from Trube[4]). Permission to publish by JPT.

The method:

(1) Determine p_{pc}, T_{pc}, p_{pr}, and T_{pr}.

(2) From Figure 3-19 or 3-20, find $c_{pr}\,T_{pr}$.

(3) $c_g = (c_{pr}\,T_{pr}) / (p_{pc}\,T_{pr})$

Problem 3-5:

For the following gas, what is its compressibility?

$$p_{pc} = 667 \text{ psia},\ T_{pc} = 390\ ^\circ\text{R},\ p = 1815 \text{ psia},\ T = 195\ ^\circ\text{F}$$

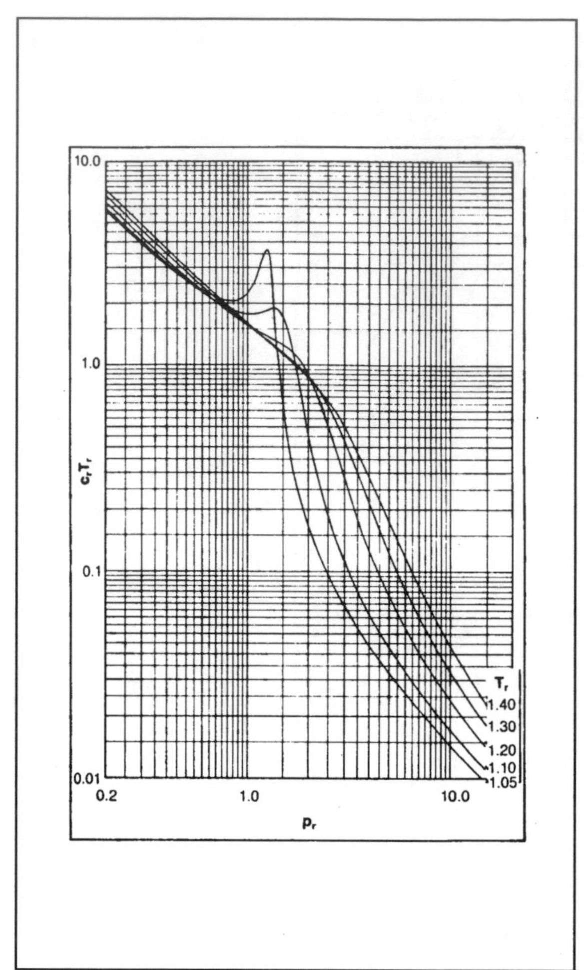

Fig. 3-19. Variation of c_rT_r with reduced temperature and pressure. ($1.05 \leq T_r \leq 1.4$; $0.2 \leq p_r \leq 15.0$). Courtesy of The Journal of Canadian Petroleum Technology, Canadian Institute of Mining and Metallurgy.

Fig. 3-20. Variation of c_rT_r with reduced temperature and pressure. ($1.4 \leq T_r \leq 3.0$; $0.2 \leq p_r \leq 15.0$). Courtesy of The Journal of Canadian Petroleum Technology, Canadian Institute of Mining and Metallurgy.

(A) Trube Method:

$$p_{pr} = 1815/667 = 2.72; \; T_{pr} = (460 + 195)/390 = 1.68$$

From Figure 3-17, $c_{pr} = 0.4$

$$c_g = c_{pr}/p_{pc} = 0.4/667 = 0.00060 \text{ psi}^{-1}$$

(B) Mattar *et al.* method:

$$p_{pr} = 2.72; T_{pr} = 1.68$$

From Figure 3-20, $c_{pr}T_{pr} = 0.66$

$$c_g = (c_{pr}\ T_{pr})/(p_{pc}\ T_{pr}) = (0.66)/[(667)(1.68)] = 0.00059\ psi^{-1}$$

D. Viscosity of Gas Mixtures

Gas viscosity can be measured in the laboratory, but usually is not. Relatively good values can be developed from published correlations. Where a gas contains an inordinately high quantity of non-hydrocarbon components, laboratory measurement could be justified.

Carr *et al.*[5] developed Figures 3-21 and 3-22 which may be used to predict gas viscosity where composition is primarily hydrocarbons. Figure 3-21 will yield μ_1, which is the viscosity at one atmosphere of pressure and the desired temperature. Correction charts are included for H_2S, N_2, and CO_2. For viscosity at higher pressure, the μ_1 from Figure 3-21 is used in conjunction with Figure 3-22 and the theorem of corresponding states. The method is illustrated with the following example.

Problem 3-6:

Find the viscosity of the following natural gas.

Gas specific gravity = 0.702; temperature = 195 °F; pressure = 1815 psia.

(1) From Figure 3-16,

$$p_{pc} = 667\ psia; \quad T_{pc} = 390°R$$

(2) Pseudoreduced properties:

$$p_{pr} = \frac{1815}{667} = 2.72; \ T_{pr} = \frac{460 + 195}{390} = 1.68$$

(3) From Figure 3-22,

$$\mu / \mu_1 = 1.28$$

(4) From Figure 3-21, for

$$\gamma_g = 0.702, \quad \mu_1 = 0.0122 \ cp$$

(5) Hence, gas viscosity at 195 °F and 1815 psia is:

$$\mu_g = (\mu_1)(\mu/\mu_1) = (0.0122)(1.28) = 0.0156 \ cp$$

Where more than trace amounts of H_2S, N_2, and/or CO_2 are present, the pseudocritical properties read from Figure 3-16 should be corrected using either the small correction charts contained in Figure 3-16 or by using the larger ones shown in Figure 3-23.

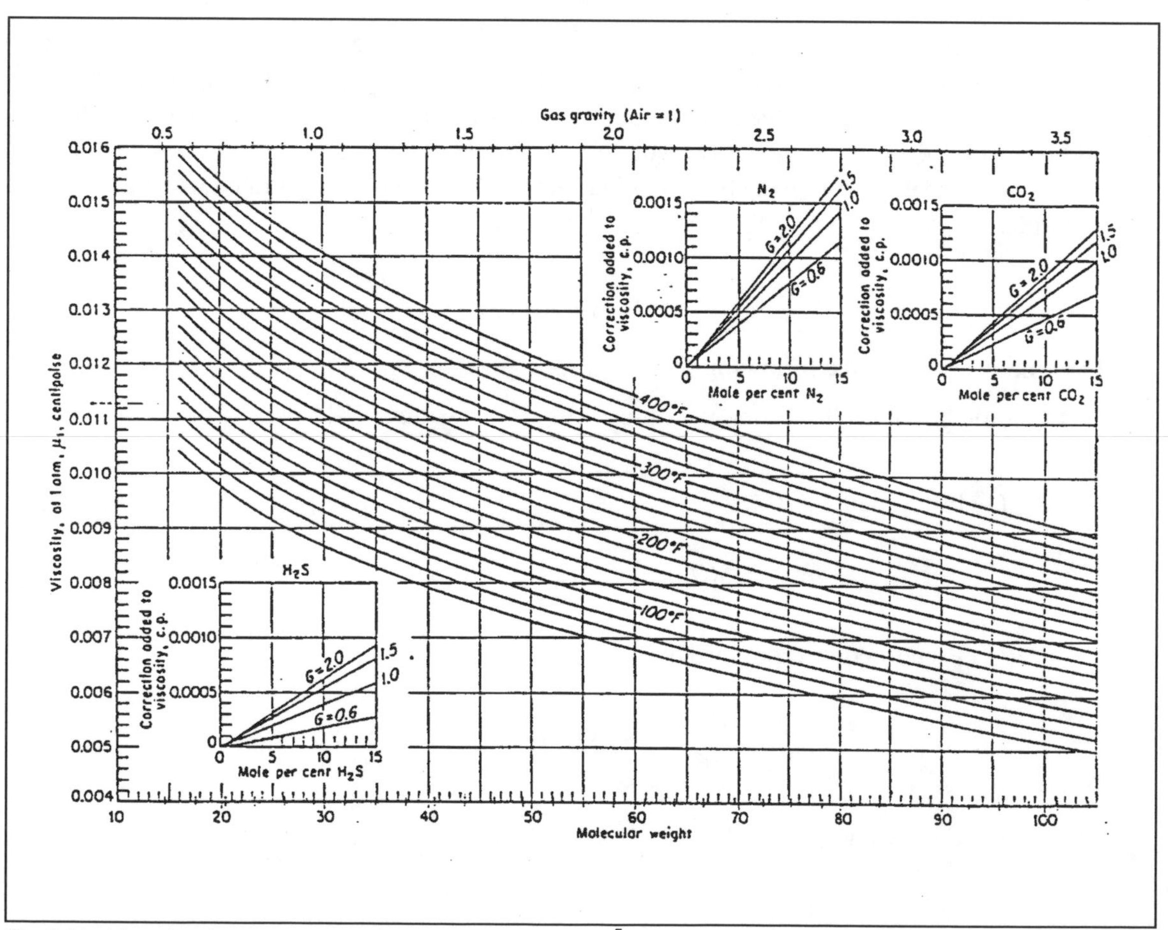

Fig. 3-21. Viscosity of gases at atmospheric pressure (Carr et al.[5]). Permission to publish by SPE.

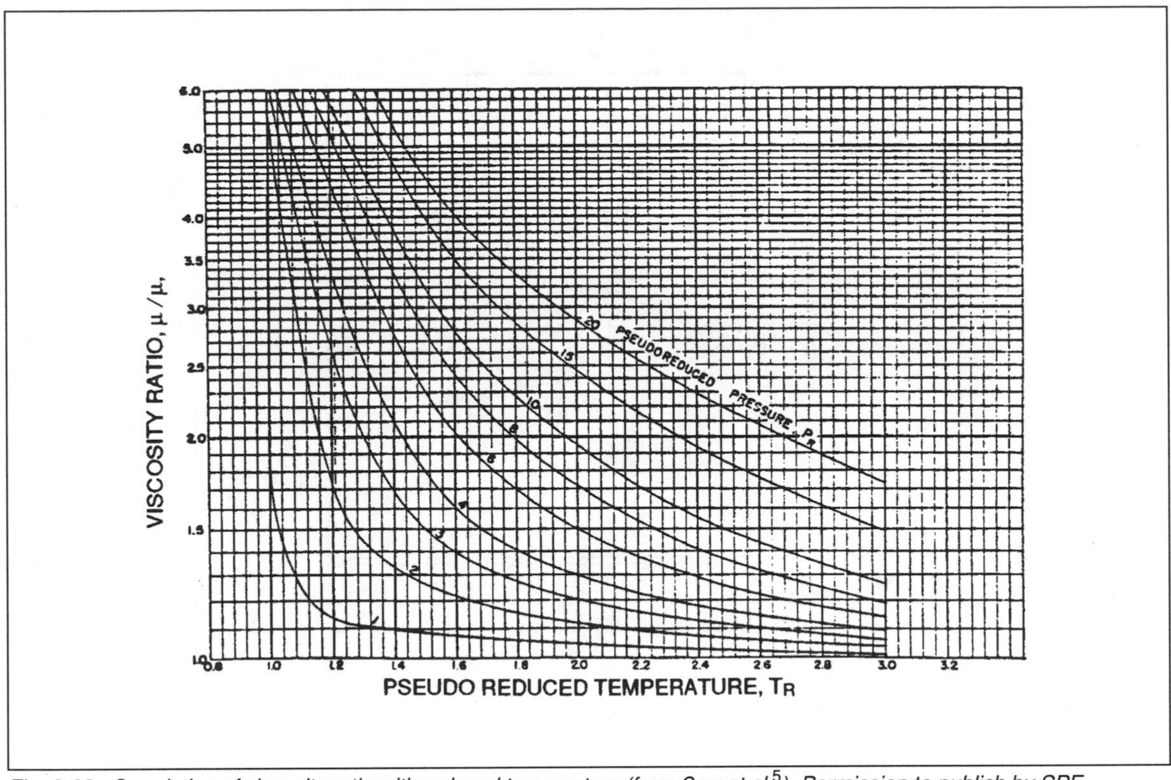

Fig. 3-22. Correlation of viscosity ratio with reduced temperature (from Carr et al.[5]). Permission to publish by SPE.

A general method for hydrocarbon gas viscosity determination that is quite easily programmed was published by Lee *et al.*[30] in 1966.

(1) Calculate molecular weight, M:

(a) With gas gravity, γ_g (relative to air): $M = (\gamma_g)(28.97)$

(b) With a gas analysis (n components):

$$M = \sum_{i=1}^{n} (y_i)(MW_i)$$

(2) Calculate K:

$$K = \left[\frac{(9.4 + 0.02\,M)\,T^{1.5}}{209 + 19\,M + T} \right] (10^{-4})$$

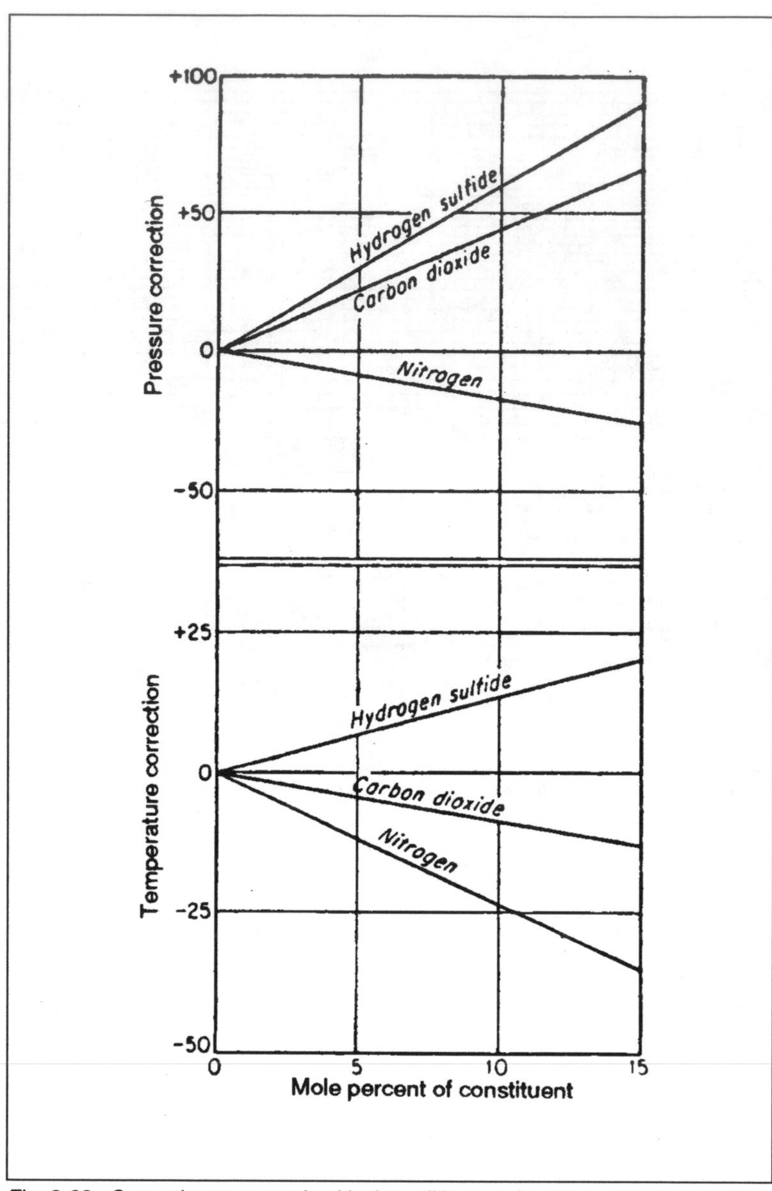

Fig. 3-23. Corrections to pseudocritical conditions as read from Fig. 3-16; to be used when finding viscosity of gases containing H_2S, N_2, and CO_2 (from Katz et al.[6]). Permission to publish by McGraw-Hill.

(3) Calculate X:

$$X = 3.5 + \frac{986}{T} + 0.01\,M$$

(4) Calculate a:

$$a = 2.4 - 0.2\,X$$

(5) Calculate gas density, ρ_g:

$$\text{(a)} \quad \rho_g = \frac{(p)(\gamma_g)}{(23.1)(z)(T)} \; gm/cm^3$$

or

$$\text{(b)} \quad \rho_g = \frac{(p)(M)}{(669.5)(z)(T)} \; gm/cm^3$$

(6) Calculate viscosity:

$$\mu_g = K \exp[(X)(\rho_g)^a] \; cp$$

Where

M = gas molecular weight,
γ_g = gas gravity relative to air, fraction,
y_i = mole fraction, component i,
MW_i = molecular weight, component i,
T = temperature, °R,
ρ_g = gas density, gm/cm^3,
p = pressure, psia
z = gas deviation factor,
μ_g = gas viscosity, cp,
K = intermediate calculation parameter,
X = intermediate calculation paramerer, and
a = intermediate calculation parameter.

Correction for sour hydrocarbon gases is made through the z factor. So, when calculating the density in Step 5, the required z factor should be corrected for nonhydrocarbon components.

Where only hydrocarbons are present in the gas, it is often convenient to use gas viscosities as provided by Figure 3-24. Linear interpolation between the graphs can be made for the specific gravity of the gas under study.

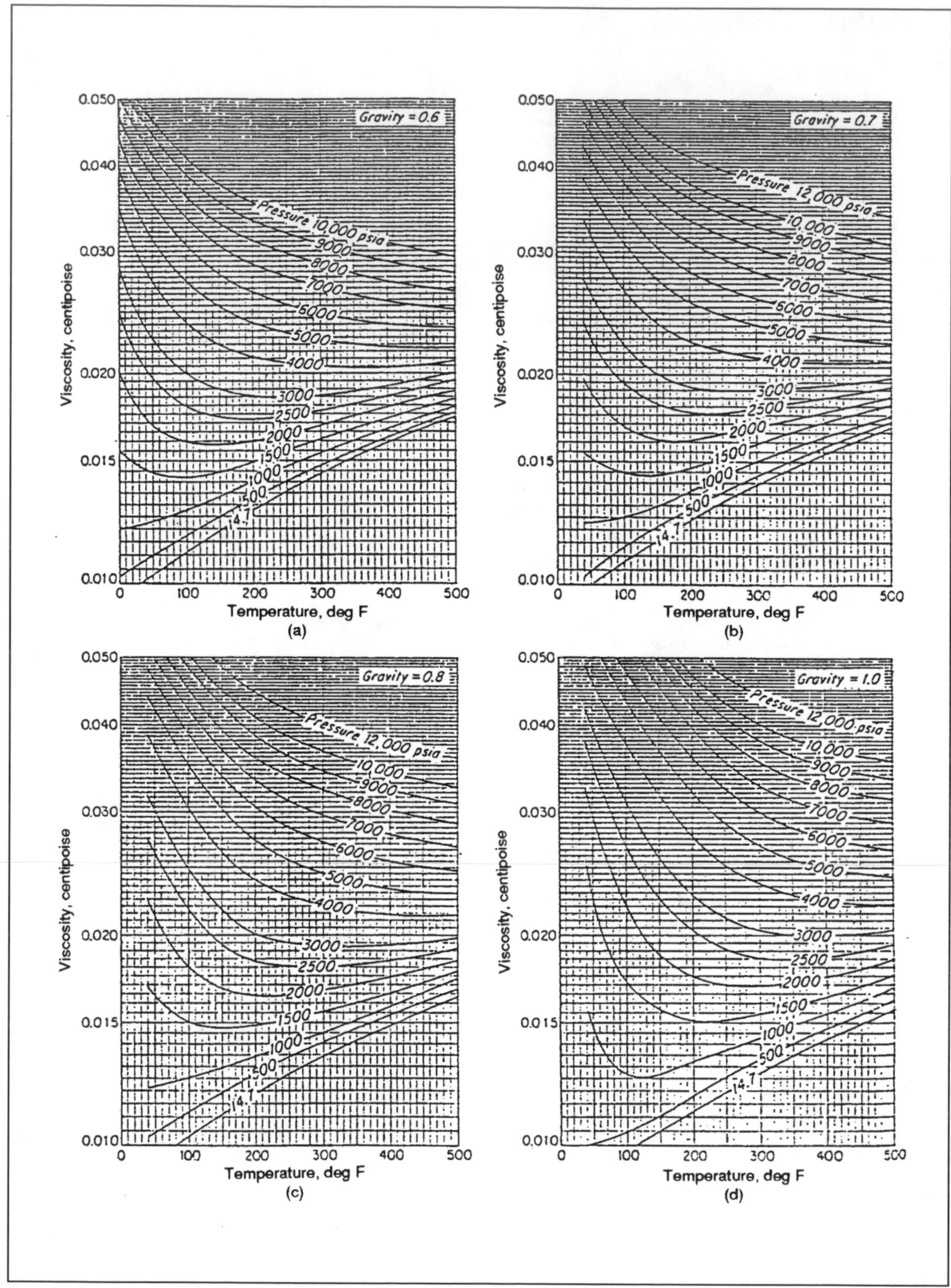

Fig. 3-24. Viscosity of natural gases. (a) 0.6 gravity; (b) 0.7 gravity; (c) 0.8 gravity; (d) 1.0 gravity. (from Katz and Cornell[7]). Permission to publish by University of Michigan.

E. Heating Value of Gas Mixtures

Today, the price of gas is related directly to heating value. In the past, cost was usually a function of the number of MSCF (1000 standard cubic ft) delivered with a heating value minimum of 1000 BTU/SCF. These days a premium is paid for gas with a heat content of more than 1000 BTU/SCF and a penalty for gas having less than 1000 BTU/SCF.

So, our present pricing structure is indicative that it is the energy contained in the gas that is important. Of course, price is also affected by other considerations such as gas gravity and amount of nonhydrocarbons such as water and sour components.

The heating value of various components may be found in Table 3-3. Where direct measurement has not been made, the heating value can be calculated for a gas mixture from a gas analysis. The following example illustrates the method.

Problem 3-7:

Component	Mole fraction y_j	Heating value of component at 14.7 psia and 60°F, Btu/cu ft, x_j	$y_j x_j$
C_1	0.887	1010	895.87
C_2	0.056	1769	99.06
C_3	0.021	2518	52.88
i-C_4	0.003	3253	9.76
n-C_4	0.006	3262	19.57
C_{5+}	0.004	4010	16.04
N_2	0.023	0	0
	1.000		1093.18

The calculated value of 1,093.18 BTU/cu ft is on a dry basis.

When a hydrocarbon gas is burned in the air, the following general reaction occurs:

$$C_nH_{2n+2} + Air \rightarrow CO_2 + CO + H_2O + NO_x + N_2 + heat$$

If the reported heating value is on a "wet" basis, then usually, this is referring to the heating value of the gas and the water vapor contained therein (before burning). Returning to the example heating value calculation above, the 1093.18 BTU/cu ft was calculated on a dry basis. Now, if the gas is saturated with water vapor at 60°F and 30 inches of H_g, then to make the adjustment required to report the heating value on a wet basis (i.e., to consider the heating value on the basis of a cubic foot of gas and associated water vapor):

$$[\,1093.18\,]\,[\,30.00 / 29.92\,]\,[\,(\,30.00 - 0.522\,) / 30\,] = 1077.0\ BTU / cu\ ft$$

At 60°F and 30 inches of Hg, equilibrium water vapor has a partial pressure of 0.522 in. of mercury. So, in the adjustment, the pressure basis is taken from 29.92 in. of H_g to 30 in. of H_g

by the term (30/29.92). Then the remainder of the correction, through the use of Dalton's law of partial pressures, has the effect of subtracting out the water vapor. With this type of adjustment, if corrections to different pressure or temperature bases are needed, Boyle's or Charles' laws may be used.

Katz[6] refers to the "gross" or "total" heating value of a gas mixture as that amount of energy obtained by cooling the products of combustion to 60°F and condensing the moisture formed. Thus the gross heating value does not subtract off the amount of heat required to vaporize the water formed in the products (latent heat of water). The "net" heating value is equal to the gross heating value minus the latent heat of water, because the heat that is used to vaporize the water in the products cannot be used for other purposes.

F. Other Properties of Gas

Many other gas properties are not basic to reservoir engineering such as enthalpy, entropy, heat content, thermal conductivity, etc. Should such data be needed, the reader is referred to Katz *et al.*[6] as the starting point. Table 3-3 also provides certain basic data.

G. Gas Concluding Remarks

We have found that gas properties are easily correlated by using the theorem of corresponding states. For gas mixtures, which all naturally occurring hydrocarbon gases are, the mole fraction of each component is usually determined from a gas chromatographic analysis. Using the analysis along with the theorem of corresponding states, most of the properties that would be of interest to the reservoir engineer can be calculated.

PROPERTIES OF LIQUID HYDROCARBONS

A. Introduction

Liquids differ from gases in that higher densities and higher viscosities are involved. Liquids take the shape of their container but do not entirely fill it as do gases. In the reservoir engineering sense, when speaking of liquid hydrocarbons, we usually mean oil; therefore, when discussing these properties, the subscript will usually be "o".

Methods to get these properties of a reservoir oil include (1) from a sample (preferable method), and (2) from published correlations.

B. Sampling

1. Bottomhole Sampling

With this method, basically, a bottomhole sampler is run in a well on a wire line, a sample is collected at the bottom of the hole, the sampler is retrieved, and the sample is taken to the laboratory for analysis.

However, some qualifications do exist. It is best to collect the sample with pressure as close to the discovery value as possible. Thus, the sample should be taken quite early in the life of the reservoir (such as day 1), and it is best to use a well with productivity as high as possible. A higher productivity well will flow with a higher pressure at a given rate than a lower productivity well, and it will build up to static pressure faster. Much better results will be obtained if the sampling pressure is above the bubblepoint.

Oil reservoirs that are discovered without an associated gas cap have all gas that is associated with that reservoir in solution in the oil. However, at some pressure below the initial pressure, gas will begin coming out of solution. The pressure at which gas first starts coming out of solution is the bubblepoint pressure. If an oil reservoir is discovered with an associated gas cap, then the initial pressure is the bubblepoint pressure, and any pressure decrease will cause gas to come out of solution in the oil. In this case, a representative sample is quite difficult to capture, either at the surface or at the perforations.

After a well has been selected to be sampled, it must be conditioned. The well is flowed at a low rate until the well stabilizes. Then the well is flowed at a series of lower and lower rates, each time to stabilized conditions. The final rate is as low a stabilized rate as possible. For each new rate or choke size, flowing conditions are maintained until a constant surface GOR is obtained. Provided the final GOR represents only solution gas, any mobile free gas saturation around the wellbore should have gone back into solution. The conditioning process may take only a matter of hours, or it could take days, depending on the reservoir maturity and how the well has been produced.

After conditioning, the well is shut-in until pressure builds up to the formation pressure around the wellbore. At this time the sampling container is run in the wellbore to the desired sampling position, which should be between the oil-water and gas-oil interfaces. This is usually just opposite the perforations or slightly above them. In some cases, after the shut-in period, the well can be put back on production at a low stabilized rate and sampled with the well flowing. However, this is only applicable if the flowing bottomhole pressure is above the bubblepoint pressure. As fresh representative reservoir fluid is entering the wellbore all of the time, this is probably the best sampling situation. After the sample is collected, it is retrieved and taken to the laboratory.

2. Recombination Sampling

Figure 3-25 is a schematic diagram of the producing formation, wellbore, high pressure separator, and the stock tank. With the recombination sample method, a sample of oil and a sample of gas are collected at the high pressure separator. These samples are recombined in the laboratory according to the producing gas/oil ratio at the separator. This requires a metering device on the separator oil stream. If the meter is not available, then the oil is sometimes collected at the stock tank with the resulting requirement that the vent gas be metered and collected for recombination with the high pressure separator gas and the stock tank oil samples.

If this method is carried out properly, the results obtained should be the same as those obtained with a bottomhole sample. Well selection and conditioning, as discussed under bottomhole sampling, are just as necessary with this method. Particularly important is a stabilized gas/oil ratio. Further discussion including the actual production test data to be collected may be found in reference 18.

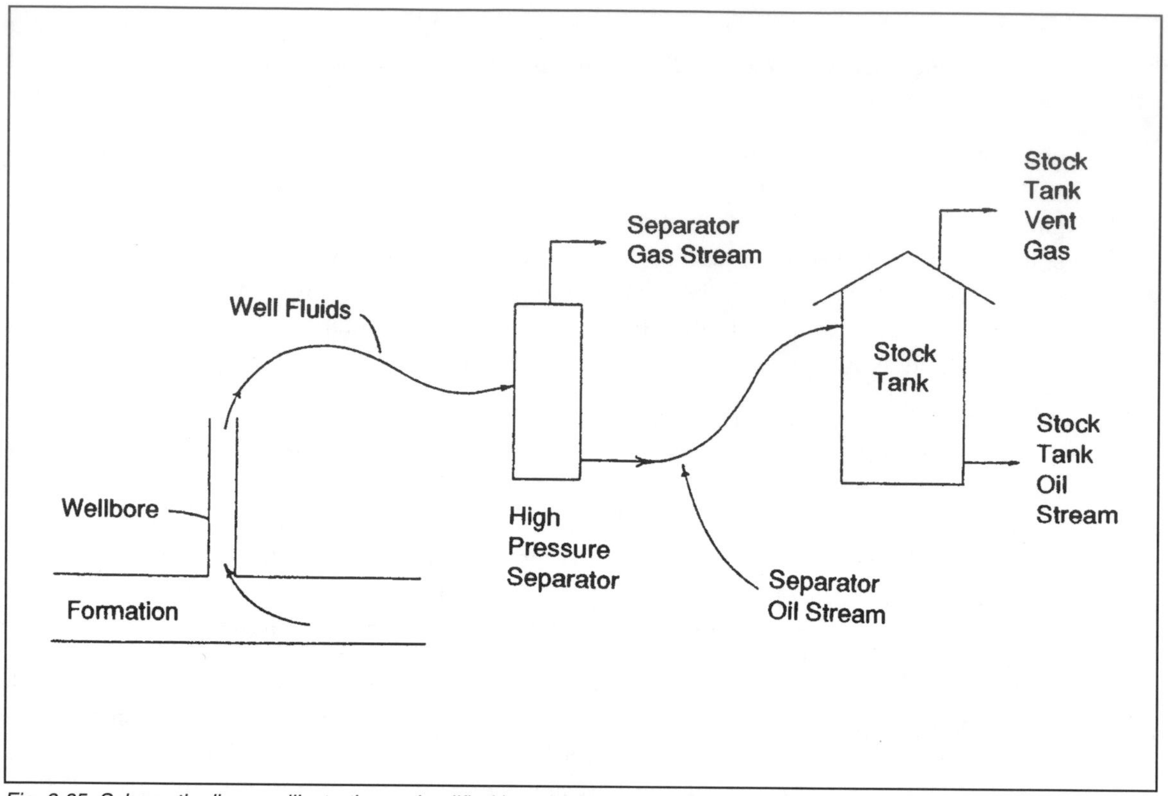

Fig. 3-25. Schematic diagram illustrating a simplified layout of the producing formation, wellbore, and surface facilities.

3. Pressure-Volume-Temperature Analysis

After the fluid sample is collected, it is taken to the laboratory where a pressure-volume-temperature (PVT) analysis is performed. This analysis can only be as good as the samples provided. An example PVT report, provided by Core Laboratories, Inc., is given in Appendix A.

Of course, if a reservoir PVT analysis is not available, the reservoir engineer will need to develop the equivalent information if a meaningful reservoir study is to be accomplished. This can be done using data from a nearby similar reservoir or via published correlations, which will be discussed later in this chapter.

C. Density, Specific Gravity and API Gravity

Density relates the mass per volume of a given substance. The density of a liquid is affected by changes in temperature and pressure, but less so than is a gas. However, the density of oil at reservoir conditions is usually quite different than at the surface.

Where stock tank liquid composition is available, the stock tank oil density can be calculated in the following manner.

Problem 3-8:

Component	Mole fraction x_j	Molecular weight, lb/lb-mol M_j	Relative weight, lb x_jM_j	Liq. Dens. at 60°F, 14.7 psia, lb/cu ft D_j	Relative Liquid Volume, cu ft/mol x_jM_j/D_j
C_1	0.0019	16.04	0.0305	(19.70)	(0.0015)
C_2	0.0098	30.07	0.2947	(23.26)	(0.0127)
C_3	0.0531	44.09	2.3412	31.64	0.0740
C_4	0.0544	58.12	3.1617	35.71	0.0885
C_5	0.0555	72.15	4.0043	39.08	0.1025
C_6	0.0570	86.17	4.9117	41.36	0.1188
C_{7+}	0.7683	263. *	202.0629	55.28	3.6553
	1.0000		216.8070 lb/mol		4.0391 ft^3/mol (4.0533)

*from laboratory analysis of stock tank liquid

$$\text{Oil density} = 216.8070 / 4.0391 = 53.68 \text{ lb/cu ft} \quad (\text{ignoring } C_1 \text{ and } C_2)$$

Oil specific gravity, γ_o, (relative density) is defined as the ratio of the density of the given liquid to the density of water, with both taken at specified conditions of temperature and pressure. When calculating specific gravity, the water density conditions are often taken to be one atmosphere of absolute pressure and 60°F. The petroleum industry uses another gravity term called API gravity that is defined as:

$$\text{API gravity (degrees)} = (141.5/\gamma_o) - 131.5 \tag{3-21}$$

and inversely,

$$\gamma_o = 141.5 / (°\text{API} + 131.5) \tag{3-22}$$

with γ_o taken at 60°F and atmospheric pressure. So, API gravity can only be related to stock tank liquid density at standard surface conditions. A widely used indicator of crude oil worth is its API gravity.

So, continuing Example 3-8, we can calculate the surface oil specific gravity and API gravity:

$$\text{Oil specific gravity} = 53.68 / 62.4 = 0.860$$

$$\text{Oil gravity} = (141.5 / 0.860) - 131.5$$

$$= 33.0\ °\text{API}$$

D. Thermal Expansion of Liquid Hydrocarbons

The most frequent application of thermal expansion is in correcting stock tank liquids to 60°F. Figure 3-26 provides a density correction that can be used in the absence of laboratory measurement.

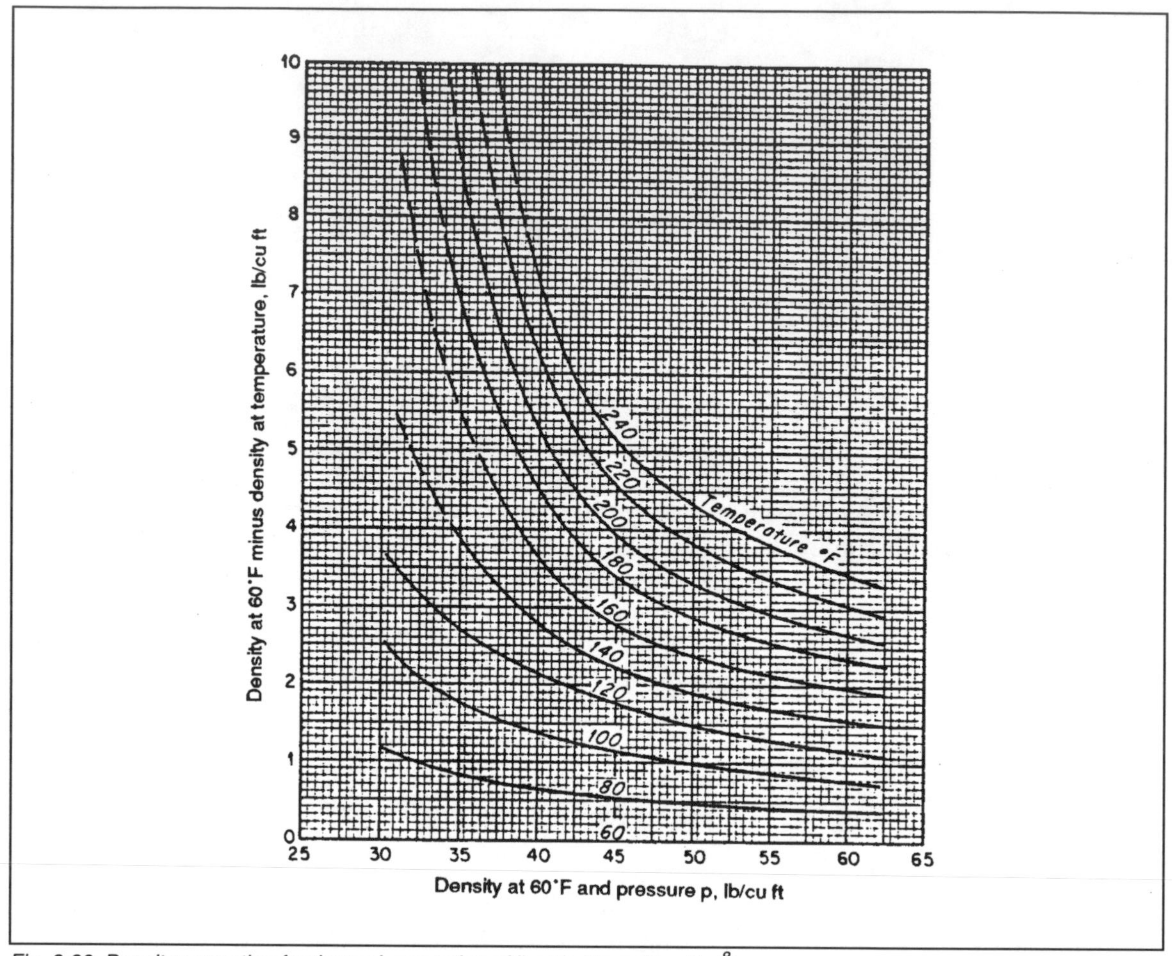

Fig. 3-26. Density correction for thermal expansion of liquids (from Standing[8]).

E. Isothermal Compressibility of Liquid Hydrocarbons

The general mathematical definition of isothermal compressibility is given as Equation 3-18. As the name "oil compressibility" indicates, this property relates how much volume change (compared to a unit volume) occurs with a change in pressure.

Oil compressibility is usually defined as:

$$c_o = -\frac{1}{v}\left[\frac{\partial v}{\partial p}\right]_T \tag{3-23}$$

where the subscript "T" indicates an isothermal process. By separation of variables, the following equation can be obtained:

$$Lnv_2 = Lnv_1 + c_o(p_1 - p_2) \qquad (3\text{-}24)$$

Here "v" refers to specific volume (1/density). If laboratory information is available on oil compressibility, c_o, and initial conditions of specific volume, v_1, and pressure, p_1, are known, then a second specific volume may be readily determined. Units for c_o are vol/vol/pressure (psi^{-1}). Information on isothermal compressibility is particularly valuable above the bubblepoint of a reservoir, since it is here that production is largely influenced by liquid expansion.

Sometimes published correlations are used to predict oil compressibility. Trube's correlation makes use of the theorem of corresponding states. Recalling that density is the reciprocal of specific volume, Equation 3-23 can be modified to be:

$$c_{pr} = \frac{1}{\rho}\left[\frac{\partial \rho}{\partial p_{pr}}\right] \qquad (3\text{-}25)$$

Here c_{pr} is the reduced coefficient of isothermal compressibility; p_{pr} and T_{pr} are reduced pressure and temperature. The reduced pressure and temperature are determined in exactly the same manner as previously described for gases. If possible, pseudocritical pressure and temperature should be determined from a laboratory analysis. Then, having calculated the pseudoreduced pressure and temperature, the pseudoreduced compressibility is obtained from Figure 3-27. Oil compressibility is calculated using:

$$c_o = c_{pr} / p_{pc} \qquad (3\text{-}26)$$

The following example outlines the calculation procedure.

Problem 3-9:

A hydrocarbon liquid has a pseudocritical temperature of 1,000 °R and a pseudocritical pressure of 300 psia. Calculate the isothermal compressibility at reservoir conditions.

Reservoir conditions: p = 2100 psia; T = 240 °F

(i) $T_{pr} = \frac{T}{T_{pc}} = \frac{(240 + 460)}{1000} = 0.70$

$p_{pr} = \frac{p}{p_{pc}} = \frac{2100}{300} = 7.0$

(ii) Read c_{pr} from Figure 3-27: $c_{pr} = 0.0046$

(iii) Calculate c_o: $c_o = \frac{c_{pr}}{p_{pc}} = \frac{0.0046}{300} = 15.3\times10^{-6}\ psi^{-1}$

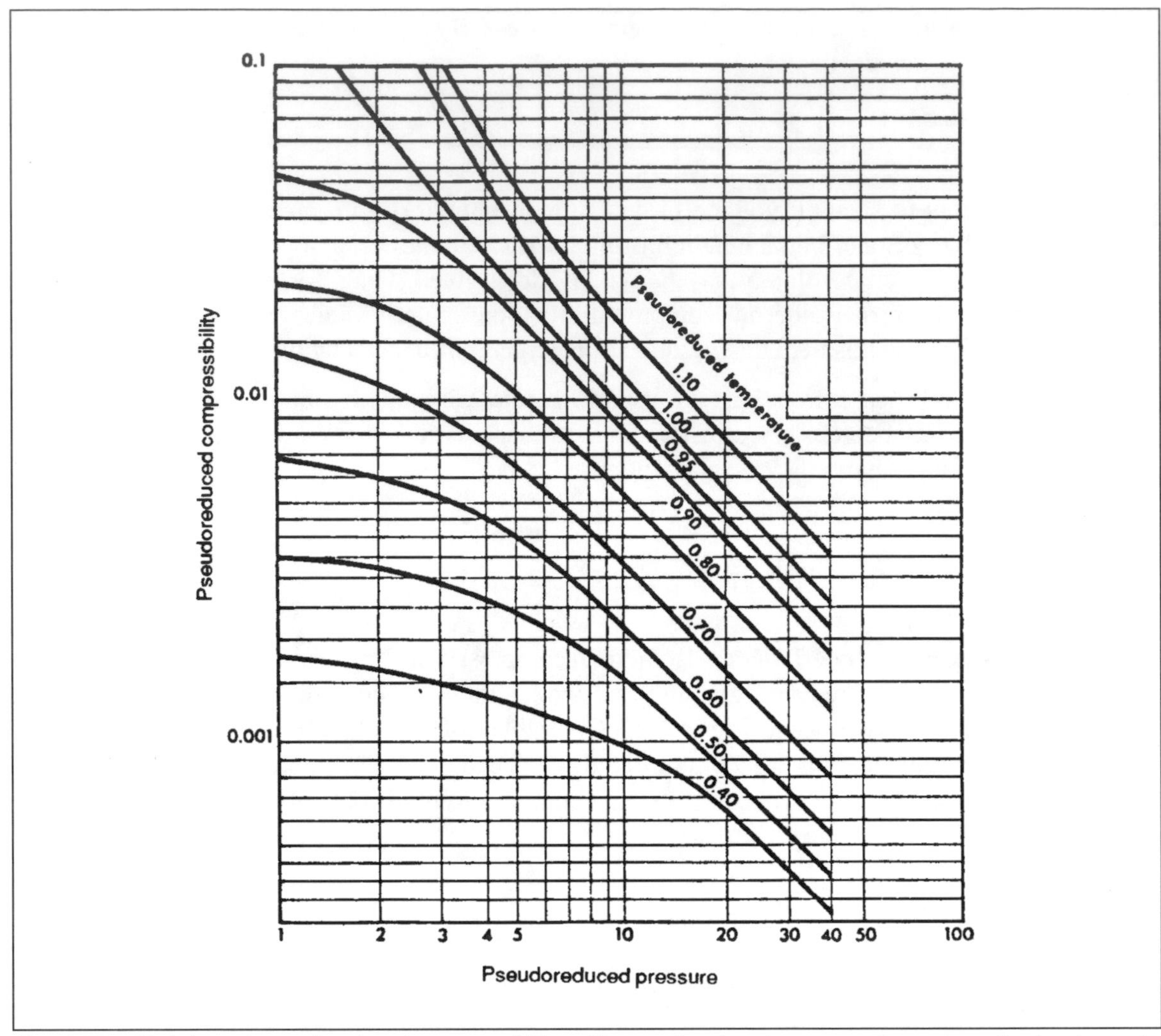

Fig. 3-27. Variation of pseudoreduced compressibility with pseudoreduced pressure for various fixed values of pseudoreduced temperature (from Trube[9]). Permission to publish by SPE.

Unfortunately, the reservoir engineer rarely has accurate values for pseudocritical properties of a reservoir oil. Trube[9] developed Figure 3-28 to estimate these properties. However, it should be used with care, as the results are only approximate.

Another method to calculate undersaturated (above the bubblepoint) isothermal oil compressibility is from Calhoun[23] and described by Burcik.[24] This method is probably not as accurate as the Trube analysis (especially if the critical conditions are determined through a laboratory analysis), but it is easier to program. First, the specific gravity of the oil at the bubblepoint is calculated. To do this, the weight of gas dissolved in one stock tank barrel of oil at the bubblepoint is computed. To this, the weight of one stock tank barrel of oil is added. This oil and solution gas occupy a volume in the reservoir at the bubblepoint of B_{ob} barrels. (See the section entitled "Formation Volume Factor for Oil" in this chapter.) The weight of an equivalent amount of water (B_{ob} barrels) is divided into the weight of the oil and gas. Of course, specific gravity is the result. These calculations may be distilled into the following formula:

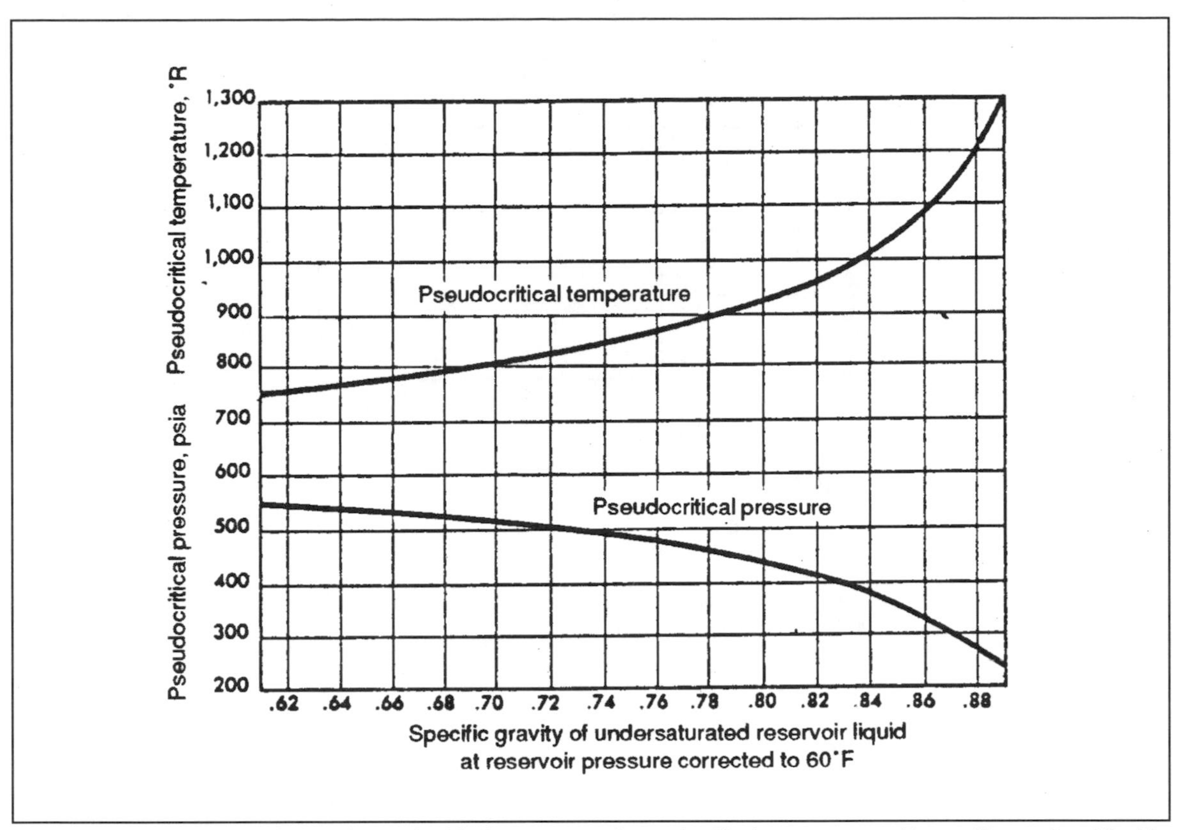

Fig. 3-28. Approximate variation of pseudocritical pressure and pseudocritical temperature with specific gravity of liquid corrected to 60°F (from Trube[9]). Permission to publish by SPE.

$$\gamma_{ob} = \frac{(2.181)(10^{-4})(R_{sb})(\gamma_g) + [(141.5)/(°API + 131.5)]}{B_{ob}} \tag{3-27}$$

Where

γ_{ob} = the reservoir oil specific gravity at the bubblepoint,
R_{sb} = the solution gas/oil ratio at p_b, SCF/STBO,
γ_g = gas specific gravity (related to air),
$°API$ = API gravity of the stock tank oil,
B_{ob} = oil formation volume factor at the bubblepoint, BBL/STB, and
p_b = bubblepoint pressure.

After the bubblepoint oil specific gravity has been calculated, then the undersaturated oil compressibility may be estimated with Figure 3-29. From Figure 3-29, it can be seen that the compressibility of reservoir oil above the bubblepoint will normally be between 5×10^{-6} and 25×10^{-6} psi^{-1}.

A newer method by Vasquez *et al.*[25] combines multiple regression analysis with an attempt to account for separator conditions. This procedure is quite easily programmed. However, additional data are needed that include separator temperature and pressure. First, the gas gravity is corrected to a value approximating that obtained with a gauge separator pressure of 100 psi.

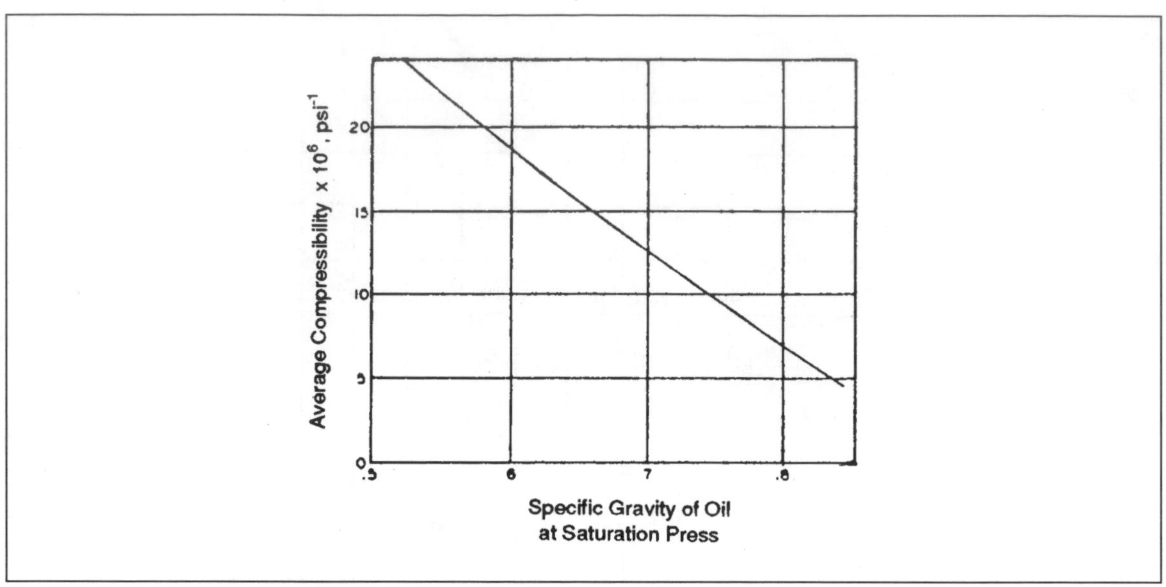

Fig. 3-29. Average compressibility of oil above the bubblepoint as a function of oil gravity at the bubblepoint (from Calhoun[23]). Permission to publish by Oklahoma University.

$$\gamma_{gs} = \gamma_{gp} \left\{ 1 + \left[(5.912)(10^{-5})(°API)(T_{sep}) \left(\log \frac{p_{sep}}{114.7} \right) \right] \right\} \tag{3-28}$$

Where

γ_{gs} = corrected gas gravity to separator pressure of 100 psig,
γ_{gp} = gas gravity with separator pressure of p_{sep},
$°API$ = stock tank oil gravity,
T_{sep} = separator temperature in °F, and
p_{sep} = actual separator pressure, psia.

Then the oil compressibility equation is used:

$$c_o = \frac{-1433.0 + (5.0)(R_{sb}) + (17.2)(T_f) - (1180.0)(\gamma_{gs}) + (12.61)(°API)}{(p)(10^5)} \tag{3-29}$$

where

c_o = isothermal undersaturated oil compressibility (psi^{-1}) at p,
R_{sb} = bubblepoint solution gas/oil ratio, SCF/STBO
T_f = reservoir temperature, °F
γ_{gs} = corrected gas gravity to separator pressure of 100 psig,
$°API$ = stock tank oil API gravity, and
p = reservoir pressure, psia (p must be greater than or equal to p_b).

F. Differential *vs.* Flash Liberation

In an oil reservoir, or in a laboratory cell, gas will break out of solution from the oil as pressure is reduced. The quantity of gas liberated, as well as its composition, is somewhat dependent on the manner in which the pressure is reduced.

Differential liberation is that process where as free gas is liberated, it is removed from the proximity of the oil. Assume that a crude oil sample is in a laboratory PVT cell at reservoir temperature and initial reservoir pressure, which happens to be higher than the bubblepoint. Now the pressure is decreased in steps with the resulting liquid volume noted at each new pressure (Fig. 3-30). Pressure in the system is maintained via a mercury pump, so a pressure decrease is accomplished by withdrawing some mercury.

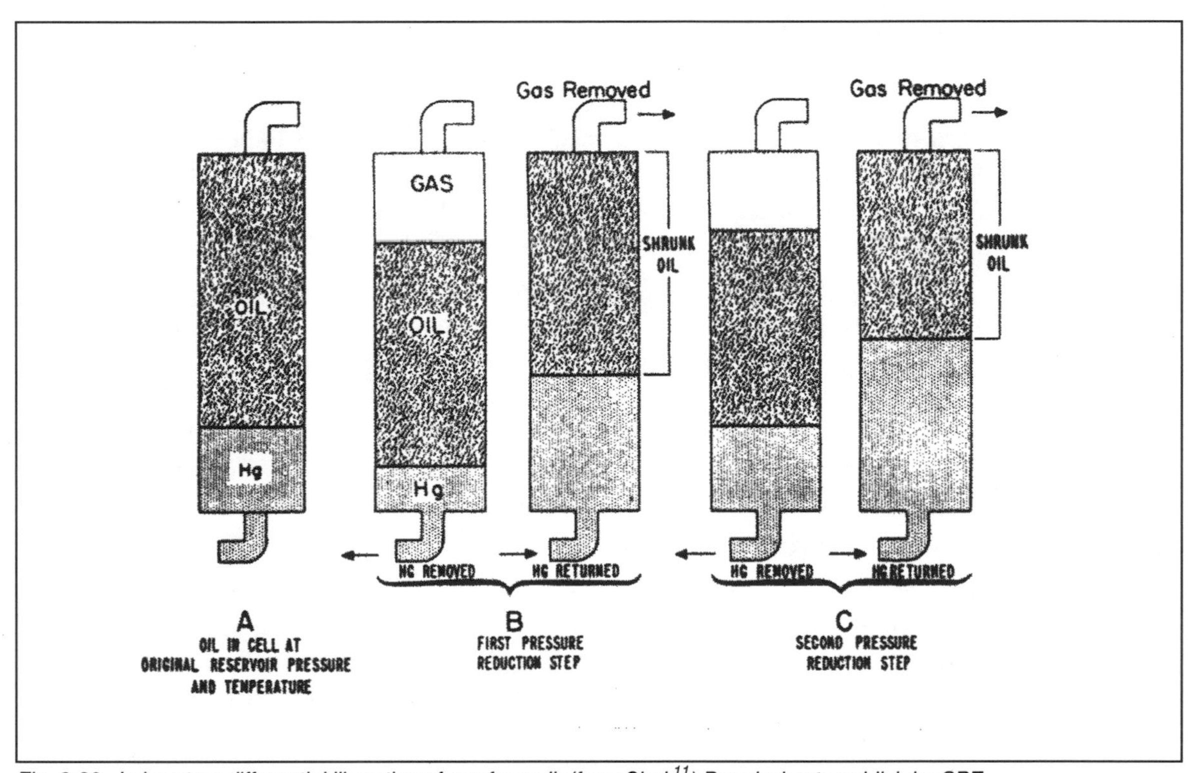

Fig. 3-30. Laboratory differential liberation of gas from oil. (from Clark[11]) Permission to publish by SPE.

When the pressure reaches the point where one bubble of gas is seen, this is the bubblepoint pressure. Now, from this point on, at each new lower pressure, the gas and oil are allowed to come to equilibrium. At this time, the gas is withdrawn from the PVT cell. Then the liquid and gas volumes are measured. It is important to note that the gas volume is removed at constant pressure, normally by injecting mercury as the gas is exiting. Differential liberation is also known as a constant volume, variable composition process.

Now, if the gas were not removed at each pressure decrement, but allowed to remain in intimate contact with the liquid, then we would have a flash or equilibrium liberation. This is also called a constant-composition, variable-volume process.

According to Clark,[11] with a normal low shrinkage (black) oil, flash conditions will cause more gas to be liberated (with resultant greater shrinkage of the liquid) down to a given pressure than will the differential process. This is caused by the attraction of the heavy liquid molecules to the light gas molecules with the resulting increased vaporization of some of these heavy molecules in the flash process.

However, with a high-shrinkage (volatile) oil,[11] this situation is usually reversed: the differential process liberates more gas. (A discussion of low-shrinkage and high-shrinkage oils appears later in this chapter under the section entitled "Reservoir Hydrocarbon Fluid Classification"). Although this is the result of complex thermodynamic behavior, a few qualitative comments will be offered.

High-shrinkage oil has a greater percentage of intermediates than does low-shrinkage oil. Through high kinetic energy of the intermediates and great attraction of the intermediate molecules to the densely spaced, light gas molecules, a large quantity of intermediates are pulled into the gas phase at high pressures. If the gas is not removed, as in flash liberation, some intermediate gas molecules may condense because their attraction to light gas molecules is reduced as the pressure drops and the gas molecules become farther apart. In the differential process, however, the vaporized intermediates are removed from the system, thereby having no further influence, and oil shrinkage remains high through the lower pressure ranges.

But what is happening in the reservoir? Flash or differential liberation? Most people believe that it is a combination of the two. After the bubblepoint pressure is reached, but before the critical gas saturation is built up, the free gas formed in the reservoir cannot flow. It therefore remains in intimate contact with the liquid, and, of course, this is more like a flash liberation process. At the lower pressures with gas saturations exceeding the critical value, because the gas is more mobile than the oil, it tends to flow away as it is liberated. This is more like a differential process. The numerical differences in gas liberated by these two processes is usually minor.

Thus far, we have compared the flash and differential processes at a constant temperature, which is formation temperature. However, the trip that the oil makes from the formation through the wellbore and flow line to the separator is not an isothermal process. This is usually regarded as a flash process, but the temperature is decreasing. At lower temperatures, gas solubility is generally increased. Therefore, the quantity of gas coming out of solution with pressure reduction is much reduced over the constant temperature case. It is common with either volatile or black oil, for this type of flash process to liberate less gas than either of the constant (reservoir) temperature processes.

Both high and low shrinkage oils will shrink less to the stock tank if they are first passed to a high pressure separator where the gas is removed from the proximity of the oil. Usually an optimum separator pressure exists where minimum shrinkage of the oil is obtained.

G. Solution Gas/Oil Ratio

The solution gas/oil ratio, R_s, is defined as the volume of gas dissolved in a unit volume of stock tank oil at reservoir temperature and pressure. Common units are standard cubic feet per stock tank barrel (SCF/STB) and standard cubic meters per stock tank cubic meter. Figure 3-31 includes a typical graph of solution gas/oil ratio *vs.* pressure. Note that in this particular case, the initial pressure, p_i, is greater than the bubblepoint pressure. For oil reservoirs with gas caps, the initial pressure and the bubblepoint pressure are equal. Note, however, that the initial pressure can never be less than the bubblepoint pressure.

It could be said that somewhere during the history of the reservoir fluid as pressure was increasing (with increasing overburden), the bubblepoint was that pressure where the fluid system "ran out of gas," or all available gas went into solution. Notice, then, that the solution gas/oil ratio curve is flat (indicating a constant amount of solution gas) for pressures above the bubblepoint. Below the bubblepoint, with decreasing pressures, the liquid can hold less and less gas, so gas comes out of solution. The bubblepoint is a point of discontinuity (abrupt break in slope) on the R_s curve.

Since PVT analyses may not always be available to the reservoir engineer, it is necessary to be able to construct reasonable approximations for the reservoir properties of interest. It is also helpful to construct values even when a PVT analysis is available, especially when concern exists as to its accuracy.

Standing[8] observed that bubblepoint pressure varied with liquid type, gas type, temperature, and solution gas/oil ratio. Although far from a perfect characterization, he decided to relate liquid type to API gravity and gas type to gas gravity.

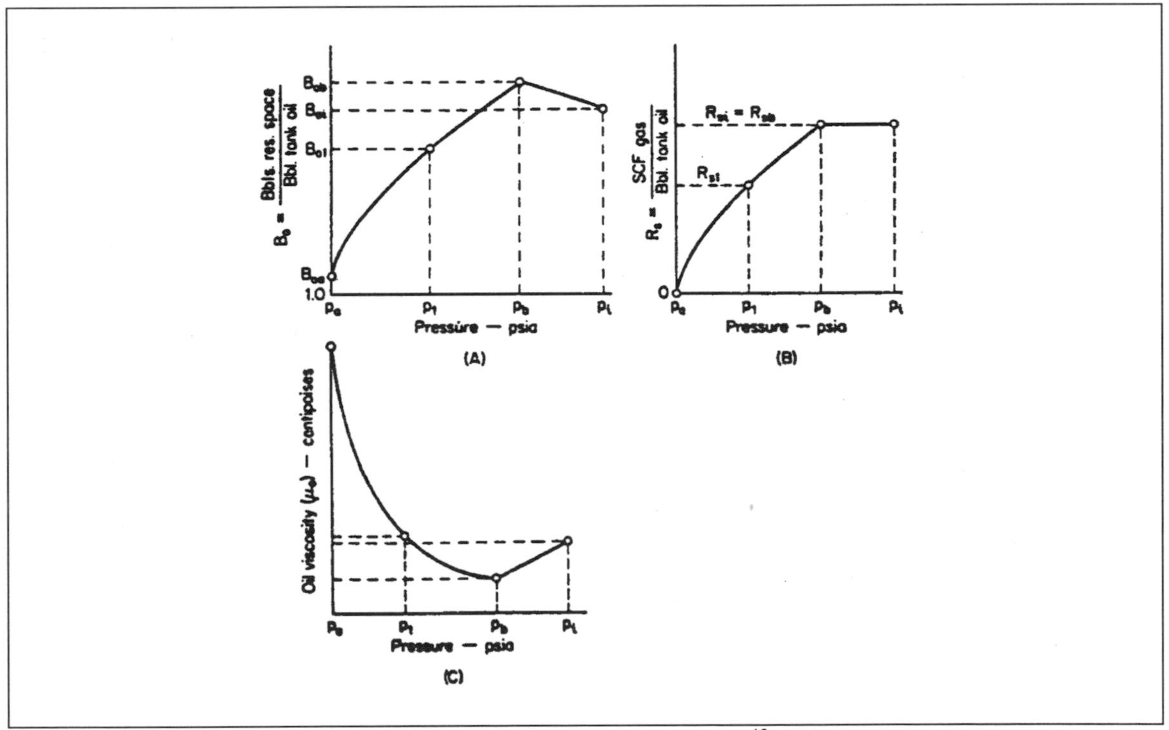

Fig. 3-31. Graphical presentation of fluid properties of a reservoir oil (from Gatlin[10]). Permission to publish by Prentice-Hall.

Standing's empirical correlation is as follows:

$$p_b = 18 \left\{ \left[\left(\frac{R_s}{\gamma_g} \right)^{0.83} \right] \left[\frac{10^{0.00091\, T_f}}{10^{0.0125\, °API}} \right] \right\} \tag{3-30}$$

where:

p_b = bubblepoint or saturation pressure, psia,
R_s = solution gas/oil ratio, SCF/STB,
γ_g = gas gravity (relative density) related to air,
T_f = reservoir temperature, °F, and
$°API$ = API gravity of the stock tank oil.

Equation 3-30 is based on 105 experimentally determined bubblepoint pressures of mostly California crude oil and gas systems. The data had the following ranges:

Bubblepoint pressures, psia	130 - 7,000
Temperature, °F	100 - 258
Solution gas/oil ratios, SCF/STB	20 - 1,425
Tank oil gravities, °API	16.5 - 63.8
Gas gravities (Air = 1)	0.59 - 0.95

The gases contained no nitrogen or hydrogen sulfide. California crudes are largely asphaltic in nature; however, this correlation should be applicable to other geographical areas unless fluid composition is substantially different. Standing's method was one of the first widely used means of estimating bubblepoint pressure.

Appendix B-1 provides a graphical means of using Standing's correlation. The chart contained in Appendix B-1 (Fig. B-1) or Equation 3-30 may be used to determine probable bubblepoint pressure where the tank oil and gas gravities, reservoir temperature, and solution gas/oil ratio at the time of discovery are known. Should the determined bubblepoint pressure essentially coincide with discovery pressure, we would expect the field to have an associated gas cap (perhaps not yet found). Provided the indicated bubblepoint pressure is less (by 10% or more) than the discovery pressure, then we could reasonably expect the reservoir fluid to be undersaturated and no gas cap present. This determination is particularly useful in checking the reasonableness of a reported bubble point pressure from a laboratory PVT analysis. Substantial disagreement would place the laboratory analysis in question.

Figure B-1 can be used in another way. The Standing correlation was prepared by using equilibrium (flash) vaporization data at surface separation conditions. We may generate the solution gas/oil ratio data from the original bubblepoint pressure down to atmospheric pressure by choosing saturation pressures below the original bubblepoint and then determining solution gas/oil ratio at each of these pressures. Equation 3-30 may also be rearranged to solve for R_s given the saturation pressure ($p \leq p_b$):

$$R_s = \gamma_g \left[\left(\frac{p}{18} \right) \left(\frac{10^{0.0125\,(°API)}}{10^{0.00091\,(T_f)}} \right) \right]^{1.20482} \qquad (3\text{-}31)$$

Thus, an R_s *vs.* pressure curve may be developed that is similar to that shown in Figure 3-31.

Two other correlations for R_s are due to Borden and Rzasa[12] and to Lasater[13]. The Borden and Rzasa correlation is somewhat restricted by the gas gravity being defined by the pentane and lighter material in the system. Therefore, a chemical analysis is required that, if you are in the market for a correlation, is usually not available.

The Lasater correlation provides a guide to use when nonhydrocarbon gases are present and would appear to be preferred to Standing in this instance. Also, Lasater used 158 samples from Canada, U.S.A., and South America which included more paraffin-base fluids than did those used for the Standing correlation. So, the Lasater correlation may be more general. In any event, it is recommended that the Lasater estimation be used for paraffin-base crudes such as those normally found in the midcontinent region of the U.S.A. Figure 3-32 provides Lasater's chart for determining bubblepoint pressure.

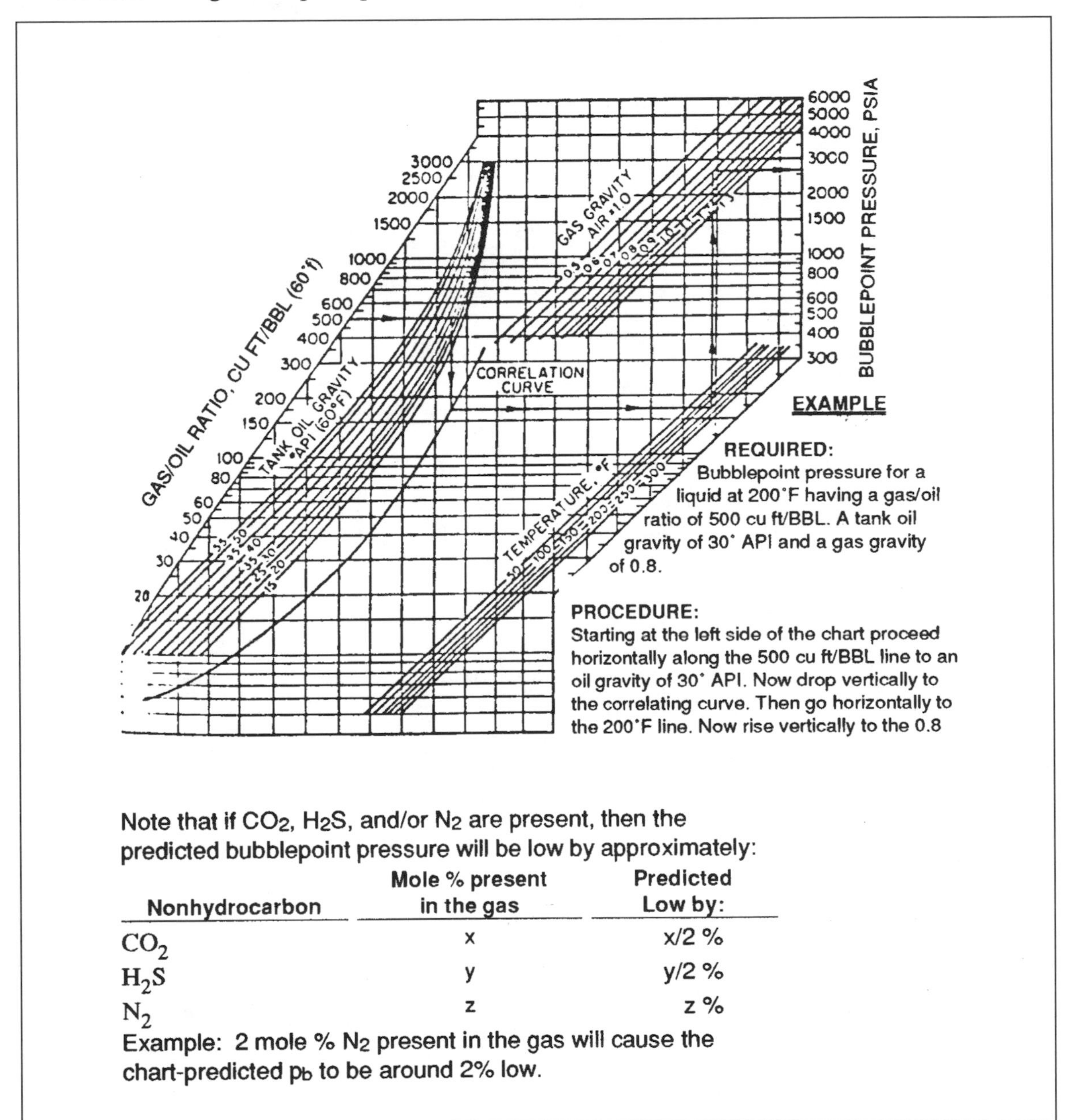

Note that if CO_2, H_2S, and/or N_2 are present, then the predicted bubblepoint pressure will be low by approximately:

Nonhydrocarbon	Mole % present in the gas	Predicted Low by:
CO_2	x	x/2 %
H_2S	y	y/2 %
N_2	z	z %

Example: 2 mole % N_2 present in the gas will cause the chart-predicted p_b to be around 2% low.

Fig. 3-32. Chart for calculating bubblepoint pressure (from Lasater[13]). Permission to publish by SPE.

Interestingly, Standing has provided the following comparisons of the accuracy of the three already-mentioned bubblepoint correlations.

Item	Standing	Borden-Rzasa	Lasater
Number of points in correlation	105	194	158
% of data points within 10% of correlation	87	65	87
% of data points more than 200 psi in error	27	12	N.A.
% mean error	4.8	9±	3.8

The Vasquez-Beggs[25] bubblepoint correlation has not been in existence as long as the Standing or Lasater methods, but results appear to be excellent, and it is easily programmed. It is meant to be used with well test data. First, gas gravity is corrected to a separator pressure of 100 psig using Equation 3-28. Then, using the following table of coefficients, the bubblepoint pressure can be calculated with Equation 3-32.

Coefficient	API ≤ 30°	API > 30°
c_1	0.0362	0.0178
c_2	1.0937	1.1870
c_3	25.7240	23.9310

$$p_b = \left[R_{sb} \frac{e^{-c_3 [°API / (T_f + 460)]}}{c_1 \gamma_{gs}} \right]^{1/c_2} \tag{3-32}$$

where:

p_b = bubblepoint pressure, psig
γ_{gs} = corrected gas gravity to separator at 100 psig,
R_{sb} = solution gas/oil ratio at p_b, SCF/STB,
$°API$ = stock tank oil gravity, and
T_f = reservoir temperature in °F.

Alternatively, for pressures at or below the bubblepoint, the above equation may be rearranged to calculate the corresponding solution gas/oil ratio:

$$R_s = (c_1)(\gamma_{gs})(p^{c_2}) \left[e^{c_3 [°API / (T_f + 460)]} \right] \tag{3-33}$$

H. Formation Volume Factor for Oil

The volume of liquid entering the stock tank is less than the volume of the same liquid plus dissolved gas in the reservoir. The main reason for this is that the liquid in the reservoir is swollen due to the solution gas. A second reason is that the reservoir fluid is in a thermally expanded state due to the higher temperature in the reservoir than in the stock tank. The reservoir engineer often has need to relate the volume of oil seen at the stock tank to its corresponding volume in the reservoir.

Hence, the definition of oil formation volume factor is:

$$B_o = \frac{\textit{volume of oil plus solution gas at reservoir pressure} \text{ and } \textit{temperature}}{\textit{volume of the oil at stock tank pressure} \text{ and } \textit{temperature}} \tag{3-34}$$

Figure 3-31 provides a typical graph of oil formation volume factor both below and above the bubblepoint. Several items are to be noted about this graph.

1. The graph is valid only at reservoir temperature. Thus, the point on the y-axis (at atmospheric pressure) illustrates the thermal expansion of stock tank oil that would occur at reservoir temperature.

2. The slope of the curve is positive between atmospheric pressure and the bubblepoint pressure. This is because more gas is in solution at higher pressures with a corresponding volume increase.

3. At the bubblepoint, a discontinuity exists. In fact, to the left of the bubblepoint, the slope is positive; while to the right the slope is negative.

4. The negative slope in the curve above the bubblepoint pressure is indicative that no further free gas is in the system. Higher pressures from this point cause liquid compression. So, crude oil has its greatest volume (or lowest density) at the bubblepoint.

Standing[8] has provided a chart (Appendix B-2 at the end of this chapter) that can be used to estimate the oil formation volume factor for saturated oil. Therefore, the chart may be used for pressures at or below the original bubble point pressure. Garb[26] suggests the following two equations may be used to closely reproduce the Standing formation volume factors obtained from the chart.

$$B_o = 0.9759 + (12)(10^{-5})(C)^{1.2} \tag{3-35a}$$

$$C = (R_s)(\gamma_g / \gamma_o)^{0.5} + (1.25)(T_f) \tag{3-35b}$$

where:

B_o = oil formation volume factor (saturated oil), Res. barrel / STB,
C = correlation number,
R_s = solution gas-oil-ratio, SCF / STB,
γ_g = specific gravity (relative density) of gas,
γ_o = stock tank oil specific gravity (relative density), and
T_f = reservoir temperature, °F

According to Garb[26], the results of the equations do not agree perfectly with those obtained from the chart; however, the equations are adequate and are within the standard deviation of the data used to develop the correlation.

The Vasquez-Beggs estimation method[25] consists of calculating gas gravity corrected to separator pressure of 100 psig using Equation 3-28. Then, the following table of coefficients is used with Equation 3-36, which is valid to calculate B_o for pressures less than or equal to the separator pressure of 100 psig using Equation 3-28. Then, the following table of coefficients is used with Equation 3-36, which is valid to calculate B_o for pressures less than or equal to the bubblepoint pressure.

Coefficient	API ≤ 30°	API > 30°
C4	4.677(10^{-4})	4.670(10^{-4})
C5	1.751(10^{-5})	1.100(10^{-5})
C6	-1.811(10^{-8})	1.337(10^{-9})

$$B_o = 1 + c_4 R_s + (T_f - 60)(°API / \gamma_{gs})(c_5 + c_6 R_s) \quad (3\text{-}36)$$

where:

B_o = oil formation volume factor, Res. barrel / STB,
R_s = solution gas/oil ratio, SCF / STB,
T_f = reservoir temperature, °F,
$°API$ = oil stock tank gravity,
γ_{gs} = gas gravity corrected to separator pressure of 100 psig.

As already mentioned, with increasing pressure above the bubblepoint, the oil formation volume factor decreases from that at the bubblepoint because no further free gas is available to go into solution. Hence, the liquid is compressed. Because this is strictly an isothermal compression, the behavior may be related in terms of the isothermal compressibility, which was described earlier in this chapter. The applicable equation is:

$$B_o = B_{ob} e^{-c_o (p - p_b)} \quad (3\text{-}37)$$

where:

B_o = oil formation volume factor, BBL/STB,
B_{ob} = oil formation volume factor, @ p_b, BBL/STB,
c_o = liquid compressibility above the bubblepoint, psi^{-1},
p = pressure (greater than or equal to p_b), psia, and
p_b = bubblepoint pressure, psia.

In certain engineering calculations, it is at times convenient to use a term called the total or two-phase formation volume factor, B_t. This factor relates the reservoir volume of liquid and initial complement of dissolved gas to the stock tank oil volume. Of course, below the bubblepoint pressure, some of the original solution gas will exist as free gas in the reservoir. The units of B_t are reservoir barrels per stock tank barrel. It is numerically equal to the volume in barrels that one stock tank barrel of liquid and its original dissolved gas occupy at reservoir temperature and at the reservoir pressure of interest. A typical B_t relationship is shown in Figure 3-33. Note that B_t and B_o are the same at and above the bubblepoint. Below the bubblepoint, however, with decreasing pressures, the two-phase volume factor rapidly increases due to the gas coming out of solution, while B_o decreases. B_t is related to oil and gas formation volume factors and to solution gas/oil ratio as:

$$B_t = B_o + B_g(R_{sb} - R_s) \quad (3\text{-}38)$$

where:

B_t = two-phase formation volume factor, BBL/STB,
B_o = oil formation volume factor, BBL/STB,
B_g = gas formation volume factor, BBL/SCF,
R_{sb} = solution gas/oil ratio at the bubblepoint, SCF/STB, and
R_s = solution gas/oil ratio at current pressure, SCF/STB.

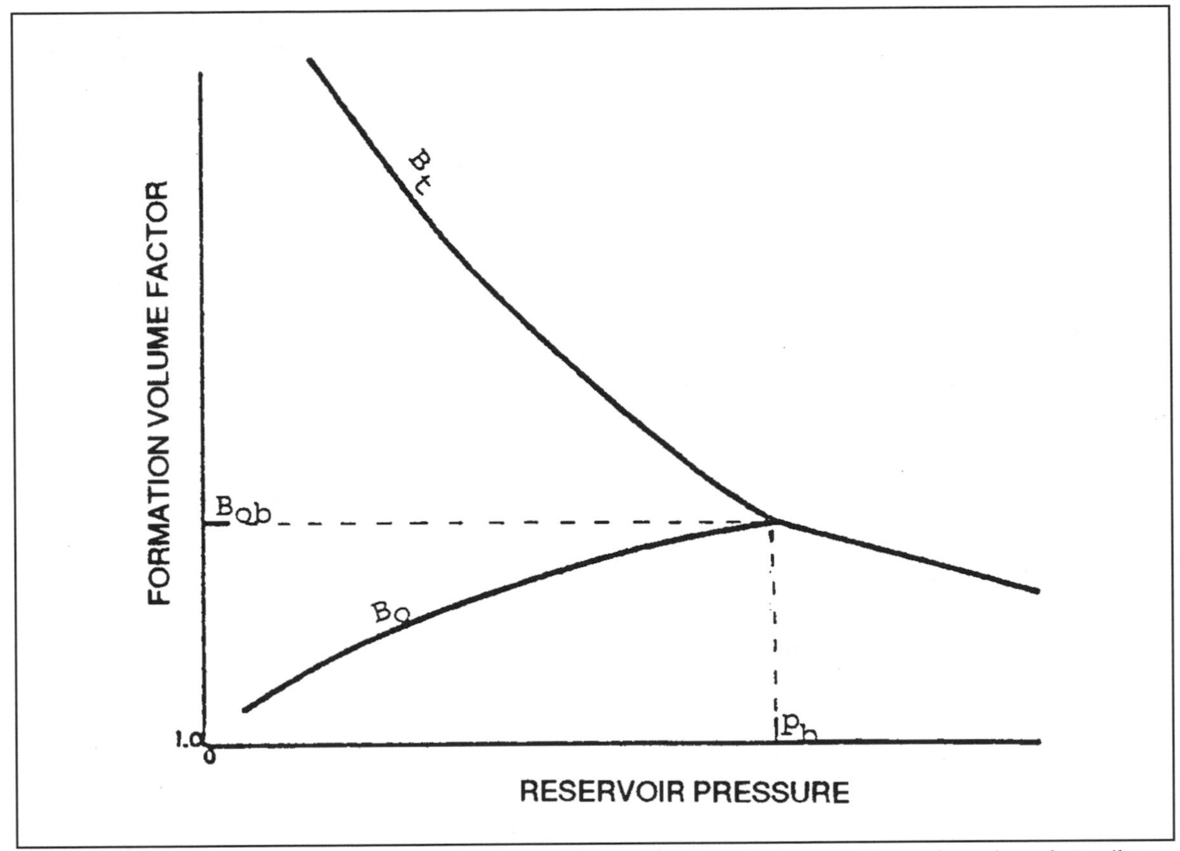

Fig. 3-33. Typical relationship between total or two-phase formation volume factor and oil formation volume factor (from McCain[1]). Permission to publish by Petroleum Publishing Company.

I. Converting Laboratory R_s and B_o Data to Actual Conditions

Appendix A contains a sample PVT analysis report on an oil sample. Typically, laboratory data for R_s and B_o are obtained under differential, isothermal (reservoir temperature) conditions. Unfortunately, such data do not allow direct correspondence between reservoir fluid volumes and surface fluid volumes. Why? Surface fluid volumes are at 60°F, which is likely to be cooler than reservoir temperature. Further, the volume of surface oil and surface gas is directly related to the type of separation process the reservoir fluids undergo on their way to the stock tank. It is essentially a nonisothermal flash process that is occurring between the reservoir and the separator and then another between the separator and the stock tank.

As seen in the PVT report in Appendix A, a group of tests that are usually performed during a laboratory fluid study are separator tests. The effects of different separator conditions are considered. In these tests in Appendix A (page 7), a sample of reservoir oil (saturated, at reservoir temperature and bubblepoint pressure) is flashed to separator conditions and then to stock tank conditions. The reported "formation volume factor" (here designated as B_{ofb}) is actually:

$$B_{ofb} = \frac{(\,volume\ \ of\ \ bubblepoint\ \ oil\ \ at\ \ T_f\,)}{(\,volume\ \ of\ \ stock\ \ tank\ \ oil\,)} \tag{3-39}$$

After choosing the laboratory separator conditions that most closely correspond to actual field conditions, the laboratory differential B_o data may be converted to more realistic formation volume factors[31]:

$$B_o = (\,B_{od}\,)\left[\frac{B_{ofb}}{B_{odb}}\right] \tag{3-40}$$

where:

B_o = formation volume factor to be used for reservoir engineering calculations,
B_{od} = laboratory differential oil formation volume factor,
B_{odb} = B_{od} at the bubblepoint pressure,
B_{ofb} = the separator formation volume factor (described with Equation 3-39).

Problem 3-10:

With the data of Appendix A, determine B_o at 2100 psig if separator conditions are 75 °F and 100 psig.

As the separator pressure is 100 psig, (page 7, Appendix A.)

$B_{ofb} = 1.474$

From page 5, Appendix A, at 2100 psig,

$B_{od} = 1.515$

At the bubblepoint pressure (2650 psig),

$B_{odb} = 1.600$

Hence: $B_o = (1.515)(1.474/1.600) = 1.396$

Along with the other results of the separator tests (Appendix A, page 7), notice that Core Laboratories, Inc., provides a determination of separator GOR and stock tank (vent gas) GOR. The total GOR, here termed R_{sfb}, associated with a given set of separator conditions is merely the sum of these two numbers. For instance, at separator conditions of 75°F and 100 psig, the total GOR in going from the bubblepoint pressure and reservoir temperature to the stock tank is:

$$R_{sfb} = 676 + 92 = 768\ SCF/STB$$

After determining R_{sfb}, then differential laboratory R_s values can be converted to more realistic solution GOR's[31]:

$$R_s = R_{sfb} - (R_{sdb} - R_{sd}) \frac{B_{ofb}}{B_{odb}} \quad (3\text{-}41)$$

where:

R_s = solution GOR to be used for reservoir engineering calculations, SCF/STB,
R_{sfb} = total separator GOR from separator test (as described in the last paragraph), SCF/STB,
R_{sd} = solution GOR determined in the laboratory under differential conditions at reservoir temperature, SCF/STB,
R_{sdb} = R_{sd} at the bubblepoint pressure.
B_{ofb} and B_{odb} were defined under Equation 3-40.

At low pressures, the R_s *vs.* pressure relationship usually must be drawn manually back to R_s = 0 at atmospheric pressure (to avoid negative R_s values at low pressure).

Problem 3-11:

Considering the data of Appendix A, with a reservoir pressure of 2100 psig and separator conditions of 100 psig and 75 °F, determine R_s.

R_{sfb} = 676 + 92 = 768 SCF/STB

From the previous example,

B_{odb} = 1.600 and

B_{ofb} = 1.474

From page 5, Appendix A: at 2100 psig,

R_{sd} = 684 SCF/STB, and

R_{sdb} = 854 SCF/STB

So, R_s = 768 - (854 - 684)(1.474 / 1.600)
= 611 SCF/STB

J. Oil Viscosity

Viscosity is the property of resistance to shear stress. Alternatively, viscosity may be viewed as a fluid's internal resistance to flow and therefore, depends greatly on density and composition. A thick, usually heavy liquid (e.g. tar) has a higher viscosity than a thin one that flows easily.

Reservoir oil viscosity, μ_o, is directly related to tank-oil gravity, gas gravity, gas in solution in the oil, pressure, and reservoir temperature. With the wide variety of compositions of crude oil, we should expect to find a large variation in oil viscosities even with oils of similar gravity, solution gas/oil ratio, and reservoir temperature. And, of the more important oil physical properties that are needed in reservoir engineering, crude oil viscosity has the poorest correlation.

Typical crude oil viscosity characteristics are graphed in Figure 3-31. This graph is at reservoir temperature. Note that the viscosity of dead oil (gas-free oil at atmospheric pressure) is much higher than the viscosity of the oil at reservoir conditions. Higher pressures up to the bubblepoint represent increased dissolved gas in the crude, which causes a viscosity reduction. Solution gas apparently has the effect of a "lubricant" and reduces the liquid's internal resistance to shear movement. Alternatively, the solution gas may be viewed as causing the oil to swell thus reducing its density. So, the more dissolved gas in the oil, the more the viscosity is lowered. Thus, the minimum viscosity (and also a point of discontinuity) is obtained at the bubblepoint pressure. Above this pressure, with no more free gas available to go into solution, the viscosity increases with increasing pressure as the liquid molecules are forced closer together.

Beal's correlation[14] is well known and is based on a large number of oil samples. Figure 3-34 shows the correlation. Notice that the correlation is for gas-free oils at 100°F, and the correlation is only fair.

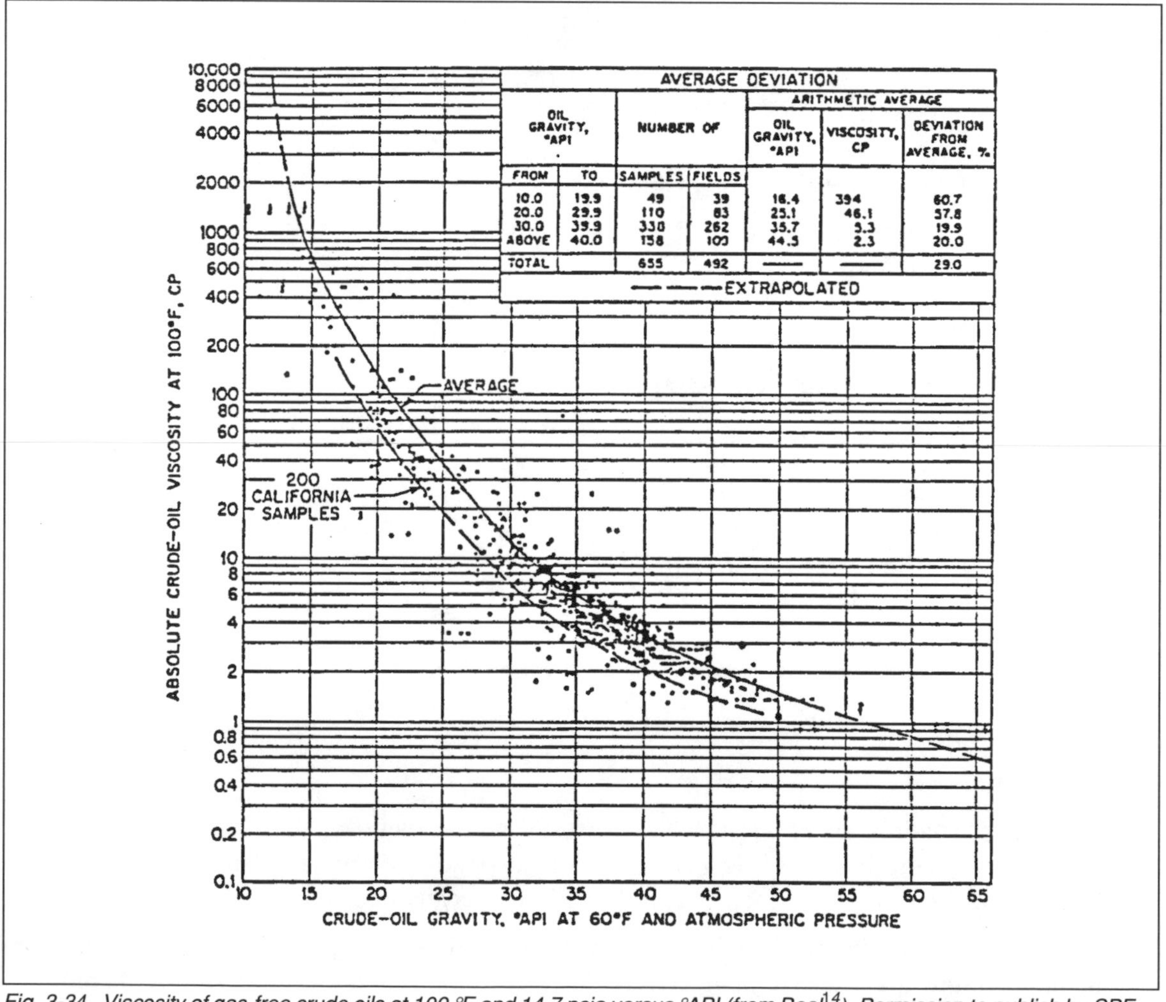

Oil gravity, °API From	Oil gravity, °API To	Number of Samples	Number of Fields	Arithmetic average: Oil gravity, °API	Arithmetic average: Viscosity, cp	Arithmetic average: Deviation from average, %
10.0	19.9	49	39	16.4	394	60.7
20.0	29.9	110	83	25.1	46.1	57.8
30.0	39.9	330	262	35.7	5.3	19.9
ABOVE	40.0	158	103	44.5	2.3	20.0
TOTAL		655	492	—	—	29.0

Fig. 3-34. Viscosity of gas-free crude oils at 100 °F and 14.7 psia versus °API (from Beal[14]). Permission to publish by SPE.

In Beal's method, Figure 3-35 is used to determine gas-free crude viscosity at reservoir temperature. With this value and a value for gas in solution, reservoir oil viscosity is read directly from Figure 3-36. If Figure 3-36 is entered with different values for gas in solution (corresponding to a range of pressures below the bubblepoint), then the complete viscosity relationship from bubblepoint to atmospheric pressure, may be generated. For pressures above the bubblepoint, viscosities may be estimated using Figure 3-37.

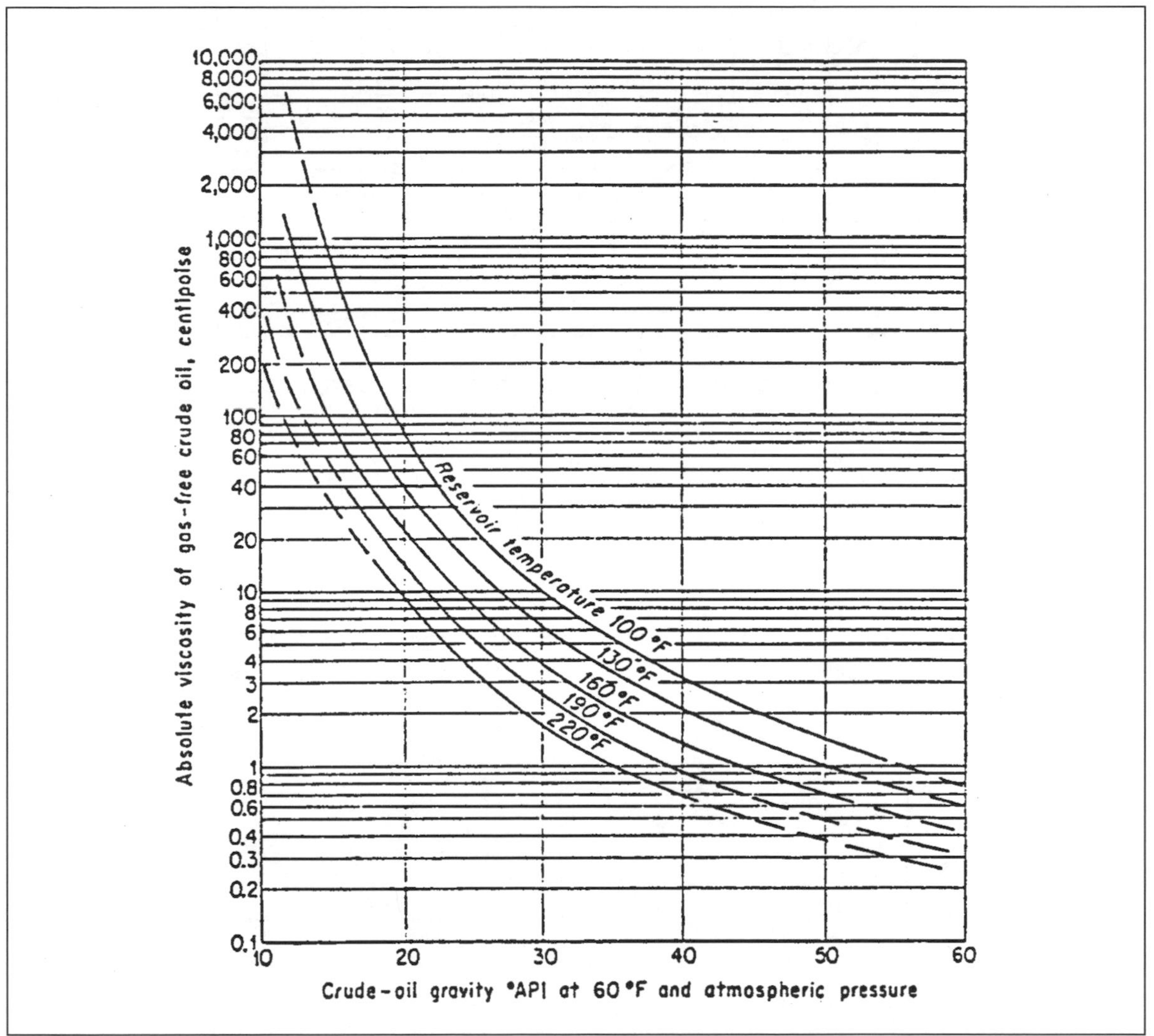

Fig. 3-35. Gas-free crude viscosity as a function of reservoir temperature and stock tank crude gravity (from Beal[14]). Permission to publish by SPE.

Another well-known viscosity correlation is that of Chew and Connally.[15] Their correlation is strictly for gas-saturated oils at reservoir temperature and pressure. A measured dead-oil (gas-free) viscosity at or near reservoir temperature is needed. If not available, one may be estimated from Figure 3-35. With the dead-oil viscosity and the appropriate solution gas/oil ratio, Figure 3-38 may be used to estimate crude oil viscosity at or below the bubblepoint.

Beggs *et al.*[27] have published an easily programmable viscosity correlation presented here as Equations 3-42, 3-43, and 3-44.

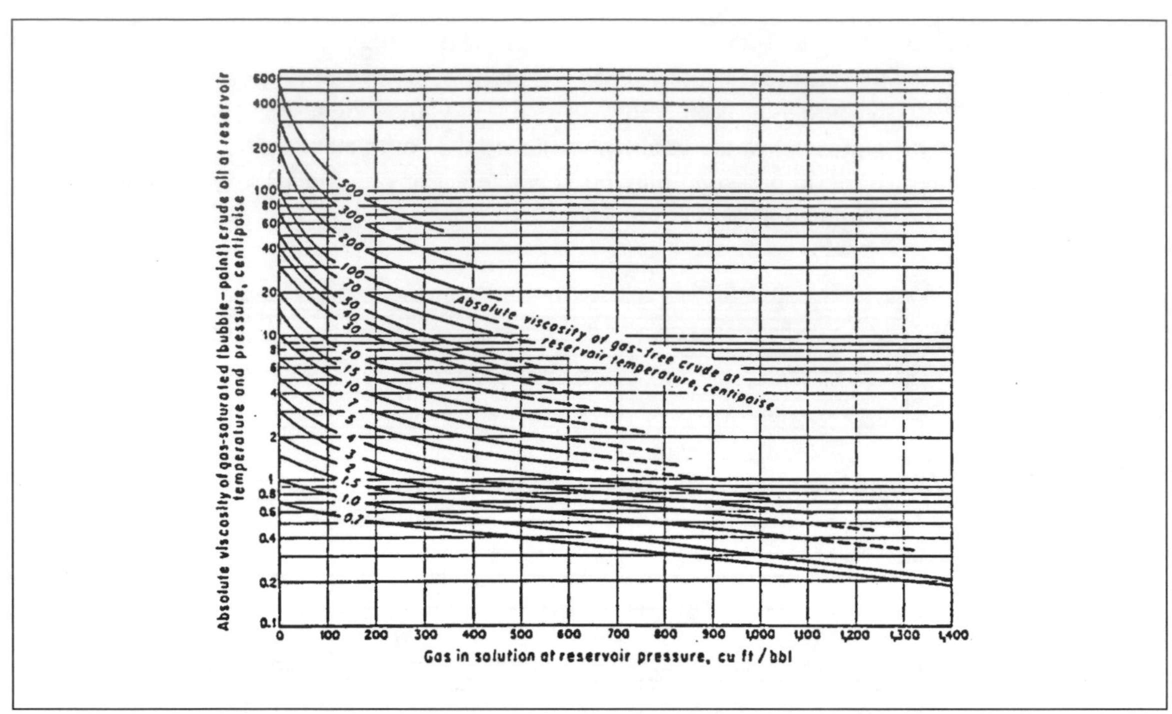

Fig. 3-36. Reservoir oil viscosity from gas-free oil viscosity and gas in solution. Correlation based on 351 viscosity observations from 41 oil samples representing average conditions for 29 oil fields. Average deviation 13.4 percent (from Beal[14]). Permission to publish by SPE.

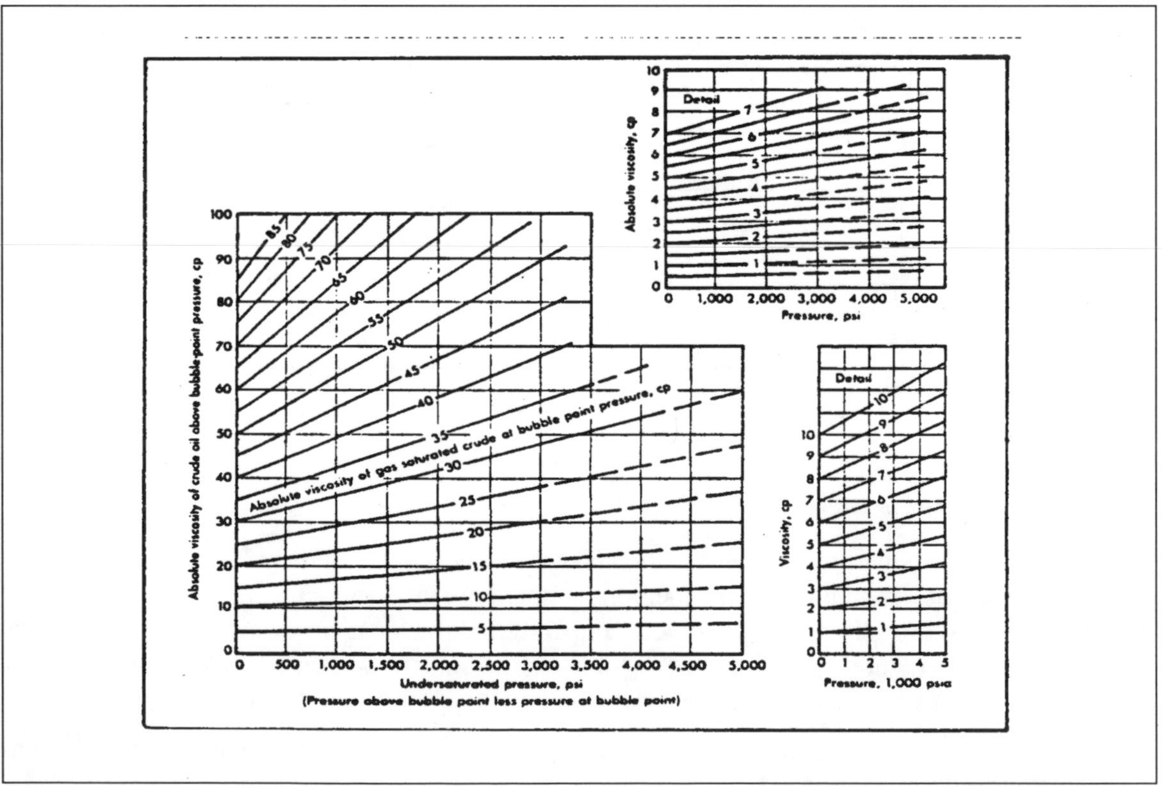

Fig. 3-37. Viscosity of oil above the bubblepoint pressure. Average deviation, 2.7 percent (from Beal[14]). Permission to publish by SPE.

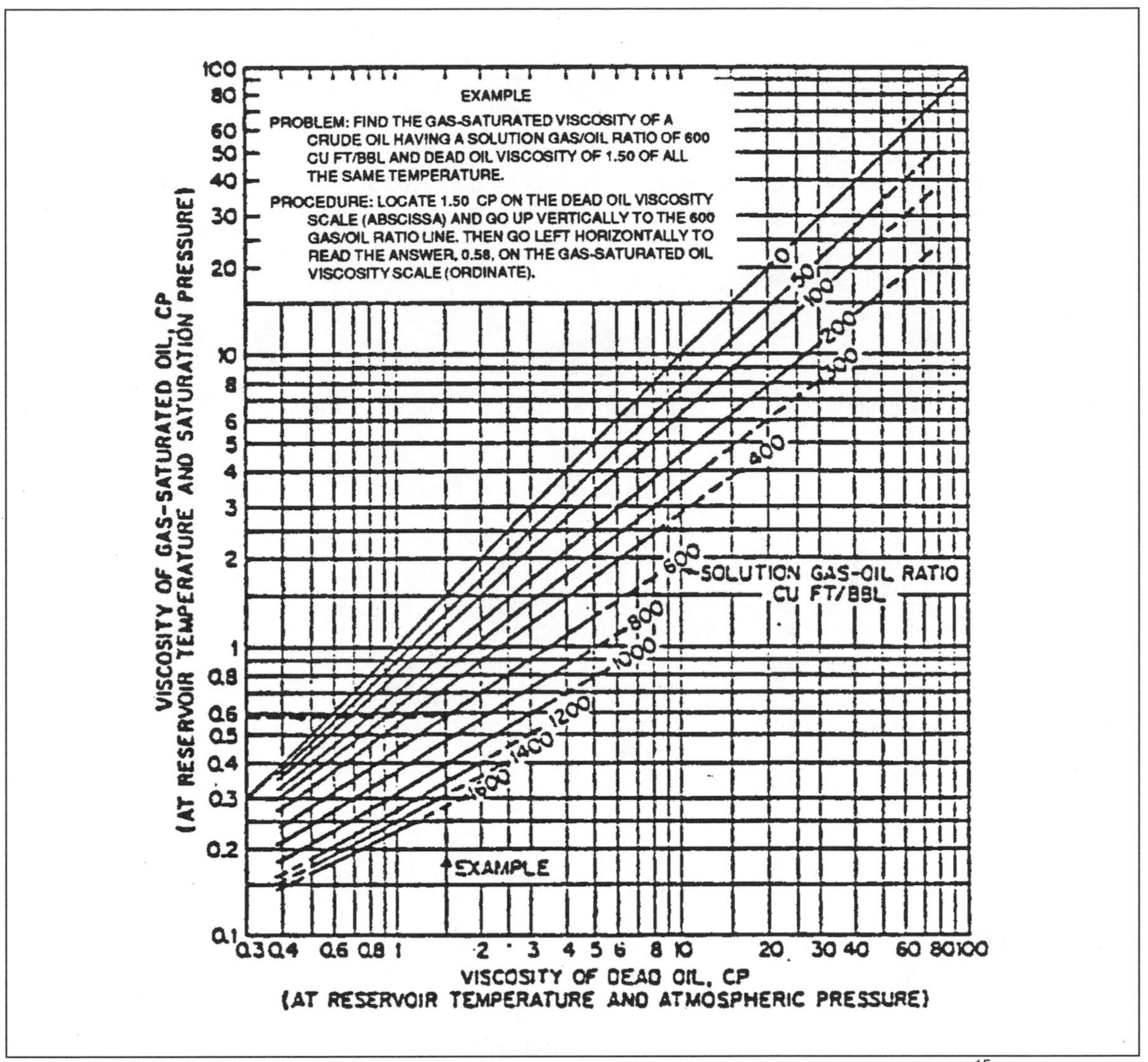

Fig. 3-38. Viscosity of gas-saturated oil at reservoir temperature and pressure (from Chew and Connally[15]). Permission to publish by SPE.

First, the dead oil viscosity is calculated.

$$\mu_{od} = 10^{x} - 1 \tag{3-42}$$

where:

μ_{od} = viscosity of gas-free oil at T_f, cp,
API = stock tank oil gravity, °API,
T_f = reservoir temperature, °F, and
$x = (y)(T_f)^{-1.163}$
$y = 10^z$
$z = 3.0324 - (0.02023)(API)$

Then, to calculate the viscosity of saturated oil ($p \leq p_b$):

$$\mu_o = (A)(\mu_{od})^B \tag{3-43}$$

where:

μ_o = viscosity of saturated oil at reservoir conditions, cp,
R_s = solution gas/oil ratio, SCF/STB, and

$$A = (10.715)(R_s + 100)^{-0.515}$$

$$B = (5.44)(R_s + 150)^{-0.338}$$

Finally, to determine viscosities for pressures above the bubblepoint:

$$\mu_o = (\mu_{ob})(p / p_b)^m \tag{3-44}$$

where:

μ_o = oil viscosity at reservoir conditions at pressure p, cp,
μ_{ob} = oil viscosity at bubblepoint conditions, cp,
p_b = bubblepoint pressure, psig,
p = reservoir pressure (which must be greater than p_b), psig, and

$$m = [(2.6)(p)^{1.187}][w]$$

$$w = 10^{\{[(-3.9)(10^{-5})(p)]-5\}}$$

RESERVOIR HYDROCARBON FLUID CLASSIFICATION

It is essential to characterize a petroleum reservoir fluid before devising a recovery scheme and proceeding with development. We will attempt this classification on the basis of API gravity. The boundaries between the following classifications are not meant to be strict. Indeed, these are "rule of thumb" divisions or are guidelines that are actually quite indistinct. Where a particular fluid belongs, when close to an API gravity boundary, is composition dependent.

1. Bitumen: $4 < °API < 10$

The authors are unaware of any known reservoirs with gravity less than 4°API that have proved to be of commercial significance. However, it is believed that at least one company has investigated the enhanced oil recovery of a reservoir fluid as low as -2°API, but commerciality is yet to be demonstrated. Other properties of a bitumen (at initial reservoir conditions) include:

$$R_{s\ initial} = R_{si} \approx negligible$$

$$B_o \approx 1.0\ Res.\ bbl / STB$$

$$1{,}000{,}000 > \mu_o > 5{,}000\ cp$$

The color is usually quite dark or even jet black, however, some are dark chocolate brown.

2. Tar or Heavy Oil: $10 < °API < 20$

$$Negligible < R_{si} < 50\ scf/STBO$$

Heavy oils from the Sparky Sandstone (Lloydminster, Alberta, Canada), have sufficient gas in solution to fuel engines on the lift equipment.

$$1.0 < B_o < 1.1\ Res.\ bbl/STB$$

$$5{,}000 > \mu_o > 100\ cp$$

3. Low-Shrinkage Oils: $20 < °API <$ *low 30's*

These liquids are sometimes called "black" oils.

$$50 < R_{si} < 500\ scf/STBO$$

$$1.1 < B_o < 1.5\ Res.\ bbl/STB$$

$$100 > \mu_o > 2\ to\ 3\ cp$$

Although, the color of these types of oils is generally lighter than that of the bitumens or heavy oils, it still tends to be rather dark. However, we do find these oils with casts of green, gold, and even light reddish brown.

As previously mentioned, two effects are causing B_o to be different than 1.0: (1) gas in solution, and (2) greater temperature at reservoir conditions.

Actually, this type of liquid is a multicomponent, low-shrinkage hydrocarbon system that may be represented on a phase diagram (pressure-temperature) as shown in Figure 3-39. By definition, this type of reservoir fluid will exist at a temperature that is less than the critical temperature. At or above the critical temperature, the hydrocarbon would not be an oil in the strict sense. Note that the dewpoint line is shown as a dashed curve. All of the action in this type of reservoir takes place to the left of this line. If the reservoir oil is at the temperature and pressure represented by point 1, it is said to be undersaturated. With isothermal pressure depletion, the oil would be fully saturated with gas at point 2, the bubblepoint for this particular reservoir-fluid system. It is worth noting that had the reservoir discovery pressure corresponded to this bubblepoint pressure, we would expect to find a gas cap in contact with the oil reservoir. With further pressure depletion, point 3 could be reached, at least in the laboratory. Here, 75 volume percent liquid and 25 percent gas would be in equilibrium. The actual reservoir case would be complicated by the presence of the porous rock that would tend to prevent total system equilibrium.

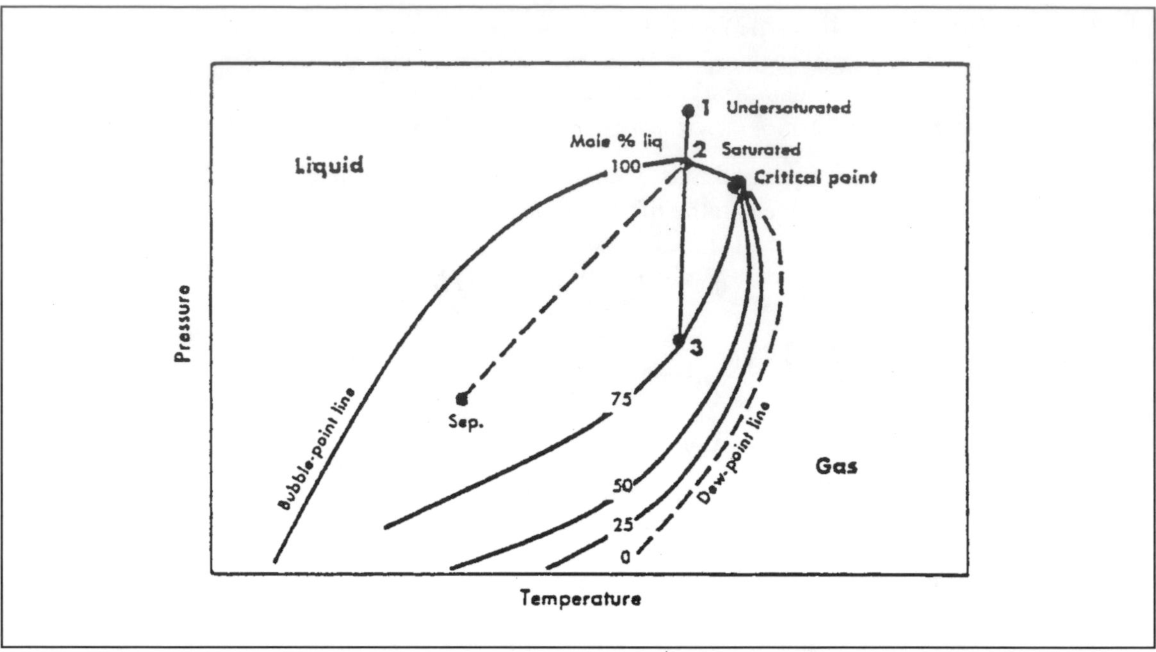

Fig. 3-39. Phase diagram of a low-shrinkage crude oil (from McCain[1]). Permission to publish by Petroleum Publishing Company.

4. High-shrinkage Oils: *low 30's < °API < low 50's*

These liquids are sometimes referred to as "volatile" oils. It should be pointed out at this point that the boundaries between fluid types are becoming more and more indistinct.

$$1.5 < B_o < 2.5 \text{ to } 3.5 \; Res.\,bbl\,/\,STB$$

$$500 < R_{si} < 2000 \text{ to } 6000 \; scf\,/\,STBO$$

$$2 \text{ to } 3 > \mu_o > 0.25 \; cp\,.$$

The upper boundary range for both B_o and R_{si} is not well defined. A fluid falling in this area ($2000 < R_{si} < 6000 \; scf\,/\,STB$) could easily fall in the next category (retrograde condensate), depending on the composition. For instance, given a liquid whose API gravity was 55 with a GOR of 5000 scf/STB, then it would be uncertain whether the fluid was a high-shrinkage oil or a condensate.

The color of these high-shrinkage liquids is much lighter than those already discussed. This type of oil could perhaps be as light as a bright gold.

The phase diagram for a typical high-shrinkage oil is shown in Figure 3-40. The formation temperature for these systems will always be less than (but possibly quite close to) the critical temperature. The envelope has a tendency to be slightly thinner than that for a low-shrinkage oil. Notice that the quality lines have shifted (from those for a "black" oil) such that a proportionately higher quantity of gas will exist at a given point within the two-phase region than at a comparable position in Figure 3-39. Volatile oils contain substantial quantities of the lighter hydrocarbon components which more easily flash into the vapor state with reduced pressure, thereby contributing to the high-shrinkage of the liquid and the higher gas/oil ratios.

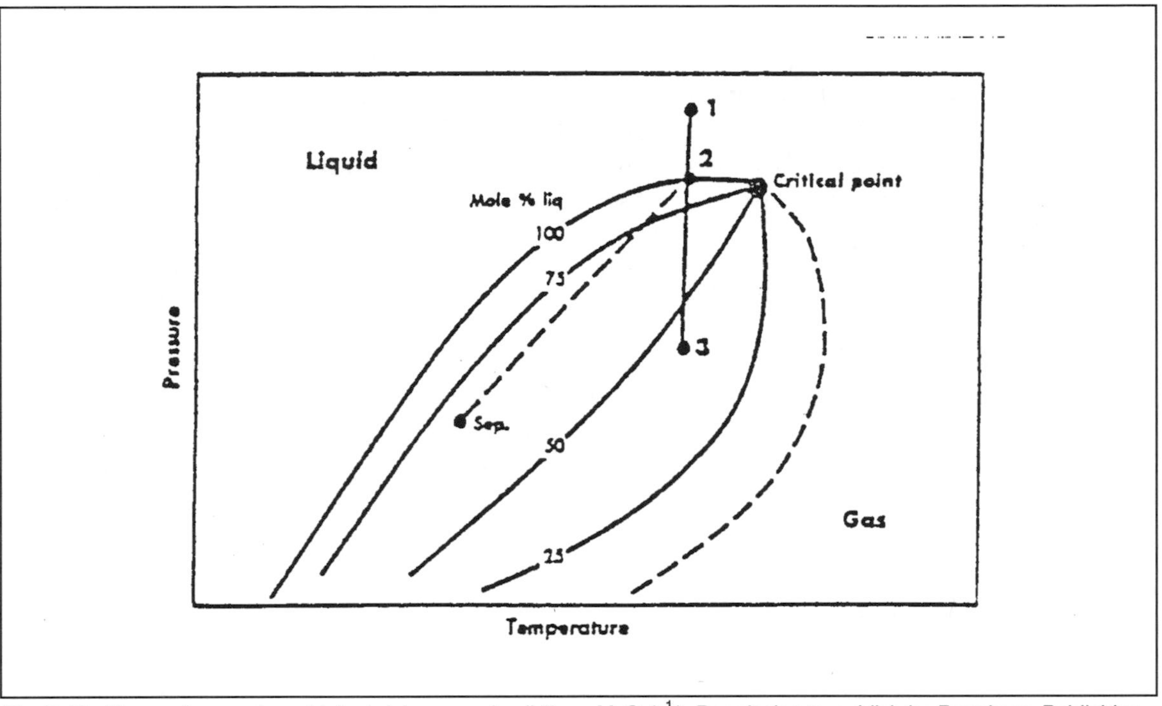

Fig. 3-40. Phase diagram for a high-shrinkage crude oil (from McCain[1]). Permission to publish by Petroleum Publishing Company.

Notice in Figure 3-40, if the volatile oil's initial conditions are at point 1, then the liquid is undersaturated and an associated gas cap will not exist. However, if the reservoir is found at point 2, then the probability of the existence of a gas cap (the size of which could range anywhere from quite small to huge) is high. Note that the pressure traverse from point 2 to separator conditions, shown as a dashed line, is not an isothermal process.

5. Retrograde Condensate Gas: *Middle 50's < °API < 70*

Actually, a practical maximum API gravity for these systems is around 68°, because above that, at surface conditions, part of the liquid would undoubtedly flash off.

B_o: not applicable because this is actually a gas system.

$$2000 \text{ to } 6000 < R_i < 15{,}000 \ scf/STBO$$

μ_o (condensate liquid) $\approx$ 0.25 cp

Note that the lower R_i boundary range is a "gray" area (2000 to 6000 scf/STB), and a fluid exhibiting GOR's in this range might be either a volatile oil or a retrograde condensate. Similarly, the upper R_i boundary is somewhat indistinct in that reports exist of some retrograde condensate reservoirs having initial producing GOR's as high as 25,000 scf/STBO.

The colors of condensate liquids range from clear to straw yellow.

The phase diagram for a retrograde condensate system is shown in Figure 3-41. Recall that the cricondentherm is the maximum temperature of the two-phase envelope, so it is the maximum temperature that the two phases can exist in equilibrium. In a retrograde condensate system, the temperature is always greater than the critical temperature but less than the cricondentherm.

In Figure 3-41, note that if the initial reservoir pressure and temperature are represented by point 1, this is above the phase envelope; and the fluid has an undersaturated, single phase state. With isothermal pressure depletion, it is possible to reach point 2, the upper dewpoint. Here the reservoir fluid is 100% saturated vapor. Any further pressure reduction (@ reservoir temperature) will cause liquids to condense as the pressure traverse enters the two-phase region. Continued pressure reduction causes more and more liquids to condense until point 3 is reached.

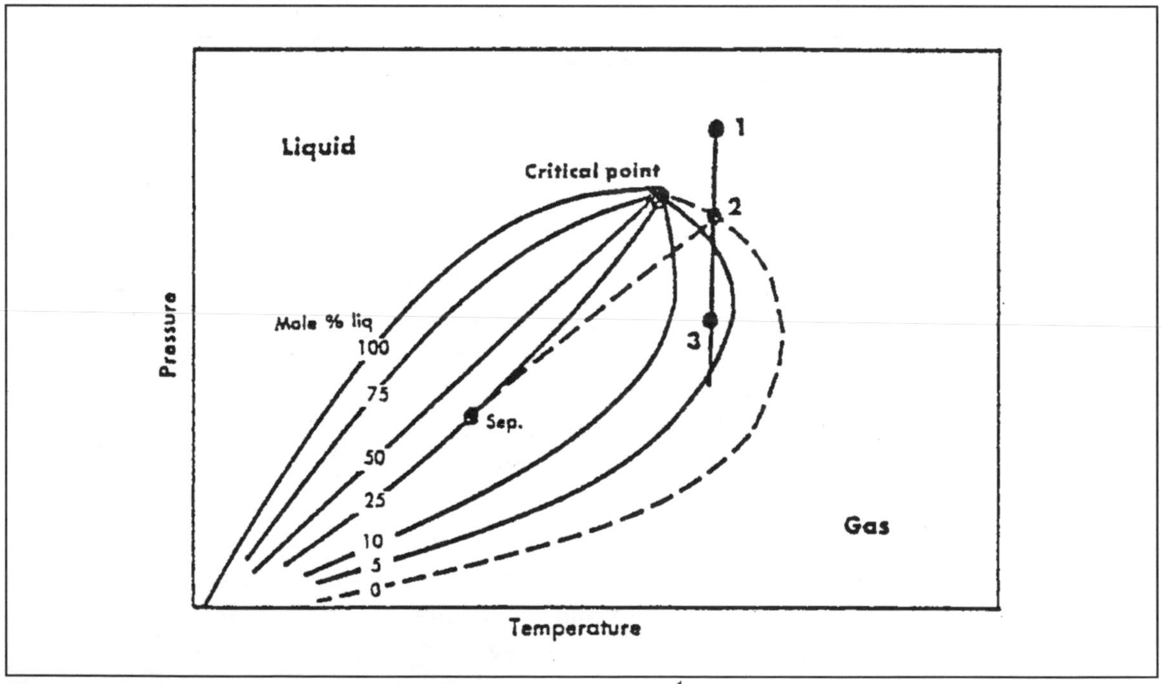

Fig. 3-41. Phase diagram for a retrograde condensate gas (after McCain[1]). Permission to publish by Petroleum Publishing Company.

With further pressure depletion, the phase diagram indicates that the liquids begin to revaporize. In fact, if the lower phase boundary can be reached (lower dewpoint), in the laboratory PVT cell, all hydrocarbon liquids will have revaporized. Although revaporization (below point 3) occurs in the laboratory, the rock in the reservoir gets in the way. The condensed liquids are held in the rock pores by capillary forces, and this prevents equilibrium of the fluid system. The unfortunate result is that significant revaporization of the liquids normally does not occur.

Also, in Figure 3-41, if initial conditions are represented by point 1, because we are above the phase envelope (indicating the fluid to be undersaturated), no possibility exists of an oil zone down structure. If the initial conditions are at point 2, then the fluid is at the dewpoint, and we have 100% gas that is saturated. In this event, an oil zone is likely to be below, but the size is indeterminate.

A retrograde condensate reservoir is an excellent candidate for “gas cycling”. With this type of reservoir management, dry gas is injected into the reservoir to keep the pressure above the phase envelope. Therefore, a single phase will be maintained in the reservoir with the benefit that no liquids will drop out in the reservoir and create an unproducible residual oil saturation.

It is necessary that the dry gas cricondentherm be less than the reservoir temperature to prevent any condensation from the injection gas. This is illustrated in Figure 3-42. Actually, the dry gas will tend to withdraw intermediates from the reservoir fluid. Thus, with time the phase diagrams for both fluids will be changing. The reservoir fluid diagram moves to the

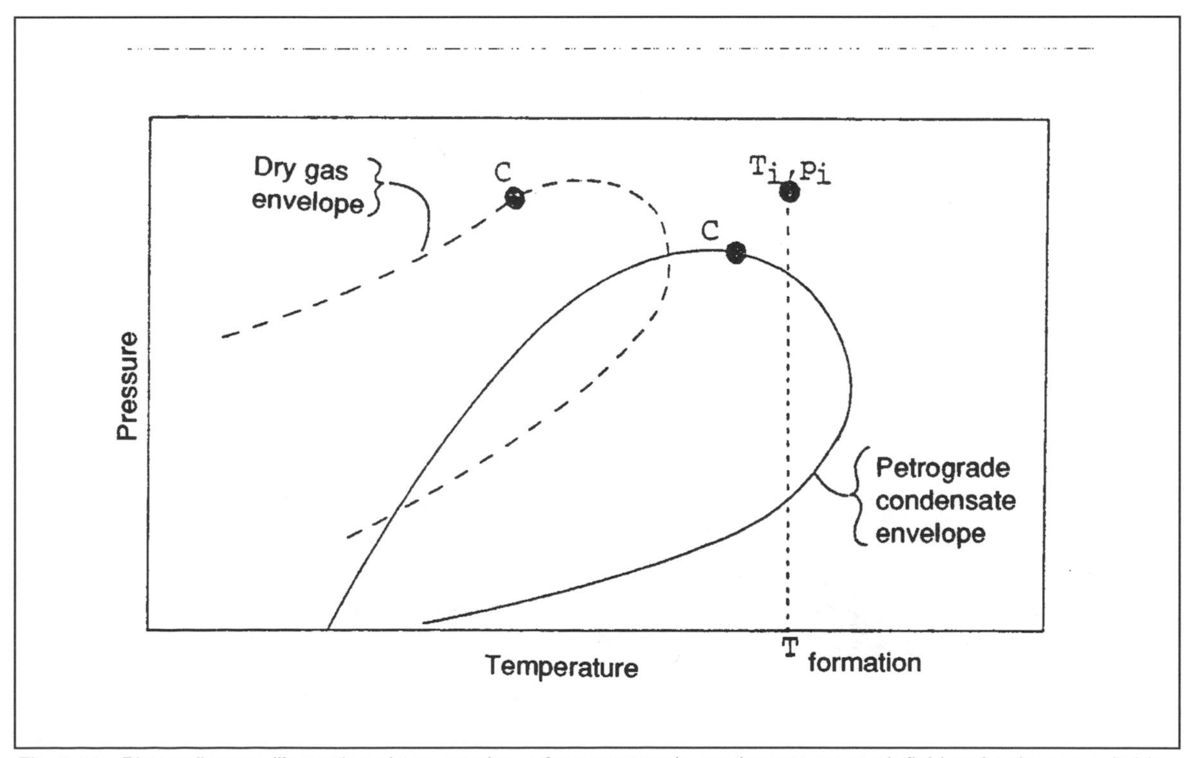

Fig. 3-42. Phase diagram illustrating phase envelopes for a retrograde condensate reservoir fluid and a dry gas suitable for “gas cycling” this reservoir.

left, and the dry gas envelope moves to the right. When the produced fluid becomes predominantly dry gas, then injection is suspended. At this point, production (blowdown) continues until the economic limit is reached.

6. Wet Gas: *°API > 60*

$$15{,}000 < R_i < 100{,}000\ scf/STBO$$

$$\mu_o\ (condensate\ liquid) \approx 0.25\ cp\,.$$

The system temperature must be greater than the cricondentherm to be a wet gas system. A typical phase diagram is shown in Figure 3-43. Note the characteristic shifting of the critical point as compared to a retrograde condensate system. It should also be observed that isothermal pressure depletion (as would occur in the reservoir) results in no hydrocarbon liquids dropping out. However, after the fluid leaves the formation (in the wellbore, flow-lines, and surface equipment), liquids may be recovered due to the accompanying temperature reduction. So, this type of reservoir does not have to be cycled to prevent liquid drop out. Reservoir engineering performance calculations for a wet gas reservoir are based on the gas laws and are essentially the same as for a dry gas reservoir.

7. Dry Gas:

The word "dry" indicates that the gas in question provides little or no liquids at surface conditions. Of course, no liquid fall out occurs in the reservoir either. Figure 3-44 provides a typical P-T diagram. Gas composition is primarily methane with small amounts of ethane, propane, and butane. Only trace amounts of heavier hydrocarbon components are present; however, certain nonhydrocarbon gases may be included.

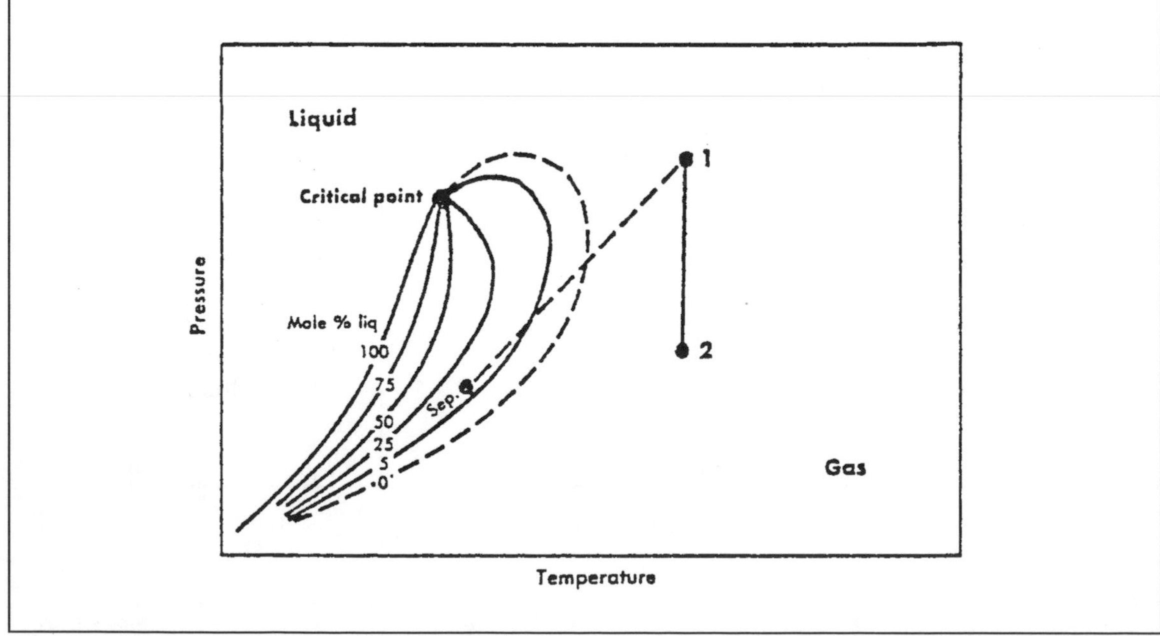

Fig. 3-43. Phase diagram of a wet gas (from McCain[1]). Permission to publish by Petroleum Publishing Company.

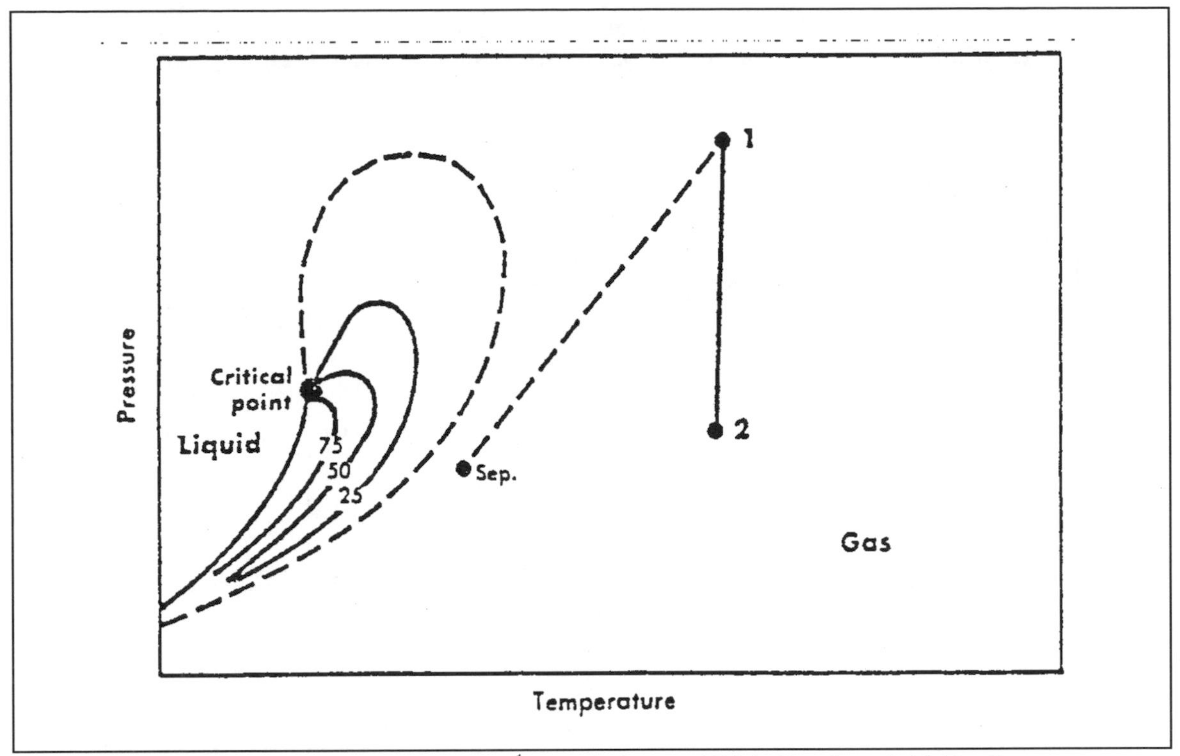

Fig. 3-44. Phase diagram for a dry gas (from McCain[1]). Permission to publish by Petroleum Publishing Company.

PROPERTIES OF FORMATION WATERS

A. Chemical Properties of Formation Water

Table 3-4 provides a summary of water characteristics from some oil-producing fields in Oklahoma.

Chemical water analysis data prepared by commercial water laboratories are often presented in graphical form. Several different methods for presenting analytical water information have been used.

Stiff[16] has presented a graphical plotting technique, whereby the effect of dilution or concentration has been reduced to a minimum. Figure 3-45 presents the essential features of the system, where the positive ions are plotted to the left and negative ions are plotted to the right of the center line in units of milliequivalents per liter. Should the results of a chemical analysis be in terms of parts per million, the conversion can be made by dividing by the equivalent weight in milligrams. Table 3-5 summarizes the appropriate conversion factors. The "milligram equivalent" value of a radical is identical with the terms "milliequivalent" or "reacting value." The parts per million of a radical, multiplied by its milligram equivalent, results in a value for the milligram equivalents per liter. For example, if a water analysis showed the sample to contain 25 ppm of carbonate, then the milligram equivalent per liter would be 25 x 0.0333, or 0.8325.

Table 3-4
Characteristics Of Some Waters Produced from Midcontinent Fields (Oklahoma)

No. of analysis*	System	Formation	Subsurface depth, ft	Constituents, mg/liter							Sp gr. 60°/60°	Total solids, mg/liter
				Ca	Mg	Na	Ba	HCO_3	SO_4	Cl		
75	Pennsylvanian	Bartlesville	4,489-5,524	1,900	910	12,100	0	0	0	24,100	1.031	39,010
				19,000	2,740	83,800	730	300	890	144,000	1.175	251,460
94	Ordovician	Wilcox	3,436-7,233	6,800	1,400	48,300	0	20	0	91,300	1.103	147,820
				18,500	3,300	80,200	130	160	720	163,000	1.178	266,010
25	Pennsylvanian	Layton	1,240-4,800	5,300	1,800	31,300	0	10	0	34,900	1.075	73,310
				18,900	4,300	79,000	380	80	510	160,000	1.179	263,170
28	Ordovician	Arbuckle	542-6,094	2,200	900	14,000	0	0	0	33,000	1.034	50,100
				18,800	2,700	63,800	110	850	1,880	127,000	1.147	214,140
19	Pennsylvanian	Cromwell	1,480-5,430	4,600	1,400	34,600	1	0	0	65,000	1.073	105,701
				11,900	4,300	51,500	20	310	1,130	113,500	1.130	182,660
12	Pennsylvanian	Burgess	1,800-2,490	5,900	2,000	42,500	1	15	0	81,600	1.091	132,016
				13,300	2,600	57,700	200	120	200	115,000	1.129	189,120
22	Mississippian	Mississippi	1,837-4,872	6,400	2,000	43,600	2	10	24	84,200	1.095	136,212
				22,400	2,500	72,000	30	80	430	157,000	1.173	254,440
18	Mississippian	Misener	3,927-5,977	4,600	1,100	29,500	0	30	60	55,400	1.066	90,600
				18,400	3,200	76,000	10	110	1,920	156,000	1.173	250,640
17	Pennsylvanian	Pennsylvanian	1,258-6,025	1,700	600	17,600	10	20	0	29,800	1.039	49,730
				15,800	3,100	61,300	280	90	2,750	121,000	1.134	201,320
10	Ordovician	Simpson	1,213-6,495	5,600	1,200	24,400	0	0	0	50,900	1.059	82,100
				17,600	3,000	71,900	2	110	440	140,000	1.159	233,042
22	Pennsylvanian	Skinner	1,030-4,567	6,200	1,500	31,700	0	20	30	64,100	1.075	103,540
				18,700	3,200	67,400	10	130	450	139,000	1.157	223,890
22	Pennsylvanian	Booch	1,876-2,300	6,600	1,500	42,500	0	50	0	90,000	1.103	141,050
				12,700	2,500	56,500	240	140	680	117,000	1.131	189,760
22	Siluro-Devonian	Hunton	3,197-5,021	300	80	4,000	0	15	0	8,200	1.012	11,995
				28,900	4,300	75,900	170	660	7,010	142,000	1.155	258,948
27	Pennsylvanian	Red Fork	2,403-4,650	9,700	1,700	42,800	5	3	0	101,000	1.115	155,208
				19,600	2,600	71,700	220	170	370	149,000	1.164	243,660
12	Ordovician	Viola	3,458-5,004	200	60	2,900	0	40	0	4,400	1.005	7,600
				16,000	2,400	62,000	10	940	980	122,000	1.137	204,380
20	Pennsylvanian	Prue	2,267-3,587	8,500	1,300	43,400	5	50	30	86,300	1.110	139,885
				11,700	3,100	72,900	20	120	480	142,000	1.158	230,320
13	Pennsylvanian	Healdton	982-3,163	740	230	10,800	0	20	2	18,600	1.022	30,392
				7,300	2,900	27,900	50	380	40	63,600	1.076	102,170
15	Pennsylvanian	Tonkawa	2,417-3,254	14,000	2,200	23,800	0	0	130	132,000	1.160	172,930
				17,400	3,100	76,400	5	50	370	156,200	1.171	253,525
24	Pennsylvanian	Burbank	790-5,000	10,900	1,800	43,200	2	20	15	99,500	1.109	155,233
				20,000	3,500	69,000	40	130	260	149,000	1.163	241,958
15	Pennsylvanian	Dutcher	1,882-3,218	5,500	900	32,000	0	50	40	45,500	1.073	88,900
				13,900	2,000	54,700	10	130	760	108,000	1.122	179,500
14	Ordovician	Bromide	2,173-7,569	700	400	11,500	10	0	0	19,500	1.024	32,110
				22,400	3,500	80,500	450	500	920	167,000	1.183	275,270

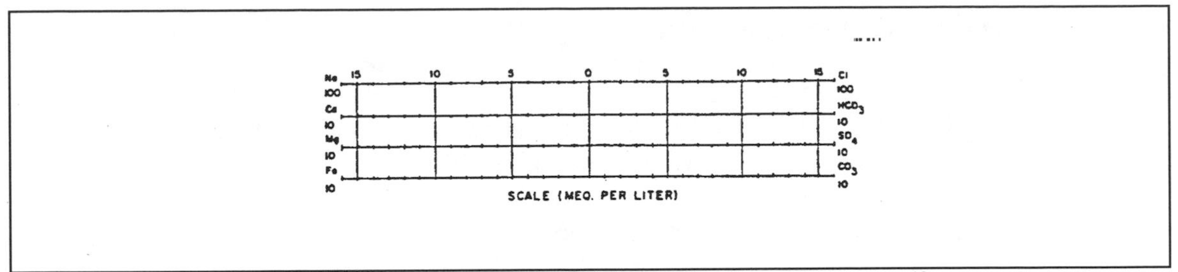

Fig. 3-45. Scaling diagram for graphical presentation of chemical water analysis data (from Stiff[16]). Permission to publish by SPE.

Table 3-5
Table of Milligram Equivalent Conversion Factors for Radicals Found in Water.

Positive Ions	Mg Equiv.*	Negative Ions	Mg Equiv.*
Aluminum, Al	0.1112	Bicarbonate, HCO_3	0.0164
Calcium, Ca	0.0499	Carbonate, CO_3	0.0333
Hydrogen, H	0.9922	Chloride, Cl	0.0282
Iron, Fe	0.0358	Hydroxide, OH	0.0588
Magnesium, Mg	0.0822	Nitrate, NO_3	0.0161
Potassium, K	0.0256	Sulfate, SO_4	0.0208
Sodium, Na	0.0435		
		Alkalinity, $CaCO_3$	0.0200
		Free CO_2, CO_2	0.0454

*The valence of a radical divided by its atomic or molecular weight.

Figure 3-46 shows some common water patterns. The actual scales used can be changed to accentuate particular properties of water in water treatment work. Water analyses taken at different points through a given system or at different times at the same sampling point can be analyzed effectively by this graphic method.

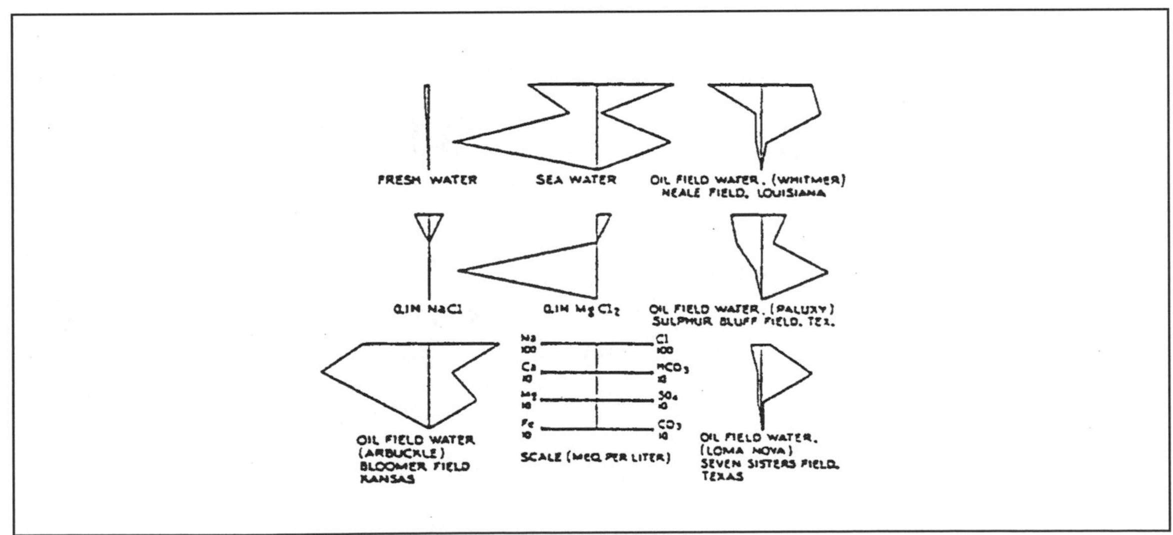

Fig. 3-46. Water analysis patterns for some common waters (from Stiff[16]). Permission to publish by SPE.

B. Physical Properties of Formation Water

1. Density or Specific Gravity

Figure 3-47 provides the density of formation water at standard conditions as a function of total dissolved solids. The density at reservoir conditions can be approximated by dividing by the water formation volume factor (see later section). The resulting number is slightly in error due to the lack of contribution of dissolved gases.

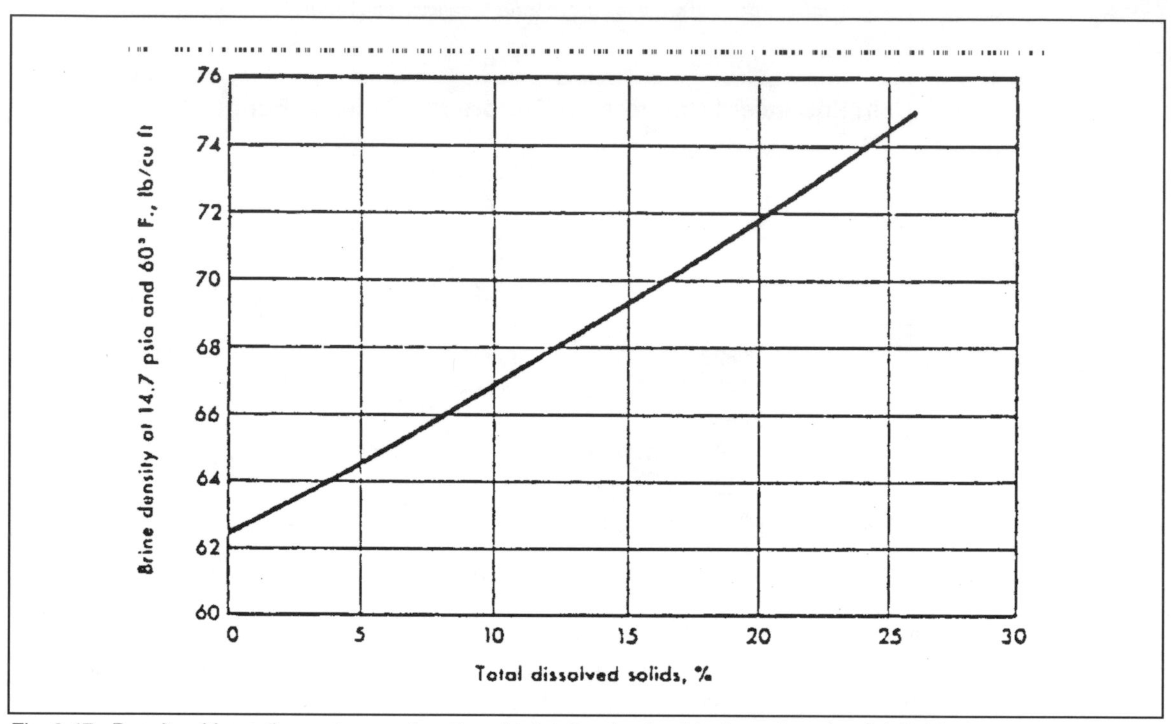

Fig. 3-47. Density of formation water as a function of total dissolved solids (from data, Int. Crit Tables).

2. Solubility of Natural Gas in Water

The solubility of natural gas in water has been determined by Dodson and Standing[17]. Figure 3-48 presents the results. The investigators used a 0.655 specific gravity gas and measured solubilities in fresh water and two brine samples.

With higher gas prices, the industry likely will give gas solubility in formation water more attention. In Japan, gas values support the production of wells for gas saturated water, which is degassed, and the processed water returned to the ocean.

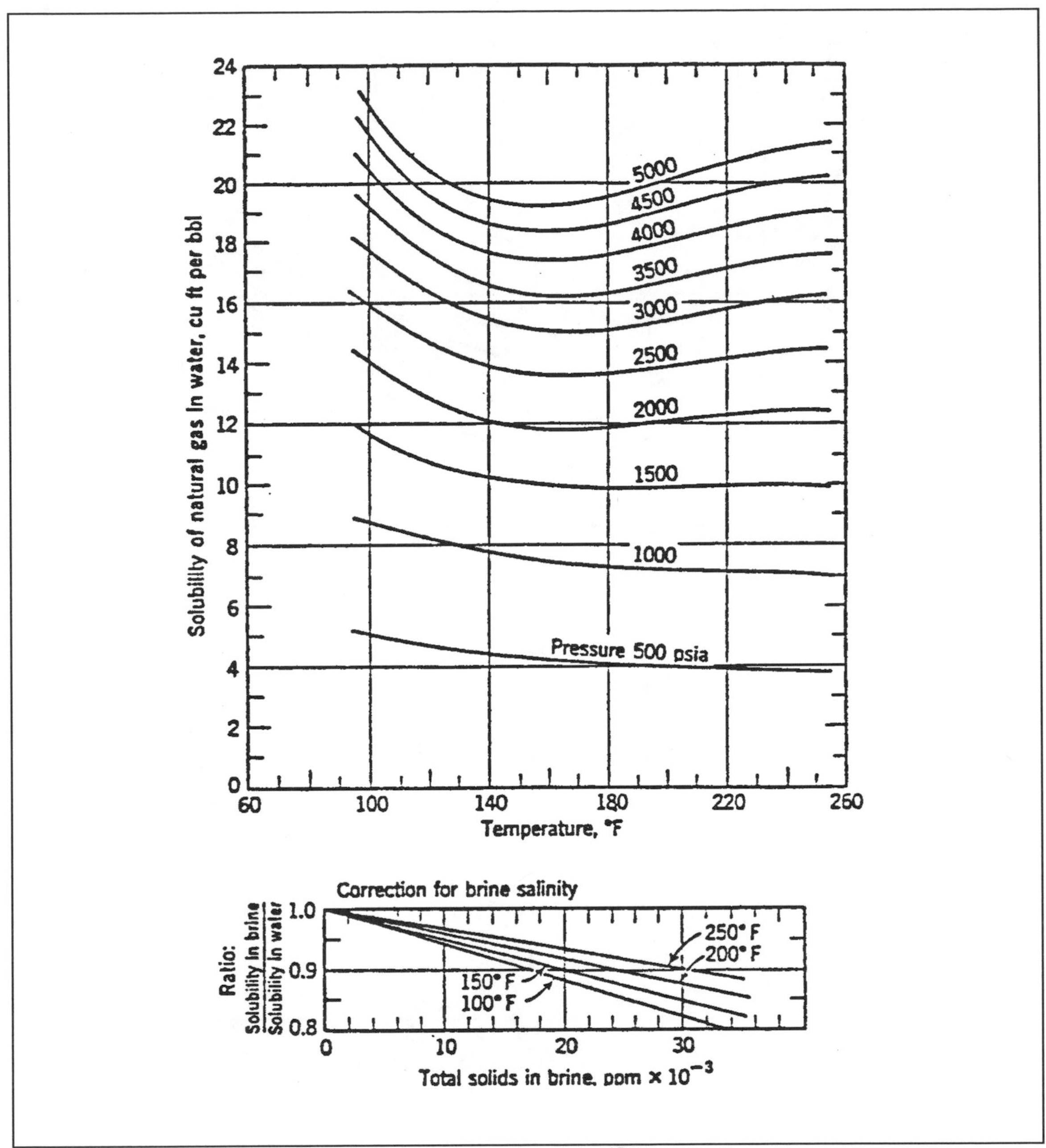

Fig. 3-48. Solubility of natural gas in water (from Dodson and Standing[17]). Permission to publish by API.

3. Compressibility of Water

The compressibility of water is influenced by pressure, temperature, gas in solution, and by dissolved solids. The compressibility of pure water is shown in Figure 3-49 with a correction provided for gas in solution. The influence of dissolved solids can be included when estimating the solubility of natural gas in water, then using this value in the (b) part of Figure 3-49.

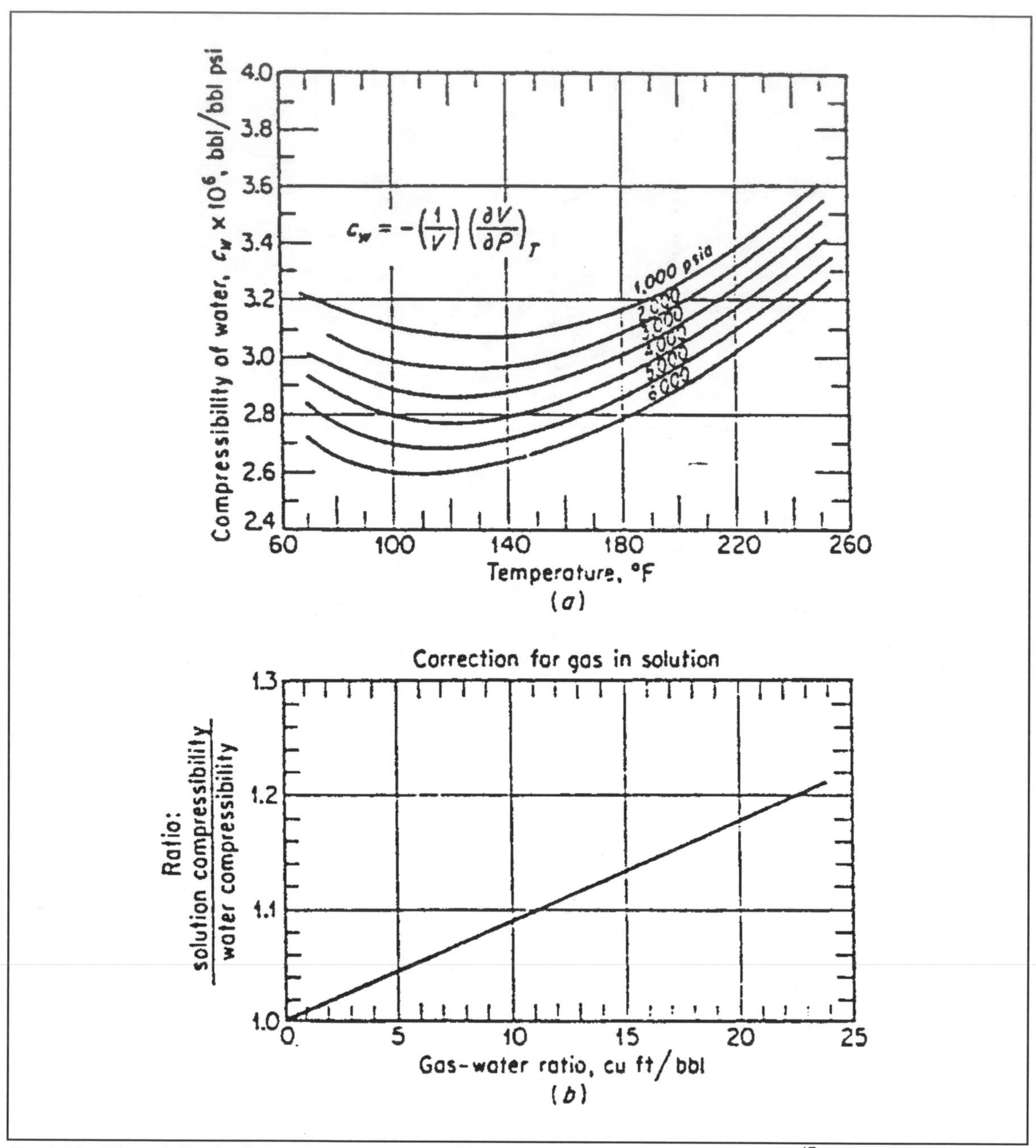

Fig. 3-49. Effect of dissolved gas upon the compressibility of water (from Dodson and Standing[17]). Permission to publish by API.

4. Thermal Expansion of Water

On occasion, the thermal expansion coefficient for water is needed. Figure 3-50 is useful in this regard. Note that with an increase in temperature from 60 to 250 °F, water volume increased by approximately 5.5 percent. In general, over the range of pressures and temperatures of interest, gas solubility and pressure have a small but definite effect upon the thermal expansion of water. The thermal expansion coefficient can be calculated as follows:

$$\beta = \frac{1}{V}\left(\frac{\Delta V}{\Delta T}\right)_p \tag{3-45}$$

where:

β = thermal expansion coefficient of water, 1/°F
V = volume of water, bbl
ΔV = change in volume of water, bbl
ΔT = change in temperature of water, °F

Note that the change of volume with temperature, a slope term, can be determined by drawing a tangent line on Figure 3-50.

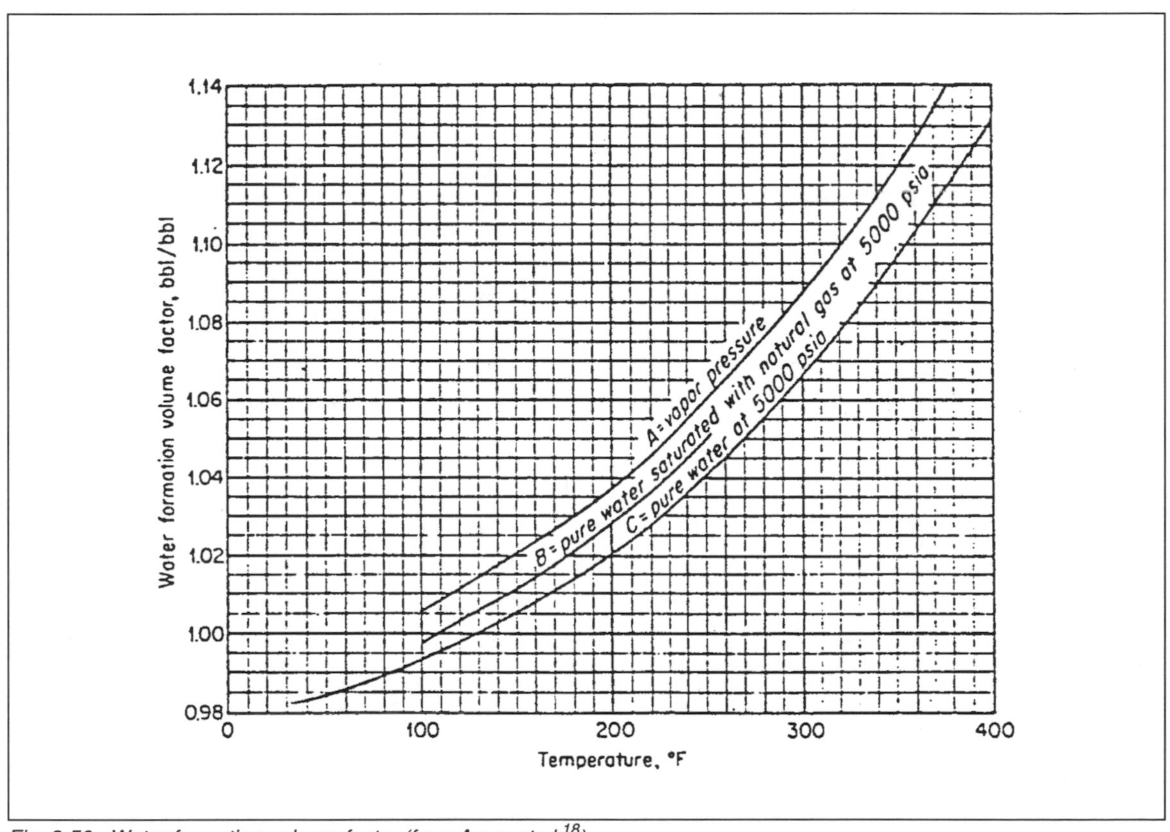

Fig. 3-50. Water formation volume factor (from Amyx et al.[18]).

5. Water Formation Volume Factor

Figure 3-51 provides a typical graph for water formation volume factor with reservoir pressure decline at constant temperature. The graphing anticipates an oil reservoir with a bubblepoint. If this were not the case, the inflection point would not be present. It is worth noting that the factor can have a value of less than unity above the bubblepoint pressure since no additional solution gas is going into solution to partially offset the effect of increased pressure.

The water formation volume factor, B_w, can be estimated using the following equation:

$$B_w = (1 + \Delta V_{wp})(1 + \Delta V_{wT}) \tag{3-46}$$

where:

B_w = water formation volume factor, bbl/STB
ΔV_{wp} = correction for reservoir pressure, dimensionless
ΔV_{wT} = correction for reservoir temperature, dimensionless

McCain[1] reports that use of Equation 3-46, in conjunction with data from Figures 3-52 and 3-53, will yield answers within 1 percent of published experimental data. This is true for oil field brines of widely varying concentrations.

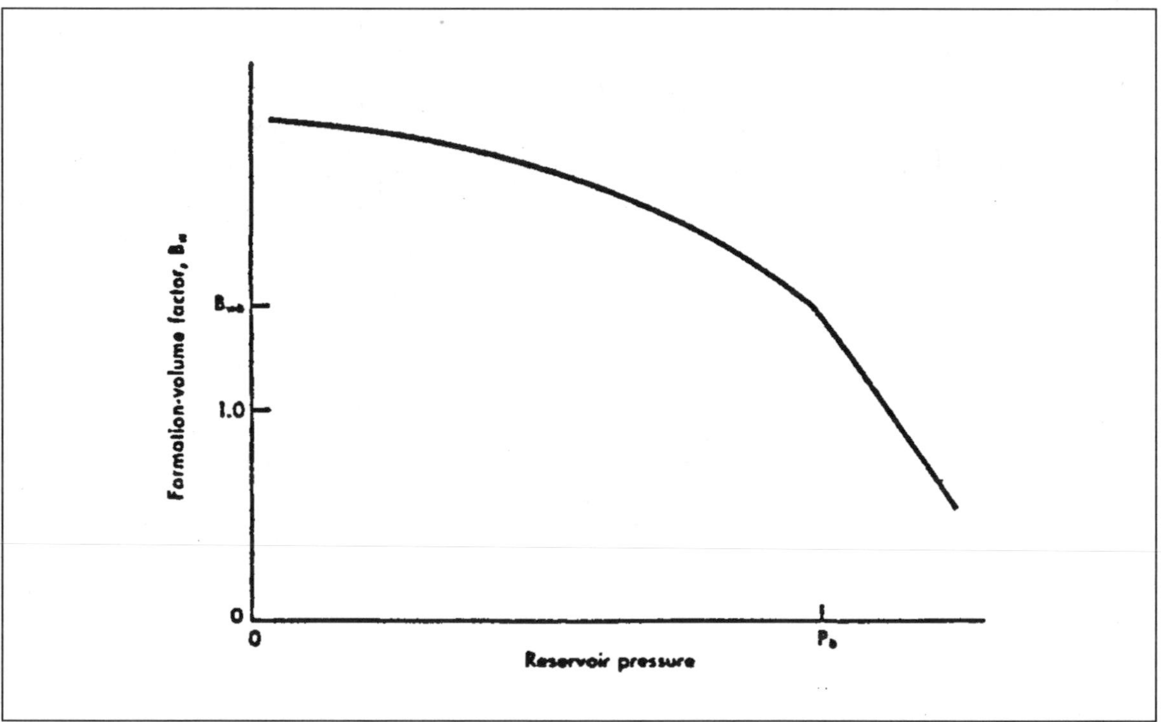

Fig. 3-51. Typical graph of water formation volume factor versus pressure, constant temperature (from McCain[1]). Permission to publish by Petroleum Publishing Company.

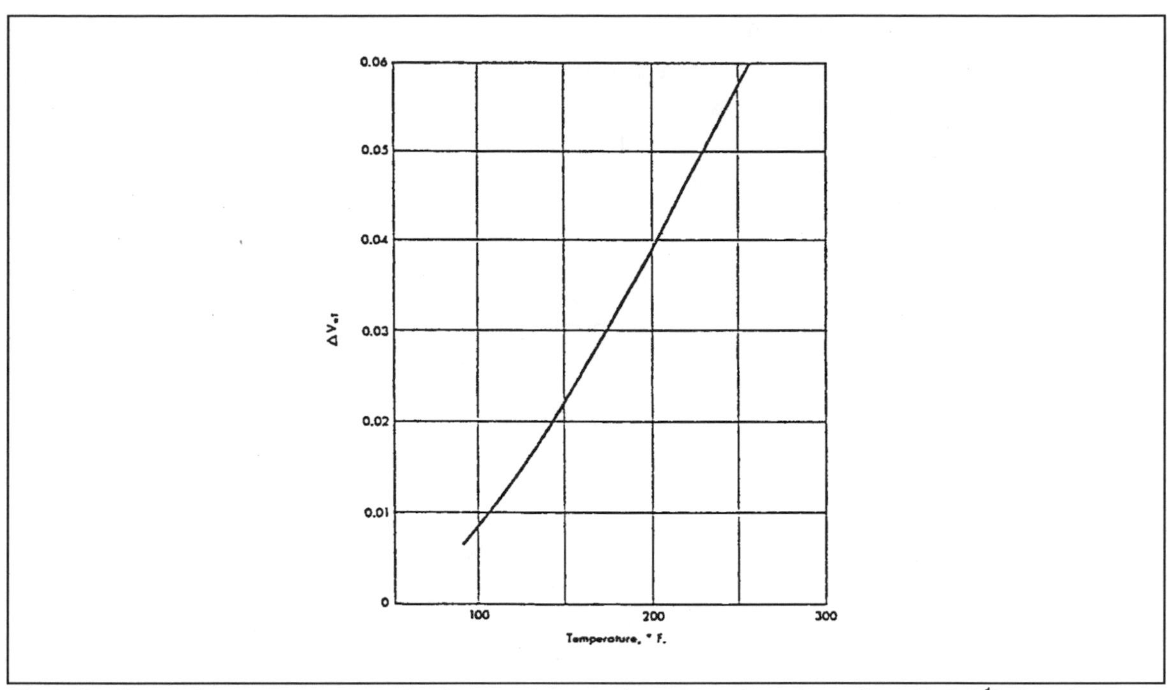

Fig. 3-52. Reservoir temperature correction factor to the water formation volume factor (from McCain[1]). Permission to publish by Petroleum Publishing Company.

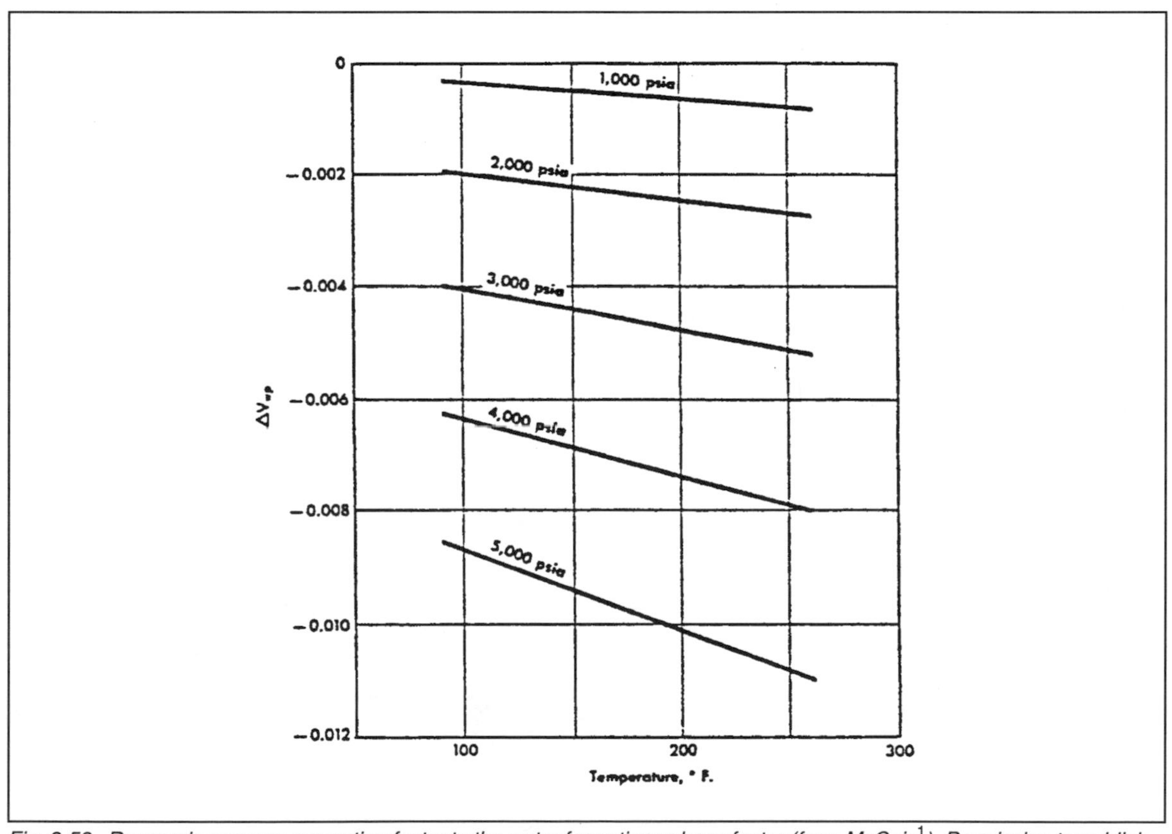

Fig. 3-53. Reservoir pressure correction factor to the water formation volume factor (from McCain[1]). Permission to publish by Petroleum Publishing Company.

6. Viscosity of Formation Water

Information has been presented by van Wingen[19] on the viscosity of a limited number of waters at varying temperatures. This is reproduced as Figure 3-54. A determination of formation water viscosity is also possible from data published by Beal[14]. This is reproduced as Figures 3-55 and 3-56. Although the change of water viscosity is not as wide ranging as that of oil, significant changes do occur over large pressure and temperature ranges.

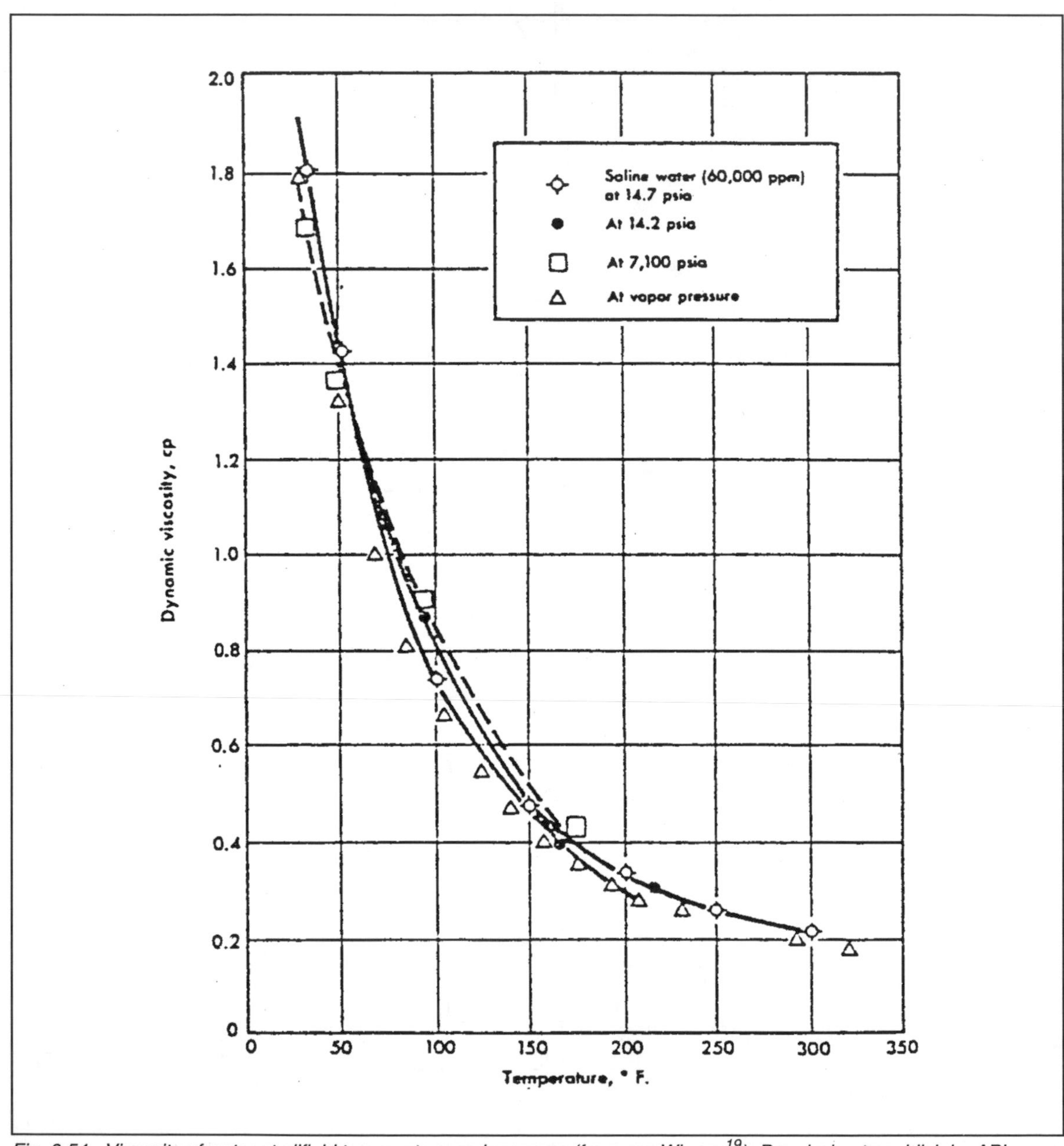

Fig. 3-54. Viscosity of water at oilfield temperature and pressure (from van Wingen[19]). Permission to publish by API.

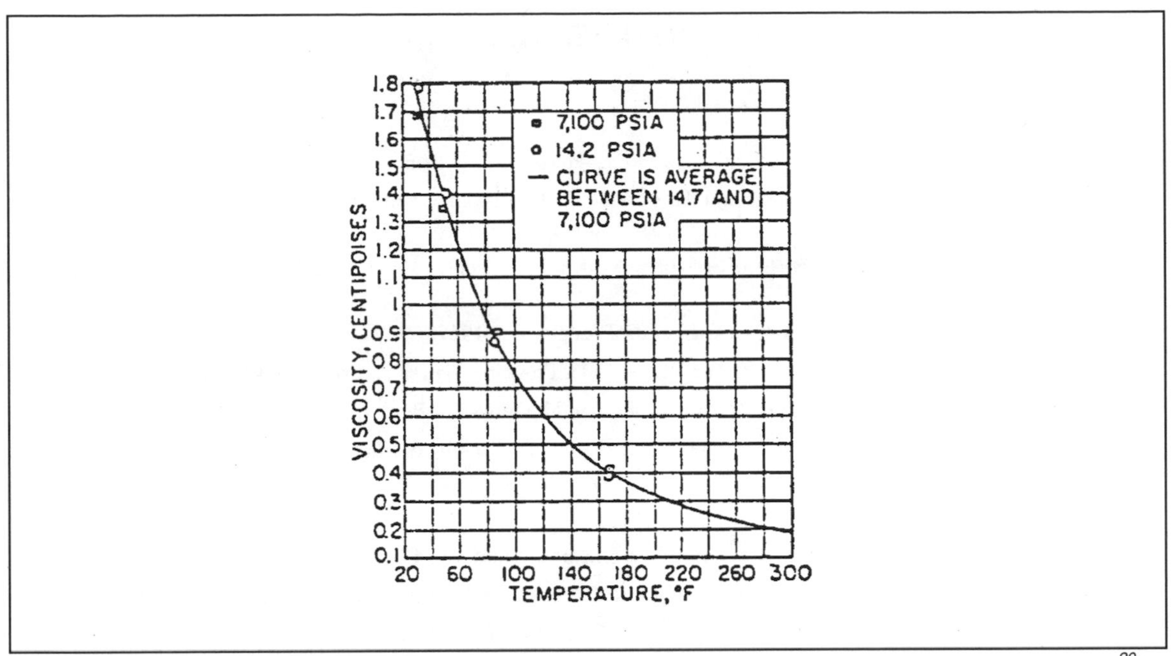

Fig. 3-55. Viscosity of pure water versus temperature. For brines, multiply viscosity by factor from Fig. 3-56 (from Frick[20]). Permission to publish by McGraw-Hill.

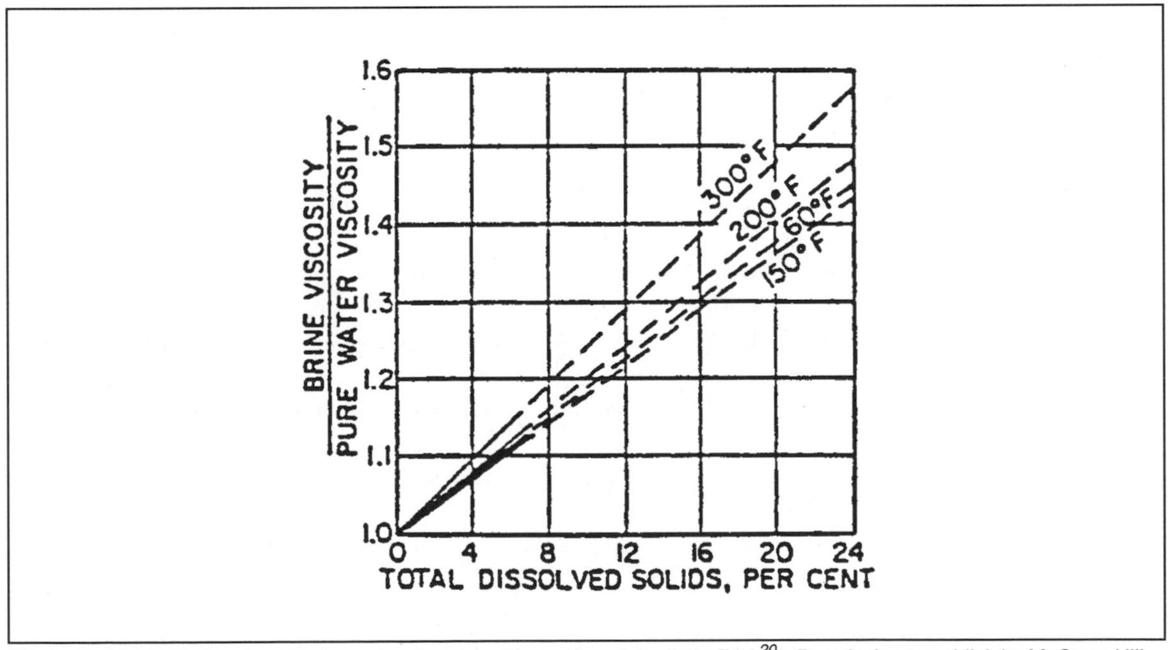

Fig. 3-56. Ratio of brine viscosity to pure water viscosity versus salinity (from Frick[20]). Permission to publish by McGraw-Hill.

REFERENCES

1. McCain, W. D. Jr.: *The Properties of Petroleum Fluids*, Petroleum Publishing Co., Tulsa, OK (1973).
2. Brown, G. G., *et al.*: "Natural Gasoline and the Volatile Hydrocarbons," Natural Gas Assoc. of America, Tulsa, OK (1948).
3. *Engineering Data Book*, Gas Processors Suppliers Association, Chapter 16, Ninth Edition (1972); Fourth Revision (1979) Tulsa, OK.
4. Trube, A. S.: "Compressibility of Natural Gases," *JPT* (September 1957) 69.
5. Carr, N. L., *et al.*: "Viscosity of Hydrocarbon Gases Under Pressure," *Trans.*, AIME (1954) **201**, 264.
6. Katz, D. L., *et al.*: *Handbook of Natural Gas Engineering*, McGraw-Hill Book Co., Inc., New York City (1959).
7. Katz, D. L. and Cornell, D.: "Flow of Natural Gas from Reservoirs," Univ. of Mich. Publ. Services, Ann Arbor (1955).
8. Standing, M. B.: *Volumetric and Phase Behavior of Oil Field Hydrocarbon Systems*, Reinhold Publ. Corp., New York (1952).
9. Trube, A. S.: "Compressibility of Undersaturated Hydrocarbon Reservoir Fluids," *Trans.*, AIME (1957) **201**.
10. Gatlin, C.: *Petroleum Engineering—Drilling and Well Completions*, Prentice-Hall, Inc., Englewood Cliffs, NJ (1960).
11. Clark, N. J.,: "Elements of Petroleum Reservoirs," SPE of AIME (1960).
12. Borden, G. Jr. and Rzasa, M. J.: "Correlation of Bottomhole Sample Data," *Trans.*, AIME (1950) **189**, 345.
13. Lasater, J. A.: "Bubblepoint Pressure Correlation," *Trans.*, AIME (1958) **213**, 379.
14. Beal, C. : "The Viscosity of Air, Water, Natural Gas, Crude Oil and Its Associated Gases at Oil Field Temperatures and Pressures," *Trans.*, AIME (1946) **165**, 94.
15. Chew, Ju-Nam and Connally, C. A.: "A Viscosity Correlation for Gas-Saturated Crude Oils," *Trans.*, AIME (1959) **216**, 23.
16. Stiff, H. A., Jr.: "The Interpretation of Chemical Water Analysis by Means of Pattern," *Trans.*, AIME (1951) **192**.
17. Dodson, C. R. and Standing, M. B.: "Pressure-Volume-Temperature and Solubility Relations for Natural Gas-Water Mixtures," *Drilling & Prod. Prac.*, API (1944).
18. Amyx, J. W., *et al.*: *Petroleum Reservoir Engineering — Physical Properties*, McGraw-Hill Book Co., Inc., New York City (1960).
19. van Wingen, N.: "Viscosity of Air, Water, Natural Gas, and Crude Oil at Varying Pressures and Temperatures," Secondary Recovery of Oil in the United States, API (1950) Chapter 6.
20. Frick, T. C.: "Petroleum Production Handbook," Vol. II, *Reservoir Engineering*, McGraw-Hill Book Co., Inc., New York City (1962).
21. Brown, G. G., Katz, D. L., Oberfell, G. G., and Alden, R. C.: "Natural Gasoline and the Volatile Hydrocarbons." Sponsored by NGAA, Tulsa, OK (1948).
22. Carr, N. L., Kobayashi, R., and Burrows, D. B.: "Viscosity of Hydrocarbon Gases under Pressure," *Trans.*, AIME (1954) **201**, 264-272.
23. Calhoun, J. C. Jr.: "Fundamentals of Reservoir Engineering", *Oil and Gas J.*, Univ. of Okla. Press., Norman, OK. (1953).
24. Burcik, E. J.: *Properties of Petroleum Reservoir Fluids*, John Wiley & Sons, Inc., New York (1957).
25. Vasquez, M., and Beggs, H. D.: "Correlations for Fluid Physical Property Prediction," paper SPE 6719 presented at the 1977 Annual Technical Conference and Exhibition, Denver (Oct. 9-12).
26. Garb, F. A.: "Waterflood Calculations for Hand-Held Computers," H. J. Gruy and Assoc., Inc. (1982).
27. Beggs, H. D. and Robinson, J. F.: "Estimating the Viscosity of Crude Oil Systems," *JPT* (September 1975) 1140-1141.
28. Wichert, E., and Aziz, K.: "Calculate Z's for Sour Gases," *Hydrocarbon Processing*, (1972) **51**.
29. Mattar, L., Brar, G. S., and Aziz, K.: "Compressibility of Natural Gases," *J. Cdn. Pet. Tech.* (October-December 1975).

30. Lee, A. L., Gonzalez, M. H., and Eakin, B. E.: "The Viscosity of Natural Gases," *JPT* (August 1966).

31. Moses, Phillip L.: "Engineering Applications of Phase Behavior of Crude Oil and Condensate Systems," *JPT* (July 1986) 715-723.

32. Novosad, Z. and Costain, T.G.: "Mechanisms of Miscibility Development in Hydrocarbon Gas Drives: New Interpretation," SPE *Reservoir Engineering*, (Aug., 1989) 341-347.

33. Meehan, N.: " Program Determines Gas Constants," *Oil and Gas Journal,* (Nov. 24, 1980) 140-141.

34. Bloomer, *et al.*, Research Bulletin 22, Institute of Gas Technology, Chicago, Ill.(1953).

Chapter 3

Appendix A

Example Hydrocarbon Liquid Fluid Properties Report

Reservoir Fluid Study for GOOD OIL COMPANY

Oil Well No. 4
Productive Field
Samson County, Texas

Courtesy of:
Core Laboratories, Inc.
Dallas, Texas

Applied Reservoir Engineering
Chapter 3, Appendix A

CORE LABORATORIES, INC.
Petroleum Reservoir Engineering
DALLAS, TEXAS

Page 1 of 15
File RFL 76000

Company	Good Oil Company	Date Sampled	
Well	Oil Well No. 4	County	Samson
Field	Productive	State	Texas

FORMATION CHARACTERISTICS

Formation Name	Cretaceous
Date First Well Completed	______, 19___
Original Reservoir Pressure	4100 PSIG @ 8692 Ft.
Original Produced Gas-Oil Ratio	600 SCF/Bbl
Production Rate	300 Bbl/Day
Separator Pressure and Temperature	200 PSIG. 75 °F.
Oil Gravity at 60° F.	°API
Datum	8000 Ft. Subsea
Original Gas Cap	No

WELL CHARACTERISTICS

Elevation	610 Ft.
Total Depth	8943 Ft.
Producing Interval	8684-8700 Ft.
Tubing Size and Depth	2-7/8 In. to 8600 Ft.
Productivity Index	1.1 Bbl/D/PSI @ 300 Bbl/Day
Last Reservoir Pressure	3954* PSIG @ 8500 Ft.
Date	______, 19___
Reservoir Temperature	217* °F. @ 8500 Ft.
Status of Well	Shut in 72 hours
Pressure Gauge	Amerada
Normal Production Rate	300 Bbl/Day
Gas-Oil Ratio	600 SCF/Bbl
Separator Pressure and Temperature	200 PSIG, 75 °F.
Base Pressure	14.65 PSIA
Well Making Water	None % Cut

SAMPLING CONDITIONS

Sampled at	8500 Ft.
Status of Well	Shut in 72 hours
Gas-Oil Ratio	SCF/Bbl
Separator Pressure and Temperature	PSIG, °F.
Tubing Pressure	1400 PSIG
Casing Pressure	PSIG
Sampled by	
Type Sampler	Wofford

REMARKS:

* Pressure and temperature extrapolated to the mid-point of the producing interval = 4010 PSIG and 220°F.

CORE LABORATORIES, INC.
Petroleum Reservoir Engineering
DALLAS, TEXAS

Page 2 of 15

File RFL 76000

Company Good Oil Company Formation Cretaceous

Well Oil Well No. 4 County Samson

Field Productive State Texas

HYDROCARBON ANALYSIS OF Reservoir Fluid SAMPLE

COMPONENT	MOL PERCENT	WEIGHT PERCENT	DENSITY @ 60° F. GRAMS PER CUBIC CENTIMETER	° API @ 60° F.	MOLECULAR WEIGHT
Hydrogen Sulfide	Nil	Nil			
Carbon Dioxide	0.91	0.43			
Nitrogen	0.16	0.05			
Methane	36.47	6.24			
Ethane	9.67	3.10			
Propane	6.95	3.27			
iso-Butane	1.44	0.89			
n-Butane	3.93	2.44			
iso-Pentane	1.44	1.11			
n-Pentane	1.41	1.09			
Hexanes	4.33	3.97			
Heptanes plus	33.29	77.41	0.8515	34.5	218
	100.00	100.00			

Applied Reservoir Engineering
Chapter 3, Appendix A

CORE LABORATORIES, INC.
Petroleum Reservoir Engineering
DALLAS, TEXAS

Page 3 of 15
File RFL 76000
Well Oil Well No. 4

VOLUMETRIC DATA OF Reservoir Fluid SAMPLE

1. Saturation pressure (bubble-point pressure) 2620 PSIG @ 220 °F.
2. Specific volume at saturation pressure: ft³/lb 0.02441 @ 220 °F.
3. Thermal expansion of saturated oil @ 5000 PSI = $\frac{V @ 220\ °F}{V @ 76\ °F}$ = 1.08790
4. Compressibility of saturated oil @ reservoir temperature: Vol/Vol/PSI:

From	PSI to	PSI =
5000	4000	13.48×10^{-6}
4000	3000	15.88×10^{-6}
3000	2620	18.75×10^{-6}

Applied Reservoir Engineering
Chapter 3, Appendix A

CORE LABORATORIES, INC.
Petroleum Reservoir Engineering
DALLAS, TEXAS

File RFL 76000
Well Oil Well No. 4

Pressure-Volume Relations at 220 °F.

Pressure PSIG	Relative Volume(1)	Y Function(2)
5000	0.9639	
4500	0.9703	
4000	0.9771	
3500	0.9846	
3000	0.9929	
2900	0.9946	
2800	0.9964	
2700	0.9983	
2620	1.0000	
2605	1.0022	2.574
2591	1.0041	2.688
2516	1.0154	2.673
2401	1.0350	2.593
2253	1.0645	2.510
2090	1.1040	2.422
1897	1.1633	2.316
1698	1.2426	2.219
1477	1.3618	2.118
1292	1.5012	2.028
1040	1.7802	1.920
830	2.1623	1.823
640	2.7513	1.727
472	3.7226	1.621

(1) Relative Volume: V/Vsat is barrels at indicated pressure per barrel at saturation pressure.

(2) $\text{Y Function} = \dfrac{(\text{Psat}-\text{P})}{(\text{Pabs})\ (\text{V/Vsat}-1)}$

CORE LABORATORIES, INC.
Petroleum Reservoir Engineering
DALLAS, TEXAS

Page 5 of 15
File RFL 76000
Well Oil Well No. 4

Differential Vaporization at 220 °F.

Pressure PSIG	Solution Gas/Oil Ratio(1)	Relative Oil Volume(2)	Relative Total Volume(3)	Oil Density gm/cc	Deviation Factor Z	Gas Formation Volume Factor(4)	Incremental Gas Gravity
2620	854	1.600	1.600	0.6562			
2350	763	1.554	1.665	0.6655	0.846	0.00685	0.825
2100	684	1.515	1.748	0.6731	0.851	0.00771	0.818
1850	612	1.479	1.859	0.6808	0.859	0.00882	0.797
1600	544	1.445	2.016	0.6889	0.872	0.01034	0.791
1350	479	1.412	2.244	0.6969	0.887	0.01245	0.794
1100	416	1.382	2.593	0.7044	0.903	0.01552	0.809
850	354	1.351	3.169	0.7121	0.922	0.02042	0.831
600	292	1.320	4.254	0.7198	0.941	0.02931	0.881
350	223	1.283	6.975	0.7291	0.965	0.05065	0.988
159	157	1.244	14.693	0.7382	0.984	0.10834	1.213
0	0	1.075		0.7892			2.039
	@ 60°F. =	1.000					

Gravity of residual oil = 35.1°API @ 60°F.

(1) Cubic feet of gas at 14.65 psia and 60°F. per barrel of residual oil at 60°F.
(2) Barrels of oil at indicated pressure and temperature per barrel of residual oil at 60°F.
(3) Barrels of oil plus liberated gas at indicated pressure and temperature per barrel of residual oil at 60°F.
(4) Cubic feet of gas at indicated pressure and temperature per cubic foot at 14.65 psia and 60°F.

Applied Reservoir Engineering
Chapter 3, Appendix A

CORE LABORATORIES, INC.
Petroleum Reservoir Engineering
DALLAS, TEXAS 75207

Page 6 of 15
File RFL 76000
Well Oil Well No. 4

Viscosity Data at 220 °F.

Pressure PSIG	Oil Viscosity Centipoise	Calculated Gas Viscosity Centipoise	Oil/Gas Viscosity Ratio
5000	0.450		
4500	0.434		
4000	0.418		
3500	0.401		
3000	0.385		
2800	0.379		
2620	0.373		
2350	0.396	0.0191	20.8
2100	0.417	0.0180	23.2
1850	0.442	0.0169	26.2
1600	0.469	0.0160	29.4
1350	0.502	0.0151	33.2
1100	0.542	0.0143	37.9
850	0.592	0.0135	43.9
600	0.654	0.0126	51.8
350	0.738	0.0121	60.9
159	0.855	0.0114	75.3
0	1.286	0.0093	137.9

Applied Reservoir Engineering
Chapter 3, Appendix A

CORE LABORATORIES, INC.
Petroleum Reservoir Engineering
DALLAS, TEXAS

Page 7 of 15

File RFL 76000

Well Oil Well No. 4

SEPARATOR TESTS OF Reservoir Fluid SAMPLE

SEPARATOR PRESSURE, PSI GAUGE	SEPARATOR TEMPERATURE, °F.	GAS/OIL RATIO (1)	GAS/OIL RATIO (2)	STOCK TANK GRAVITY, °API @ 60° F.	FORMATION VOLUME FACTOR (3)	SEPARATOR VOLUME FACTOR (4)	SPECIFIC GRAVITY OF FLASHED GAS
50	75	715	737			1.031	0.840
to							
0	75	41	41	40.5	1.481	1.007	1.338
100	75	637	676			1.062	0.786
to							
0	75	91	92	40.7	1.474	1.007	1.363
200	75	542	602			1.112	0.732
to							
0	75	177	178	40.4	1.483	1.007	1.329
300	75	478	549			1.148	0.704
to							
0	75	245	246	40.1	1.495	1.007	1.286

(1) Gas/Oil Ratio in cubic feet of gas @ 60° F. and 14.65 PSI absolute per barrel of oil @ indicated pressure and temperature.
(2) Gas/Oil Ratio in cubic feet of gas @ 60° F. and 14.65 PSI absolute per barrel of stock tank oil @ 60° F.
(3) Formation Volume Factor is barrels of saturated oil @ 2620 PSI gauge and 220 ° F. per barrel of stock tank oil @ 60° F.
(4) Separator Volume Factor is barrels of oil @ indicated pressure and temperature per barrel of stock tank oil @ 60° F.

Applied Reservoir Engineering
Chapter 3, Appendix A

CORE LABORATORIES, INC.
Petroleum Reservoir Engineering
DALLAS, TEXAS

Page 8 of 15

File RFL 76000

Company: Good Oil Company — Formation: Cretaceous

Well: Oil Well No. 4 — County: Samson

Field: Productive — State: Texas

HYDROCARBON ANALYSIS OF Separator GAS SAMPLE

COMPONENT	MOL PERCENT	GPM
Hydrogen Sulfide	Nil	
Carbon Dioxide	1.62	
Nitrogen	0.30	
Methane	67.00	
Ethane	16.04	4.265
Propane	8.95	2.449
iso-Butane	1.29	0.420
n-Butane	2.91	0.912
iso-Pentane	0.53	0.193
n-Pentane	0.41	0.155
Hexanes	0.44	0.178
Heptanes plus	0.49	0.221
	100.00	8.793

Calculated gas gravity (air = 1.000) = 0.840

Calculated gross heating value = 1405 BTU
per cubic foot of dry gas at 14.65 psia at 60° F.

Collected at 50 psig and 75 ° F. in the laboratory.

Applied Reservoir Engineering
Chapter 3, Appendix A

CORE LABORATORIES, INC.
Petroleum Reservoir Engineering
DALLAS, TEXAS

Page 9 of 15

File RFL 76000

Company Good Oil Company — Formation Cretaceous

Well Oil Well No. 4 — County Samson

Field Productive — State Texas

HYDROCARBON ANALYSIS OF Separator GAS SAMPLE

COMPONENT	MOL PERCENT	GPM
Hydrogen Sulfide	Nil	
Carbon Dioxide	1.67	
Nitrogen	0.32	
Methane	71.08	
Ethane	15.52	4.127
Propane	7.36	2.014
iso-Butane	0.92	0.299
n-Butane	1.98	0.621
iso-Pentane	0.33	0.120
n-Pentane	0.26	0.094
Hexanes	0.27	0.110
Heptanes plus	0.29	0.131
	100.00	7.516

Calculated gas gravity (air = 1.000) = 0.786

Calculated gross heating value = 1321 BTU
per cubic foot of dry gas at 14.65 **psia at 60° F.**

Collected at 100 **psig and** 75 **° F.** in the laboratory.

CORE LABORATORIES, INC.
Petroleum Reservoir Engineering
DALLAS, TEXAS

Page 10 of 15

File RFL 76000

Company Good Oil Company — Formation Cretaceous

Well Oil Well No. 4 — County Samson

Field Productive — State Texas

HYDROCARBON ANALYSIS OF Separator GAS SAMPLE

COMPONENT	MOL PERCENT	GPM
Hydrogen Sulfide	Nil	
Carbon Dioxide	1.68	
Nitrogen	0.36	
Methane	76.23	
Ethane	13.94	3.707
Propane	5.31	1.453
iso-Butane	0.57	0.185
n-Butane	1.21	0.379
iso-Pentane	0.20	0.073
n-Pentane	0.16	0.058
Hexanes	0.16	0.065
Heptanes plus	0.18	0.081
	100.00	6.001

Calculated gas gravity (air = 1.000) = 0.732

Calculated gross heating value = 1236 BTU
per cubic foot of dry gas at 14.65 psia at 60° F.

Collected at 200 psig and 75 ° F. in the laboratory.

Applied Reservoir Engineering
Chapter 3, Appendix A

CORE LABORATORIES, INC.
Petroleum Reservoir Engineering
DALLAS, TEXAS

Page 11 of 15

File RFL 76000

Company Good Oil Company — Formation Cretaceous

Well Oil Well No. 4 — County Samson

Field Productive — State Texas

HYDROCARBON ANALYSIS OF Separator GAS SAMPLE

COMPONENT	MOL PERCENT	GPM
Hydrogen Sulfide	Nil	
Carbon Dioxide	1.65	
Nitrogen	0.39	
Methane	79.42	
Ethane	12.48	3.318
Propane	4.21	1.152
iso-Butane	0.43	0.140
n-Butane	0.90	0.282
iso-Pentane	0.15	0.055
n-Pentane	0.12	0.043
Hexanes	0.12	0.049
Heptanes plus	0.13	0.059
	100.00	5.098

Calculated gas gravity (air = 1.000) = 0.704

Calculated gross heating value = 1192 **BTU**
per cubic foot of dry gas at 14.65 **psia at 60° F.**

Collected at 300 **psig and** 75 **° F.** in the laboratory.

Core Laboratories, Inc.

Manager
Reservoir Fluid Analysis

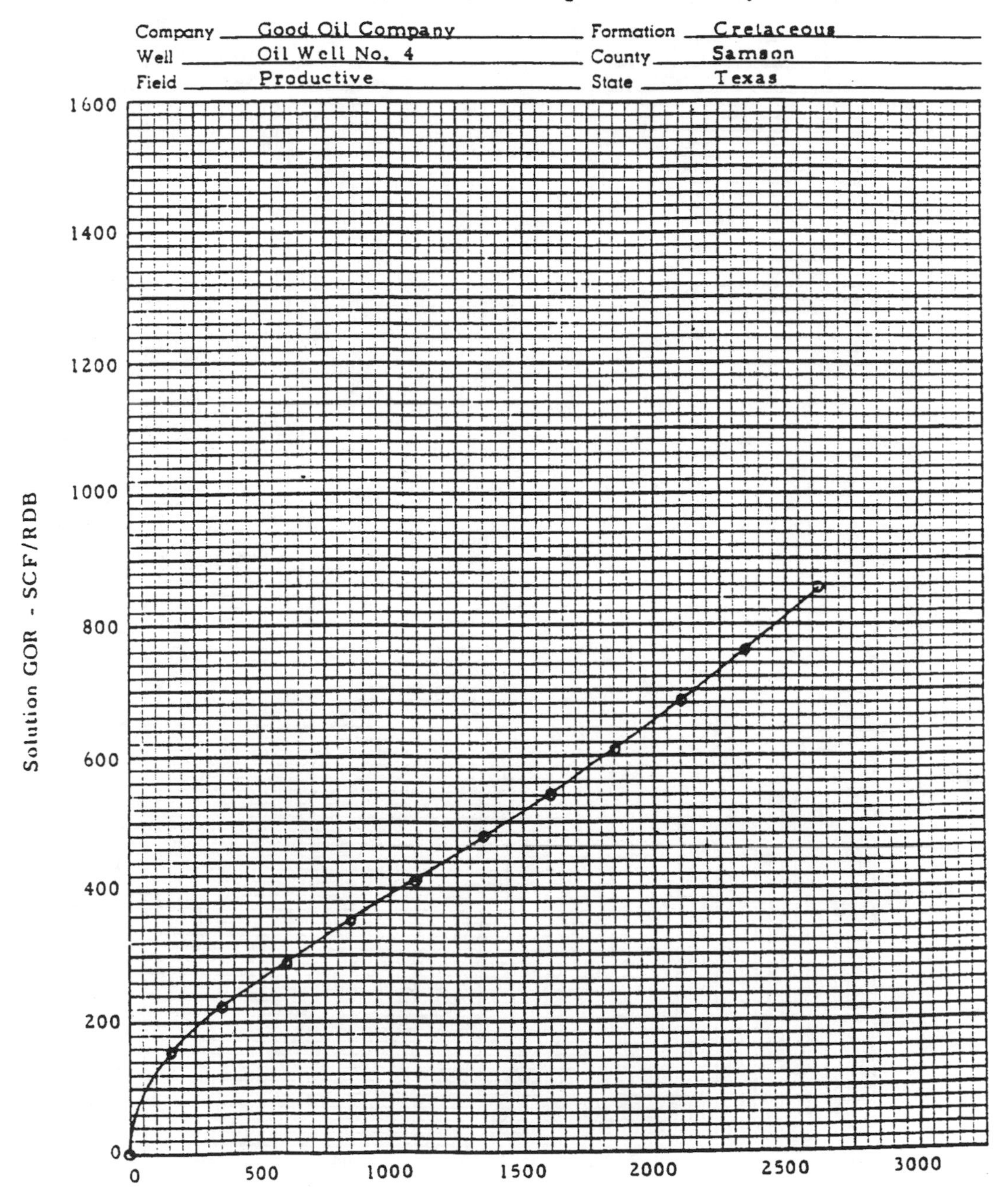
Applied Reservoir Engineering
Chapter 3, Appendix A
CORE LABORATORIES, INC.
Petroleum Reservoir Engineering
DALLAS, TEXAS
Page 12 of 15
File RFL 76000
Solution Gas/Oil Ratio During Differential Vaporization
Company Good Oil Company
Well Oil Well No. 4
Field Productive
Formation Cretaceous
County Samson
State Texas
Solution GOR - SCF/RDB
1600
1400
1200
1000
800
600
400
200
0
Pressure - PSIG
0
500
1000
1500
2000
2500
3000

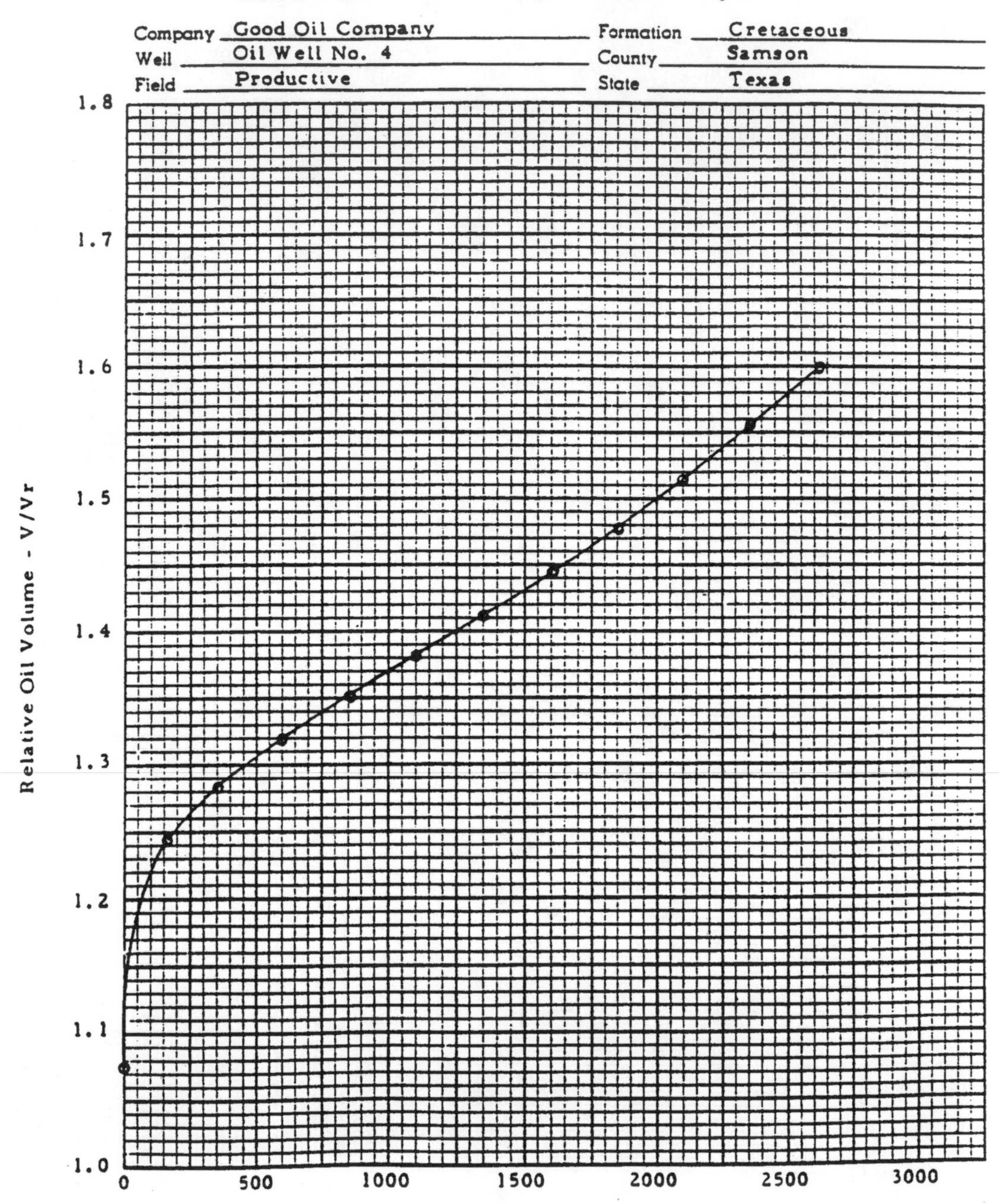
Applied Reservoir Engineering
Chapter 3, Appendix A
CORE LABORATORIES, INC.
Petroleum Reservoir Engineering
DALLAS, TEXAS
Page 13 of 15
File RFL 76000
Relative Oil Volume During Differential Vaporization
Company Good Oil Company
Formation Cretaceous
Well Oil Well No. 4
County Samson
Field Productive
State Texas
Relative Oil Volume - V/Vr
1.8
1.7
1.6
1.5
1.4
1.3
1.2
1.1
1.0
0
500
1000
1500
2000
2500
3000
Pressure - PSIG

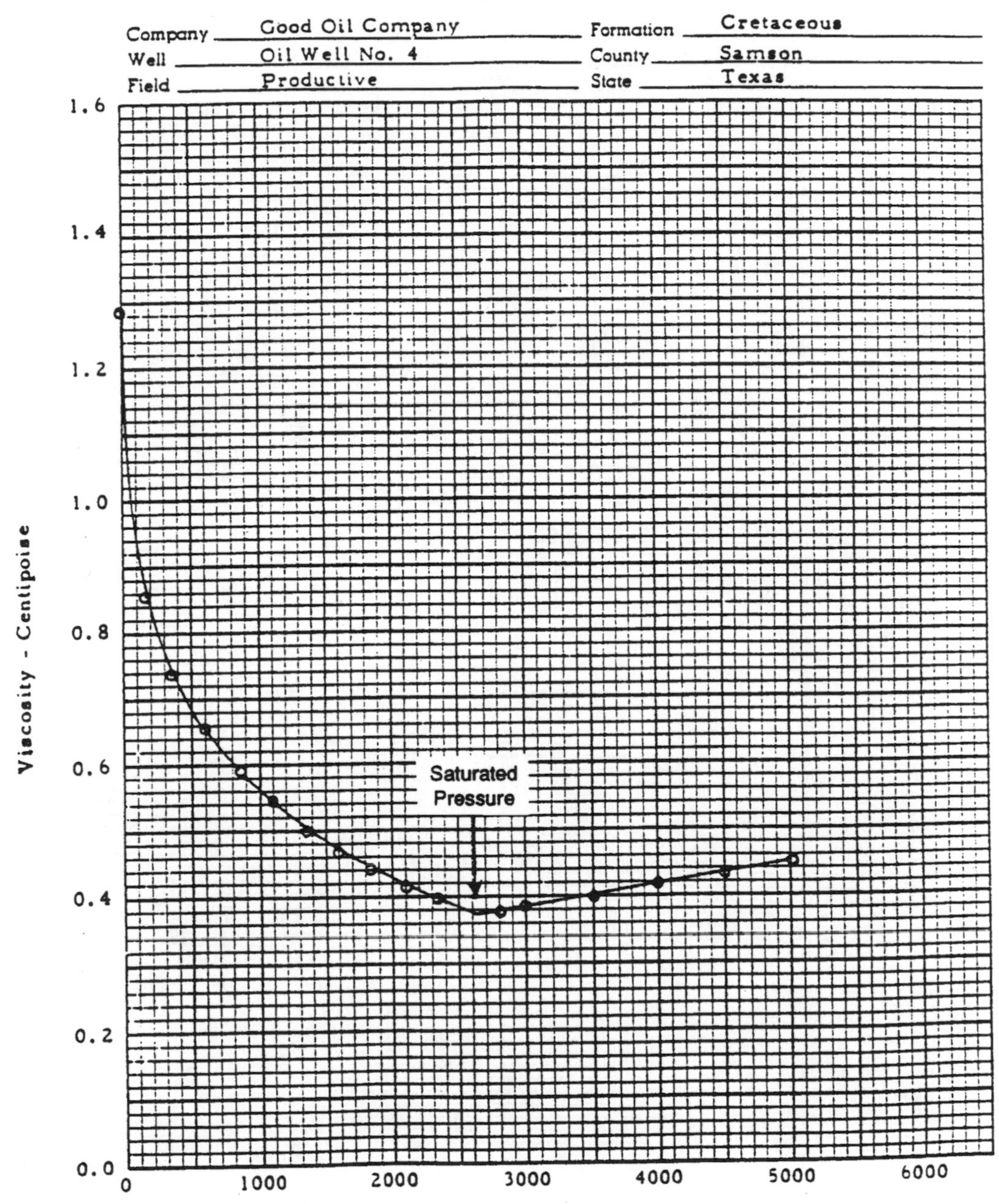
Applied Reservoir Engineering
Chapter 3, Appendix A
CORE LABORATORIES, INC.
Petroleum Reservoir Engineering
DALLAS, TEXAS
Page 14 of 15
File RFL 76000
Viscosity of Reservoir Fluid
Company Good Oil Company
Formation Cretaceous
Well Oil Well No. 4
County Samson
Field Productive
State Texas
Saturated Pressure
Viscosity - Centipoise
1.6
1.4
1.2
1.0
0.8
0.6
0.4
0.2
0.0
0
1000
2000
3000
4000
5000
6000
Pressure - PSIG

Applied Reservoir Engineering
Chapter 3, Appendix A

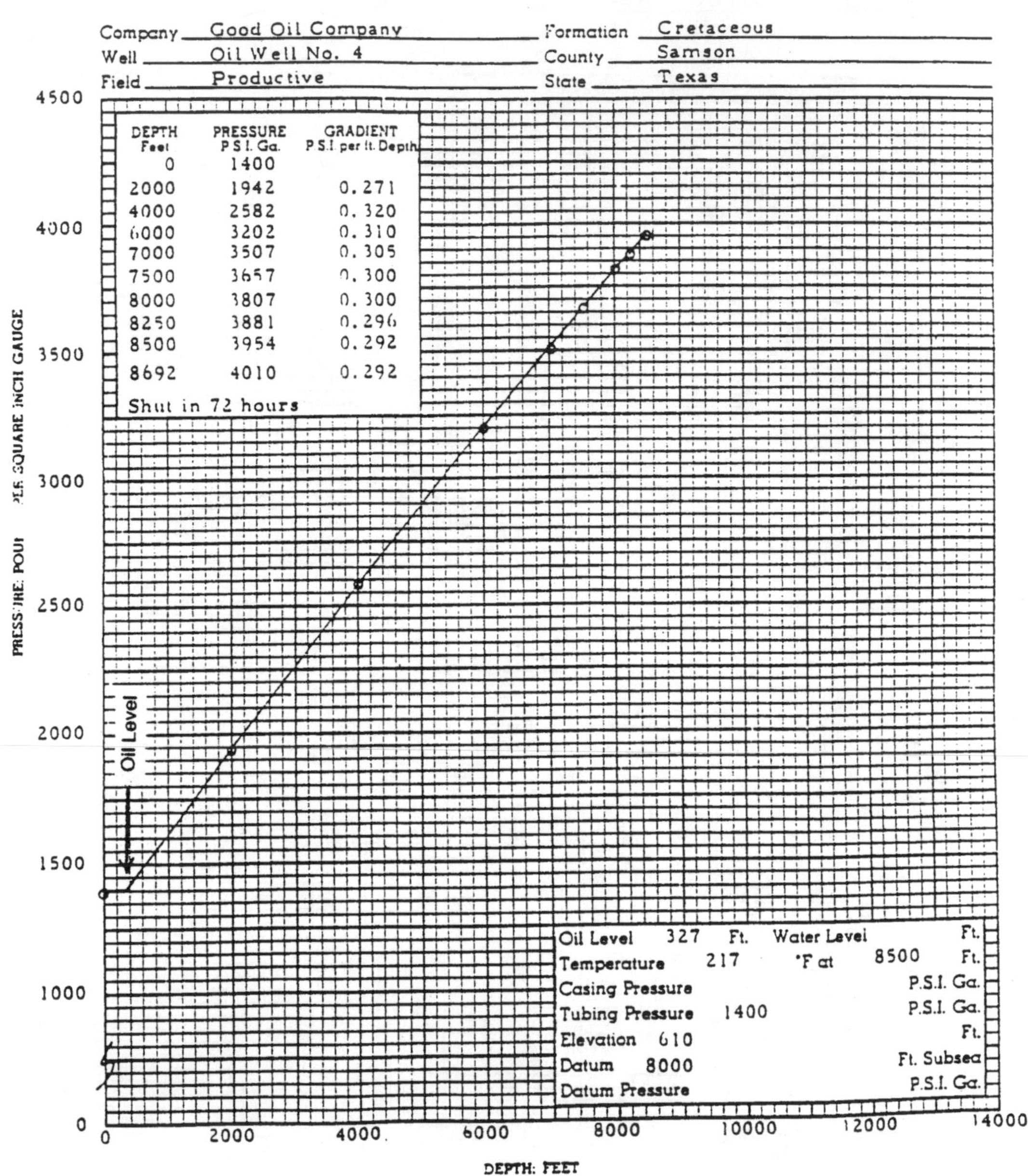

Chapter 3

Appendix B

Standing Oil Fluid Property Charts

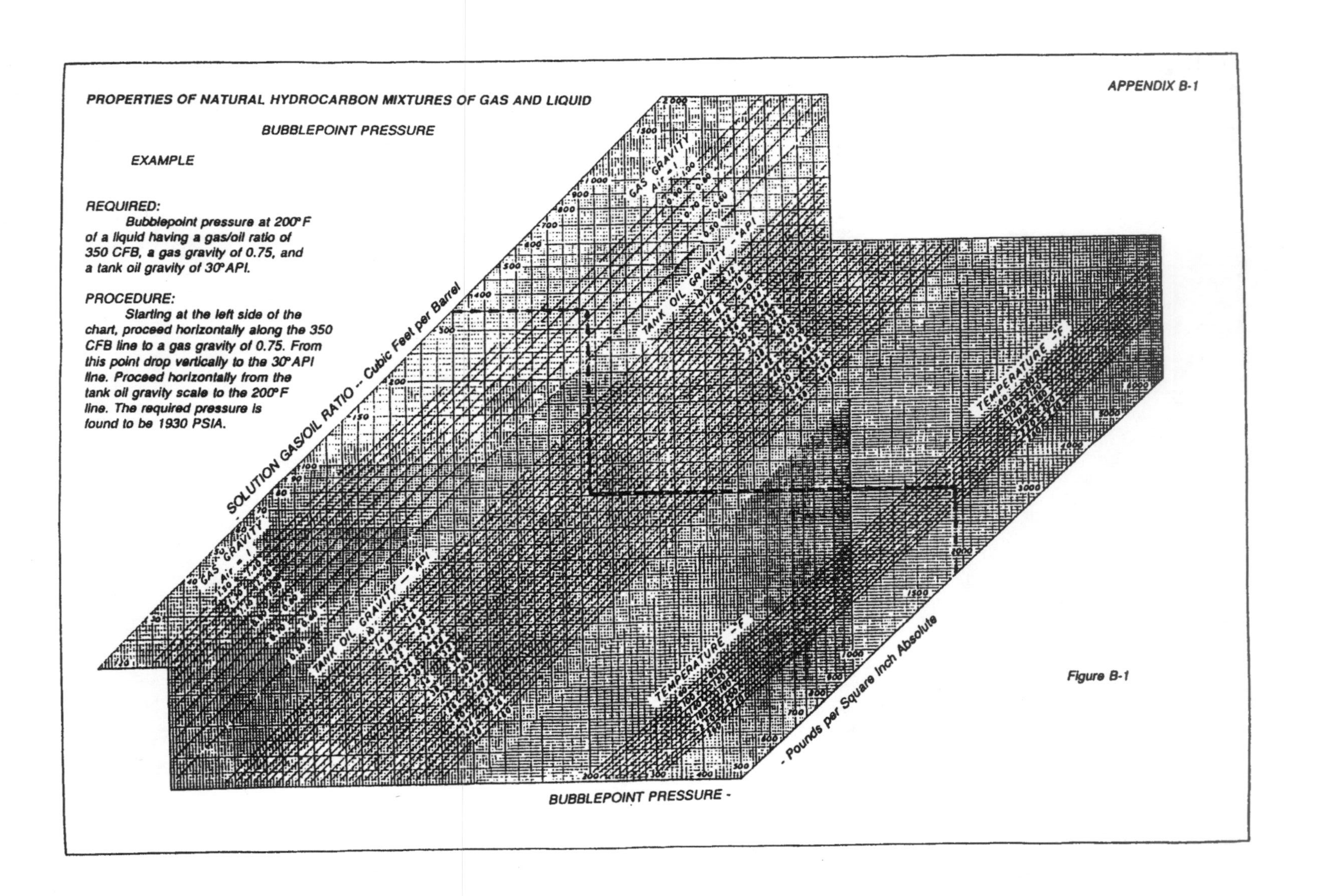
APPENDIX B-1
PROPERTIES OF NATURAL HYDROCARBON MIXTURES OF GAS AND LIQUID
BUBBLEPOINT PRESSURE
EXAMPLE
REQUIRED:
Bubblepoint pressure at 200°F of a liquid having a gas/oil ratio of 350 CFB, a gas gravity of 0.75, and a tank oil gravity of 30°API.
PROCEDURE:
Starting at the left side of the chart, proceed horizontally along the 350 CFB line to a gas gravity of 0.75. From this point drop vertically to the 30°API line. Proceed horizontally from the tank oil gravity scale to the 200°F line. The required pressure is found to be 1930 PSIA.
SOLUTION GAS/OIL RATIO -- Cubic Feet per Barrel
GAS GRAVITY Air = 1
TANK OIL GRAVITY - °API
TEMPERATURE - °F
BUBBLEPOINT PRESSURE - Pounds per Square Inch Absolute

Figure B-1

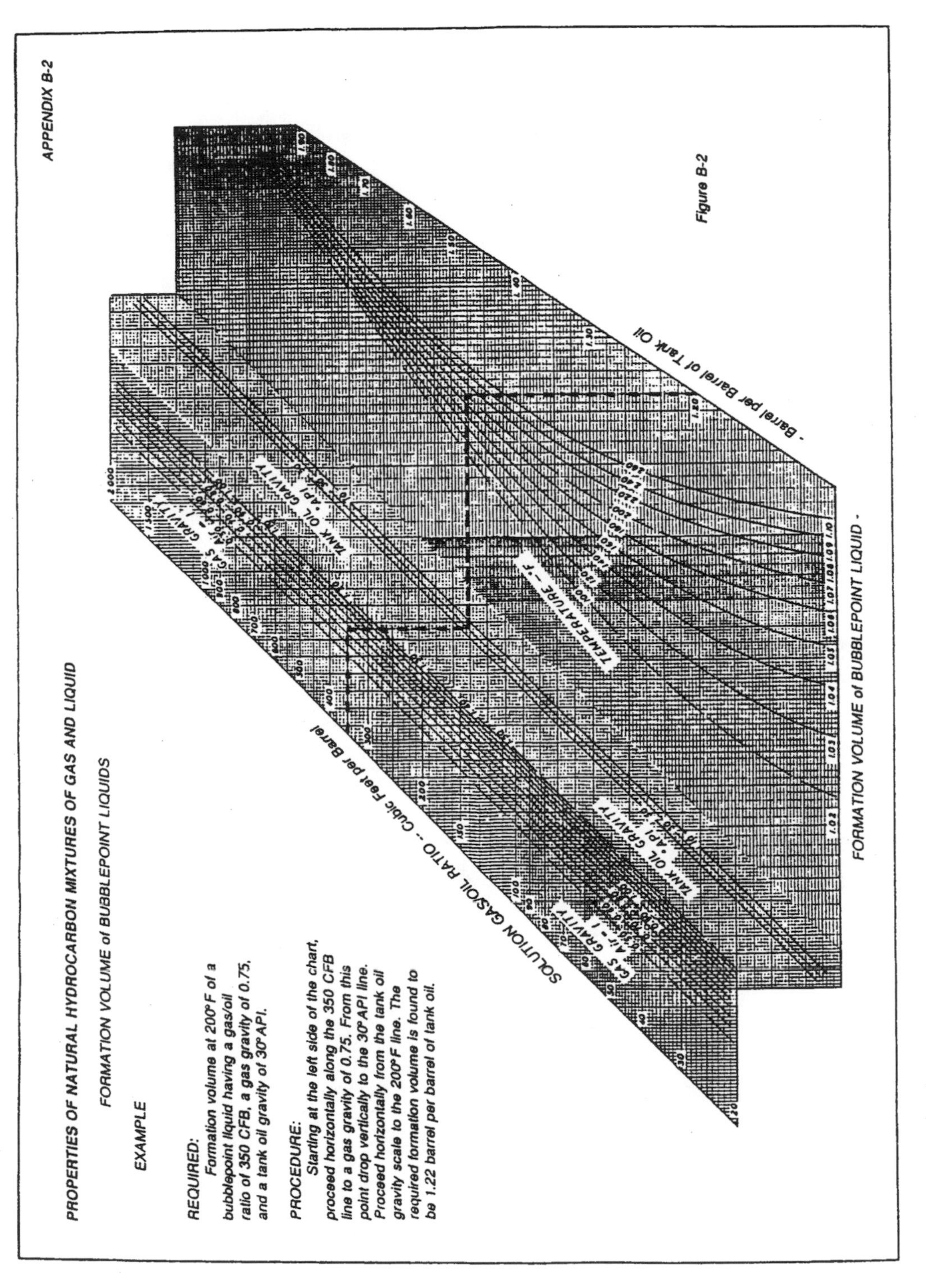
APPENDIX B-2
PROPERTIES OF NATURAL HYDROCARBON MIXTURES OF GAS AND LIQUID
FORMATION VOLUME of BUBBLEPOINT LIQUIDS
EXAMPLE
REQUIRED:
Formation volume at 200°F of a bubblepoint liquid having a gas/oil ratio of 350 CFB, a gas gravity of 0.75, and a tank oil gravity of 30°API.
PROCEDURE:
Starting at the left side of the chart, proceed horizontally along the 350 CFB line to a gas gravity of 0.75. From this point drop vertically to the 30°API line. Proceed horizontally from the tank oil gravity scale to the 200°F line. The required formation volume is found to be 1.22 barrel per barrel of tank oil.
SOLUTION GAS/OIL RATIO -- Cubic Feet per Barrel
GAS GRAVITY Air = 1
TANK OIL GRAVITY °API
TEMPERATURE - °F
FORMATION VOLUME of BUBBLEPOINT LIQUID -
- Barrel per Barrel of Tank Oil

Figure B-2

4 RESERVOIR VOLUMETRICS

INTRODUCTION

Today, the drilling of a well is an extremely expensive endeavor. Therefore, the management of an oil company is much more careful than in the past to justify the drilling of a well that may eventually cost millions of dollars. Typically this justification is given in the form of an "expected" oil in place calculation that is normally based on a volumetric estimate made by the petroleum engineer or geologist. Of course, the expected oil in place is only the first part of the justification, which also includes an economic investigation of the hypothetical development plan.

If other reservoirs are in the vicinity of a prospect, then performance data from these reservoirs are often used either directly or on an analogy basis to arrive at estimates in terms of recoverable barrels of oil or thousands of standard cubic feet (Mscf) of gas per acre. Surrounding wells sometimes offer subsurface control as to thickness, areal extent, and reservoir quality of the target reservoir.

In a new area, volumetric estimates made before the drilling of the first well are usually based on geophysical maps that may have no subsurface well control. These maps are used to get an estimate of possible productive size that, together with estimates of recoverable barrels or MCF per acre, will allow expected total reserves (recoverable hydrocarbons) to be calculated.

All this, of course, assumes that a reservoir exists. The chance of failure is denoted by the term "expected."

So, before a reservoir performance prediction can be made, an estimate of the volume of original oil in place is needed. To calculate this volume, we must establish the geologic boundaries of the reservoir. To this end, some definitions are needed.

"Gross formation thickness" is the total thickness of the formation. "Gross pay" (for an oil reservoir) is the total thickness of the oil-bearing portion of the formation or reservoir. At a well, the interval of the formation below the oil/water contact is included in the gross formation thickness but is excluded from gross pay. "Net pay" or "effective pay" is that part of the gross pay that contributes to hydrocarbon recovery and is defined by lower limits of porosity and permeability and upper limits of water saturation[1]. Figure 4-1 illustrates these concepts. Analogous definitions for gross and net pay exist for gas reservoirs or gas caps. The most common exclusion from gross pay (to arrive at net pay) is shale. Shale intervals are usually determined from well logs such as the self-potential or the gamma ray surveys.

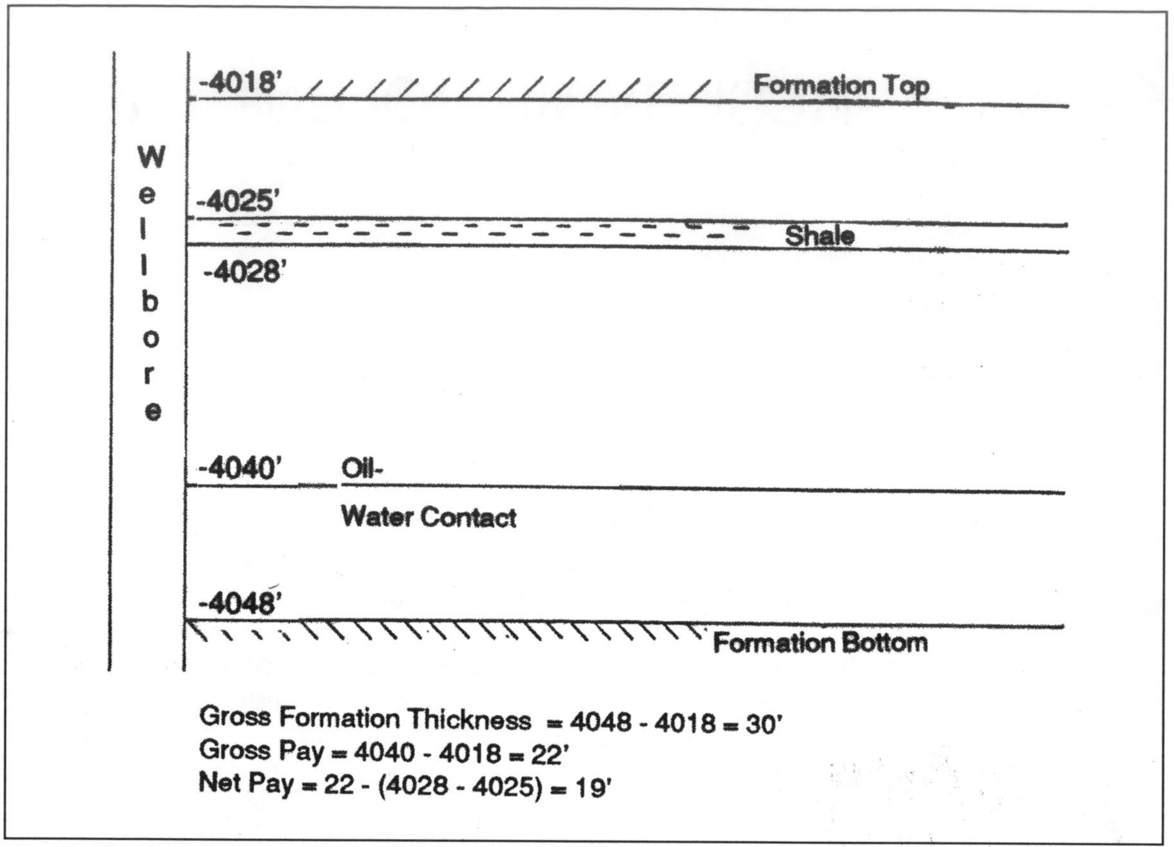

Fig. 4-1. Hypothetical oil-bearing formation illustrating the calculation of gross formation thickness, gross pay and net pay.

Often times, the lower boundary of a reservoir is the "oil/water contact." Because usually there is really not a sharp surface where the fluid saturation changes from 100% oil to 100% water (see Capillary Pressure in the "Rock Properties" chapter), we define the oil/water contact to be that level where, just below it, 100% water production would be obtained. This determination is normally made via conventional electric logs.

Similarly, the upper boundary of an oil reservoir is sometimes the "gas/oil contact." Here again, a transition zone is between the oil and gas. But, the gas/oil contact is defined as that level above which 100% gas production would be obtained.

If we had a homogeneous, isotropic reservoir, then it would be valid to obtain volumetric estimates of the original hydrocarbon in place with the following equations:

$$Oil,\ STB = \frac{(7758)(A)(h)(\phi)(1 - S_w)}{B_o} \tag{4-1}$$

$$Gas,\ Mscf = \frac{(43{,}560)(A)(h)(\phi)(1 - S_w)}{(1000)(B_g)} \tag{4-2}$$

where:

7758 = conversion factor: barrels per acre-ft,
$43{,}560$ = conversion factor: square ft per acre,
A = reservoir area, acres,
h = net thickness, ft,
ϕ = porosity, fraction,
S_w = average water saturation, fraction,
B_o = oil formation volume factor, res. barrels / STB, and
B_g = gas formation volume factor, res. cu ft / scf.

Notice that the unit of volume employed on the right hand side of these equations is an "acre-foot." As you might expect, one acre-foot is equal to the volume having one acre in area and one foot in thickness.

These equations are also applicable for the case of a new reservoir with no wells or perhaps only one or two wells drilled. For this situation, only unrefined estimates will be available for the needed parameters. The petroleum engineer or geologist who is using either Equations 4-1 or 4-2 will usually determine a range of values for each parameter: minimum, maximum, and most likely. Thus, estimates for oil and/or gas in place may be bracketed.

With the drilling of more wells, it is usually found that the average porosity (for the reservoir in question) differs in each well. Similarly, net thickness, average water saturation, and possibly even formation volume factor are changing with position in the reservoir. Normally, the formation factor will be constant for the reservoir unless a large vertical distance exists between the reservoir top and bottom.

With additional information available from multiple wells, we normally prepare maps to keep track of and display these data. These maps include: structure top, structure bottom, gross thickness, net-to-gross (thickness) ratio, iso-porosity, and iso-water saturation maps.

The volumetric estimation of oil in place is an on-going project. Each time that new information becomes available, usually from additional wells, then all the maps should be updated and a new volumetric calculation made. In this manner, as the field is drilled, the reserves estimate becomes more accurate.

SUBSURFACE MAPPING

Contour maps are used extensively for the determination of hydrocarbon in place and reserves. All the maps mentioned previously may be prepared as contour maps. According to Bishop[2], contours are lines drawn on a map to connect points of equal value compared to some chosen reference. Figure 4-2 is an example from Pirson[1] that illustrates a contour map. In this case, the map is a gross thickness isopach. Each contour represents a given constant gross thickness in the reservoir.

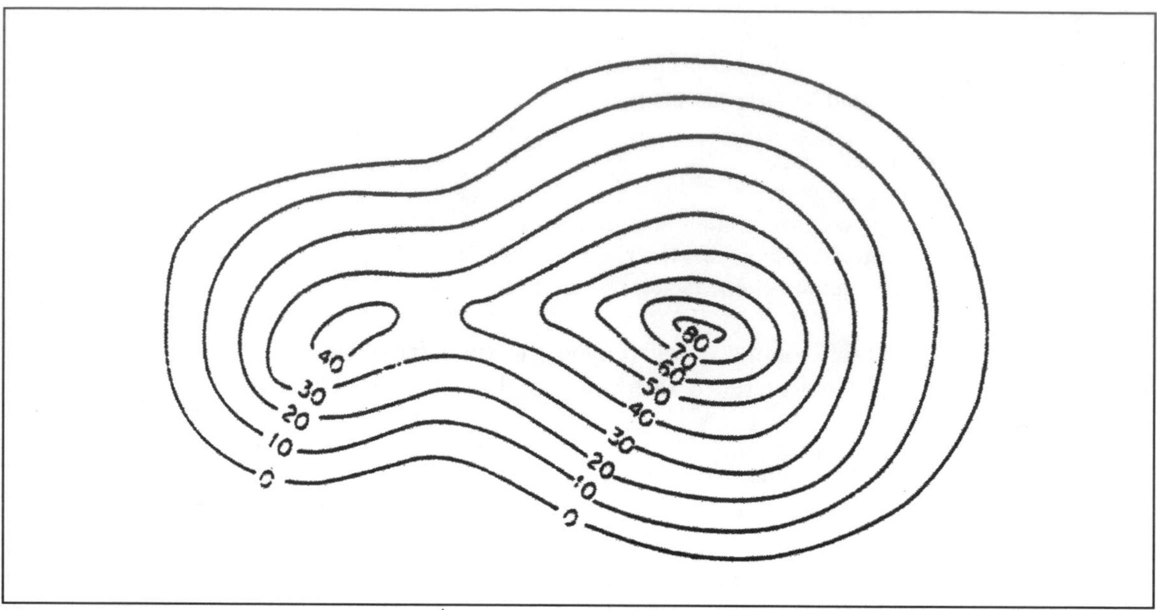

Fig. 4-2. Reservoir gross isopach (from Pirson[1]). Permission to publish by McGraw-Hill.

Bishop[2] has presented some rules concerning contour maps:

1. Contour lines cannot cross one another. (In the special case of an overhanging cliff or fault, the contours appear to cross. In space, these lines would not be in contact but would be one above the other.)
2. Contour lines may not merge with contours of different values or of the same value. (When a vertical plane is projected upon a map, the contours appear to be merging. In space, those lines would not be in contact but would be one above the other.)
3. Contours must always close or end at the edge of the map.
4. Contours of the same value must be repeated to indicate a reversal of direction of slope.
5. The contour interval, or unit upon which the map is drawn, should be a function of (a) the scale of the map, (b) the amount of variation between the values being contoured, and (c) the amount of detail which is desirable for the special purpose of the map.

A. Structure Maps

Structure maps are drawn to show the geometric shape of a reservoir or formation. Traditionally, if the term "structure" map is used; this indicates a structure map of the top of the zone. It is probably better terminology to use "structure top" map if the top of the zone is being mapped because "structure bottom" maps also exist. An example structure top map is shown in Figure 4-3.

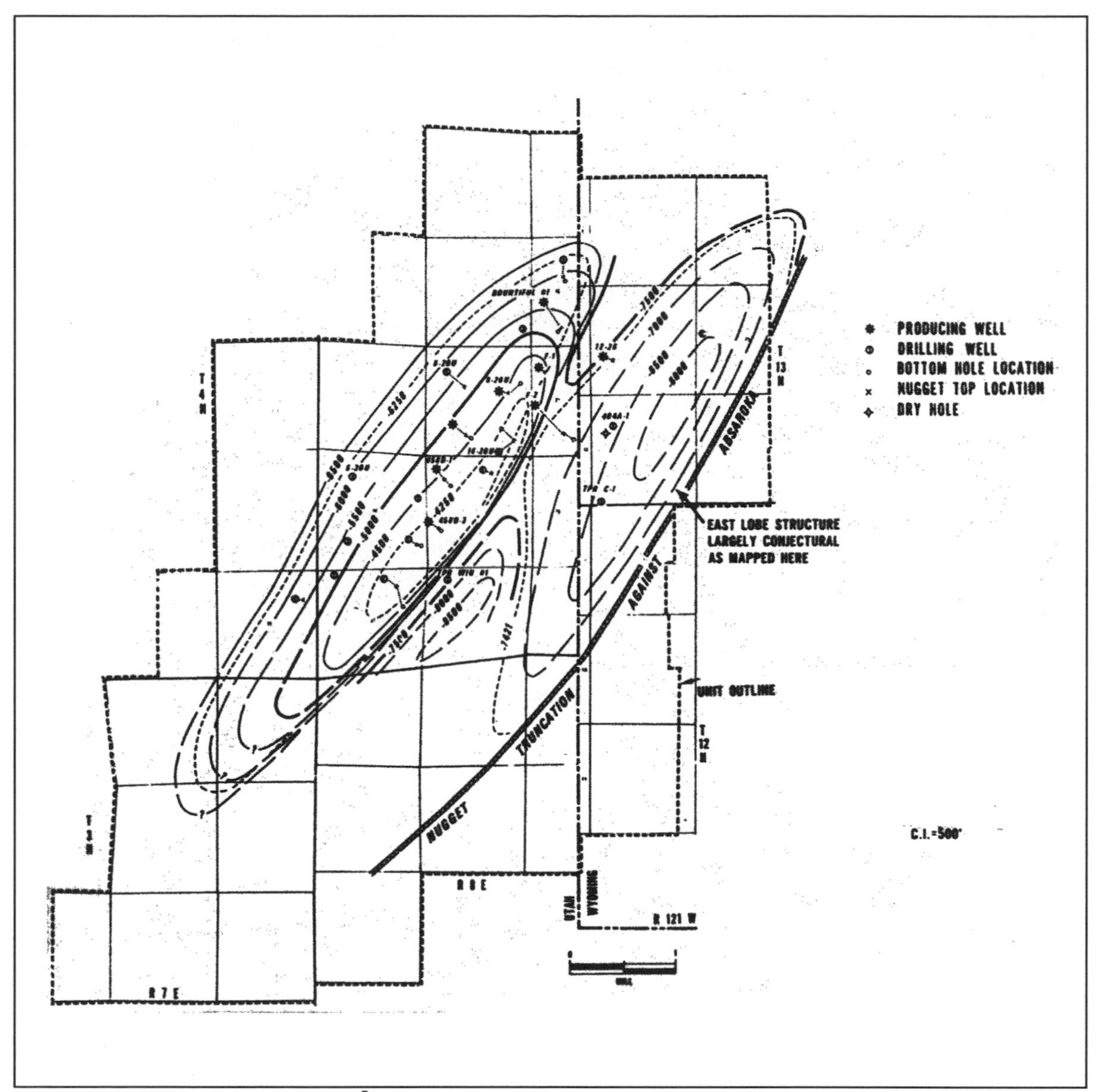

Fig. 4-3. Structure top map (from Lelek[5]). Permission to publish by JPT.

Often the top of the formation is different for the geologist and the reservoir engineer. The aim of the geologist is normally to map the top of a lithologic or stratigraphic unit, regardless of quality. However, the aim of the reservoir engineer is to map the surface that is the highest point where "reservoir"-quality rock exists. The reservoir engineer is interested in mapping the top of the oil within the formation; the geologist is interested in mapping the top of the formation. Sometimes these tops are the same, but sometimes they are not. When a formation has a caprock with no or very little porosity, it will act as an impermeable barrier. The engineer will probably map the bottom of this; the geologist may choose to map the top. So, it pays to be careful and make sure what is actually being mapped.

It is common practice for the contours of a structure map to be solid. If an oil/water or gas/oil contact is shown, these are often dashed lines.

The contours represent lines of intersection between a series of horizontal planes (separated by a constant vertical interval) and the surface (usually the top but could be the bottom) of the formation being mapped. The configuration of the surface is interpreted from the pattern made by the contour lines. Faulting and folding may be shown by means of contours.

Strike is defined as the compass direction of the line of intersection between a horizontal plane and the mapping horizon.[2] Because a contour line is also the line of intersection between a horizontal plane and the mapping horizon, each structure contour line is also a strike line.

Dip is defined as the angle between the bedding plane and a horizontal plane measured at 90° in a compass direction from the strike of the bed. Dip has both magnitude and direction, with the direction being that of the maximum downward slope. So, dip may be determined from a structure map by considering the change in slope along a line at right angles to the contours.

The first step in making a subsurface structure map is to determine the surface elevation of the well with reference to mean sea level. Then, the structural position of the formation surface is found by subtracting the depth at which the surface was encountered from the surface elevation. The result will be a structure map whose contours are referred to mean sea level. Thus, if the mapping surface lies below sea level, the contour values are negative.

Only in completely drilled fields on relatively close spacing will there be enough data to permit accurate contouring by merely mechanically connecting points of equal elevation. With the limited data normally on hand, "interpretative" contouring is probably the most acceptable method. With this procedure[2], contours are drawn to illustrate possible structural patterns that are consistent with the given set of data as well as with the known (or supposed) trends of the region.

Usually the contouring of structure maps is left in the hands of the geologist. These data for this purpose are basically from three sources: (1) well control, (2) geophysical data usually in the form of "Time" maps, and (3) geological hypotheses concerning deposition and post-depositional events.

B. Isopach Maps

An isopach map shows by means of contour lines the distribution and thickness of a chosen mapping unit. The contour lines connect points of equal vertical interval. Isopach maps illustrate the size and shape of a given horizon.

Two common types of isopach maps are used by reservoir engineers: gross isopachs and net isopachs. For the moment, comments will be restricted to oil reservoirs. The gross oil thickness isopach map contours gross pay; i.e., the depth of the top of the oil minus the depth of the bottom of the oil. The net oil thickness isopach map has contours that relate only to the zone thickness contributing to oil recovery; i.e., net pay. Therefore, for example, a gross oil isopach would include any shale sections within the oil-bearing interval of the formation; a net isopach would not. Note that the top of the oil interval in a well with a gas/oil contact would not be the top of the formation, but would be the gas/oil contact. Similarly, in a well with an oil/water contact, the lower limit of the oil would be the oil-water contact. Figure 4-1 illustrates these concepts in a single borehole. The situation is similar for a gas zone (whether it is a gas reservoir or the gas cap of an oil reservoir) with the exception that if a gas/oil contact is present, it will be the lower thickness limit, not the upper. Data for isopach maps may be obtained directly from well logs.

For the sake of calculating hydrocarbons in place, it is more geologically and mathematically consistent to prepare gross isopach maps for subsequent use with net-to-gross-ratio maps than it is to use net isopach maps. The mathematical reasons for this will be touched on later. However, many oil companies prefer to use net oil isopach maps for volumetric calculations.

C. Net-to-Gross (Pay) Ratio Maps

In a well, considering a single formation, the net-pay to gross-pay ratio relates the fraction of the total hydrocarbon interval that is effectively contributing to recovery. Therefore, the contours of a net-to-gross ratio map will illustrate at a glance how clean the formation is and how it is distributed. From well logs, the gross pay section is determined. Then, within this interval (again using logs); zones of shale, low porosity, and high water saturation are located. The thicknesses of these zones are subtracted from gross pay, which leaves net pay. Then, for that well location, the net-to-gross ratio is merely the net pay divided by the gross pay.

D. Iso-Porosity Maps

To prepare these maps, for each well the average porosity over the net pay portions of the desired formation is calculated. So one number is representing the average porosity at each well location. Then, contours are drawn which illustrate the net pay porosity trends in the reservoir.

E. Iso-Water Saturation Maps

For each well, considering the net pay portions of the desired formation, average water saturation is calculated, usually from conventional electric logs. Then, similar to porosity mapping, each well location will have one average water saturation. With enough wells, contours can be drawn which illustrate how average water saturation is distributed in the reservoir.

F. Determining Reservoir Volume from Contour Maps

1. Limited Data Available

As we have discussed, during the early development of a reservoir (such as zero, one, or two wells drilled), only one or two (perhaps crude) estimates for each of the parameters in Equations 4-1 or 4-2 are available. However the geologist may have provided a structure map primarily based on geophysical data from which a rough net pay isopach can be generated. In this case, it would be appropriate to use Equations 4-1 or 4-2 together with estimates of average values over the reservoir for all the parameters except "A" and "h." Net hydrocarbon volume for this situation is determined by numerical integration of the net pay isopach. This result in acre-ft is substituted into the appropriate equation, 4-1 or 4-2, in place of the product: (A)(h).

A planimeter is a device for integrating surface areas. This instrument has an arm that is used to trace the boundary of the area, such as one of the contours on the net isopach map. In this way, the area inside a particular contour can be obtained.

Two equations are commonly used to determine the approximate volume from the planimeter readings[3] of an isopach of the productive zone. Consider the volume between two contour lines (see Figure 4-4.)

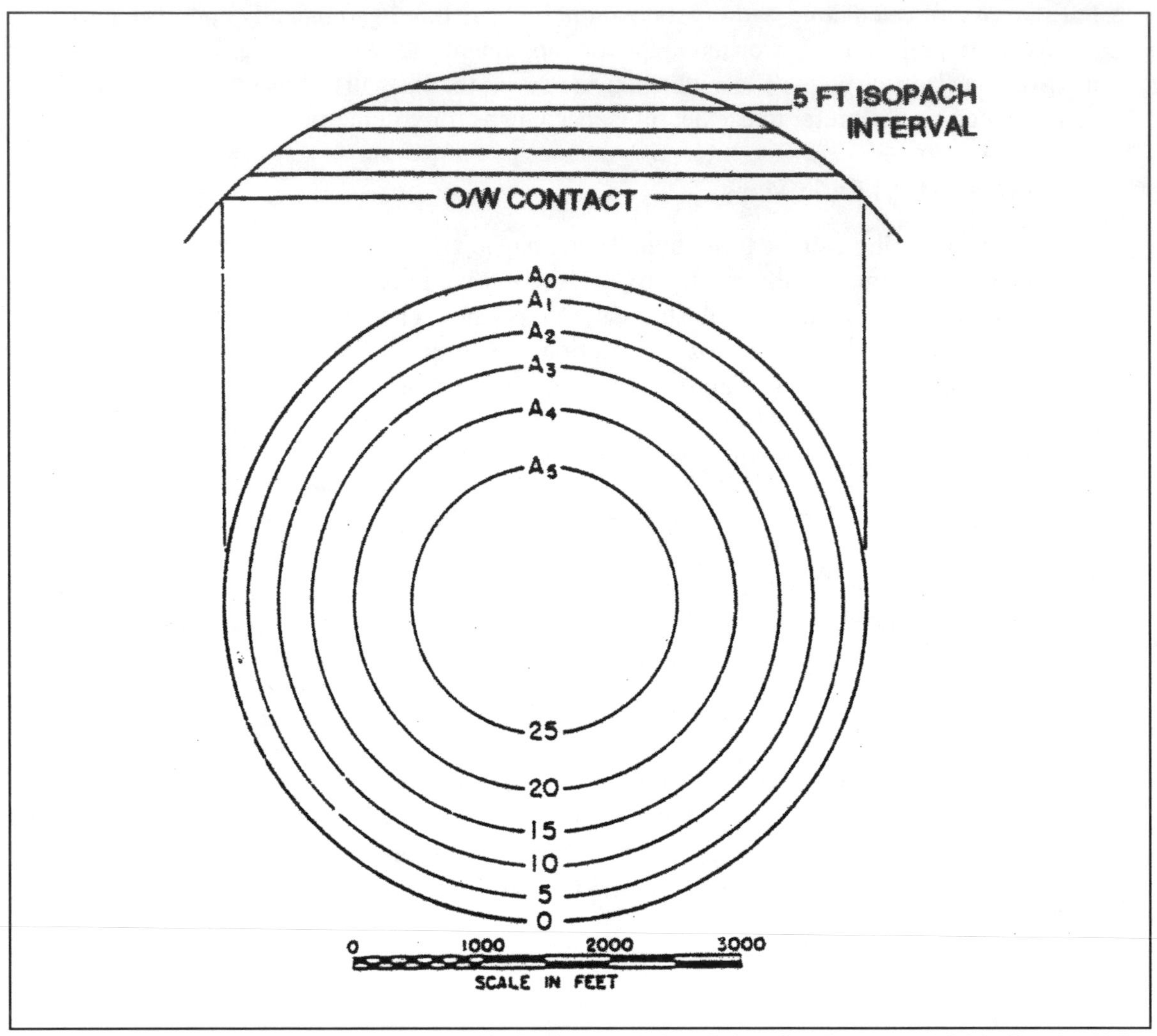

Fig. 4-4. Cross section and isopachous map of an idealized reservoir (from Ref. 3). Permission to publish by Prentice-Hall.

This may be regarded approximately as the frustrum of a pyramid, the volume of which is given by:

$$\{\Delta V_b = \frac{h}{3}\left(A_j + A_{j+1} + \sqrt{A_j A_{j+1}}\right) \tag{4-3}$$

where:

ΔV_b = the bulk volume in acre-ft between contours "j" and "j+1",
h = the interval between isopach contours in ft,
A_j = area enclosed by lower isopach contour, "j", acres (see Figure 4-4),
A_{j+1} = area enclosed by upper isopach contour, "j+1", in acres.

Thus, if this formula is used by itself, it must be applied repeatedly between each set of contours, and then the total volume is the sum of all the ΔV_b's.

The trapezoidal volume formula is:[3]

$$\Delta V_b = \frac{h}{2}\,(A_j + A_{j+1}) \tag{4-4}$$

For a series of successive trapezoids with constant "h":

$$\Delta V_b = \frac{h}{2}\,(A_o + 2A_1 + 2A_2 + \ldots + 2A_{n-1} + A_n) + t_{avg}\,A_n \tag{4-5}$$

where:

A_o = the area enclosed by the zero isopach line in acres,
A_1 = the area of the contour just above A_o, acres,
$A_2 \ldots A_n$ = areas enclosed by successive isopach lines, acres, and
t_{avg} = average thickness above the top contour, ft.

Notice in Equation 4-5 that the uppermost volume element of the isopach has been calculated as $(A_n)(t_{avg})$. Alternatively, Pirson[1] suggests that this upper volume could be calculated pessimistically as the volume of a pyramid:

$$(\Delta V_b)_{upper} = \frac{A_n\,h_n}{3} \tag{4-6}$$

where:

h_n = vertical distance between top contour and highest reservoir point, ft, and
A_n = area of highest contour, acres

Pirson[1] goes on to say that some surface curvature is in the upper part of most geologic structures. An optimistic formula for this upper volume element is:

$$(\Delta V_b)_{upper} = \frac{\pi\,h_n^{\,3}}{6} + \frac{A_n\,h_n}{2} \tag{4-7}$$

which is the formula for the volume of a spherical segment. Equations 4-6 and 4-7 should probably be considered as lower and upper limits, respectively, for this upper volume element. Actually, not much volume is usually involved.

Normally, because of its ease, the successive trapezoidal formula is used. However, the best accuracy is obtained with the pyramidal formula. In fact, an error of about 2% is introduced[3] by using the trapezoidal formula when the ratio of successive areas is 0.50.

According to Craft and Hawkins[3], a commonly adopted rule is to use the pyramidal formula whenever the ratio of any two successive areas is 1/2 or smaller, and the trapezoidal rule is used otherwise.

It should be emphasized that Equations 4-1 and 4-2 should only be used when average values of the needed parameters are available. This is also the only instance that it is appropriate to calculate reservoir acre-ft with the procedure just outlined. A better method exists.

2. ***Better Method***

This method assumes that enough wells have been drilled to be able to prepare each of the following types of maps for this reservoir:

(a) structure top map,

(b) gross isopach map,

(c) net-to-gross ratio map,

(d) iso-porosity map,

(e) iso-water saturation map,

(f) water/oil contact map (if WOC is not constant), and

(g) gas/oil contact map (if the GOC is not constant).

Numerical Integration by Hand

When integrating by hand, the reservoir will need to be divided into a grid, i.e., a number of small blocks, elements or samples. Thus, a block size must be selected. This size is generally based on (1) the desired precision of the results and (2) the variability of the contoured surfaces. So, it is somewhat related to the number of wells that have been drilled. Results tend to be more accurate if a greater number of elements are used; however, it does no good to use a large number of blocks if not enough data are available to justify them.

Once the block size has been selected, then a transparent overlay is usually made which contains the grid. Figure 4-5 illustrates such a grid. The elements do not have to be rectangular, although they usually are. This grid can then be placed over each of the contour maps for the purpose of selecting a representative value from each map for each block. It is probably a good idea to keep track of all these data and subsequent calculations by using a columnar worksheet.

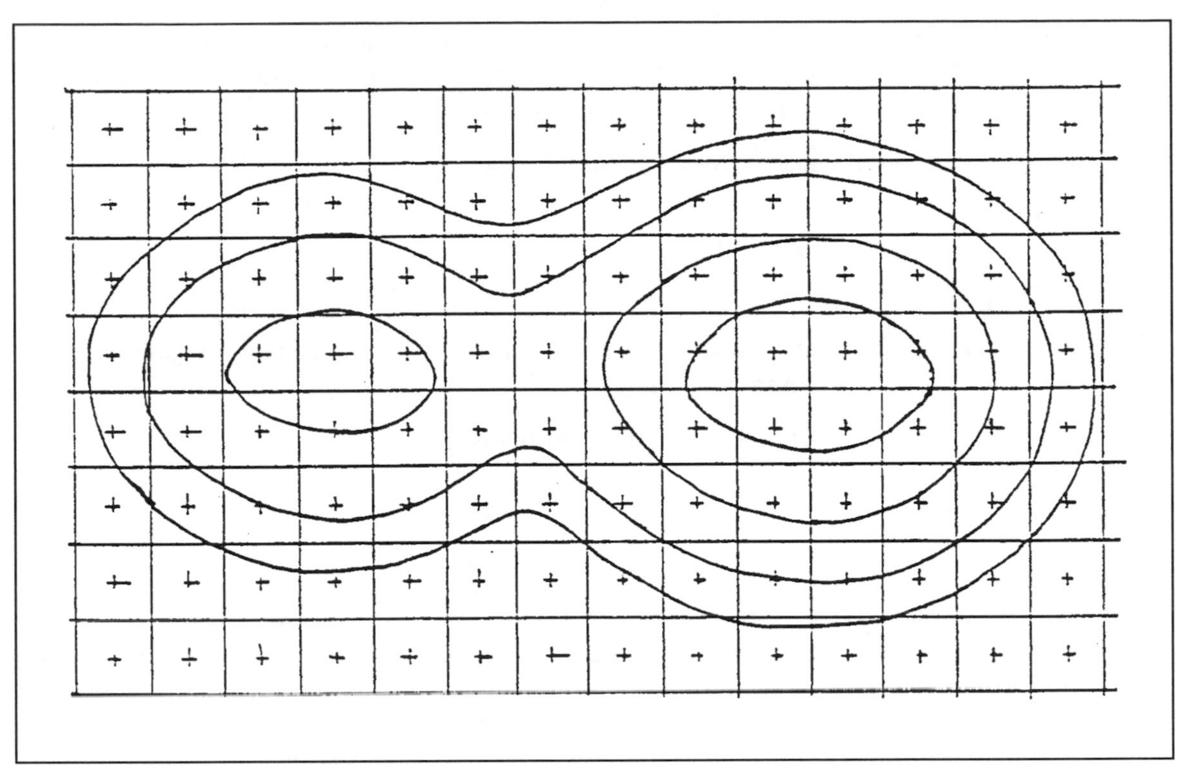

Fig. 4-5. Example grid for hand integration of contour maps.

The equations to be used are:

$$Oil,\ STB = 7758 \sum_{j=1}^{n} \frac{(A_j)(h_{gj})(N/G)_j(\phi_j)(1 - S_{wj})}{B_{oj}} \tag{4-8}$$

$$Gas,\ Mscf = 43.56 \sum_{j=1}^{n} \frac{(A_j)(h_{gj})(N/G)_j(\phi_j)(1 - S_{wj})}{B_{gj}} \tag{4-9}$$

where:

j = subscript indicating value from the jth element,
A = area, acres,
h_g = gross pay thickness, ft,
N/G = net-to-gross ratio, fraction,
ϕ = porosity, fraction,
S_w = average water saturation, fraction,
B_o = oil formation volume factor, res. barrels / STB, and
B_g = gas formation volume factor, res. cu ft / scf.

Equations 4-8 and 4-9 are better approximations than are Equations 4-1 and 4-2 to the actual volume of hydrocarbons in place. The reasons come from the calculus and are beyond the scope of this work, but some comments will be offered.

The actual rigorous equation for oil in place is a volume integral:

$$Oil, STB = 7758\int_V \frac{(N/G)(\phi)(1 - S_w)}{B_o}\, dV \tag{4-10}$$

where dV indicates that the integration is in terms of gross volume, and V signifies that the integration is over the entire reservoir volume.

According to Widder[4], the following relationship is true:

$$(\bar{A})(\bar{B})(\bar{C})\int_V dV = \int_V (A)(B)(C)\, dV \tag{4-11}$$

only with functions (A, B, and C) that are well behaved. The bar over a function indicates the average value for that function over the volume being integrated. The functions within the volume integral of Equation 4-10 (porosity, water saturation, etc.) are not well behaved. Equation 4-8 is an approximation to Equation 4-10 which is needed because the integral in 4-10 is too difficult to be evaluated for an actual reservoir. This approximation becomes more accurate as the number of elements increases. Therefore, if these data are available, Equation 4-8 or 4-9 should be used instead of Equation 4-1 or 4-2.

Computer Integration

When integrating using the computer, the first task is to contour (by computer) all the individual surfaces that affect the integration. Do not contour combinations of independent variables such as $\phi(1-S_w)$ or ϕh . Computer routines may select interior points for interpolation and functions are usually extrapolated out to the boundaries. Thus, if a combined surface, such as ϕh is input, then the values will be unreliable at the interpolated and extrapolated points unless the variables are dependent, e.g. porosity is a function of thickness. The zero isopach on the net pay map will not occur at the same place as the zero isopach on the ϕh map. This is easily seen on the following two-dimensional sketch. The same principle applies for three-dimensional surfaces.

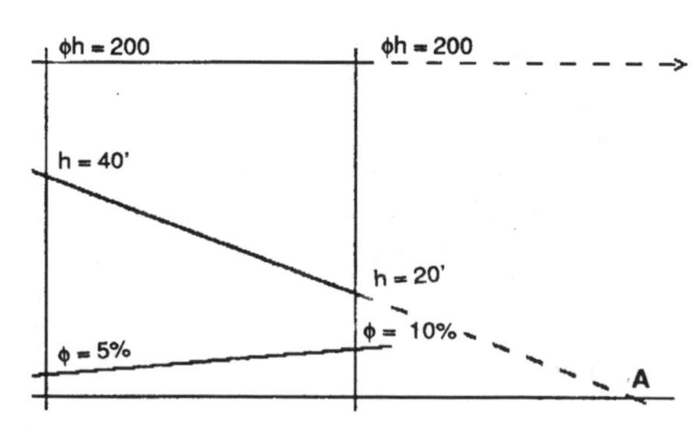

In this instance, pay is thinning and porosity is increasing. However, ϕh is constant. Reservoir thickness would extrapolate to zero at point A, but ϕh would extrapolate at a constant value to infinity.

Advantages of Computer Integration

Some of the advantages of computer contouring and integration, particularly where diverse interests are involved, are as follows:

1. Contouring is objective. Even though the contours might not be precisely as each person would draw them, they were drawn with the same mathematical objectivity.
2. If the various parties can agree on pay picks, methods of averaging porosities, ways to calculate water saturation, etc., then the computer can contour and integrate the data methodically and quickly.
3. If changes have to be made due to new wells, new test data, etc., then these changes are easy to incorporate in computer contoured maps. The new maps can be integrated quickly and objectively, whereas changes are disruptive and time consuming when done by hand.
4. Reservoirs can be separated into zones easily when the contouring and integration are done by computer.
5. Much smaller elements can be justified.

REFERENCES

1. Pirson, S. J.: *Oil Reservoir Engineering*, McGraw-Hill Book Co., Inc., New York City (1958).
2. Bishop, M. S.: *Subsurface Mapping*, John Wiley & Sons, New York (1960).
3. Craft, B. C. and Hawkins, M. F.: *Applied Petroleum Reservoir Engineering*, Prentice-Hall, Inc., Englewood Cliffs, NJ (1959).
4. Widder, D. V.: *Advanced Calculus*, Prentice-Hall, Inc., Englewood Cliffs, NJ (1961).
5. Lelek, J. J.: "Geologic Factors Affecting Reservoir Analysis, Anschutz Ranch East Field, Utah and Wyoming," *JPT*, (August 1983).

5 GAS RESERVOIRS

INTRODUCTION

In this chapter, discussion will be limited to single phase non-associated gas reservoirs (see Figure 5-1). Gas could also be produced from gas caps overlying oil zones (so-called associated gas), from partially or totally gas-saturated oil formations or from gas-saturated water reservoirs. Only those reservoirs that remain single phase regardless of the state of depletion shall be discussed. Gas condensate reservoirs will be covered in the next chapter.

Gas fluid properties have already been reviewed, and gas processing is beyond the scope of this work. This chapter will discuss calculations involving gas in place, reserves, reservoir performance, and gas well deliverability testing.

It would be most helpful for the reader to review the section of the "Fluid Properties" chapter entitled "Reservoir Hydrocarbon Fluid Classification." Specifically, the subsections "Wet Gas" and "Dry Gas" concern the gas systems that will be studied here.

For these gas reservoirs, the reservoir fluid is always a single phase gas for the life of the reservoir because in these systems the formation temperature is greater than the cricondentherm (maximum temperature of the phase envelope). The phase diagrams for wet and dry gases are shown in Figures 5-2 and 5-3.

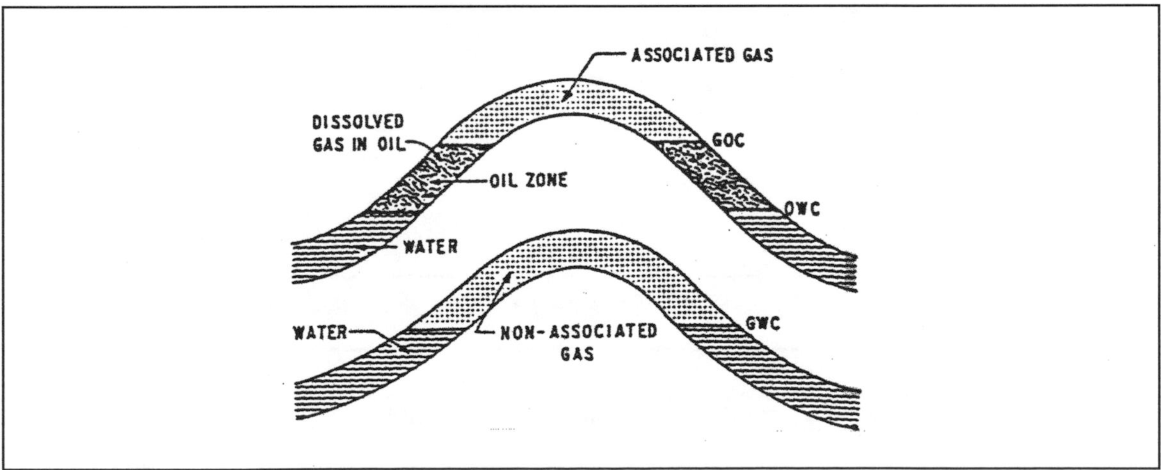

Fig. 5-1. Classification of gas based on source in reservoir (from Clark[1]). Permission to publish by SPE.

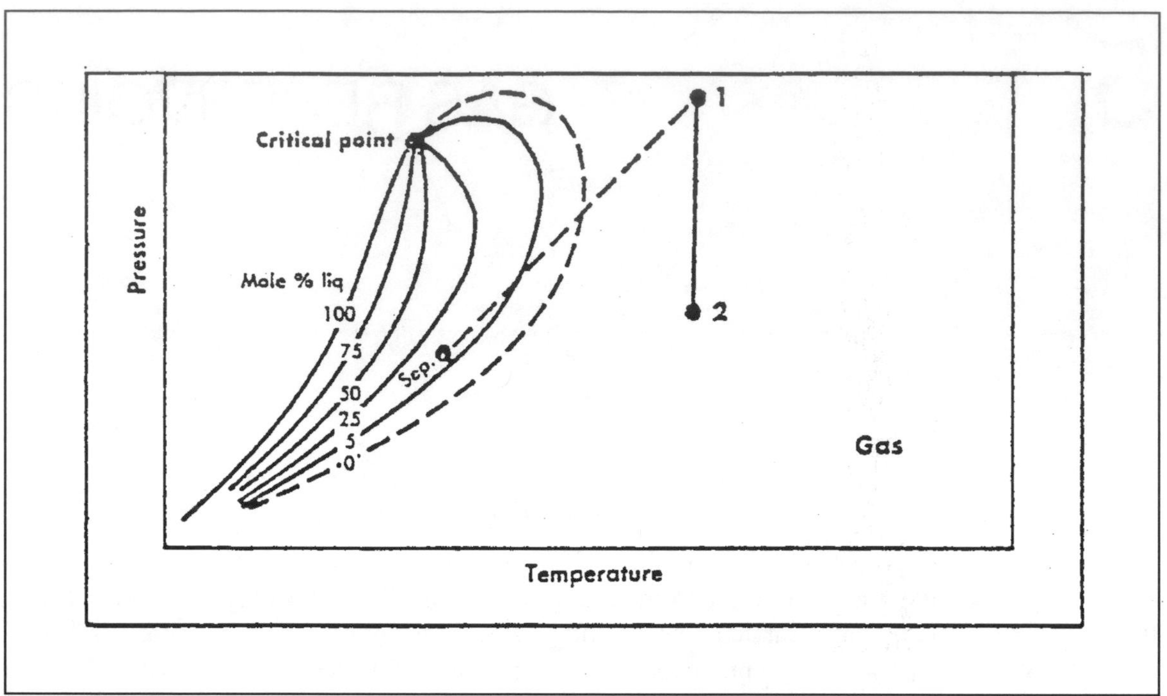

Fig. 5-2. Phase diagram of a wet gas (from McCain[26]). Permission to publish by Petroleum Publishing Company.

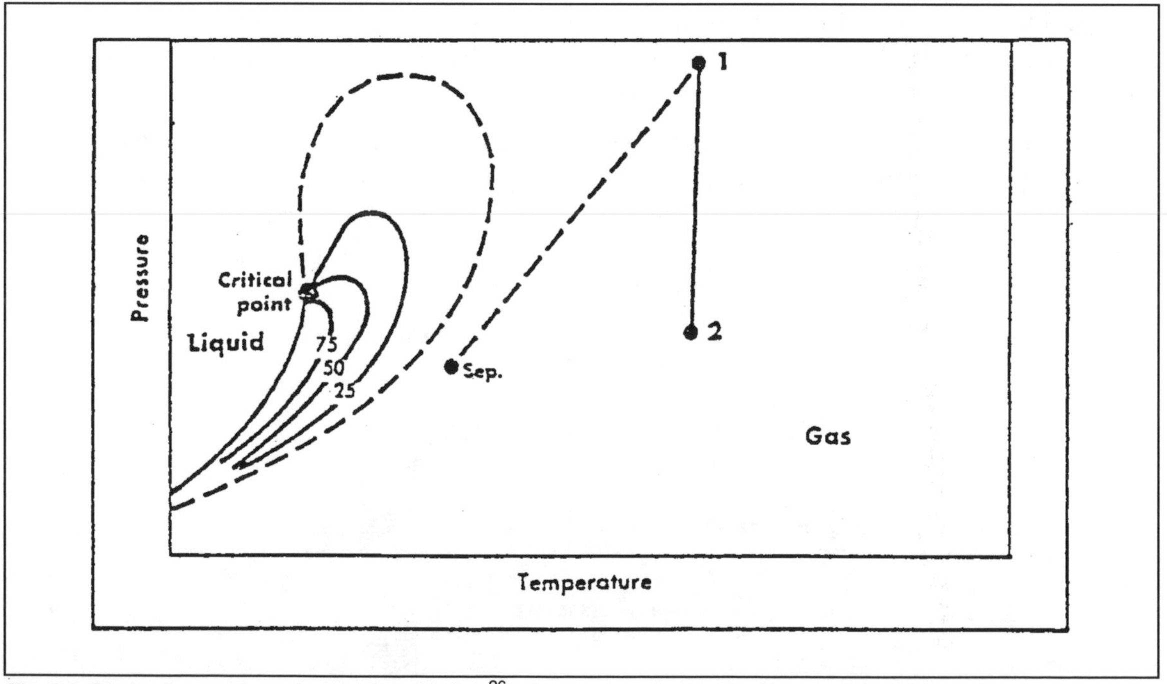

Fig. 5-3. Phase diagram for a dry gas (from McCain[26]). Permission to publish by Petroleum Publishing Company.

DETERMINATION OF ORIGINAL GAS IN PLACE

As with oil reservoirs, there are two main ways to estimate the original gas in place: the volumetric method and material balance.

A. Volumetric Method

This approach is used early in the life of the reservoir (for instance, before 5% of the reserves have been produced). However, this method is also often used to determine equity for sales and unitization rather than using performance data on individual leases. A complete treatment of this method is given in the "Reservoir Volumetrics" chapter. The simplest form of the equation for the volumetric (or pore volume) method that assumes a homogeneous, isotropic reservoir is:

$$G = \frac{(43{,}560)(A)(h)(\phi)(1 - S_w)}{B_{gi}} \tag{5-1}$$

where:

G = original gas in place, scf,
$43{,}560$ = conversion factor: square ft per ac,
A = reservoir productive area, ac,
h = net thickness, ft,
ϕ = porosity, fraction,
S_w = average water saturation, fraction, and
B_{gi} = initial gas formation volume factor, res. cu ft / scf

The significance and derivation for B_{gi} is found in the "Fluid Properties" chapter.

$$B_{gi} = \frac{(p_{sc})(T_f)(z_i)}{(p_i)(T_{sc})}\ ft^3 / scf$$

Notice here that B_{gi} is the initial or discovery gas formation factor. It is necessary to know the pressure base, p_{sc}, and the temperature base, T_{sc}. These items are not always 14.7 psia and 520°R.

For a reservoir having more than one or two wells drilled, the method outlined in the "Reservoir Volumetrics" chapter will yield more accurate results than will Equation 5-1. Where some production history is available, then a material balance calculation can be made.

B. Material Balance Method

In general, the use of pressure decline as a means to calculate the original gas in place assumes that the space occupied by the gas is constant. This means that the expansion of the rock and water is negligible and that no subsidence or collapse of the reservoir rock exists. It is further assumed that there is no net migration of gas into or away from the volume of interest. For the gas systems under consideration, assume that gas composition is constant. Therefore, we can make the material balance in terms of moles of gas:

$$n_p = n_i - n_f \tag{5-2}$$

where:

n_p = moles of gas produced,
n_i = original moles of gas in place,
n_f = moles remaining in the reservoir.

Figure 5-4 helps visualize the three corresponding volumes involved. Notice that if there is a water drive, then the final hydrocarbon pore volume is not equal to the original volume. In fact, the final volume is equal to:

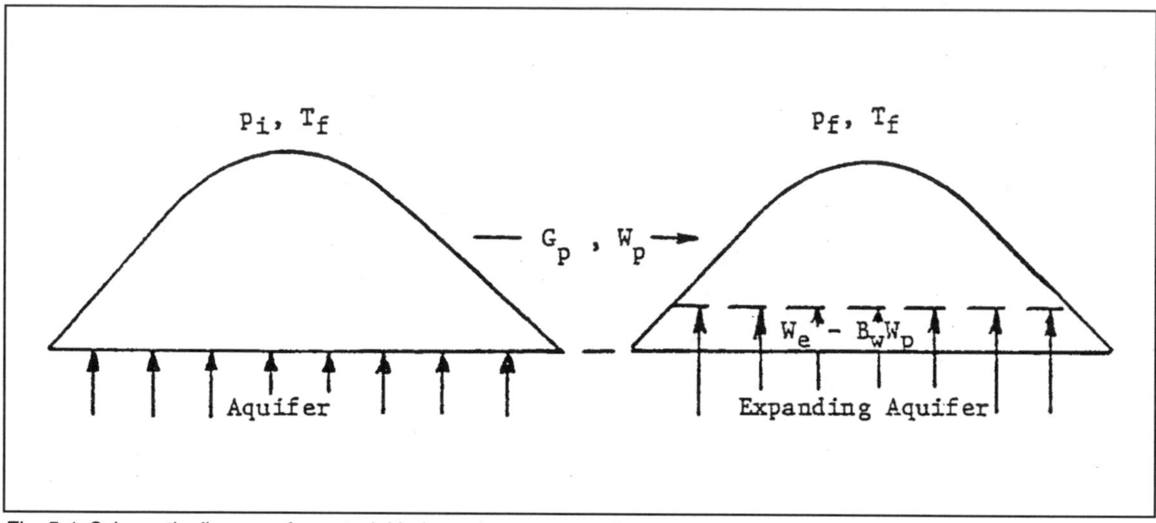

Fig. 5-4. Schematic diagrams for material balance in gas reservoirs.

$$V_f = V_i - (W_e - B_w W_p) \tag{5-3}$$

where:

V_f = gas occupied reservoir volume after G_p, W_p production, ft^3,
V_i = initial reservoir volume occupied by gas, ft^3,
W_e = volume of encroached water, reservoir ft^3,
W_p = volume of produced water, stock tank ft^3, and
B_w = water formation volume factor, res. ft^3/(stock tank ft^3).

Substituting the real gas law, pV = znRT, into Equation 5-2 and simplifying, then:

$$\frac{p_{sc} G_p}{T_{sc}} = \frac{p_i V_i}{z_i T} - \frac{p_f V_f}{z_f T} \tag{5-4}$$

where:

p_{sc} = pressure at standard conditions, psia

T_{sc} = temperature at standard conditions, °R,
G_p = cumulative gas production, scf
p_i = original pressure, psia,
z_i = z factor at p_i and T,
T = reservoir temperature, °R,
p_f = reservoir pressure after G_p production, psia,
z_f = z factor at p_f and T.

If the reservoir has no aquifer present, or if it is small, and if there is no meaningful water production, then $V_f = V_i$, and Equation 5-4 may be written as:

$$\frac{p_{sc} G_p}{T_{sc}} = \frac{p_i V_i}{z_i T} - \frac{p_f V_i}{z_f T} \tag{5-5}$$

It is convenient to get Equation 5-5 into the form y = mx + b. This may be done with the resulting equation:

$$\frac{p}{z} = \frac{p_i}{z_i} + m G_p \tag{5-6}$$

where:

$$m = -\frac{T p_{sc}}{V_i T_{sc}}$$

and where the subscript "f" has been dropped. Note that "p" and "z" still refer to static values after "G_p" production. After cumulative production of G_{p1}, the reservoir can be shut in (by shutting in all the wells), the reservoir pressure may be allowed to build up to static pressure, and $(p/z)_1$ measured. At a later time, or after cumulative production, G_{p2}, the corresponding shut-in value for $(p/z)_2$ may be obtained. It is important to note that the pressure value required in Equation 5-6 is not the surface or flowing bottomhole pressure for a well, but the reservoir static pressure at the datum elevation.

Equation 5-6 behaves as a straight line relationship. A typical graph is seen in Figure 5-5. For a constant volume reservoir, the plot of p/z vs. G_p does indeed yield a linear plot. When this line is extrapolated to p/z = 14.7 psia (atmospheric pressure; $z_{sc} = 1$), then the initial gas in place, G, may be determined as shown. If the abandonment pressure and corresponding z factor are known, then reserves may be determined from this plot. In Figure 5-5, reserves are denoted as $(G_p)_{abd}$.

Similarly, if one particular cumulative production, G_p, on the straight line and the corresponding p/z are known, then:

$$G = \left\{ \frac{[(p_i/z_i) - (p_{ref}/z_{ref})]}{[(p_i/z_i) - (p/z)]} \right\} (G_p)$$

where (p_{ref}/z_{ref}) normally equals 14.7 psia. However, this is a merely a definition; i.e., G has been defined here to be the amount of cumulative gas that would be produced if it were possible to lower the static reservoir pressure to 14.7 psia (reference pressure). A different pressure reference could be used (and often is) such as 15.025 or 0 psia.

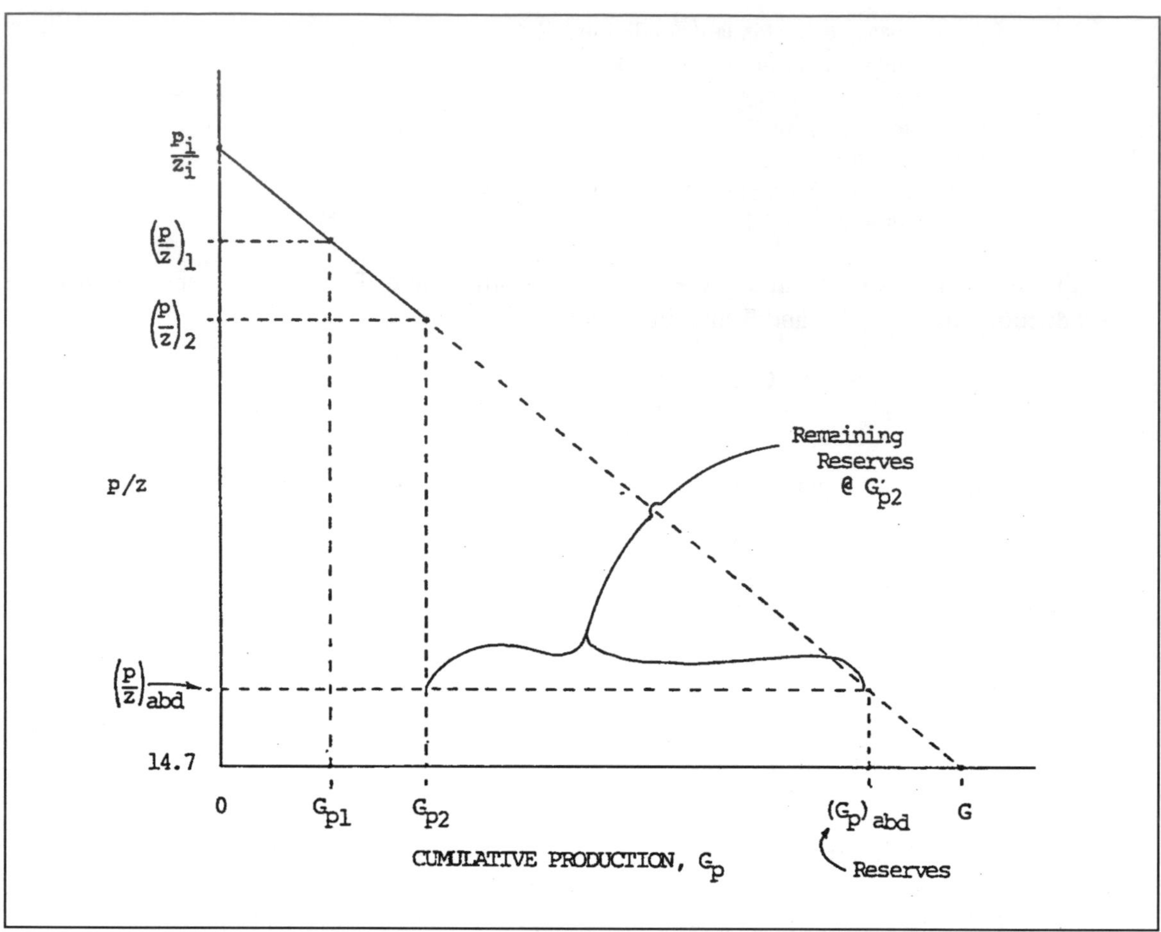

Fig. 5-5. Example plot of p/z vs. G_p for a volumetric gas reservoir.

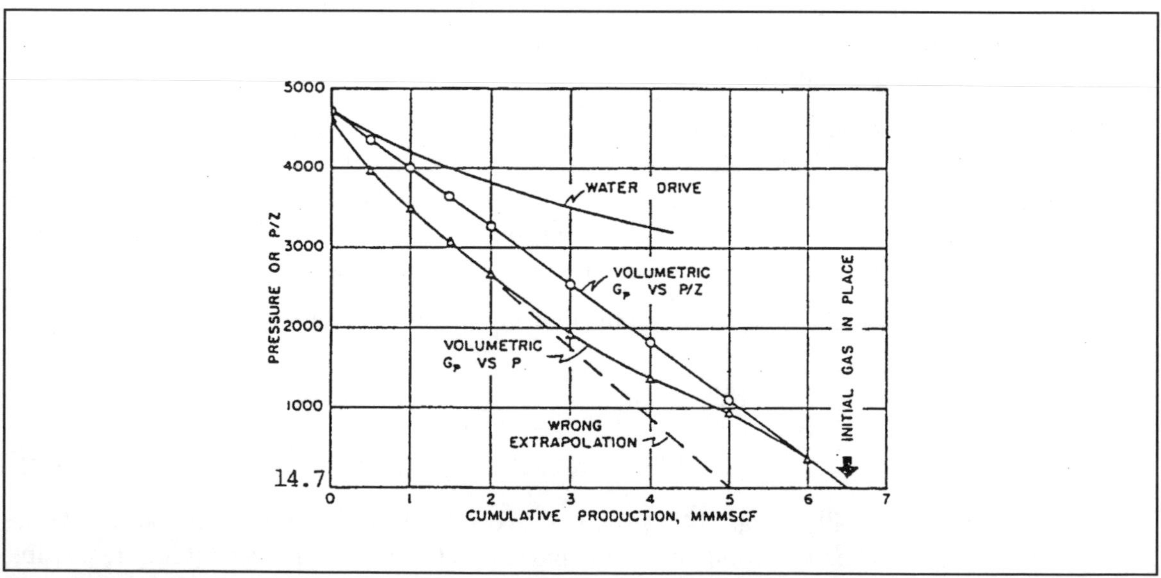

Fig. 5-6. Comparison of theoretical values of p/z and p plotted versus cumulative production for a volumetric-type gas reservoir (from Craft & Hawkins[2]). Permission to publish by Prentice-Hall.

Some reservoir engineers simply graph p vs. G_p. Figure 5-6 illustrates the pitfall involved if such a graph is extrapolated in a linear fashion. If an active water drive is present, then the constant volume assumption used to obtain Equation 5-6 is invalidated. A linear curve will not be obtained if p/z vs G_p is plotted. Figure 5-6 shows the plot obtained for a water-drive reservoir curving upward.

For efficient reservoir management, volumetrically calculated initial gas in place should be checked with a material balance calculation. If the volumetric result agrees reasonably well with that calculated via material balance, then there is probably a good estimate. If, however, production indicates a larger reservoir than is mapped, an extension of the reservoir may exist. Alternatively, the initial estimates of S_w, ϕ, or h are in error.

A more common situation is to have the early material balance calculations indicate less original gas in place than the volumetric computation. Usually, this discrepancy arises due to improperly conducted pressure buildup tests. Especially with low permeability, layered systems, or dual porosity reservoirs; if pressure buildup wells are not shut in long enough, a low estimate of static pressure is obtained. The result is a low material balance estimate of G. Fortunately, the difference between the true static pressure and the pressure buildup indicated static pressure lessens with years of production.

Where water influx is occurring, the material balance equation may be obtained by substituting Equation 5-3 into Equation 5-4 and dropping the subscript "f":

$$\frac{p_{sc}\,G_p}{T_{sc}} = \frac{p_i\,V_i}{z_i\,T} - \frac{p\,(V_i - W_e + B_w\,W_p)}{z\,T} \tag{5-7}$$

With this equation, there are multiple unknowns: V_i and W_e. Also, W_e is changing as production continues. A mathematical model can be used that characterizes the aquifer performance with water influx constants, differential pressure across the water/oil contact, and time. Solution is obtained by writing Equation 5-7 in combination with the model for each reservoir pressure at which production data are available. Then this set of equations is solved simultaneously in a fashion similar to that discussed in the "Water Drive" chapter. Guerrero[6] has illustrated his approach to this problem with example calculations.

In Figure 5-7, some of the different possibilities are shown for the p/z vs. G_p plot. The linear relationship, already discussed, is the expression of a true constant volume reservoir.

If, however, the curve is bending upwards, then the constant volume assumption is invalid and is probably due to one of the following possibilities:

1. water influx into the system;
2. bad data;
3. subsidence or compaction drive is occurring;
4. communication or leakage into the reservoir along faults or a leakage due to operations problems (such as channeling behind the casing);
5. retrograde phenomena occurring (this is usually not visible, and if it is seen, the curve usually bends downward, not upward); and
6. an oil zone underneath (perhaps undiscovered yet) that is expanding.

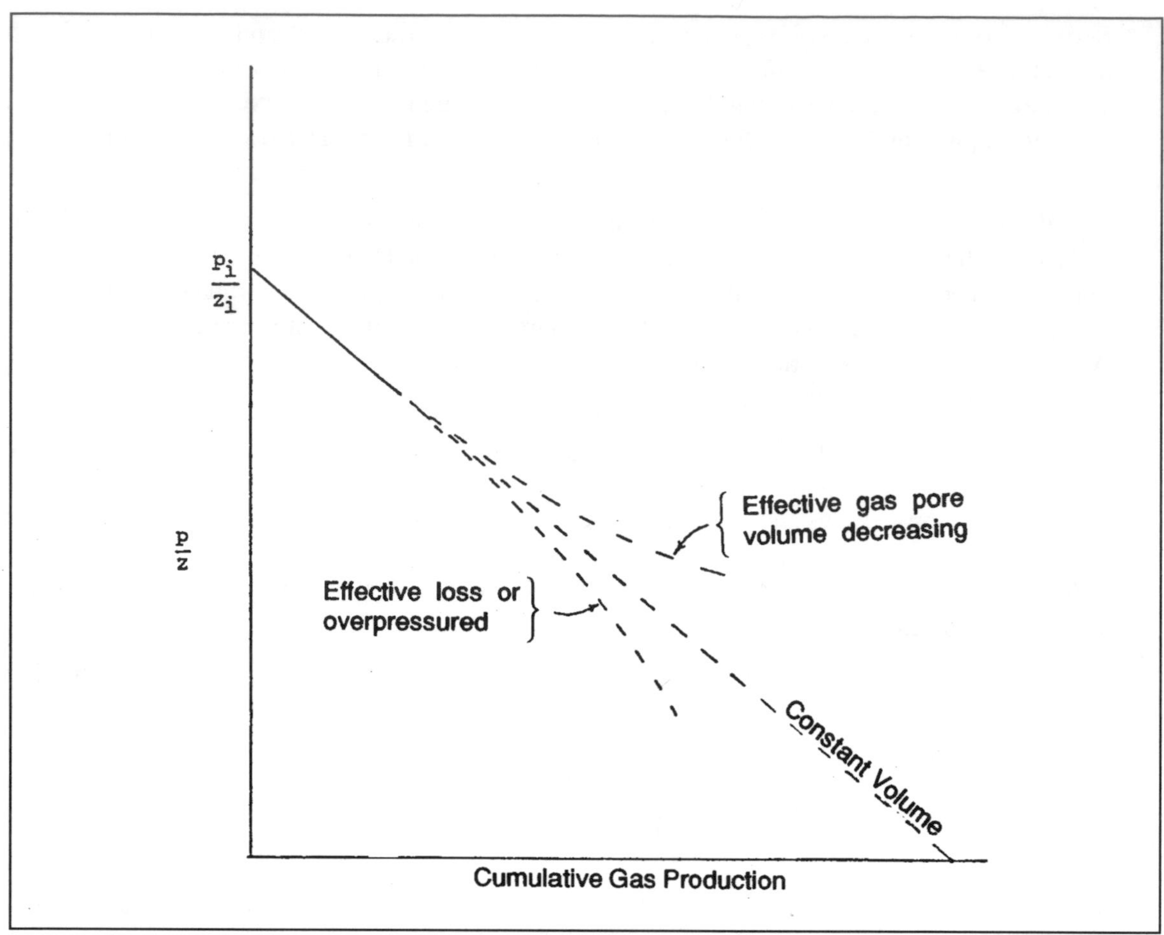

Fig. 5-7. Three main possible curve types for a gas reservoir performance plot of p/z vs. cumulative production.

If the p/z curve is bending downward, the possibilities include:

1. bad data;
2. retrograde condensation occurring (usually not significant enough to be seen);
3. drainage or leakage out of the reservoir (similar to (4) in the previous list, but in the opposite direction);
4. the company on the other side of the lease line (if graph is on a lease basis) is taking more reserves competitively than you are (they are draining your side of the reservoir as well as their own); and
5. the reservoir is overpressured, and a downshift of the curve is normal.

C. Wet Gas Reservoirs

If we consider Figure 5-2, point 1 represents either initial or current conditions.The solid vertical line (constant temperature) represents the pressure traverse taken in the reservoir caused by production. Notice that this solid line is to the right of the phase envelope, and therefore, the fluid state in the reservoir remains 100% gas with pressure depletion. However, the pressure-temperature traverse from the reservoir pressure at point 1 through the tubulars to the separator (represented by the dashed line) does enter the two-phase region with resultant liquid dropping out. Because (1) there is associated liquid production and (2) single phase gas in the reservoir, these types of systems are sometimes referred to as single-phase gas condensate reservoirs. The authors prefer to call them by the more appropriate name, "wet gas" reservoirs. What is important is that even though some produced liquids are seen at the surface, there is 100% gas in the reservoir.

Usually single-phase gas reservoirs will produce with gas/oil ratios exceeding 15,000 scf/STB of condensed liquids (condensate). As long as these liquids do not condense in the reservoir, the calculations of this chapter can be used. However, the cumulative gas production should be modified to include the "gas equivalent" of these condensed liquids. If liquids do drop out in the reservoir, then the methods of the "Gas Condensate" chapter can be used.

The produced liquid or condensate can be converted to its gas equivalent (if the specific gravity, γ_o, is known) by assuming that it behaves as an ideal gas when vaporized in the produced gas.

The gas equivalent, GE, is:

$$GE = V = \frac{nRT_{sc}}{p_{sc}} = \frac{(350.5)(\gamma_o)(R)(T_{sc})}{(M_o)(p_{sc})} \quad scf/STB \tag{5-8a}$$

If standard conditions are 14.7 psia and 520 °R, and the gas constant, R = 10.73, then the gas equivalent is:

$$GE = 133{,}000\ \gamma_o / M_o \quad scf/STB \tag{5-8b}$$

Cragoe[3] has provided the following formula as an estimate of the molecular weight of the condensate:

$$M_o = \frac{44.29\ \gamma_o}{1.03 - \gamma_o} = \frac{6084}{°API - 5.9} \tag{5-9}$$

It is common for a small amount of water to be produced as a condensate from the gas phase. If it is actually from the gas phase, it will be fresh and should be added to the gas production. Since the specific gravity of fresh water is 1.00 and its molecular weight 18, its gas equivalent is:

$$GE_w = \frac{nRT_{sc}}{p_{sc}} = \frac{(350.5)(1.00)}{18} \times \frac{RT_{sc}}{p_{sc}} \quad scf/STB \tag{5-10}$$

At the same conditions as for Equation 5-8b, the gas equivalent for water produced as condensate is 7,390 scf/STB of water.

Figure 5-8 is a convenient source of data for the water content of natural gas in equilibrium with liquid water. The use of gas equivalents for condensed liquid hydrocarbons and water is best shown by example.

Problem 5-1:

Determine the total daily gas production from a reservoir, including the gas equivalents of liquid hydrocarbons (condensate) and water. The available data are:

Separator-gas production	1.0 MMscf/D
Condensate production	20.0 STB/D
Stock-tank gas production	3.0 Mscf/D
Fresh-water production	3.0 STB/D
Initial reservoir pressure	3500 psia
Current reservoir pressure	1000 psia
Reservoir temperature	200 °F
Condensate gravity	60 °API (0.739 sp. gr.)

Solution:

Condensate gas equivalent production:

$M_o = (44.29)(0.739)/(1.03 - 0.739) = 112.5$ lb/mole

$GE = (133{,}000)(0.739)/112.5 = 873.7$ scf/STB

or $GE = (873.7)(20) = 17{,}473$ scf/D

Total hydrocarbon gas equivalent production =

$1{,}000{,}000 + 17{,}473 + 3{,}000 = 1{,}020{,}473$ scf/D

Water equivalent production:

According to Slider[27], for material balance calculations, any water production in excess of the original equilibrium (saturated) water vapor should be treated as produced water. Note that Figure 5-8 indicates increased water content of natural gas at lower pressures. But, if this is happening in the reservoir, it is connate water that is being vaporized. The production of some interstitial (connate) water acts to increase the pore volume available to gas. When calculating gas equivalent water production, use the water content (from Figure 5-8) at the original reservoir pressure.

From Figure 5-8,

@ 3500 psia and 200 °F, the equilibrium water vapor is 260 lb/MMscf.

Bbl/MMscf = 260/350 = 0.743 Bbl/MMscf

Gas equivalent water production = $(0.743)(7390)(1.020473) \approx 5600$ scf/D

Hence, the total daily vapor (equivalent gas) production is:

$$G_p = 1{,}020.5 + 5.6 = 1{,}026.1 \text{ Mscf/D}$$

And @ 1000 psia, the reservoir liquid water production (to be used with the material balance equation) is: $Q_w = 3 - (0.743)(1.02) = 2.24$ Bbl/D.

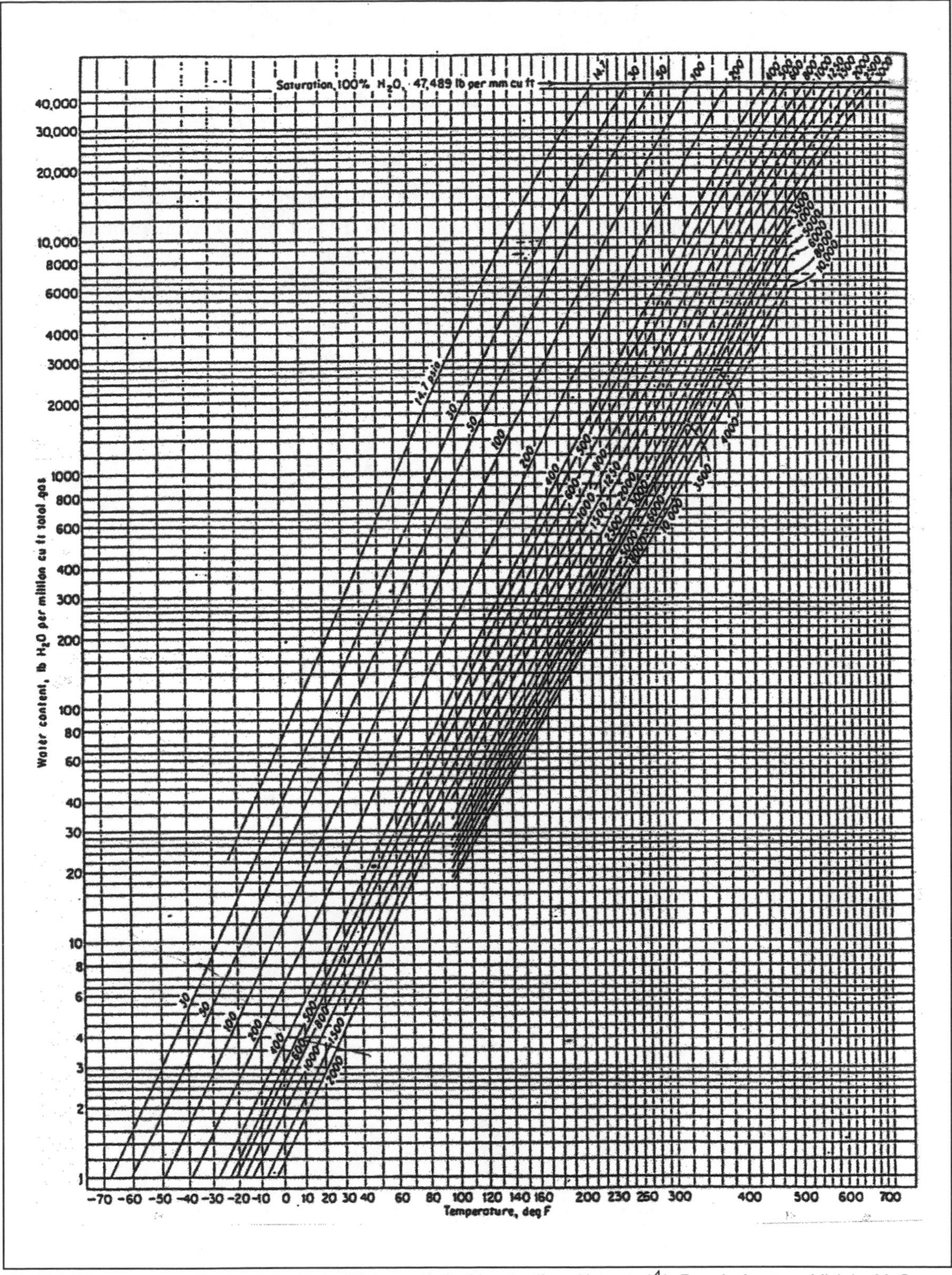

Fig. 5-8. Water content of natural gas in equilibrium with liquid water (from Katz et al.[4]). Permission to publish by McGraw Hill.

CALCULATION OF INITIAL IN-PLACE GAS AND OIL FROM FIELD DATA—WET GAS RESERVOIR

The method[3] can also be used for a retrograde gas condensate reservoir if the measured data have been obtained early in the life of the reservoir when the reservoir fluid is in a single phase. Gas-cap gas and condensate can be determined in the same manner.

It is important to note that there is gas production at the separator and at the stock tank. Additional gas escapes from the oil (condensate) at stock tank conditions. Thus, to get the average produced gas gravity:

$$\gamma_{g,avg} = \frac{(q_{g,sep})(\gamma_{g,sep}) + (q_{g,tank})(\gamma_{g,tank})}{(q_{g,sep} + q_{g,tank})} \tag{5-11}$$

where:

$\gamma_{g,avg}$ = average produced gas gravity (air = 1.0),
$q_{g,sep}$ = producing gas rate measured at the separator, Mscf/D,
$\gamma_{g,sep}$ = separator gas gravity (air = 1.0),
$q_{g,tank}$ = stock tank vent gas rate, Mscf/D, and
$\gamma_{g,tank}$ = vent gas gravity (air = 1.0).

Then, the actual producing gas/oil ratio is:

$$R = \left[\frac{q_{g,sep} + q_{g,tank}}{q_o}\right] \times 1000 \tag{5-12}$$

where:

R = producing gas/oil ratio, scf/STB, and
q_o = oil (condensate) producing rate, STB/D

Now, consider one stock-tank barrel of produced condensate and its associated gas, R (scf). Then, the total pounds of well fluid is:

$$m_w = (R)(\gamma_{g,avg})(28.97)/379.4 + (\gamma_o)(350) \tag{5-13}$$

where:

m_w = total pounds of well fluid associated with one STB of condensate,
γ_o = specific gravity of the stock tank condensate (water = 1.0),
28.97 = the molecular weight of air,
379.4 = number of cu ft of gas in 1 mole at 60 °F and 14.7 psia,
350 = number of pounds that one barrel of water weighs.

Notice in Equation 5-13, that (R/379.4) x ($\gamma_{g,avg}$) x (28.97) represents the weight of the gas, and (γ_o)(350) is the weight of the oil.

Recall that the tank oil specific gravity, γ_o, (60°F and 14.7 psia) is related to API gravity as:

$$\gamma_o = 141.5 / (131.5 + °API) \tag{5-14}$$

Then, the total moles of gas and oil associated with one barrel of condensate is:

$$n_t = (R / 379.4) + (350) (\gamma_o) / M_o \tag{5-15}$$

where:

n_t = total moles of gas and oil associated with 1 STB of condensate,
γ_o = tank oil specific gravity (water = 1.0), and
M_o = is the molecular weight of the condensate.

Equation 5-9 is repeated here for ease of reference:

$$M_o = \frac{44.29\,\gamma_o}{1.03 - \gamma_o} = \frac{6084}{°API - 5.9} \tag{5-9}$$

The total well fluid molecular weight can be calculated:

$$M_w = m_w / n_t \tag{5-16}$$

where:

M_w = total well fluid molecular weight.

Total well fluid gas gravity is therefore:

$$\gamma_w = M_w / 28.97 \tag{5-17}$$

where:

γ_w = total well fluid gas gravity (air = 1.0)

Assume that specific gravity from Equation 5-17 is representative of the reservoir vapor. Now determine approximate pseudocritical pressure and temperature from the condensate well fluids curves of Figure 3-16 of the "Fluid Properties" chapter. With the reservoir temperature and discovery pressure, pseudoreduced temperature and pressure are calculated:

$$p_{pr} = p_i / p_{pc} \tag{5-18}$$

$$T_{pr} = (T_f + 460) / T_{pc} \tag{5-19}$$

where:

p_{pr} = pseudoreduced pressure, discovery conditions,
p_i = reservoir static discovery pressure, psia,
p_{pc} = pseudocritical pressure, psia,
T_{pr} = pseudoreduced reservoir temperature,

T_f = formation temperature, °F, and
T_{pc} = pseudocritical temperature, °R.

The z factor is obtained from Figure 3-15 of the "Fluid Properties" chapter.

If we consider Equation 5-1, it is apparent that $(43{,}560)(\phi)(1-S_w)$ is equal to the volume in cubic feet available for reservoir gas in one net acre-foot of reservoir rock.

At this point, let "G" represent the total initial standard cubic feet of gas (one phase reservoir vapor including gas, condensate vapor, and water vapor) in place per net acre-foot of reservoir rock. Then, the total moles in "G": n = G/379.4.

Substitute both the volume and mole results into the real gas law:

$$pV = znRT$$

$$(p_i)(43{,}560)(\phi)(1 - S_w) = (z_i)(G/379.4)(R)(T_f + 460)$$

Simplifying and using R = 10.73, obtain:

$$G = 1{,}540{,}230 \frac{p_i \phi (1 - S_w)}{z_i (T_f + 460)} \tag{5-20}$$

where:

G = total initial scf of reservoir vapor per net ac-ft of res. rock,
p_i = discovery pressure, psia
ϕ = porosity, fraction,
S_w = water saturation, fraction,
z_i = reservoir gas discovery conditions z factor, and
T_f = formation temperature, °F.

The surface gas mole fraction is: $f_g = n_g / (n_g + n_o)$, which is equal to:

$$f_g = \frac{(R / 379.4)}{(350\, \gamma_o / M_o) + (R / 379.4)} \tag{5-21}$$

where f_g is equal to the gas mole fraction related to gas and condensate. This is also equal to the mole fraction of reservoir fluid that will be surface gas. Then:

$$GIP = (f_g)(G) \tag{5-22}$$

where:

GIP = initial hydrocarbon in place (scf/net ac-ft) that will be gas at the surface, or we usually say initial gas in place.

$$OIP = GIP / R \tag{5-23}$$

where:

OIP = initial oil (condensate) in place, STB / net ac-ft.

$$q_f = (q_{g,sep} + q_{g,tank})/f_g \tag{5-24}$$

where:

q_f = total daily reservoir fluid (gas + condensate) production, Mscf/D.

Then, to calculate total daily reservoir voidage:

$$Voidage = \left[\frac{q_{g,sep} + q_{g,tank}}{f_g}\right]\left[\frac{T_f + 460}{520}\right]\left[\frac{14.7}{p}\right][z] \tag{5-25}$$

where:

$voidage$ = total daily reservoir volume of hydrocarbon fluid withdrawn, thousands of reservoir cubic feet per day, (Mcf/D, not Mscf/D),

T_f = formation temperature, °F,

p = formation pressure, psia, and

z = z factor corresponding to pressure, p.

Problem 5-2:

Determine the initial gas and condensate in place per net acre-foot of reservoir rock for the following data:

Discovery pressure	3000 psia
Reservoir temperature	200 °F
Average porosity	20%
Average water saturation	30%
Oil (condensate) producing rate	200 STB/D
Oil gravity, 60 °F	55 °API
Gas producing rate	3,000 Mscf/D
Separator gas gravity	0.70
Vent gas produced at the stock tank	100 Mscf/D
Vent gas gravity	1.10

Solution:

(1) Average gas gravity = [(3,000 x 0.70) + (100 x 1.10)] / (3,000 + 100) = 0.713

$\gamma_o = (141.5) / (55 + 131.5) = 0.759$

$M_o = (6084) / (55 - 5.9) = 123.9$

$R = [(3,000 + 100) / 200] (1,000) = 15,500$ scf/STB

$m_w = [(15,500 \times 0.713 \times 28.97 / 379.4] + [350 \times 0.759] = 1,109.5\ lb$

$n_t = [15,500 / 379.4] + [(350 \times 0.759) / 123.9] = 43.00\ moles$

$M_w = 1,109.5/43.00 = 25.80$

$\gamma_w = 25.80/28.97 = 0.891$

(2) Using an average fluid (here taken to be a gas) gravity compared to air of 0.891, reference to Figure 3-16, in the "Fluid Properties" chapter, for condensate well fluids, yields a pseudocritical pressure of 651 psia and a pseudocritical temperature of 426 °R. Then, T_{pr} = (200 + 460)/426 = 1.549 and p_{pr} = 3000/651 = 4.608, and Figure 3-15 of the same chapter yields a supercompressibility factor, z = 0.813, at 3,000 psia and 200 °F.

(3) Initial hydrocarbon vapor in place per acre-foot of reservoir rock is:

$$G = 1{,}540{,}230 \times \frac{(3000)(0.20)(1-0.3)}{(0.813)(660)} = 1{,}205{,}592 \; scf / acre\text{–}ft$$

$$= 1{,}205.6 \; Mscf / acre\text{–}foot$$

(4) The mole fraction of the total initial hydrocarbon fluid in place that is residue or sales gas is:

$$f_g = \frac{15{,}500 / 379.4}{15{,}500 / 379.4 + 350 \times 0.759 / 123.0} = 0.9502$$

And,

Initial gas in place = 0.9502 × 1,205.6 = 1,145.5 Mscf/ac-ft
Initial oil in place = 1,145,500/15,500 = 73.90 STB/ac-ft

(5) The total daily hydrocarbon vapor production is:

= Daily gas/0.9502 = (3,000 + 100)/0.9502 = 3,262 Mscf/D

(6) The total daily reservoir voidage is

= 3,262,000 x [(200 + 460) / (60 + 460)] x [14.7/3,000] x 0.813 = 16,493 cu ft/day

Notice that the answers are as good as the data used. Poorly chosen data will yield unreliable answers. Particular care should be taken in including the stock tank vapors (and their gravity) with the separator gas when calculating the compressibility factor (z factor) for the reservoir gas. If carefully taken data and accurate analyses of the produced gas and hydrocarbon liquid are available, then a meaningful calculation (similar to the previous problem) can be accomplished.

GAS RESERVES

A. Introduction

The term "gas reserves" refers to the fraction or portion of the original gas in place that is producible. This is dependent on factors such as initial pressure, abandonment pressure, and reservoir drive mechanism.

Similar to gas in place, reserves may be calculated based on volumetric considerations or via material balance. Basically, the volumetric methods are used early in the life of the reservoir before much production data exist. Later in the reservoir life, it is usually more accurate to predict overall reservoir performance on the basis of material balance calculations.

Generally speaking, the higher the discovery pressure, the higher will be the recovery, if all other factors are held constant.

B. Abandonment Pressure

Abandonment pressure is usually a function of economic considerations. Basically, it is that static pressure at which the gross profits from the produced gas are approximately equal to the operating costs of the reservoir.

When considering a newly discovered reservoir, often an estimate of abandonment pressure is needed, say to estimate reserves. To determine abandonment pressure, p_a, two pieces of data must be known (or estimated): (1) the minimum economic rate from a single well, q_{abd}, and (2) a surface pressure, p_{surf} (sales line pressure or compressor inlet pressure). Using q_{abd}, calculate the surface pressure drop, Δp_{surf} (from the wellhead to p_{surf}). Next, Δp_{tub} (pressure drop through the wellbore) and Δp_{res} (through the reservoir) are calculated. Then:

$$p_a = p_{surf} + \Delta p_{surf} + \Delta p_{tub} + \Delta p_{res} \tag{5-26}$$

Methods for calculating Δp_{tub} and Δp_{res} will be discussed later in this chapter.

C. Reservoir Drive Mechanism

Constant Volume Reservoir

It has been estimated that between one-half and two-thirds of all gas reservoirs are constant volume (so-called "volumetric") reservoirs. These reservoirs essentially maintain a constant pore volume for the life of the reservoir. Actually, the rock and connate water are slightly compressible, and therefore, a small volume change occurs in the reservoir with pressure depletion. This volume differential is usually small and can be ignored. If, however, the reservoir in question is overpressured, has mobile connate water, has a matrix system undergoing compaction, or has an active water drive, then the reservoir volume available to the gas is not constant. In this case, the volumetric calculation methods should not be used.

The performance of a volumetric reservoir may be represented as shown in Figure 5-5. Here "G" is defined to be that "G_p" obtained by extrapolating the curve to a p/z of 14.7 psia; or the standard cubic feet of gas that would be produced if it were possible to reduce the reservoir static pressure to atmospheric pressure. Then, if abandonment pressure and z factor are known, reserves may be determined from the linear plot as shown.

Recovery efficiency is defined to be the ratio of reserves to gas initially in place expressed as a percent:

$$R.\,E. = (100)(G_{pa}/G) \tag{5-27}$$

where:

$R.\,E.$ = recovery efficiency, %,
G_{pa} = cumulative gas produced at abandonment (reserves), scf,
G = gas initially in place, scf.

Notice that if G is defined as in Equation 5-27 (p/z extrapolated to 14.7 psia), then if Equation 5-27 is used on any reservoir that has an abandonment pressure below atmospheric, there would be a problem with a recovery efficiency greater than 100%. In this case, it would be more appropriate to define G as the amount of gas that would be produced if the reservoir static pressure could be lowered to 0 psia.

Equation 5-1 is the volumetric equation for a homogeneous, isotropic gas reservoir. During the development stage, the reservoir bulk volume may not be known. In this instance, we could modify Equation 5-1 and calculate the original gas in place per net acre-foot:

$$G = \frac{(43{,}560)(\phi)(1 - S_w)}{(B_{gi})} \; scf/\; ac\text{–}ft \tag{5-28}$$

where:

$$B_{gi} = [\,(p_{sc})(T_f)\,(z_i)\,]/\,[\,(p_i)(T_{sc})\,]$$

Recall that these temperatures have units of degrees Rankine. If a reasonable estimate can be made of the abandonment pressure, then the gas (per acre-foot) remaining in the reservoir at abandonment can be determined:

$$G_a = \frac{(43{,}560)(\phi)(1 - S_w)}{B_{ga}} \; scf/\; ac\text{–}ft \tag{5-28a}$$

where:

$$B_{ga} = [\,(p_{sc})(T_f)(z_a)\,]/[\,(p_a)(T_{sc})\,]$$

The subscript “a” refers to abandonment conditions. The recovery efficiency can be written as:

$$Recovery\ efficiency = (100)(G - G_a)/\,G = 100\left(1 - \frac{B_{gi}}{B_{ga}}\right)\% \tag{5-29}$$

If we denote that pressure to which the p/z vs G_p plot is extrapolated to obtain G as the base pressure (often 14.7 psia), then a third recovery efficiency equation can be written:

$$R.E. = \left[\frac{\left(\frac{p}{z}\right)_i - \left(\frac{p}{z}\right)_a}{\left(\frac{p}{z}\right)_i - \left(\frac{p}{z}\right)_b} \right] (100)\ \% \tag{5-30}$$

where:

b = subscript referring to base pressure such as 14.7 psia or 0 psia or whatever, and
a = subscript indicating abandonment conditions

Then,

reserves = (R.E./100)(G)

Of course, the units of these "reserves" would depend on the units of G: normally either Bscf or scf/(net ac-ft).

The term "unit recovery"[2,17] refers to recoverable scf per net acre-foot:

$$G - G_a = 43{,}560\,(\phi)\,(1 - S_w)\left(\frac{1}{B_{gi}} - \frac{1}{B_{ga}}\right) \tag{5-31}$$

If reserves calculations are being made for less than the total reservoir then implicit in the use of Equations 5-29, 5-30, and/or 5-31 is that no migration is occurring across property lines.

It should be mentioned that there are some points of confusion in the petroleum industry when reporting reserves. Recovery efficiency is sometimes referred to as recovery factor[2]. Other times, recovery factor refers to unit recovery[18]. In this work, recovery factor and recovery efficiency will be used interchangeably.

Another point of possible confusion is that some authors report formation volume factor as the ratio of surface volume to reservoir volume. This is the reciprocal of the formation volume factor defined in this work.

The highest recovery efficiencies, normally 70 to 90 percent, are obtained in depletion-type (volumetric) reservoirs having no water encroachment. In the United States, current gas prices often permit the installation of compression facilities so that flowing pressures can be reduced to low levels, yet sales made to much higher pressure gas lines. Those depletion-type reservoirs that are abandoned at relatively high pressures, such as 30 percent or more of discovery pressure, either have low gas prices, gas permeability problems due to sensitive swelling clays or reservoir compaction, or some regulatory or economic constraint.

WATER-DRIVE GAS RESERVOIRS

Many gas reservoirs are associated with sizeable underlying aquifers, which, as reservoir pressure drops, allow water encroachment. This serves to maintain reservoir pressure to a degree, depending on the amount of encroachment. Thus, for a time, wellhead pressures and, therefore, producing rates are maintained. This may result in lower operating cost since compression equipment may not be needed.

Many people think intuitively that since a water-drive oil reservoir has a better recovery efficiency than does a depletion-type oil reservoir, then the same should be true for gas reservoirs. Unfortunately, the encroaching water is inefficient in displacing gas from porous media. Consider the water-invaded zone of Figure 5-4. For a gas reservoir, the remaining gas saturation in the water-invaded zone is normally in the range of 30 to 50 percent. Why is so much gas left behind?

Residual hydrocarbon saturation, or that saturation at which hydrocarbon permeability goes to zero, is directly related to interfacial tension between the hydrocarbon and the water which is normally in the range of 40 to 60 dynes/cm for a gas reservoir. This high interfacial tension causes a high relative attraction between the water and the rock (compared to that between the gas and the rock). The resultant effect is to allow the water to bypass pockets of gas that then remain "trapped." This is discussed further in the capillary pressure section of the "Rock Properties" chapter.

Due to heterogeneity, sweep efficiency (fraction of the reservoir rock contacted by the water in the invaded zone) is normally less than 100%. Therefore, the remaining hydrocarbon saturation is generally higher than the residual hydrocarbon saturation.

Table 5-1 provides values for residual gas saturations after waterflood in core plugs. The values given here should be considered a minimum of what will happen in an actual reservoir because heterogeneity and anisotropy are not considered in core floods.

If the encroaching water in a water-drive gas reservoir results in gas displacement and abandonment at a stabilized pressure, then the unit recovery can be written as:

$$Unit\ recovery = (43{,}560)(\phi)\Big[((1 - S_w)/B_{gi}) - (S_{gr}/B_{ga})\Big] \tag{5-32}$$

where:

Unit recovery = reserves, scf/net ac-ft,
S_{gr} = residual gas saturation, fraction, and
B_g = gas formation volume, ft^3/scf

The corresponding recovery efficiency is:

$$R.E. = \frac{(100)\ [((1 - S_w)/B_{gi}) - (S_{gr}/B_{ga})]}{(1 - S_w)/B_{gi}} \tag{5-33}$$

where:

R. E. = recovery efficiency, percent

Table 5-1
Residual gas saturation after waterflood as measured on core plugs
(from Geffen *et al.*[5])

Porous Material	Formation	S_{gr}, percent	Remarks
Unconsolidated sand		16	13-ft column
		21	1-ft core
Slightly consolidated sand (synthetic)			
Synthetic consolidated materials	Selas Porcelain	17	1 core
	Norton Alundum	24	1 core
Consolidated Sandstones	Wilcox	25	3 cores
	Frio	30	1 core
	Nellie Bly	30-36	12 cores
	Frontier	31-34	3 cores
	Springer	33	3 cores
	Frio	30-38 (Ave 34.6)	14 cores
		34-37	
	Torpedo	40-50	6 cores
	Tensleep		4 cores
Limestone	Canyon Reef	50	2 cores

The obvious question, "How can recovery efficiency be improved in a water-drive gas reservoir?", has several possible answers.

(1) Lower the interfacial tension between the water and the gas. This idea is sound but not easy to implement.

(2) In some cases, it may be possible to produce the wells at very high rates. The idea is to "get ahead" of the advancing gas/water contact. This rapid depletion, if possible, will result in higher-percentage gas recoveries since the trapped gas exists at a lower pressure. Such an operating strategy assumes that there is a market for all the gas and that coning will not be a problem.

(3) Cycling with nitrogen, flue gas, or some other low value (probably inert) gas is a possible solution if such a gas is available. In this case, we would cycle the reservoir with the low value gas until a limiting percentage of the injected gas is reached in the produced gas. At this point (determined by economic considerations), cycling is terminated, and blowdown of the reservoir begins. Produced gas will need to be processed to separate the reservoir fluid from the injected gas. A sizeable portion of the gas entrapped by the advancing aquifer during blowdown will be the less valuable injected gas.

(4) Produce the water from the aquifer at the same time as producing the gas. The idea is to balance withdrawals so that the gas/water contact does not move. As far as actual field application, this is quite a new idea. A different strategy has been reported to have been implemented successfully in the Texas gulf coast area. In several large water-drive gas fields, the gas / water contact had advanced much of the way through the reservoir, and the operation was at the economic limit. High volume downhole centrifugal pumps were installed in wells near the original gas/water contact, and water withdrawals made. The pressure was reduced due to the water production, and the residual gas saturation expanded such that mobile gas

resulted. This allowed the gas to be produced. Such a strategy might be termed "secondary gas recovery." One big problem is the disposal of the produced water.

As has been mentioned, Equation 5-7 may be used to predict the performance of a water-drive gas reservoir when used in combination with an appropriate aquifer mathematical model. The method is discussed and illustrated in the "Water Drive" chapter.

COMPACTION DRIVE

Compaction drive can constitute an important drive mechanism in some hydrocarbon reservoirs. This type of drive, which is usually associated with unconsolidated formations, may not be apparent until reservoir pressure declines below a certain threshold pressure[19]. At this point, the formation can no longer support the weight of the overburden and actually begins to compact with resulting decrease in pore volume. This helps to maintain formation pressure, but may cause some loss in system permeability.

If the compaction is appreciable, there may be accompanying visible surface subsidence. There is evidence[19] that this subsidence can be analyzed using elastomechanical theory to predict future compaction and future withdrawal versus pressure.

CALCULATION OF BOTTOMHOLE PRESSURES

A. Introduction

While it is necessary to determine original gas in place and producible gas reserves, it is also necessary to estimate productivity or absolute open-flow potential of gas wells. For this, the bottomhole pressure will need to be known, either by actual measurement with a bottomhole pressure gauge or by a calculation from wellhead pressure measurements. Often measurement of pressure opposite the producing formation is impractical due to wellbore conditions, cost, and time considerations.

The published methods for calculation of bottomhole pressures have been developed from the first law of thermodynamics, which is a basic energy balance. For steady state flow, the equation is:

$$\frac{144}{\rho}\, dp + \frac{u}{2\, \alpha\, g_c}\, du + \frac{g}{g_c}\, dX + \frac{2 f u^2}{g_c D}\, dL + W_s = 0 \qquad (5\text{-}34)$$

where:

ρ = density of the fluid, lb_m/ft^3,
p = pressure, psia,
u = average velocity of the fluid, ft/sec,
α = correction factor to compensate for the variation of velocity over the tube cross section (It varies from 0.5 for laminar flow to 1.0 for fully developed turbulent flow),
X = distance in the vertical downward direction,
f = Fanning friction factor,
D = inside diameter of the pipe, ft,

L = length of the flow string, ft. For a vertical flow string, L = X.

$\frac{u}{2\,\alpha\, g_c}\, du$ = pressure drop due kinetic energy effects,

$\frac{2 f u^2}{g_c D}\, dL$ = pressure drop due friction effects,

W_s = mechanical work done on or by the gas. Here, $W_s = 0$.

The real gas law can be used to determine the density of the reservoir gas by recognizing that the moles of gas in one cubic foot of reservoir gas pore space are p/zRT. Also, the molecular weight of gas is equal to (28.97)(γ_g) pounds per mole. Therefore, the reservoir gas density is:

$$\rho_g = \frac{(28.97)(\gamma_g)(p)}{z\ R\ T}\ lb\,/\,ft^3$$

This can be readily converted to gas gradient by dividing the density in pounds per cubic foot by 144 sq in/sq ft:

$$Gas\ gradient = \frac{(0.01875)(\gamma_g)(p)}{z\ T}\ psi\,/\,ft \tag{5-35}$$

Note that the gas constant, R, was replaced by 10.73.

B. Static Bottomhole Pressure—Single Phase Gas

For a shut-in gas well, the pressure at the producing depth is the sum of the static wellhead pressure and the pressure due to the weight of the gas column in the wellbore. Because this is a no-flow situation, the kinetic energy, friction, and work terms of Equation 5-34 are zero. Therefore, we may write:

$$dp = (Gas\ Gradient)\ dX \tag{5-36}$$

where dX is equal to the change in vertical distance in the wellbore.

Now, if we substitute Equation 5-35 into Equation 5-36, separate variables and integrate (from surface to reservoir depth, D; and from surface pressure, p_s, to bottomhole pressure, p_w), we arrive at the "Bureau of Mines" equation:

$$p_w = p_s \exp\left\{[(0.01875)(\gamma_g)(D)]\ /\ [\ (z_{avg})(T_{avg})\,]\right\} \tag{5-37}$$

This equation requires that average values in the wellbore for "z" and "T" be used.

To illustrate the use of this equation, its terms and units, an example problem will be worked. First, however, an equation will be given that is used to estimate the pressure due to the weight of the column of wellbore gas.

$$\Delta p = (0.25)\left(\frac{p_s}{100}\right)\left(\frac{D}{100}\right) psi \tag{5-38}$$

where:

Δp = approximate [(bottomhole pressure) - (wellhead pressure)], psi.

This approximation is used to obtain an average wellbore pressure so that an initial z_{avg} can be calculated.

Problem 5-3:

Determine the sandface pressure for single phase gas well where the shut-in tubing pressure is 1000 psia, gas gravity is 0.65, depth is 3000 ft, formation temperature is 115 °F, and surface temperature is 70 °F.

Solution:

Assume that the gas is sweet and the theorem of corresponding states allows calculation of z factor directly.

Pressure caused by gas column weight:

= (0.25)(1000/100)(3000/100)

= 75 psi

Approximate p_w = 1000 + 75 = 1075 psia

p_{avg} = (1000 + 1075)/2 = 1037.5 psia

$\bar{p}_{pr} = p_{avg}/p_{pc}$ = 1037.5/670 = 1.549

T_{avg} = (70 + 115)/2 = 92.5 °F = 552.5 °R

$T_{pr} = T_{avg}/T_{pc}$ = 552.5/372 = 1.485

z_{avg} = 0.85

Now, using the "Bureau of Mines" equation, Equation 5-37:

$$p_w = 1000\ \exp\left\{[(0.01875)(0.65)(3000)]/\ [(0.85)(552.5)]\right\} = 1081\ psia$$

Notice that this is a little higher than the bottomhole pressure that was estimated. So, to complete the solution, a new average pressure is calculated, and a better value for p_w is determined. In this instance, the correction would be minor.

C. Flowing Bottomhole Pressure—Single-Phase Gas

The flowing bottomhole pressure in a gas well is the sum of the flowing wellhead pressure, the weight of the flowing gas column in the wellbore, and the kinetic energy change and friction losses that occur. Studies have shown that the kinetic energy term (the second one in Equation 5-34) is negligible in all practical gas well calculations. So, if we disregard the term and assume an average temperature and average compressibility factor for the wellbore, then we can separate variables and integrate between the appropriate limits. Although several limiting assumptions have been made, good results can often be obtained from the resulting equation:

$$(p_w)^2 = (p_s)^2 (e^S) + \frac{100\ \gamma_g\ T\ \bar{z}\ \bar{f}\ X\ (e^S - 1)\ Q^2}{d^5\ S} \tag{5-39}$$

where:

p_w = flowing bottomhole pressure, psia,
p_s = flowing wellhead pressure, psia,
γ_g =gas gravity (air = 1.0),
X = distance in the vertical downward direction, ft,
T = arithmetic mean of bottomhole and wellhead temperatures, °R,
$\bar{z}$ = compressibility factor at T and $(p_w + p_s)/2$,
$\bar{f}$ = Fanning friction factor,
d = internal pipe diameter, in,
Q = gas production rate in MMscf/D (14.65 psia, 60 °F),and
$S = (2)(\gamma_g)(X) / (53.34)(T)(\bar{z})$.

The Fanning friction factor is a function of Reynolds number and relative roughness of the wellbore wall.

For steady state flow, the Reynolds number is equal to:

$$R_e = (20{,}011)(\gamma_g)(Q) / (\mu)(d) \tag{5-40}$$

where:

R_e = Reynolds number, dimensionless, and
μ = viscosity, cp, at the average wellbore temperature and pressure.

The relative roughness, δ/d, is equal to the ratio of the absolute roughness δ (the distance, perpendicular to the pipe wall, between peak highs and valley lows in the pipe wall irregularities) to the internal pipe diameter, d.

Figure 5-9 presents experimental data for the relative roughness of pipe. For new pipe or tubing used in most wells, the absolute roughness has been found to be on the order of 0.00060 to 0.00065 in. This corresponds to the lower line in Figure 5-9, which will be satisfactory unless better information is available. For old tubing, Brown[20] suggests using 0.018 in., the roughness of commercial steel, which is represented by the middle line of Figure 5-9. For extremely dirty or rough pipe, the top line should be used.

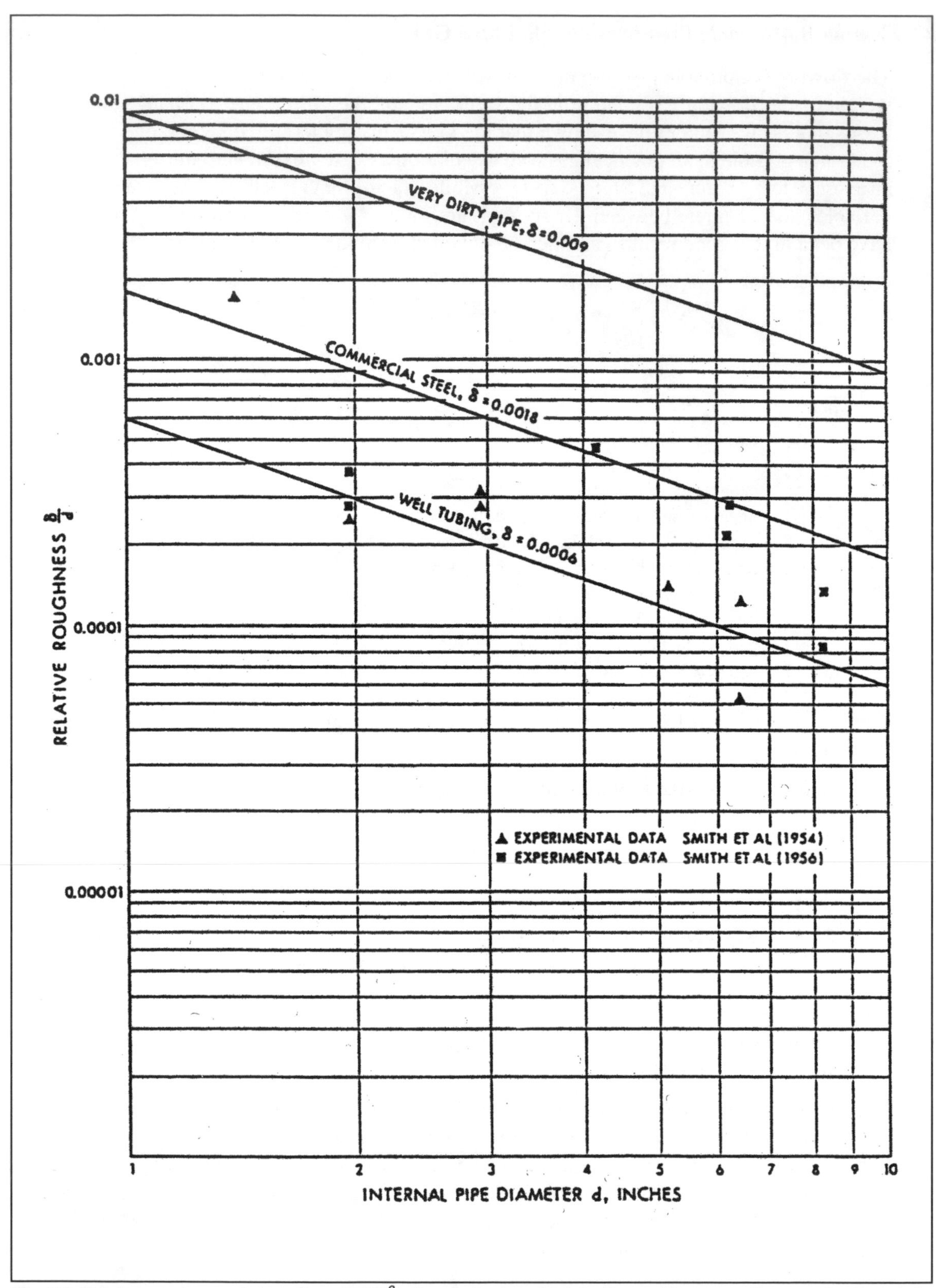

Fig. 5-9. Relative roughness of pipes (from ERCB[8]). Permission to publish by Energy Resources Conservation Board of Canada.

If the Reynolds number and relative roughness are known, then the Fanning friction factor can be determined from Figure 5-10.

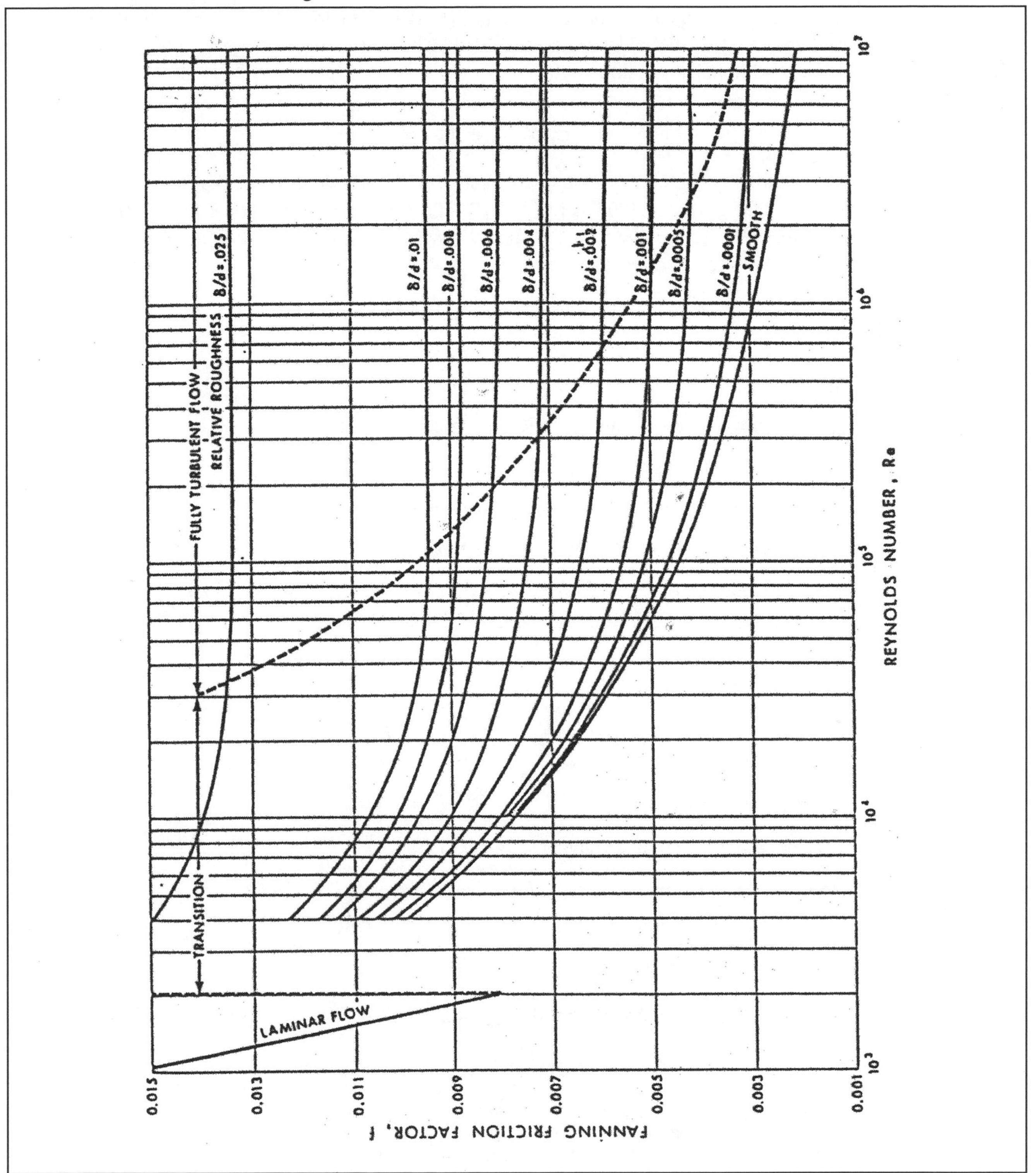

Fig. 5-10. Friction factors for fluid flow in pipes (from ERCB[8]). Permission to publish by Energy Resources Conservation Board of Canada.

It is now possible to solve Equation 5-39 and obtain the flowing sandface pressure in a gas well where a single phase is flowing. If the well is relatively shallow (less than 3000 ft), a one-step calculation from the wellhead to the sandface is usually satisfactory. However, as in problem 5-3, a trial-and-error calculation is the normal procedure using mean temperature, compressibility factor and, additionally, friction factor.

Problem 5-4:

Determine the flowing sandface pressure for a single-phase gas well where the flowing tubing pressure is 500 psia, gas gravity is 0.65, depth is 3000 ft, formation temperature is 115 °F, flowing surface temperature is 90 °F, and the flow rate is 500 Mscf/D at standard conditions of 14.65 psia and 60 °F through 2 in. internal diameter tubing.

Solution:

Assume that the gas is sweet.

(1) Average flowing temperature = (550 + 575)/2 = 562.5 °R.

(2) Average gas compressibility factor:

First, estimate bottom hole flowing pressure with Equation 5-38:

Approximate p_w = 500 + (0.25)(500/100)(3000/100) = 537.5 psia.

However, we need to allow for friction loss in the tubing. So, arbitrarily, this will be increased to 575 psia. Therefore, the first estimate of average flowing pressure is 537.5 psia.

$p_{pr} = 537.5/670 = 0.802$

$T_{pr} = 562.5/372 = 1.512$

$\bar{z} = 0.921$

(3) Average friction factor:

From Figure 5-9, relative roughness is 0.000298.

Average viscosity may be obtained from a gas viscosity correlation such as Figure 3-24 (since the gas is sweet), "Fluid Properties" chapter: $\mu = 0.011$ cp.

The Reynolds number is:

$$R_e = (20{,}011)(0.65)(0.5)/(0.011)(2) = 295{,}617 \text{ or } 2.95617 \times 10^5$$

Then, from Figure 5-10, the friction factor is : $\bar{f} = 0.0042$.

(4) Flowing sandface pressure:

$S = (2)(0.65)(3000)/(53.34)(562.5)(0.921) = 0.1411$

$e^S = 1.1515$

And,

$$p_w^2 = (500)^2 (1.1515) + \frac{(100)(0.65)(562.5)(0.921)(0.0042)(3000)(0.1515)\ (0.5)^2}{(2)^5\ (0.1411)}$$

$$= 287{,}875 + 3559.1$$

$$p_w = 540 \text{ psia}$$

To complete the solution, a new average pressure is calculated, and a better value for the flowing sandface pressure is determined. In this case, the correction would be small.

Problem 5-4 illustrates a simple flowing sandface pressure calculation. Where wells are deeper, flow rates higher, and pressure differences in the flow string larger, this method can still be used, but the pressure traverse should be divided into several vertical intervals.

Other more sophisticated methods are available, the best known being those of Sukkar and Cornel[9] and Cullender and Smith.[7] Both of these methods make fewer simplifying assumptions, but do provide more accurate answers for adverse well conditions. The Sukkar and Cornell method requires considerably more complex calculations than either the method just presented or the Cullender and Smith procedure, which follows.

METHOD OF CULLENDER AND SMITH

In 1956, Cullender and Smith[7] published a method for calculating bottomhole pressure of a gas well. The procedure normally is used by starting with a known surface pressure and flow rate. The well may be either static or flowing and either vertical or slanted. For the flowing case, the friction factor equation assumes turbulent flow with an absolute roughness of 0.0006 inches. However, Peffer *et al.*[28] have illustrated how other friction factor can be used. The Cullender and Smith method may be used for single phase gas wells that are not making significant amounts of liquids. Practically speaking, it may be used as long as the liquids are not dropping out in the wellbore onto the surface of the tubing. There must be a steady stream with any contained liquids well entrained in the gas flow stream. An energy balance was used to derive the basic equation, which for vertical flow is:

$$\frac{G\,L}{53.34} = \int_{p_s}^{p_w} \frac{\left[\dfrac{p}{T\,z}\right]}{1000\,(F_r Q)^2 + \left[\dfrac{p}{T\,z}\right]^2}\,dp \tag{5-41}$$

For the flowing case, a friction factor is needed. Due to the assumption of full turbulent flow with an absolute roughness of 0.0006 inches, the relationship is a function of pipe internal diameter:

$$F_r = \frac{0.10797}{d^{2.612}} \quad for \; d \leq 4.277 \; in. \tag{5-42}$$

$$F_r = \frac{0.10337}{d^{2.582}} \quad for \; d > 4.277 \; in.$$

where:

G = gas gravity relative to air, fraction,
L = length of pipe, ft,
53.34 = combination of R and other constants,
Q = flow rate in MMscf/D,
z = z factor,
p = pressure, psia,
T = temperature, °R,
p_s = surface pressure, psia,
p_w = bottomhole pressure, psia,
d = inside diameter of tubing, in.

Notice that for a static wellbore, Equation 5-41 simplifies to:

$$\frac{G\,L}{53.43} = \int_{p_s}^{p_w} \frac{T\,z}{p}\,dp \tag{5-43}$$

For inclined flow:

$$\frac{G\,H}{53.34} = \int_{p_s}^{p_w} \frac{\left[\frac{p}{T\,z}\right]}{1000\,(L/H)\,(F_r Q)^2 + \left[\frac{p}{T\,z}\right]^2}\,dp \tag{5-44}$$

where:

H = vertical elevation difference, ft,
L = length of tubing, ft.

The method will be illustrated for a vertical flowing well. Thus, Equation 5-41 will be used. Notice that this equation involves an integral that has as its upper limit the bottomhole well pressure. Thus, the procedure will be a trial-and-error numerical evaluation process. The wellbore must be broken up into a number of segments, such as:

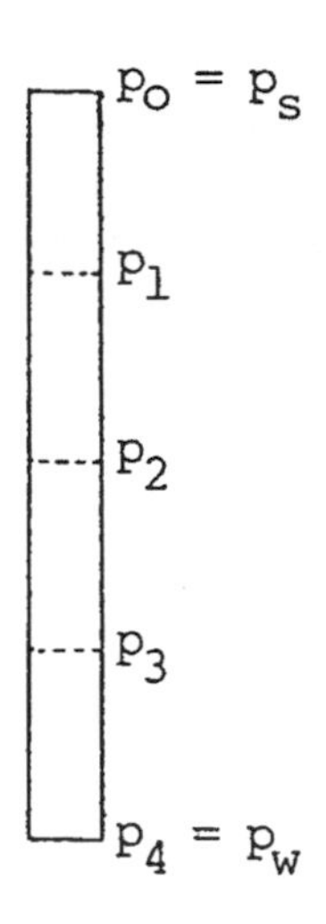

As illustrated, typically four segments are used. However, a general rule of thumb is that no segment should be longer than 2500 ft; and normally segments shorter than 1000 ft are not used. But even in shallow wells, it is best to use at least two segments. Especially with several segments, this technique is more suitable for computer solution than for a hand calculation.

Let "I" represent the integrand of the main equation (which for the vertical flowing case is Equation 5-41):

$$I = \frac{\left[\dfrac{p}{T\,z} \right]}{1000\,(F_r\,Q)^2 + \left[\dfrac{p}{T\,z} \right]^2} \tag{5-45}$$

To begin, the first segment is considered; so, with four segments, Equation 5-41 becomes:

$$\frac{G\,(L\,/\,4)}{53.34} = \int_{p_s}^{p_1} I\,dp \tag{5-46}$$

This equation is represented numerically as:

$$\frac{G\,L}{(4)\,(53.34)} = \left[\frac{I_s + I_1}{2} \right] (p_1 - p_s) \tag{5-47}$$

Thus, p_1 can be solved:

$$p_1 = p_s + \left[\frac{GL}{(4)(53.34)} \right] \left[\frac{2}{I_s + I_1} \right] \tag{5-48}$$

A trial-and-error solution is indicated because the evaluation of I_1 requires knowledge of the value of p_1. Hence, to start, let $I_1 = I_s$. Then, p_1 can be calculated, and I_1 can then be computed. The previous equation can be solved once again for p_1. In this manner, continue iterating until the next value of p_1 agrees with the last value within one psi or so. When the pressures agree closely, convergence has been obtained.

Next, we move to the second segment to solve for p_2. The procedure is almost identical to that used for the first segment. To begin, let the first guess for p_2 be equal to: Assumed $p_2 = (2)(p_1) - p_s$. With this value of p_2, evaluate I_2, and begin iteratively solving:

$$p_2 = p_1 + \left[\frac{GL}{(4)(53.34)} \right] \left[\frac{2}{I_1 + I_2} \right] \tag{5-49}$$

In this manner, work down the wellbore until finally obtaining the bottomhole pressure. Notice that the number of pressures to be solved for is equal to the number of segments used. For n segments, a general formula to be used on each segment to iteratively obtain the pressure at the bottom of the segment is:

$$p_i = p_{i-1} + \frac{(G)(L)(2)}{(n)(53.34)(I_{i-1} + I_i)} \tag{5-50}$$

where:

i = subscript indicating the "ith" segment,
n = the number of segments,
p_n = the bottomhole pressure, and
p_o = the surface pressure

Note that for all segments except the first, a reasonable formula for the initial guess for p_i (to begin iteratively solving Equation 5-50) is:

$$p_i \cong (2)(p_{i-1}) - p_{i-2}, \ for \ i > 1 \tag{5-51}$$

To calculate z factor at the bottom of each segment, a temperature will need to be assumed. Typically, a linear temperature gradient is used for these calculations; however, the Cullender and Smith method has no such restriction. Any temperature gradient may be used.

Problem 5-5:

Rework Problem 5-4 using the Cullender and Smith method.

Solution:

Because the well is shallow, only two segments will be used. A linear temperature distribution will be assumed.

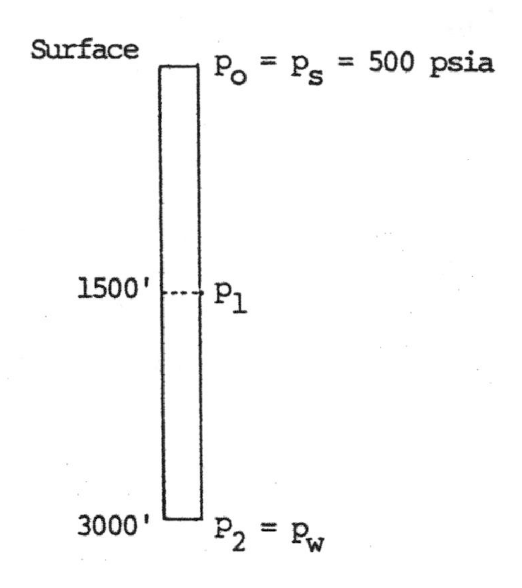

Friction factor: $F_r = (0.10797/2^{2.612}) = 0.01766$

Top segment: 0 ft to 1500 ft

Trial 1:

For a 0.65 gravity dry gas with T = 90 °F and p = 500 psia ==> z = 0.921

$$I_s = \frac{\left[\dfrac{500}{(90 + 460)(0.921)}\right]}{1000\,[\,(0.01766)(0.5)\,]^2 + \left[\dfrac{500}{(90 + 460)(0.921)}\right]^2} = 0.938$$

$$\text{Now } p_1 = p_s + \left[\frac{(G)(L)}{(2)(53.34)}\right]\left[\frac{2}{I_s + I_1}\right]$$

Assume $I_1 = I_s$

$$\text{So, } p_1 = 500 + \left[\frac{(0.65)(3000)}{(2)(53.34)}\right]\left[\frac{2}{0.938 + 0.938}\right]$$

$p_1 = 519.49$

Second trial:

$$T_1 = 90 + 460 + \frac{115 - 90}{2} = 562.6°R$$

with $p_1 = 519.49$ psia and $T_1 = 562.5$ °R ==> $z_1 = 0.925$

$$\text{So,} \quad \frac{p_1}{(T_1)(z_1)} = \frac{(519.49)}{(562.5)(0.925)} = 0.9984$$

$$\text{Hence,} \quad I_1 = \frac{0.9984}{1000\,[\,(0.01766)(0.5)\,]^2 + (0.9984)^2}$$

$$= \frac{0.9984}{0.07797 + (0.9984)^2} = 0.929$$

$$\text{So,} \quad p_1 = 500 + \left[\frac{(0.65)(3000)}{(2)(53.34)}\right]\left[\frac{2}{0.938 + 0.929}\right]$$

$$= 519.58 \text{ psia}$$

This is quite close to the last p_1, so we have convergence, and $p_1 = 519.58$ psia.

Bottom segment:

Trial 1:

Initial guess, $p_2 \cong 2p_1 - p_o = (2)(519.58) - 500 = 539.16$

$T_2 = 115 + 460 = 575$ °R

With this temperature and pressure, $z_2 = 0.928$

$$\text{So,} \quad \frac{p_2}{(T_2)(z_2)} = \frac{(539.16)}{(575)(0.928)} = 1.0104$$

$$\text{Hence,} \quad I_2 = \frac{1.0104}{0.07797 + (1.0104)^2} = 0.9195$$

$$p_2 = p_1 + \left[\frac{(G)(L)}{(2)(53.34)}\right]\left[\frac{2}{I_2 + I_1}\right]$$

$$= 519.58 + \left[\frac{(0.65)(3000)}{(2)(53.34)}\right]\left[\frac{2}{0.929 + 0.9195}\right]$$

$$= 539.36 \text{ psi}$$

Convergence. So, $p_w = 539.4$ psia.

As can be seen from this problem, convergence is usually achieved quite rapidly with the Cullender and Smith method: typically within three iterations for the first segment and within two iterations for the remaining segments.

D. Flowing Bottomhole Pressure—More Complex Situations

The preceding two sections dealt with calculation of bottomhole pressure from wellhead measurements for the case where only the gas phase is present in the wellbore. With the deeper drilling of the last two decades, many gas wells now produce from deep, high-pressure, wet-gas or retrograde-condensate reservoirs. It is common for these wells to have both gas and liquid present in the wellbore. The liquid may be condensate (oil) or water (either produced water or condensed water vapor from the reservoir gas). Several flow regimes are possible: mist flow, slug flow, annular flow, ripple flow, and combinations of these. Each regime has different pressure-drop characteristics, the calculation of which is somewhat complicated. Such methods are beyond the scope of this book. However, for those interested, references 10, 11, 12, and 20 provide multiple phase flow techniques.

One simple method that treats a gas well with small amounts of condensate is the Arnondin *et al.*[21] procedure. Their formulation accounts for wells that deviate from the true vertical direction. The application of the method should be restricted to situations where the gas/oil ratio is 75,000 scf/STB or higher. However[21], if the velocity in the tubing is high enough to keep the liquid dispersed in the flow stream, then lower GOR's can be used. No provisions have been made for water; only gas and condensate (oil) are considered. However, the thoughtful engineer can incorporate a small amount of produced water into the calculation by including a term for the gas equivalent of the water in the "equivalent flow rate" equation and by considering the water effect on the well fluid gravity equation.

The equations used in the Arnondin *et al.*[21] method follow. Although, many of these equations (or least similar formulations) have been discussed previously, they will be presented together for convenience and completeness.

$$\gamma_o = \frac{141.5}{°API + 131.5} \tag{5-14}$$

Condensate molecular weight:

$$M_o = 6084/(°API - 5.9) \tag{5-9}$$

Condensate gas equivalent:

$$GE = \left[\frac{(350.5)(\gamma_o)}{M_o}\right]\left[\frac{(10.73)(T_{sc})}{p_{sc}}\right] \tag{5-8}$$

Well stream (or total fluid) gravity:

$$\gamma_w = \frac{(GOR)(\gamma_g) + (4584)(\gamma_o)}{GOR + (133{,}000)(\gamma_o)/M_o} \tag{5-52}$$

(Gas) equivalent flow rate:

$$q_{wg} = (q_g)\left(1 + \frac{GE}{GOR}\right) \tag{5-53}$$

Reynolds number:

$$R_e = \frac{(20{,}011)(\gamma_w)(q_{wg})}{(d)(\mu_g)} \tag{5-54}$$

Moody friction factor (not Fanning friction factor):

$$f = \left\{\frac{1}{\left[1.14 - 2\left\{\log\left(\frac{\delta}{d} + \frac{21.25}{R_e^{0.9}}\right)\right\}\right]}\right\}^2 \tag{5-55}$$

Bottomhole pressure equation:

$$p_w^2 = (p_s^2)(e^{X_{tvd}}) + (667)(f)(q_{wg}^2)(T_a^2)(z_a^2)(e^{X_{md}} - 1)/d^5 \tag{5-56}$$

where:

γ_o = tank oil (condensate) specific gravity (water = 1.0),
$°API$ = tank oil API gravity, degrees,
M_o = estimated oil molecular weight,
GE = condensate gas equivalent, scf/STB,
T_{sc} = temperature at standard conditions, °R,
p_{sc} = pressure at standard conditions, psia,
γ_w = well stream (gas and oil vapor) gravity (air = 1.0),
GOR = producing gas/oil ratio, scf/STB,
γ_g = specific gravity of dry gas (air = 1.0),
q_{wg} = equivalent flow rate, MMscf/D,
q_g = surface dry-gas flow rate, MMscf/D,
R_e = Reynolds number,
d = tubing inside diameter, in.,
μ_g = gas viscosity (average wellbore conditions), cp,
f = Moody friction factor,
δ = absolute roughness of tubing, in.,
p_w = bottomhole pressure, psia,
p_s = tubing head pressure, psia,
T_a = average wellbore temperature, °R,
z_a = average compressibility factor,
TVD = true vertical depth, ft,
MD = measured depth, ft,
X_{tvd} = $(0.0375)(\gamma_w)(TVD)/(T_a)(z_a)$ and
X_{md} = $(X_{tvd})(MD)/TVD$

DELIVERABILITY (RATE) TESTING OF GAS WELLS

A. Introduction

While it is one thing to determine gas in place and reserves, it is quite another to predict productivity of gas wells. A complete analysis of a flowing well test should allow the determination of (1) the stabilized shut-in reservoir pressure, (2) the rate at which a well will flow against a particular pipeline "backpressure," and (3) an estimate of the manner in which flow rate will decrease with reservoir pressure depletion. Some types of tests will also estimate reservoir flow characteristics. However, the purpose here is to discuss tests that allow well deliverability estimation, not those designed to characterize reservoir parameters.

Well tests are widely used by regulatory bodies in setting maximum gas withdrawal rates and by producing and transporting companies in projecting well deliveries. These rate forecasts are needed in the preparation of field development programs, in the design of processing plants, and in the negotiation of gas sales contracts.

In early gas-well testing, the wells were opened to the atmosphere and the "absolute open-flow" potential determined, usually with "impact" pressure gauges. Predicting rate (deliverability) based on such tests is useful for shallow wells, but is highly inaccurate for deeper wells producing through small tubing strings. Because of this and also due to the waste of gas and the possible danger involved, this testing practice has been largely discontinued. Most modern gas-well tests utilize controlled and reasonable rates of flow, which can also yield the equivalent of an "absolute open-flow" potential. The most well-known gas-well test is the "conventional backpressure test" (also called the "flow-after-flow" test[22]). Two other quite popular deliverability-type tests are the "isochronal" and "modified isochronal" tests.

B. Basic Theory

For a time after flow from a well begins, the reservoir is said to be "infinite-acting" because the pressure disturbance created by the well production has not reached the reservoir boundaries. Tests designed to determine reservoir information during this infinite-acting period are termed "transient" tests. Examples of these are the well known pressure "buildup" and "drawdown" tests. Such procedures will be discussed in the Well Testing chapter.

For deliverability prediction purposes, the behavior of the reservoir under "stabilized flow" conditions is needed. Stabilized flow occurs in the reservoir when all pressure disturbances (usually due to production or injection) have equalized. In this case, either steady state or pseudosteady state flow conditions will exist in the reservoir. For pseudosteady state flow, the pressure at every point in the reservoir is changing linearly with time. Steady state refers to the special case where at any point in the reservoir, the pressure is not changing with time. Steady state is not obtained in many reservoirs, but for the reservoir with a strong water drive, it is sometimes closely approximated.

For stabilized flow, Lee[22] reports the following relationship:

$$\psi(p_{wf}) = \psi(\bar{p}) - 50{,}300 \frac{p_{sc}}{T_{sc}} \frac{q_g T}{k h} \left[\ln\left(\frac{r_e}{r_w}\right) - 0.75 + s + D \,\middle|\, q_g \,\middle|\, \right] \tag{5-59}$$

where:

p_{wf} = flowing bottomhole pressure, psia
$\bar{p}$ = static drainage area reservoir pressure, psia,
p_{sc} = standard condition pressure, psia (frequently 14.7 psia),
T_{sc} = standard condition temperature, °R (usually 520 °R),
q_g = gas flow rate, Mscf/D
T = reservoir temperature, °R,
k = reservoir rock permeability, md,
h = net formation thickness, ft,
r_e = external reservoir drainage radius, ft,
r_w = wellbore radius, ft,
s = skin factor (relating to wellbore damage), dimensionless,
D = non-Darcy flow constant (relates to turbulence), 1/Mscf/D

and pseudopressure[13,23,24] is defined by the integral:

$$\psi(p) = 2\int_{p_b}^{p} \frac{p}{\mu z}\, dp \tag{5-60}$$

where:

p = pressure, psia,
p_b = some arbitrary low base pressure (such as 0), psia,
μ = viscosity at pressure p, cp, and
z = compressibility factor at pressure p, dimensionless.

The term $D|q_g|$ in Equation 5-59 reflects a non-Darcy flow pressure loss. This term accounts for an extra pressure loss occurring near the wellbore due to turbulence with large gas production rates.

During the drilling and completion process, a damaged zone is normally created around the wellbore from such things as clay swelling, fine drilled particles, etc. This creates an extra pressure drop near the wellbore. Alternatively, the well could be fractured or acidized. This hopefully creates a stimulated zone of higher permeability around the wellbore, thereby causing less pressure drop. The skin factor, s, takes into account the effects of the altered (damaged or stimulated) zone. A positive value of skin indicates damage, while a negative skin is a sign of stimulation.

Although Equation 5-59 is complex, it normally only needs to be used between approximately 2000 to 3000 psia. For pressures below 2000 psia, or where the $(\mu)(z)$ product is very close to a constant, Equation 5-59 can be written in terms of p^2 as:

$$p_{wf}^2 = \bar{p}^2 - 1422\,\frac{q_g\,\bar{\mu}\,\bar{z}\,T}{kh}\left[\ln\left(\frac{r_e}{r_w}\right) - 0.75 + s + D\left|q_g\right|\right] \tag{5-61}$$

where:

$\bar{\mu}$ = viscosity at $\bar{p}$, cp, and
$\bar{z}$ = compressibility factor at $\bar{p}$.

For pressures above 3000 psia, where the (μ)(z) product may be approximated as a linear function of pressure, Equation 5-59 may be written in terms of p.

Here the procedures will be based on the p^2 formulation because these equations illustrate the general method and permit easier comparison with older methods of gas-well test analysis that are commonly still used. The most generally applicable and accurate formulation is the pseudo-pressure Equation 5-59, which can be used for any pressure.

The gas rate used in Equations 5-59 and 5-61 and in the methods to follow should include all fluids that are flowing in the vapor phase in the reservoir, i.e., gas, condensate (oil), and that part of the produced water that was vapor in the reservoir. This can be accomplished by using methods discussed earlier in this chapter.

C. Conventional Backpressure Tests

In this method, a well is put on production at a selected constant rate until bottomhole flowing pressure stabilizes. The stabilized rate and bottomhole pressure are recorded, and then the rate is changed (usually increased). (See Figure 5-11). The well is flowed at the new rate until pseudosteady state again is attained. The pressure may be measured by using a bottomhole pressure gauge (preferred) or by calculation from carefully measured surface values.

This process is repeated, each time recording the stabilized rate and pressure, for a total of four rates.
Two fundamentally different techniques exist for analyzing such tests.

Classic Method

In 1936 Rawlins and Schellhardt[12], presented the following equation:

$$q_g = C(\bar{p}^2 - p_{wf}^2) \tag{5-62}$$

where q_g, $\bar{p}$, and p_{wf} are defined as in Equation 5-59. This is Darcy's law for a compressible fluid, and "C" contains all the terms other than those of pressure; i.e., gas viscosity, permeability to gas flow, net pay thickness, formation temperature, etc. Rawlins and Schellhardt found that Equation 5-62 did not account for the turbulence usually present near gas wells, so they modified the equation with an exponent "n" on the right hand side:

$$q_g = C(\bar{p}^2 - p_{wf}^2)^n \tag{5-63}$$

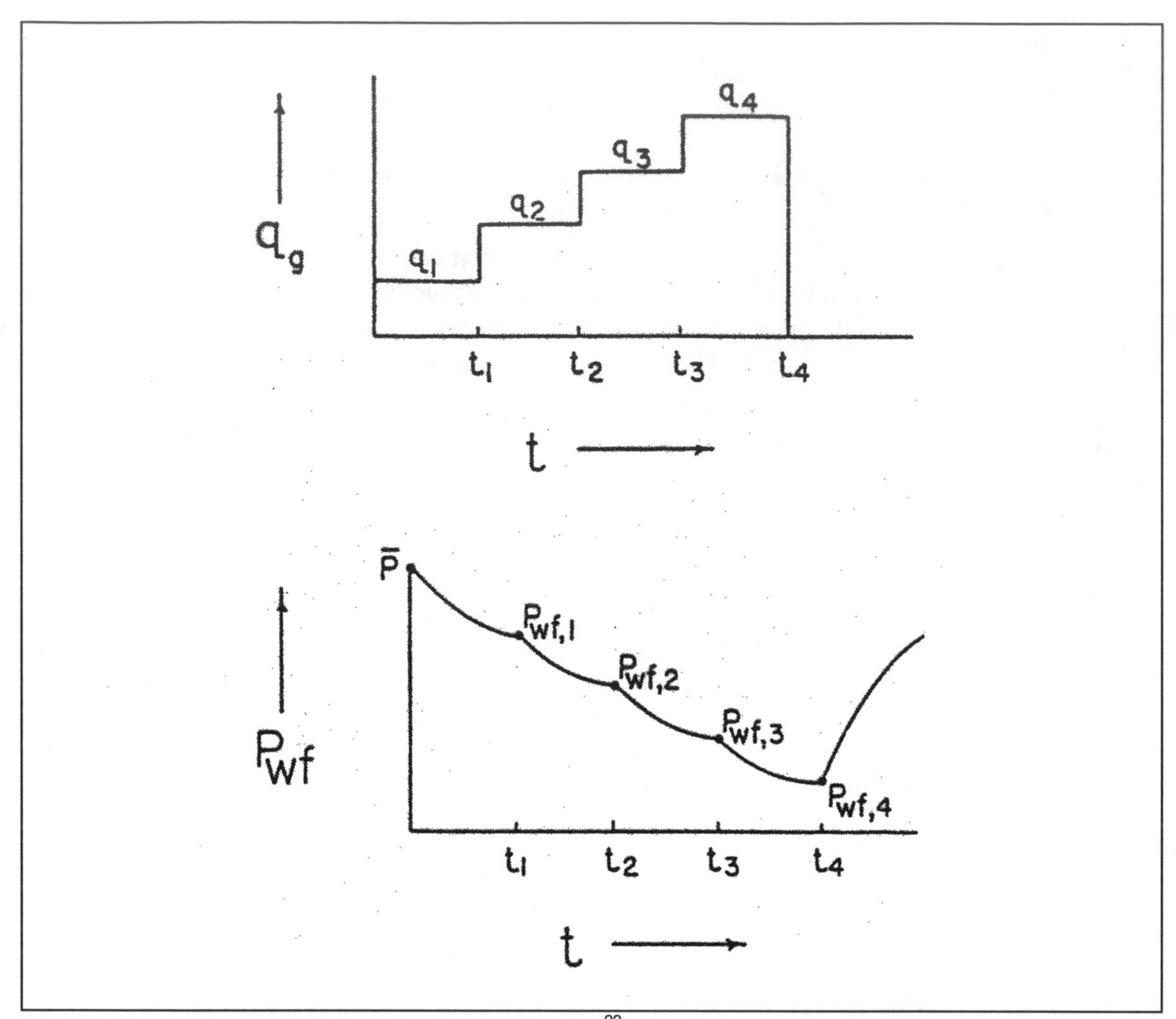

Fig. 5-11. Rates and pressures in flow-after-flow test (from Lee[22]). Permission to publish by SPE.

The exponent "n" may vary from 1.0 for completely laminar flow to 0.5 for completely turbulent flow.

According to Lee,[22] who calls this approach the "empirical method," the theoretical basis for Equation 5-63 is rather tenuous. However, due to its ease and half century of use, methods based on Equation 5-63 are more widely used in the industry today than those based on Equation 5-59 or 5-61.

Equation 5-63 can be written in the form:

$$log\,(\bar{p}^2 - p_{wf}^2) = \left(\frac{1}{n}\right)(\,log\,q_g - logC\,) \tag{5-64}$$

The implication is that a log-log plot of $(\bar{p}^2 - p_{wf}^2)$ versus q_g will be a straight line (Figure 5-12). The slope of this line is 1/n. As in Figure 5-12, a plot of the four flow rates will be approximately a straight line for many wells, provided stabilized flow conditions prevailed.

Equations 5-62, 5-63, and 5-64 are subject to the following assumptions[8].

1. Isothermal conditions prevail throughout the reservoir.
2. Gravitational effects are negligible.
3. The flowing fluid is single phase.
4. The medium is homogeneous and isotropic.
5. Permeability is independent of pressure.
6. Fluid viscosity and compressibility are constant.
7. Pressure gradients and compressibility are small.
8. The radial-cylindrical flow model is applicable.

These factors may not be even closely approximated, especially in tight gas formations.

By definition, "absolute open flow" occurs when the sandface backpressure has been reduced to atmospheric pressure (14.7 psia). Given a deliverability curve plot such as Figure 5-12, extrapolation of the straight line yields this value. In many areas regulatory agencies will limit gas-well producing rates to 25 percent of the initial, absolute open-flow potential (AOFP). Also, note that the backpressure or deliverability curve allows the determination of gas-flow rate given a specific back-pressure (flowing sandface pressure). Thus, if a sales gas pipeline pressure is known, then through flowline

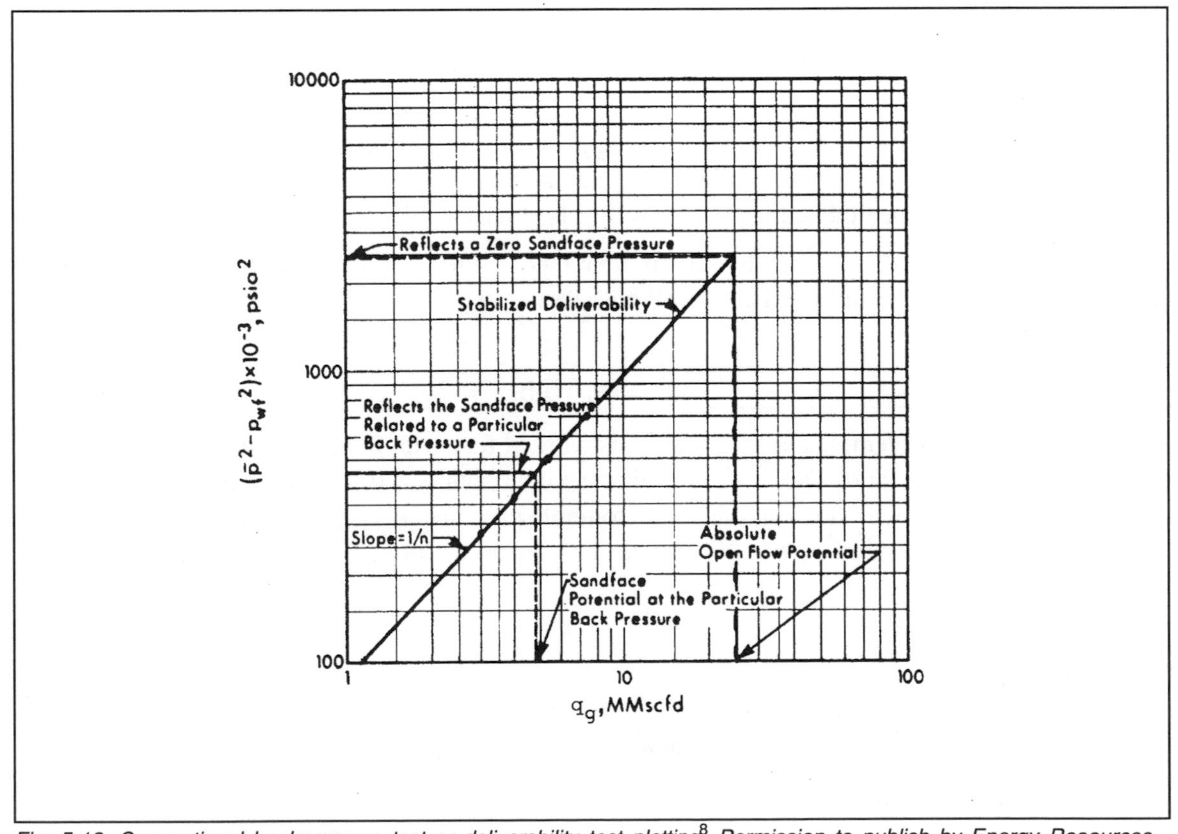

Fig. 5-12. Conventional backpressure test or deliverability test plotting[8] Permission to publish by Energy Resources Conservation Board of Canada.

and wellbore pressure drop calculations,[20] the well flowing bottomhole pressure can be determined. Then the well deliverability can be read from the backpressure curve.

As long as the factors that comprise "C" in Equation 5-63 do not change appreciably, the same stabilized deliverability plot can be used. However, the factors in "C" do change during well operating life, thus requiring a retesting of the well from time to time. Factors that can change (affecting "C") are z factor, gas compressibility, viscosity, permeability to gas flow, well damage, external boundary radius, and possibly wellbore radius. The exponent "n," which is related to the nature of the turbulence around the wellbore, can also change. Thus, retesting permits the determination of a new deliverability plot and new values for "C" and "n".

Theoretical Method

Equation 5-61 has the form:

$$\bar{p}^2 - p_{wf}^2 = a\,q_g + b\,q_g^2 \qquad (5\text{-}65)$$

where:

$$a = 1422\,\frac{\bar{\mu}\,\bar{z}\,T}{k\,h}\left[\ln\left(\frac{r_e}{r_w}\right) - 0.75 + s\right] \qquad (5\text{-}66)$$

and

$$b = 1422\,\frac{\bar{\mu}\,\bar{z}\,T}{k\,h}\,D \qquad (5\text{-}67)$$

Equation 5-65 suggests that a plot of $(\bar{p}^2 - p_{wf}^2)/q_g$ vs. q_g will result in a straight line with slope "b" and intercept "a" (see Figure 5-13). Again, this relationship will be valid only for stabilized flow rates. The constants "a" and "b" can be determined from flow tests with at least two stabilized rates. Notice that these constants are pressure dependent and probably also time dependent. Therefore, they will need to be updated with new well tests at reasonable intervals, perhaps annually.

According to Lee[22], this plot has a stronger theoretical basis than the plot of the classical (empirical) method. It should yield an AOFP with less error.

Problem 5-6:

The following problem, taken from Lee[22], is worked first with the classical approach and then with the theoretical method. Table 5-2 contains the data obtained from a four-point conventional backpressure test. At each rate, a stabilized bottomhole pressure was obtained. The shut-in bottomhole pressure before the test, $\bar{p}$, was 408.2 psia. To prepare the plots needed in the analysis, Table 5-3 was constructed.

Classical Approach:

A plot of $(\bar{p}^2 - p_{wf}^2)$ vs. q_g on log-log paper was prepared (see Figure 5-14). This plot was extrapolated to $(\bar{p}^2 - 14.7^2) = 166{,}411$ to determine the:

AOFP = 60 MMscf/D. To determine the slope of the curve, 1/n:

$$1/n = \frac{\log(\bar{p}^2 - p_{wf}^2)_2 - \log(\bar{p}^2 - p_{wf}^2)_1}{\log q_{g,2} - \log q_{g,1}}$$

$$= (\log 10^5 - \log 10^3)/(\log 42.5 - \log 1.77) = 1.449$$

So,

n = 0.690

Therefore,

$$C = \frac{q_g}{(\bar{p}^2 - p_{wf}^2)^n} = \frac{42.5}{(10^5)^{0.690}} = 0.01508$$

Thus, the deliverability equation is:

$$q_g = 0.01508\,(\bar{p}^2 - p_{wf}^2)^{0.690} \; MMscf/D$$

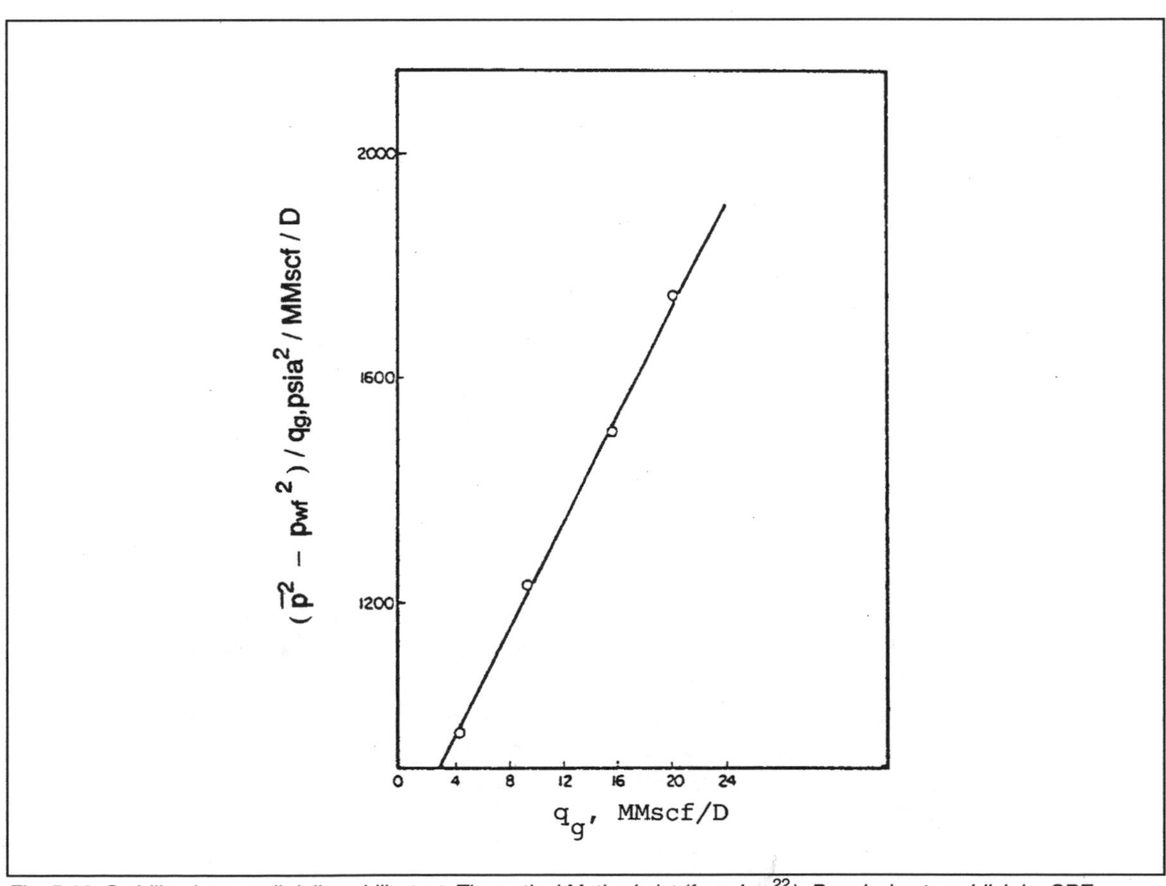

Fig. 5-13. Stabilized gas-well deliverability test. Theoretical Method plot (from Lee[22]). Permission to publish by SPE.

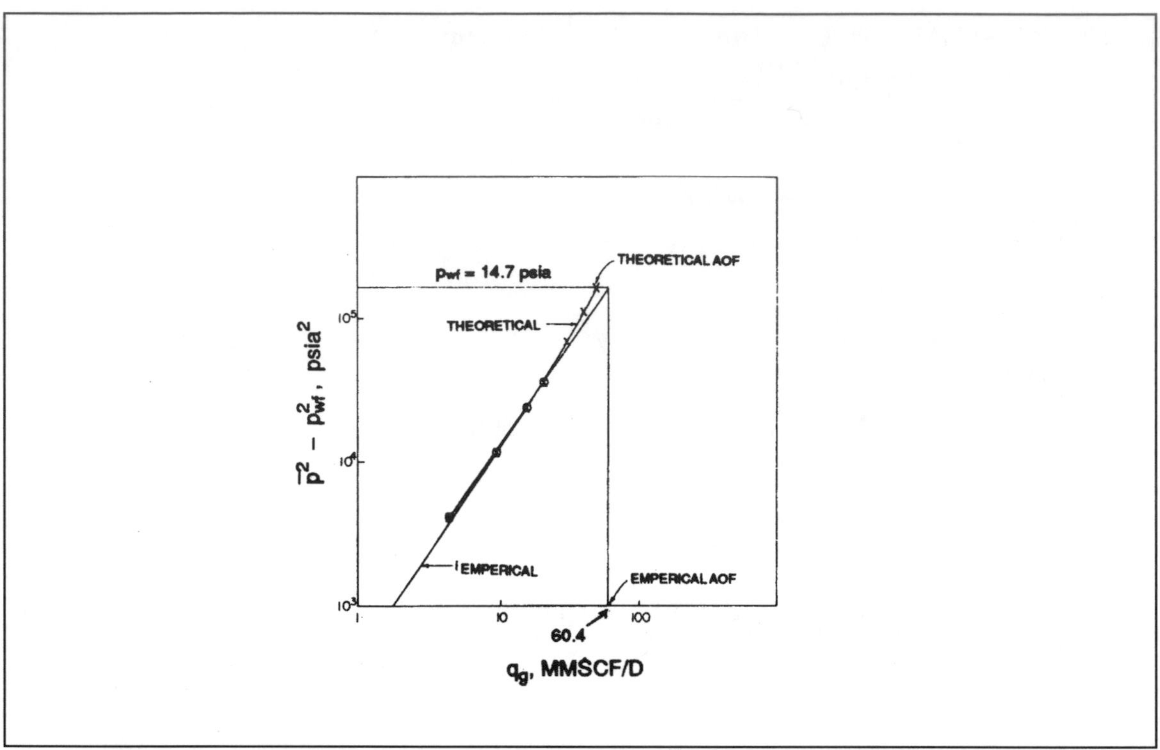

Fig. 5-14. Stabilized deliverability test. Classical method plot (from Lee[22]). Permission to publish by SPE.

Table 5-2
Stabilized-flow test data.

Test	p_{wf} (psia)	q_g (MMscf/D)
1	403.1	4.288
2	394.0	9.265
3	378.5	15.552
4	362.6	20.177

Table 5-3
Stabilized Flow Test Analysis.

p_{wf} (psia)	q_g (MMscf/D)	$\bar{p}^2 - p_{wf}^2$ (psia^2)	$(\bar{p}^2 - p_{wf}^2)/q_g$ (psia^2/MMscf/D)
408.2	0	—	—
403.1	4.288	4,138	964.9
394.0	9.265	11,391	1,229.
378.5	15.552	23,365	1,502.
362.6	20.177	35,148	1,742.
14.7	AOF	166,411	—

Theoretical Approach:

The theoretical deliverability equation is:

$$(\bar{p}^2 - p_{wf}^2) / q_g = a + b\,q_g$$

Figure 5-13 is a plot of $(\bar{p}^2 - p_{wf}^2) / q_g$ vs. q_g for the test data. Two points on the best straight line through the data are (2.7, 900) and (23.9, 1900). So,

900 = a + (2.7)b

1900 = a + (23.9)b

Solving these two equations simultaneously yields a = 772.6 and b = 47.17.

So, the theoretical deliverability equation for this well is:

$$47.17\,q_g^2 + 772.6\,q_g - (\bar{p}^2 - p_{wf}^2) = 0$$

If 14.7 psia is substituted for the flowing bottomhole pressure, the equation can be solved for the AOFP:

$$47.17\,q_g^2 + 772.6\,q_g - 166{,}411 = 0$$

Solving, using the quadratic equation:

$$q_g = AOFP = \frac{-a + \sqrt{a^2 + 4\,b\Delta(p^2)}}{2\,b}$$

= 51.8 MMscf/D

Notice that the AOFP determined with the theoretical method is less than that determined using the classical approach. Because, maximum well rates are often determined as a fraction of the initial AOFP, this may well be another reason for the continuing popularity of the classical method.

Field Procedure: Conventional Backpressure Test

1. Shut the well in until a stabilized bottomhole shut-in pressure, $\bar{p}$, is obtained.
2. Open the well on a small choke size, such as 6/64 inches, and let stabilize. Record and plot the stabilized bottomhole flowing pressure and the stabilized rate.
3. Change to a slightly larger choke size, such as 8/64 inches, and let the flowing well stabilize. Record and plot the stabilized pressure and rate.
4. Repeat step 3 using two larger choke sizes to give a total of 4 rates.

An important must here is that stabilized flow is achieved at each choke size.

D. Isochronal Tests

A rate change at a well causes a "pressure transient" (pressure wave or disturbance) to propagate out from the well. The distance this pressure transient has moved at a particular time is known as the "radius of investigation."

A conventional backpressure test uses stabilized flow rates. Therefore, the flow times must be sufficient to permit the radius of investigation to reach the limit of the reservoir or to the point of interference with offsetting wells. The effective drainage radius is constant.

In a lower permeability reservoir, it is frequently impractical to flow the well long enough to reach stabilization, especially if pseudosteady state conditions are needed at more than one rate. The objective of isochronal testing, proposed by Cullender,[15] is to obtain data to establish a stabilized deliverability curve without flowing the well long enough to achieve stabilized conditions at each rate. The principle is that the radius of investigation achieved at a given time in a flow test is independent of flow rate. Therefore, if a series of flow tests are performed on a well, each for the same period of time (isochronal), the radius of investigation will be the same at the end of each test. Consequently, the same portion of the reservoir is being drained at each rate. For an extensive discussion of the theory behind isochronal testing, the reader is referred to "Theory and Practice of the Testing of Gas Wells."[8]

Figure 5-15 provides a schematic flow rate and pressure diagram for an isochronal flow test on a gas well. Notice that the shut-in period after each flow period must be long enough for the static reservoir pressure to be reached (or at least approached). Also note that it is necessary to have one stabilized flow period at the end of the test.

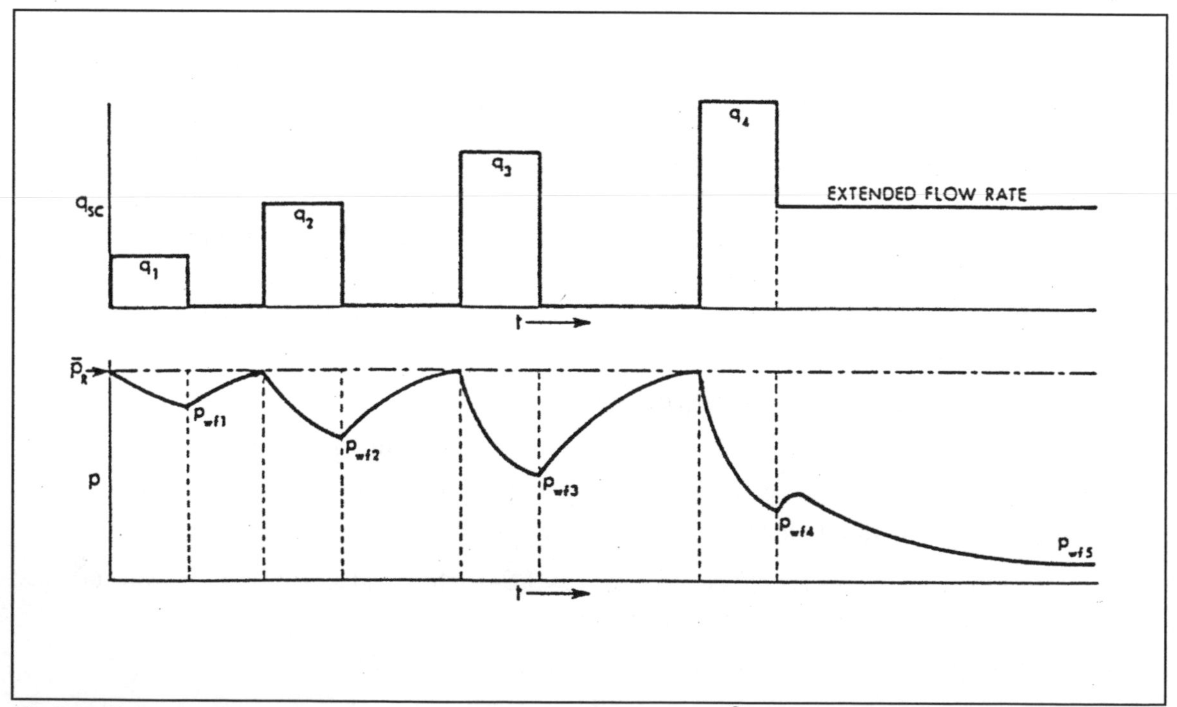

Fig. 5-15. Flow-rate and pressure diagrams for an isochronal test of a gas well.[8] Permission to publish by Energy Resources Conservation Board of Canada.

Classical Analysis

Considering the classical method, there are two constants to determine: "C" and "n." Theory indicates that "C" is a function of the radius of investigation, which means that if two flow periods have the same radius of investigation, then they will have the same "C." Flow rates having the same time interval will have the same radius of investigation and therefore the same "C." For stabilized flow periods, the "C" will be the stabilized "C", which is what we are trying to determine. For a series of equal flow periods that are not long enough to achieve stabilization, the "C's" of each test will be the same, but will not be the stabilized "C."

Because "n" relates to the nature of turbulence around the well, "n" is assumed to be the same for transient conditions or pseudosteady state conditions. Therefore, after four isochronal (equal time) flow periods, a log-log plot of $(\bar{p}^2 - p_{wf}^2)$ vs. q_g can be made and the points should lie on a straight line with slope of "1/n." (The "p_{wf}" to be used for each flow rate is that at the very end of the flow period. Similarly, if the flow rate is not remaining quite constant, the rate to be used is that at the very end of the flow period.)

Then, if the well is flowed at one flow rate until it reaches stabilized conditions, this one point on the log-log plot can be plotted. As shown in Figure 5-16, draw a line through the stabilized point that is parallel to the line of the equal time transient points. So, "n" has been obtained via transient behavior and "C" with the one stabilized point. The line through the stabilized point is the stabilized deliverability curve.

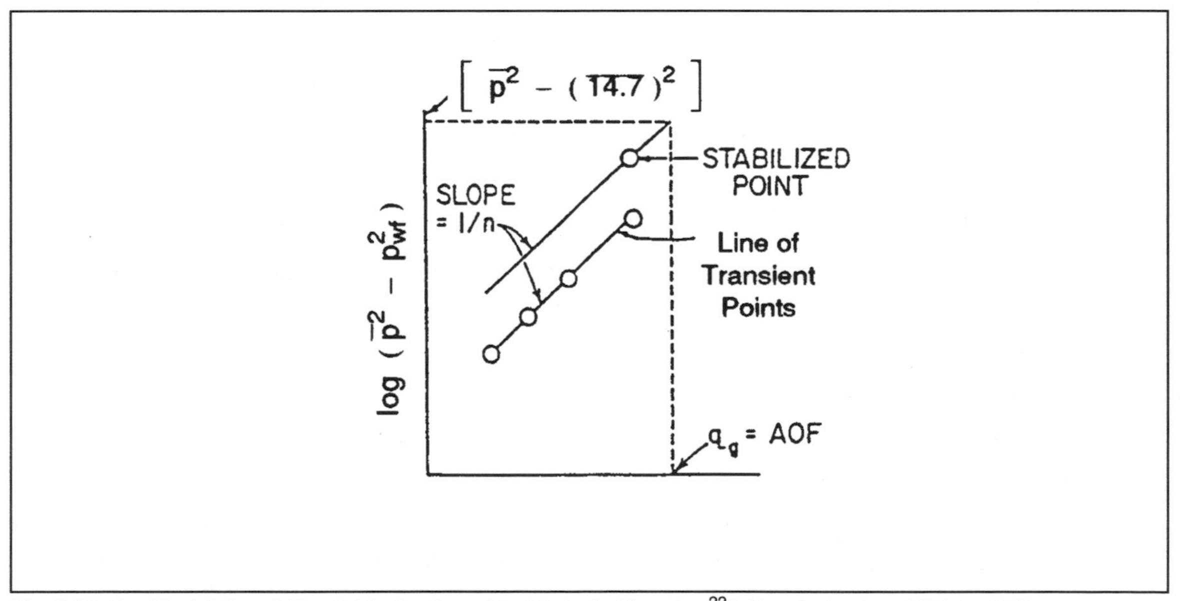

Fig. 5-16. Empirical deliverability plot for isochronal test (adapted from Lee[22]). Permission to publish by SPE.

Theoretical Analysis

In the theoretical analysis of the data, the approach used with isochronal testing is quite similar to that described previously with the classical method. Using the four isochronal nonstabilized points, make a plot of $(\bar{p}^2 - p_{wf}^2) / q_g$ versus q_g. Determine "b" from this plot as shown earlier under conventional backpressure tests or by merely measuring the slope. This constant "b" is a function of pressure and the nature of turbulence around the well. It is not a function of drainage radius, so it can be determined from the isochronal transient data. Then, using data from the one stabilized point and the "b" just calculated, "a" can be determined with the following equation:

$$a = \left[(\bar{p}^2 - p_{wf}^2) / q_g \right] - bq_g$$

These constants "a" and "b" are used in the stabilized deliverability curve (or equation):

$$\bar{p}^2 - p_{wf}^2 = aq_g + bq_g^2$$

Then absolute open flow potential may be determined as shown in the earlier example.

Field Procedure: Isochronal Test

1. Shut the well in for a stabilized shut-in bottomhole pressure.
2. Open the well on a small choke, such as 6/64 inches, and flow for eight hours.
3. At the end of the eight-hour flow period, record bottomhole flowing pressure and flow rate.
4. Shut the well in and let the bottom hole pressure build up to the beginning static pressure.
5. Open the well on a slightly larger choke, such as 8/64 inches, and let the well flow for eight hours.
6. At the end of the 8 hour flow period, record bottomhole flowing pressure and flow rate.
7. Shut the well in and let the bottomhole pressure build up to the stabilized shut-in bottomhole pressure.
8. Repeat steps five, six, and seven using two progressively larger choke sizes.
9. Ensure that the recorded flowing pressures are taken just before shut-in. Also, if the rate is varying on a flow test, record the rate just before shut-in.
10. These four transient points should be plotted just as described under the conventional backpressure test (either classical or theoretical methods).
11. Open the well for a fifth flow period (using a previous choke size or a new one), and let it flow until stabilization occurs. Record this stabilized rate and bottomhole pressure.
12. Plot this stabilized point. The stabilized deliverability curve passes through this stabilized point and is parallel to the line of the four transient points.

Note that the exact length of time of the flow periods is not important as long as they are all the same. For instance, 12-hour flow periods could be used instead of eight hour ones. Notice that the shut-in periods are not necessarily equal. Each shut-in period lasts until the bottomhole pressure has built up to the stabilized shut-in pressure.

E. Modified Isochronal Testing

The objective of modified isochronal tests is to obtain the same information as in an isochronal test without the sometimes lengthy required shut-in periods. In fact, the true isochronal test has proved to be impractical as a means of testing many wells.

Katz *et al.*[16] suggested that a modified isochronal test using shut-in periods equal to the flow periods may give satisfactory results provided that the associated, unstabilized shut-in pressure is used in the analysis instead of the average reservoir pressure, $\bar{p}$. Figure 5-17 provides a schematic diagram of the flow rates and pressures resulting from this kind of a test. Notice that for the first flow period, $(\bar{p}^2 - p_{wf,1}^2) = (p_{ws,1}^2 - p_{wf,1}^2)$ should be used; for the second flow period, use $(p_{ws,2}^2 - p_{wf,2}^2)$. Otherwise the analysis procedure is the same as for the isochronal test.

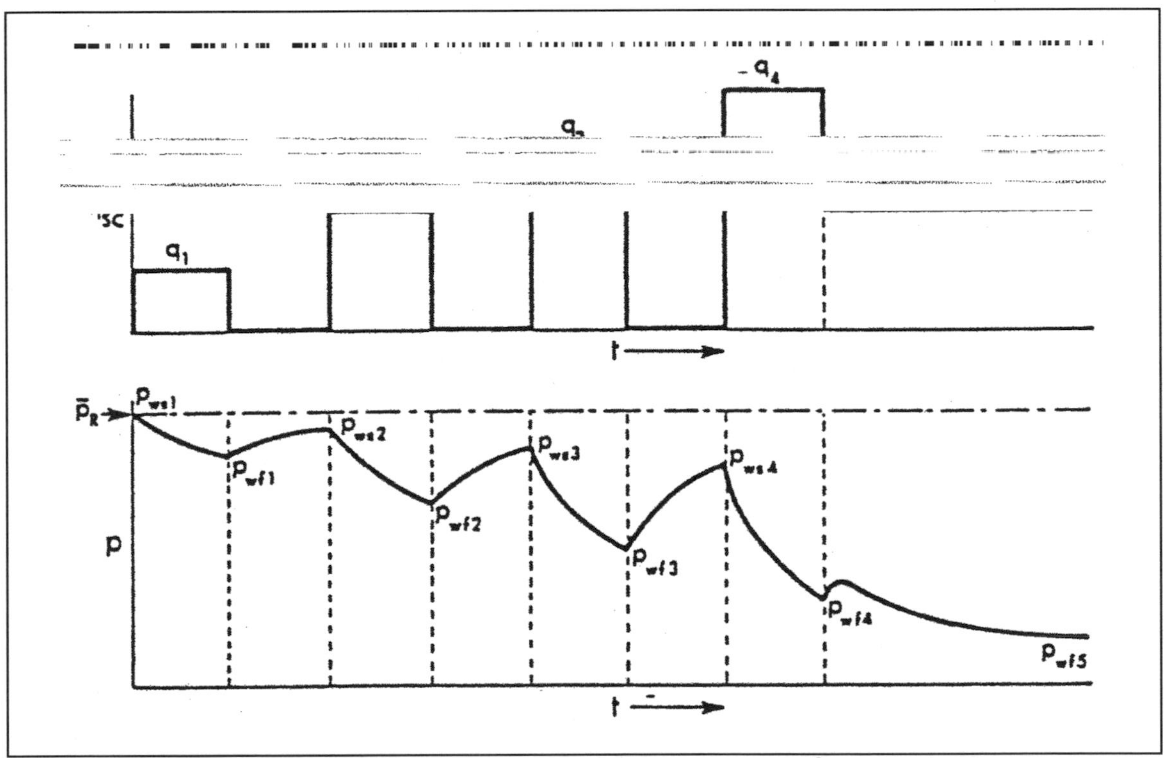

Fig. 5-17. Flow rate and pressure diagrams for modified isochronal tests on gas wells[8]. Permission to publish by Energy Resources Conservation Board of Canada.

The modified isochronal procedure involves approximations; whereas, "true" isochronal tests conform more closely to the theory behind such tests. Modified isochronal tests are used extensively in low permeability reservoirs because they save time and money. They have also proved to be excellent approximations of true isochronal tests.

Field Procedure: Modified Isochronal Test

1. Shut the well in for a stabilized shut-in pressure (or for as long as practically possible to obtain a good estimate of reservoir static pressure).
2. Open the well on a small choke, such as 6/64 inches, and flow for 12 hours.
3. At the end of this flow period, record the flow rate and the flowing bottomhole pressure.
4. Shut the well in for 12 hours.
5. At the end of the shut-in period, record bottomhole pressure. This shut-in pressure will be used in the analysis as the estimate for static pressure for the second flow period.
6. Open the well on a slightly larger choke, such as 8/64 inches, and flow for 12 hours.
7. At the end of this flow period, record the flow rate and the flowing bottomhole pressure.
8. Shut the well in for 12 hours, and then record the bottomhole pressure (to be used as the approximate static pressure for the next flow period).
9. Repeat steps six, seven, and eight using two progressively larger choke sizes. For each flow period, approximate static pressure to be used in the analysis is the shut-in pressure that existed just before the flow period began. The flowing bottomhole pressure is the pressure at the very end of the flow period, even though stabilization may not have occurred.
10. These four points are plotted in the same manner as described under conventional backpressure tests.
11. Now, the well is flowed for a fifth flow period until stabilization occurs. The choke size may be a new one or one previously used. For the analysis, the stabilized, flowing bottomhole pressure is used as well as the rate at the end of the flow period. The shut-in pressure to be used for this stabilized point is not the shut-in pressure just before this flow period, but the true, stabilized shut-in pressure.
12. Plot the stabilized point, and then draw a line through this point parallel to the line through the four transient points. This line through the stabilized point is the stabilized deliverability curve for this well.

Note that the flow and shut-in periods do not have to be 12 hours, but could be some other time such as eight or 16 hours.

F. Use of Pseudopressure in Gas-well Test Analysis

The generalized equation for stabilized flow in gas reservoirs was presented in Equation 5-59 in terms of gas pseudopressure which was defined in Equation 5-60. The usual gas-well analysis in the petroleum industry involves approximations in terms of p^2, which can be quite accurate for pressures up to 2000 psia. However, over all pressure ranges, and especially between 2000 to 3000 psia, if the pseudopressure $\psi(p)$ is used, the best accuracy will be obtained. The assumption that the $(\mu)(z)$ product is constant, implicit in the p^2 approach, is not needed with the pseudopressure analysis. Detailed discussion, including development of the pertinent equations, concerning deliverability and transient gas well tests may be found in reference 8.

For convenience, Equation 5-60 is reproduced here:

$$\psi(p) = 2\int_{p_b}^{p} \frac{p}{\mu z}\, dp \qquad (5\text{-}60)$$

where:

p = pressure, psia,
p_b = some arbitrary low base pressure, psia,
μ = viscosity at pressure p, cp, and
z = compressibility factor at pressure p, dimensionless.

Calculation of Pseudopressure

Notice in Equation 5-60 that the definition of pseudopressure involves the evaluation of an integral with a lower limit of p_b and an upper limit of p, the pressure of interest. This p_b is an arbitrary low base pressure, the value of which is often in the range of 0 to 200 psia, depending on the needs and wishes of the analyst.

The normal approach is to evaluate the integral in Equation 5-60 numerically. For the specific gas under consideration, a table is developed of μ and z versus p for pressures from the base pressure to the maximum system pressure which is usually the initial reservoir pressure. Table pressure increments are normally in the range of 50 to 200 psi for hand calculations, or 10 to 25 psi with a computer. For low pressures, a smaller increment may be chosen. Then, for each pressure, the quantity $p/\mu z$ is calculated. At this point, the integral is evaluated at each pressure by using a numerical integration scheme such as Simpson's rule or the Trapezoidal rule. In this way a graph or table of ψ versus p is developed for a particular gas at reservoir temperature.

Problem 5-7:

In this problem, a base pressure, $p_b = 0$ will be used. In the development of the ψ vs. p table, the numerical integration will be done with the trapezoidal rule. In this example, the pressure increments have been chosen to be 400 psi, which is larger than would normally be used in practice.

To use the trapezoidal rule, the integral in Equation 5-60 is represented as a sum of integrals with each integral being over one of the table pressure increments:

$$2\int_{p_b}^{p} \frac{p}{\mu z}\, dp = \sum_{i=1}^{n} 2 \int_{\Delta p_i} \frac{p}{\mu z}\, dp$$

where there are "n" pressure increments up to pressure "p," and $\int_{\Delta p_i}$ represents the integral over pressure increment "i." Each integral within the sum on the right-hand side is approximated as:

$$2 \int_{\Delta p_i} \frac{p}{\mu z} dp \approx 2 \left[\left(\frac{p}{\mu z} \right)_{i-1} + \left(\frac{p}{\mu z} \right)_i \right] [\Delta p] \ / 2$$

where $(p / \mu z)_{i-1}$ and $(p / \mu z)_i$ represent the beginning and ending $(p / \mu z)$ values, respectively, of pressure increment "i." Δp is the pressure increment which here equals 400 psi.

Problem Statement

Develop a pseudopressure relationship for a sweet gas having the following properties:

(1) p, psia	(2) z	(3) μ, cp
0	1.000	—
400	0.955	0.0118
800	0.914	0.0125
1200	0.879	0.0134
1600	0.853	0.0145
2000	0.838	0.0156

Solution:

The equation to be solved at each pressure level is as follows:

$$\psi = 2 \int_o^p \frac{p}{\mu z} dp$$

In tabular form, the calculation is as shown:

i	(1) p	(2) z	(3) μ	(4) $(p / \mu z)10^{-3}$	(5) $(4)_i + (4)_{i-1}$	(6) Δp	(7) (5) x (6) x (10^{-3})	(8) = Σ(7) $\psi \times 10^{-6}$
0	0			0.0				0.0
1	400	0.955	0.0118	35.5	35.5	400	14.2	14.2
2	800	0.914	0.0125	70.0	105.5	400	42.2	56.4
3	1200	0.879	0.0134	101.9	171.9	400	68.8	125.2
4	1600	0.853	0.0145	129.4	231.3	400	92.5	217.7
5	2000	0.838	0.0156	153.0	282.4	400	113.0	330.7

Figure 5-18 provides a graph of this pseudopressure vs. pressure data.

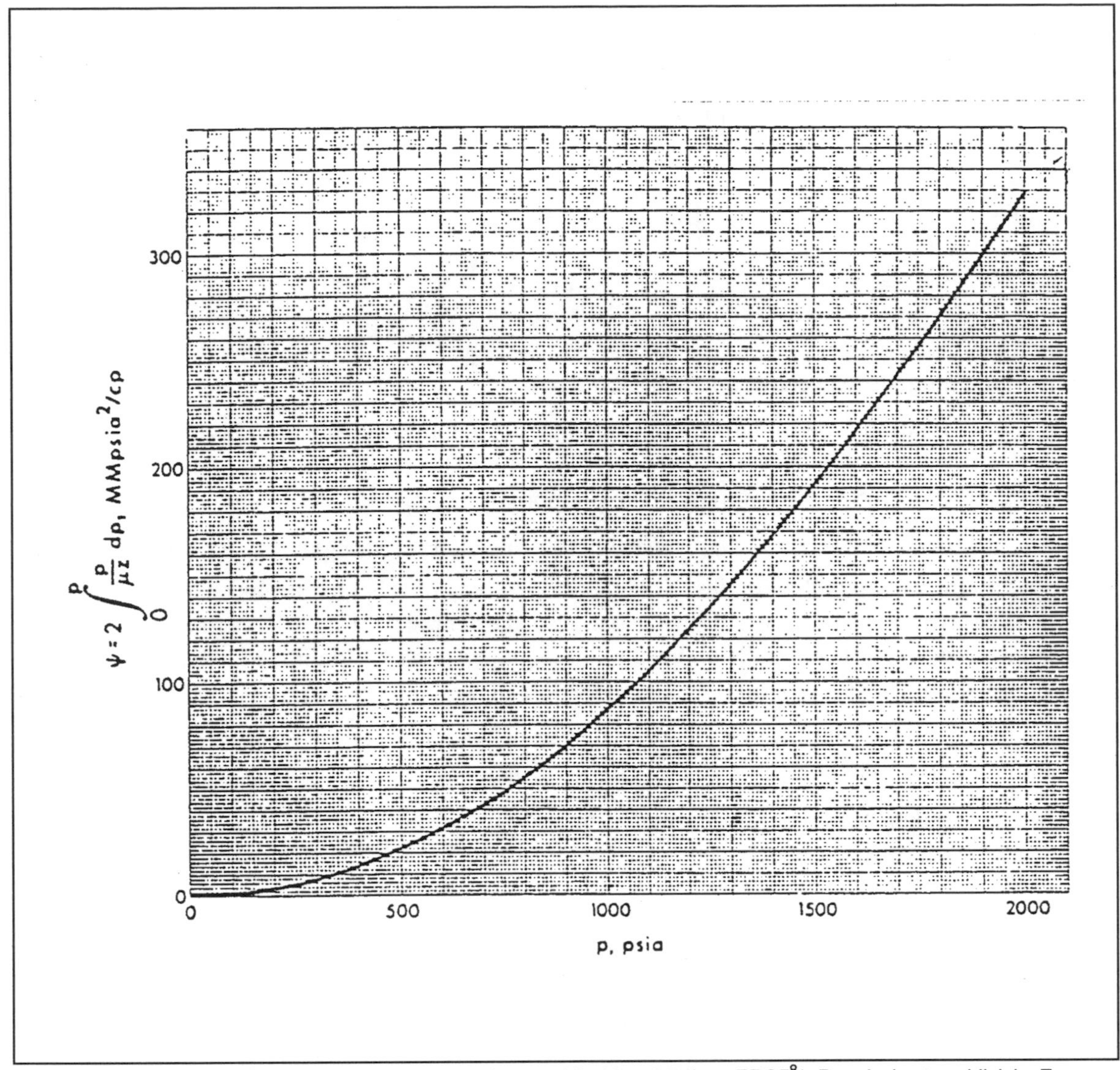

Fig. 5-18. Pseudopressure versus reservoir pressure for gas of Problem 5-7 (from ERCB[8]). Permission to publish by Energy Resources Conservation Board of Canada.

APPLICATION OF PSEUDOPRESSURE TO DELIVERABILITY TESTING

Equation 5-59, the stabilized gas flow equation, has the form:

$$\psi(\bar{p}) - \psi(p_{wf}) = a\,q_g + b\,q_g^2 \qquad (5\text{-}68)$$

where:

$\psi(\bar{p})$ = pseudopressure corresponding to $\bar{p}$,
$\psi(p_{wf})$ = pseudopressure corresponding to p_{wf},
$a\,q_g$ = pseudopressure drop due to laminar flow and well conditions,
$b\,q_g^2$ = pseudopressure drop due to inertial - turbulent flow effects.

Note that:

$$a = 1422 \frac{T}{kh} \left[\ln \left(\frac{r_e}{r_w} \right) - 0.75 + s \right] \tag{5-69}$$

and

$$b = 1422 \frac{T D}{k h} \tag{5-70}$$

The remaining terms were defined with Equation 5-59.

To apply Equation 5-68 in deliverability testing, the recommended method is analogous to that presented under Conventional Back Pressure Tests, the "Theoretical Method." That analysis was based on Equation 5-65, which employed the p^2 approximation.

Here, a plot of $(\overline{\psi} - \psi_{wf}) / q_g$ versus q_g will result in a straight line with slope "b" and intercept "a" for truly stabilized rates. Note that both scales of this plot are linear. This method of analysis may also be used with isochronal and modified isochronal testing to plot the transient points with equal flow times. Then, the stabilized point is plotted with the stabilized deliverability curve being drawn through this point and parallel to the line of the transient points.

Using a method similar to that shown in Problem 5-6, "Theoretical Approach," the stabilized deliverability curve may be analyzed to calculate "a" and "b".

If we have n data points (n flow periods) from a deliverability test, then Kulczycki[14] has shown that "a" and "b" may be calculated by using a least squares approach with the following relationships:

$$a = \frac{\sum \frac{\Delta\psi}{q_g} \sum q_g^2 - \sum q_g \sum \Delta\psi}{n \sum q_g^2 - \sum q_g \sum q_g} \tag{5-71}$$

$$b = \frac{n \sum \Delta\psi - \sum q_g \sum \frac{\Delta\psi}{q_g}}{n \sum q_g^2 - \sum q_g \sum q_g} \tag{5-72}$$

Then, using these constants in Equation 5-68, we have the stabilized deliverability equation for the specific well being tested. This equation may be solved to yield the absolute open-flow potential.

These concepts are illustrated with the following example.

Problem 5-8:

Here, Problem 5-6 will be reworked using pseudopressures. Note that with the low pressures involved (p = 408.2 psia), the pseudopressure approach is really not needed. The p^2 approximation method used in Problem 5-6 is adequate.

To begin, construct a table of the data and calculate pseudopressure versus pressure as in Problem 5-7. The reference pressure, p_b, is chosen as 0.

(1) p	(2) μ	(3) z	(4) $p/\mu z$	(5) $2(p/\mu z)_{mean}$	(6) Δp	(7) (5) x (6)	(8) ψ
0	—	—	0				
25	.01225	.9977	2,045.52	2,045.52	25	51,138	51,138
50	.01227	.9953	4,094.22	6,139.74	25	153,494	204,632
75	.01230	.9929	6,141.16	10,235.38	25	255,885	460,516
100	.01233	.9904	8,188.91	14,330.07	25	358,252	818,768
125	.01235	.9880	10,244.39	18,433.30	25	460,833	1,279,600
150	.01238	.9856	12,293.34	22,537.73	25	563,443	1,843,044
175	.01241	.9833	14,341.03	26,634.37	25	665,859	2,508,903
200	.01243	.9810	16,401.74	30,742.77	25	768,569	3,277,472
225	.01246	.9787	18,450.79	34,852.53	25	871,313	4,148,785
250	.01249	.9763	20,501.91	38,952.70	25	973,818	5,122,603
275	.01251	.9740	22,569.21	43,071.12	25	1,076,778	6,199,381
300	.01254	.9717	24,620.20	47,189.41	25	1,179,735	7,379,116
325	.01257	.9694	26,671.35	51,291.55	25	1,282,289	8,661,405
350	.01259	.9671	28,745.57	55,416.92	25	1,385,423	10,046,828
375	.01262	.9647	30,802.05	59,547.62	25	1,488,691	11,535,518
400	.01265	.9624	32,855.94	63,657.99	25	1,591,450	13,126,968
425	.01267	.9601	34,937.82	67,793.76	25	1,694,844	14,821,812

The well-test data were obtained from a conventional backpressure test where all four flow periods achieved stabilized flow. A second table is constructed using the static well pressure and the test data points.

p_{wf}	ψ	$\Delta\psi$	q_g	$\Delta\psi / q_g$	q_g2
408.2	13,682,877	0	0	—	0
403.1	13,337,129	345,748	4.288	80,632	18.387
394.0	12,745,020	937,857	9.265	101,226	85.840
378.5	11,758,321	1,924,556	15.552	123,750	241.865
362.6	10,797,128	2,885,749	20.177	143,022	407.111
		6,093,910	49.282	448,630	753.203
14.7	30,069	13,652,808	AOFP		

Notice that none of the test pressures coincided with any of the p vs. ψ table pressures. Therefore, for a particular p_{wf}, the corresponding pseudopressure is determined using linear interpolation. For example, consider p = 408.2 psia. The table values for pressure and pseudopressure are needed, which bracket the pressure of interest.

p	ψ
400	13,126,968
425	14,821,812

Then:

$$\psi\,(p = 408.2) = 13{,}126{,}968 + \left[\left(\frac{408.2 - 400}{425 - 400}\right) \times (14{,}821{,}812 - 13{,}126{,}968)\right]$$

$$\psi = 13{,}682{,}877$$

This type of approximation for pseudopressures falling between table values is sufficiently accurate as long as the table pressure increment is not too large.

Figure 5-19 is a plot of $(\overline{\psi} - \psi_{wf}) / q_g$ versus q_g for the test data. Two points lying on the straight line are (6.75; 90,000) and (14.5; 120,000). Using Equation 5-68:

$$120{,}000 = a + 14.5\ b$$
$$90{,}000 = a + 6.75\ b$$

Solving for "a" and "b," a = 63,871 and b = 3,871.

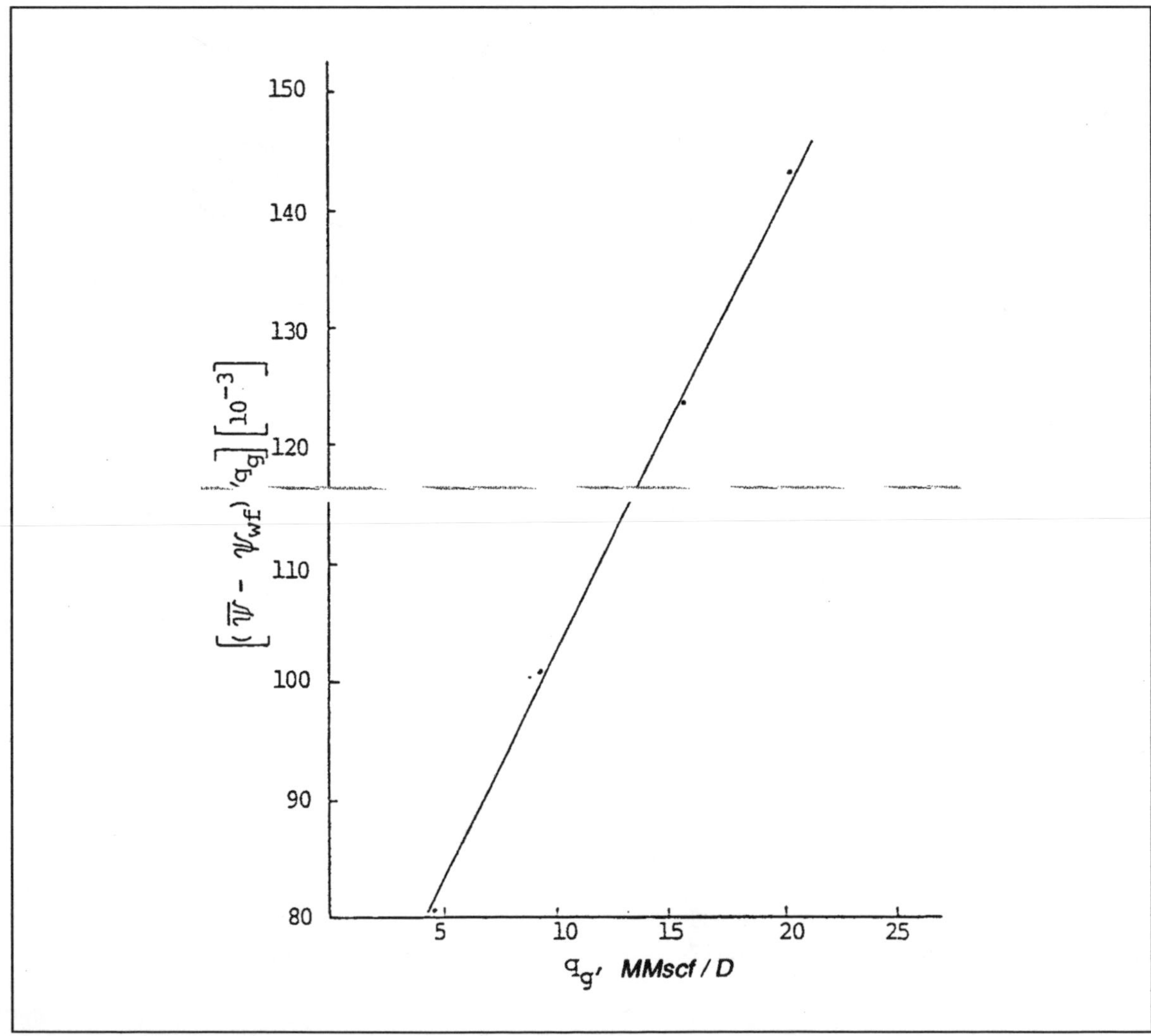

Fig. 5-19. Stabilized deliverability test. Pseudopressure formulation, theoretical method.

A more rigorous way to calculate "a" and "b" is to use the method of least squares; i.e., Equations 5-71 and 5-72. To do this, we use the data from the second table of this example. So:

$$a = \frac{(448{,}630)(753.203) - (49.282)(6{,}093{,}910)}{(4)(753.203) - (49.282)^2} = \frac{37{,}589{,}389.30}{584.0965} = 64{,}354.8$$

and

$$b = \frac{(4)(6{,}093{,}910) - (49.282)(448{,}630)}{584.0965} = \frac{2{,}266{,}256.34}{584.0965} = 3{,}879.9$$

These values are not much different than those obtained with the "eyeball" method, but they should be more accurate. Using the least squares values of "a" and "b," we write the stabilized deliverability equation:

$$3{,}879.9\, q_g^2 + 64{,}354.8\, q_g = (\overline{\psi} - \psi_{wf})$$

To solve for the AOFP, we need $[\psi(\overline{p}) - \psi(14.7)]$, which is obtained from the last table. Thus, for the AOFP:

$$3{,}879.9\, q_g^2 + 64{,}354.8\, q_g - 13{,}652{,}808 = 0$$

With the use of the quadratic equation:

$$q_g = \frac{-a + \sqrt{(a^2 + 4\,b\,\Delta\psi)}}{2\,b}$$

where $\Delta\psi$ is the particular value at the conditions of the desired flow rate.

$$AOFP = \frac{-64{,}354.8 + \sqrt{(64{,}354.8)^2 + (4)(3{,}879.9)(13{,}652{,}808)}}{(2)(3{,}879.9)} = 51.6\ MMscf/D$$

This is almost exactly the same AOFP that was obtained in Problem 5-6 using the p^2 approximation and the theoretical approach.

PSEUDOPRESSURE LOG-LOG DELIVERABILITY PLOT

After the constants "a" and "b" have been determined, then a log-log deliverability curve may be drawn by plotting $(\Delta\psi - b\,q_g^2)$ versus q_g. Figure 5-20 illustrates such a deliverability plot. This diagram is convenient for those who do not like to use equations such as Equation 5-68 to predict deliverabilities. Of course, to use this curve, another table or curve relating pseudopressure to pressure is required. This deliverability graph should be applicable for a longer period of reservoir life than the corresponding plot using the p^2 approximate solution.

PSEUDOPRESSURE WITH ISOCHRONAL AND MODIFIED ISOCHRONAL TESTING

With isochronal or modified isochronal tests, the transient points are analyzed in the same manner as previously outlined to calculate "b" only. Then, using the data from the one stabilized point, "a" is calculated:

$$a = \left\{ [\psi(\bar{p}) - \psi(p_{wf})] / q_g \right\} - b\, q_g \tag{5-73}$$

These constants "a" and "b" are used in the stabilized deliverability relationship:

$$a\, q_g + b\, q_g^2 = \psi(\bar{p}) - \psi(p_{wf}) \tag{5-68}$$

Then, AOFP may be determined as outlined in problem 5-8.

G. Stabilization Time

In high permeability reservoirs, stabilized flowing pressures are quickly reached. In tight reservoirs, stabilized pressures may not be attained for months or even years. The time to reach stabilization represents the amount of time that it takes, after a rate change, for the reservoir to reach pseudosteady state. For a regular-shaped drainage area with the well in the center, the approximate stabilized time (hours) is:

$$t_s = 381 \frac{\phi\, \bar{\mu}\, A}{k \bar{p}} \tag{5-74}$$

Where the drainage area is not regular shaped and/or the well is not in the center, then Equation 5-74 will under estimate the time to stabilization. In this case, a better estimate is:

$$t_s = 3810 \frac{\phi\, \bar{\mu}\, A}{k \bar{p}} (t_{da})_s \tag{5-75}$$

where:

t_s = time to stabilization, hr,
A = drainage area, ft^2,
$\bar{\mu}$ = gas viscosity at $\bar{p}$, cp,
ϕ = gas-filled porosity, fraction (i.e., actual porosity times gas saturation),
k = effective permeability to gas, md,
$\bar{p}$ = stabilized shut-in bottomhole pressure, psia, and
$(t_{da})_s$ = dimensionless stabilization time referred to area.

Note that $(t_{da})_s$ values for various shapes and well locations may be found in Table 16-1 of the "Well Testing" chapter. The drainage geometry is indicated on the left, and the $(t_{da})_s$ values are given in the fourth column of numbers under the heading, "Exact For t_{DA} > ". For a circle or square with the well in the center, $(t_{da})_s = 0.1$.

Where the data are known, Equation 5-74 or 5-75 may be used in designing well tests.

H. Radius of Investigation

During the infinite-acting period (before a pressure disturbance caused by a rate change at a well has reached the exterior radius of the drainage area), the radius of investigation in feet may be calculated approximately by:

$$r_{inv} = 0.032 \sqrt{\frac{k \bar{p} t}{\phi \bar{\mu}}} \tag{5-76}$$

where the nomenclature and units are the same as with Equation 5-74.

For instance, if a well has just been put on production then the distance that the pressure transients have propagated out into the reservoir (assuming that the pressure disturbance has not yet reached the outer boundary) may be estimated with Equation 5-76.

Note that the actual magnitude of the flow rate change has no effect on Equation 5-76. Theory indicates that the gradient of the pressure transients will vary with the magnitude of the change in rate, but the velocity of the transients will be the same (approximately, for the same well) whether the rate change is large or small. Those interested in pursuing this matter further are referred to reference 8.

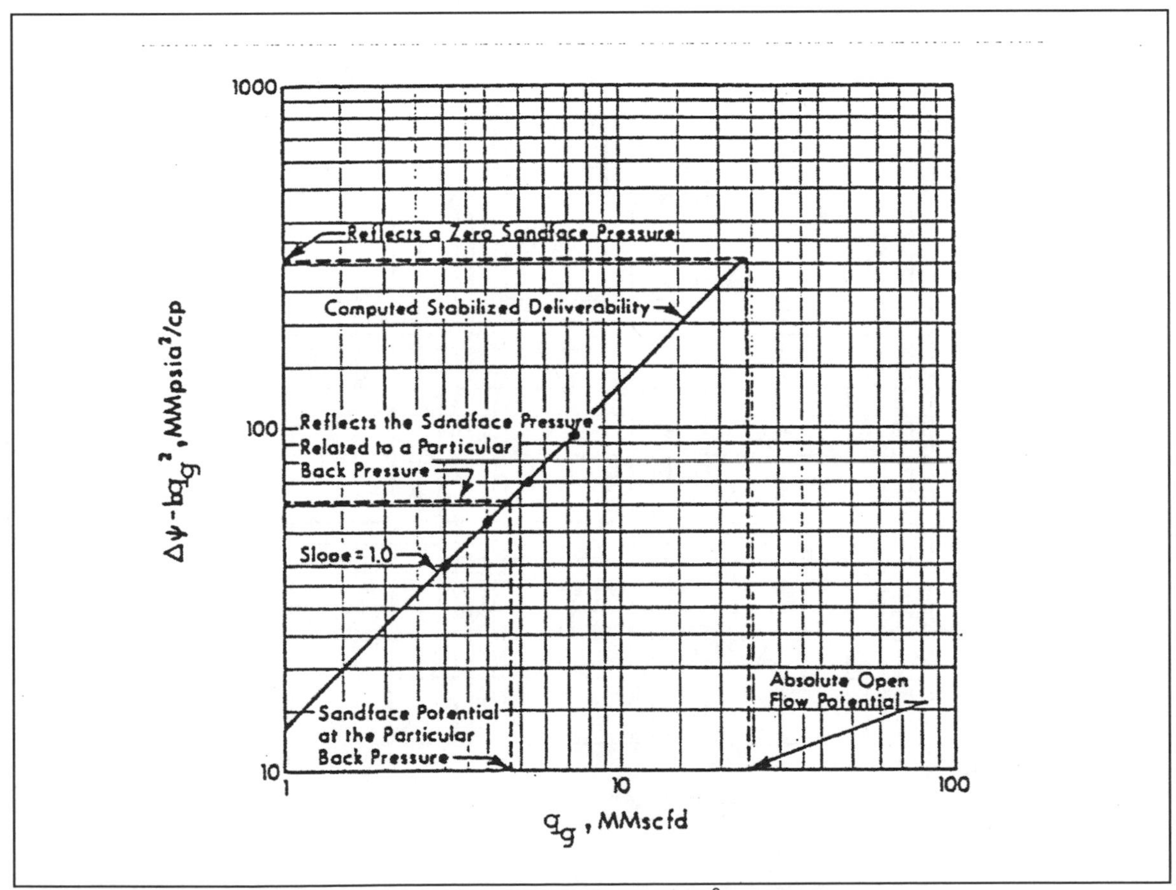

Fig. 5-20. Gas deliverability flow analysis using pseudopressure approach[8]. Permission to publish by Energy Resources Conservation Board of Canada.

Figure 5-21 provides a convenient graphical determination of the radius of investigation.

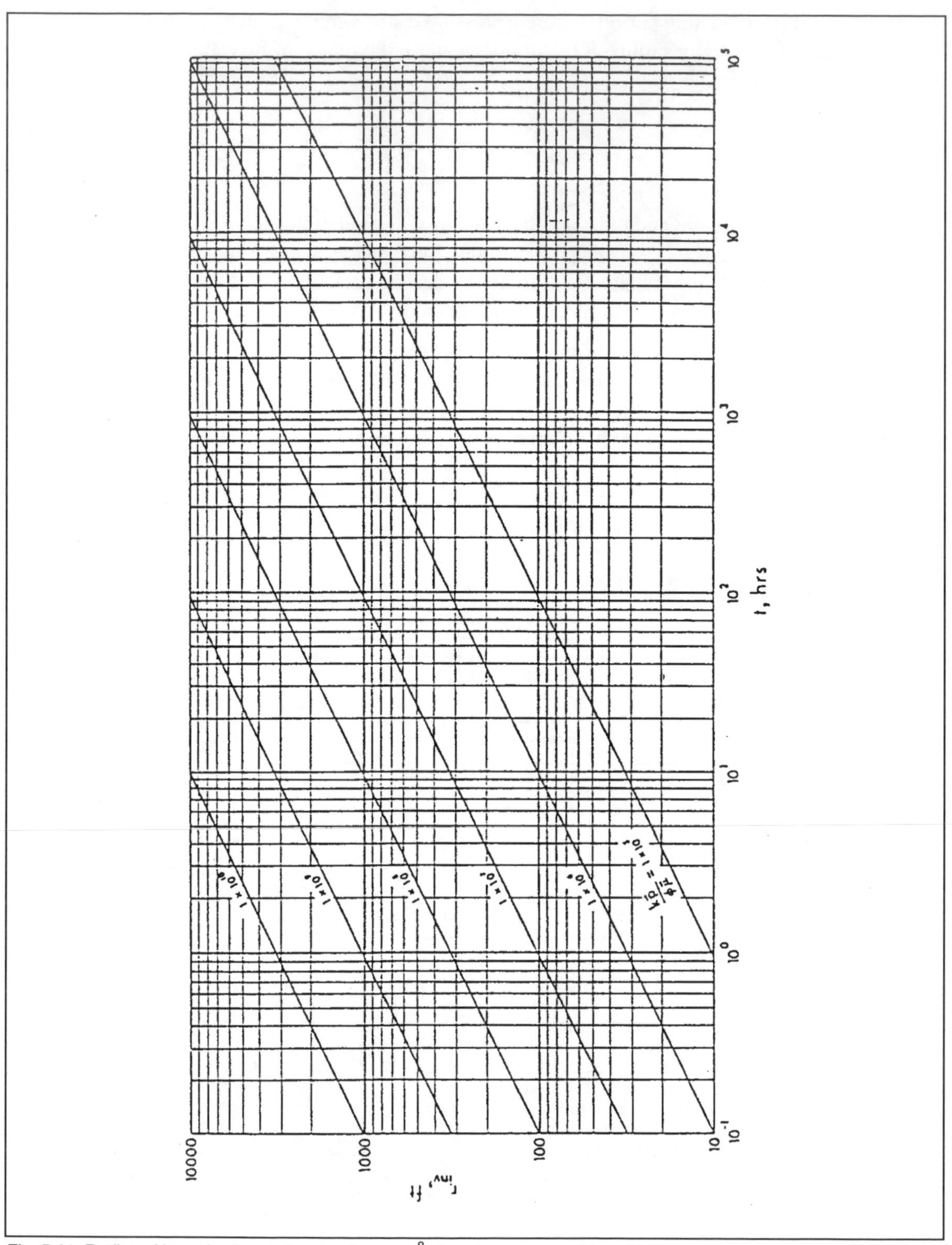

Fig. 5-21. Radius of investigation during gas flow tests[8]. Permission to publish by Energy Resources Conservation Board of Canada.

REFERENCES

1. Clark, N. J.: "Elements of Petroleum Reservoirs," SPE of AIME, Dallas (1960).

2. Craft, B. C. and Hawkins, M. F.: *Applied Reservoir Engineering*, Prentice-Hall, Inc., Englewood Cliffs, N.J. (1959).

3. Cragoe, C. S.: "Thermodynamic Properties of Petroleum Products," Bureau of Standards, U. S. Department of Commerce, Misc. Publ. No. 97 (1929) 22.

4. Katz, D. L., *et al.*: *Handbook of Natural Gas Engineering*, McGraw-Hill Book Co., Inc., New York City (1959).

5. Geffen, T. M, Parrish, D. R., Haynes, G. W., and Morse, R. A.: "Efficiency of Gas Displacement from Porous Media by Liquid Flooding," *Trans.*, AIME (1952) **195**, 37.

6. Guerrero, E. T.: *Practical Reservoir Engineering*, The Petroleum Publishing Co., Tulsa (1968) 213-218.

7. Cullender, M. H. and Smith, R. V.: "Practical Solution of Gas-Flow Equations for Well and Pipelines with Large Temperature Gradients," *Trans.*, AIME (1956) **207**.

8. Energy Resources Conservation Board: "Theory and Practice of the Testing of Gas Wells," Calgary, Alberta, Canada, Third Ed., (1975), Second Printing (1978).

9. Sukkar, Y. K. and Cornell, D.: "Direct Calculation of Bottomhole Pressures in Natural Gas Wells," *Trans.*, AIME (1955) **204**, 43-48.

10. "Manual of Back Pressure Testing of Gas Wells," Interstate Oil Compact Commission (1962).

11. "Manual of Back Pressure Testing of Gas Wells," Kansas State Corporation Commission (1959).

12. Rawlins, E. L. and Schellhardt, M. A.: "Back-Pressure Data on Natural Gas Wells and Their Application to Production Practices," USBM Monograph 7 (1936).

13. Al-Hussainy, R. and Ramey, H. J., Jr.: "Application of Real Gas Flow Theory to Well Testing and Deliverability Forecasting," *JPT* (May 1966) 637-42.

14. Kulczycki, W.: "New Method of Determination of the Output and the Absolute Open-Flow of Gas Wells," *Nafta, 11* (10) (1955) 233-237.

15. Cullender, M. H.: "The Isochronal Performance Method of Determining the Flow Characteristics of Gas Wells," *Trans.*, AIME (1955) **204**, 137-142.

16. Katz, D. L., *et al.*: *Handbook of Natural Gas Engineering*, McGraw-Hill Book Co., Inc., New York City (1959) 448.

17. Pirson, S. J.: *Oil Reservoir Engineering*, McGraw-Hill Book Co., Inc. New York City (1958).

18. *Bull. D14, A Statistical Study of Recovery Efficiency*," API, Dallas (October 1967).

19. Merle, H. A., Kentie, C. J. P., van Opstal, G. H. C., and Schneider, G. M. G.: "The Bachaquero Study - A Composite Analysis of the Behavior of a Compaction/Solution Gas Drive Reservoir," *JPT* (September 1976) 1107-1115.

20. Brown, K. E.: *The Technology of Artificial Lift Methods*, Petroleum Publishing Company, Tulsa (1977).

21. Arnondin, M., van Poollen, M., and Farshad, F. F.: "Predicting Bottomhole Pressure for Gas and Gas Condensate Wells," *Pet. Eng. Intl.* (November 1980) 90-110.

22. Lee, J.: "Well Testing," SPE Textbook Series, Vol. 1., SPE of AIME, New York (1982) Dallas.

23. Al-Hussainy, R., Ramey, H. J., Jr., and Crawford, P. B.: "The Flow of Real Gases Through Porous Media," *JPT* (May 1966) 624-636; *Trans.*, AIME, 237.

24. Wattenbarger, R. A. and Ramey, H. J., Jr.: "Gas Well Testing with Turbulence, Damage and Wellbore Storage", *JPT* (August 1968) 877-887; *Trans.*, AIME, 243.

25. Ikoku, Chi U.: *Natural Gas Engineering, A Systems Approach*, PennWell Publishing Company, Tulsa (1980).

26. McCain, W. D., Jr.: *The Properties of Petroleum Fluids*, Petroleum Publishing, Tulsa (1973).

27. Slider, H. C.: *Practical Petroleum Reservoir Engineering Methods*, Petroleum Publishing Company, Tulsa (1976).

28. Peffer, J. W., Miller, M. A., and Hill, A. D.: "An Improved Method for Calculating Bottomhole Pressure in Flowing Gas Wells with Liquid Present," paper SPE 15655 presented at the 1986 Annual Technical Conference and Exhibition, New Orleans, (Oct. 5 - 8).

6 GAS CONDENSATE RESERVOIRS

INTRODUCTION

The reader is referred to the chapter on Fluid Properties for gas properties and for a discussion based on phase diagrams that outlines the difference between volatile oil, retrograde gas condensate, wet gas, and dry gas reservoirs. In this chapter, only those reservoirs that exhibit retrograde condensation behavior in the reservoir rock will be discussed. Although, wet gas reservoirs are sometimes treated under the topic "Gas Condensates," the authors here felt that it was more appropriate to discuss such systems in the "Gas Reservoirs" chapter.

Dry gas reservoirs normally yield little or no surface liquid recovery with processing through normal lease separation equipment. In Texas, a statutory gas well is one with a gas/oil ratio exceeding 100,000 SCF/STB of recovered hydrocarbon liquid (or less than 10 STB/MMscf of gas).

To field operating personnel, a gas is "wet" if hydrocarbon liquids are dropped out in surface separation equipment. To the reservoir engineer, a "wet" gas may be produced from either a single-phase gas reservoir, a retrograde-condensate gas reservoir, or from an "associated" oil reservoir (gas produced with oil). Confusion can usually be avoided by classifying reservoirs by conditions present at the time of discovery. If the gas exhibits producing gas/oil ratios exceeding 15,000 scf/STB by early testing in reservoirs with discovery pressures not exceeding 8,000 psia and temperatures less than or equal to 225°F, then the gas probably exists in the reservoir in a single phase. With pressure depletion due to production, the gas in the reservoir will remain in a single-phase vapor state, and no liquids will be lost. Again, this type of system (wet gas or dry gas) was discussed in the "Gas Reservoirs" chapter.

Where producing gas/oil ratios fall between 6,000 and 15,000 scf/STB, retrograde behavior should be suspected, but this is not always the case. Prudent operations suggest that a representative reservoir fluid sample be obtained and a laboratory pressure-volume-temperature (PVT) analysis be made. This will permit planning for the most efficient and profitable development of the resource.

Where the initial gas/oil ratio is between 3,000 and 6,000 scf/STB, it is possible for the reservoir to contain either a volatile oil or a retrograde gas condensate. A representative sample should be captured, and PVT studies performed to make the distinction between the two types of fluid. Pressure depletion of the sample at constant temperature in the laboratory would allow either a dewpoint or a bubblepoint to occur. A dewpoint would indicate the presence of a retrograde gas condensate reservoir.

The concern is not simply academic, since optimum exploitation for the two different types of reservoirs may be substantially different. Where spacing regulations exist, spacing would normally be wider for a "gas" reservoir as opposed to an "oil" reservoir. These regulations recognize the increased mobility of gas as contrasted with oil, and the corresponding greater migration capability of gas during producing operations.

Below an initial producing gas/oil ratio of 3,000 scf/STB, the authors are unaware of any bona fide reservoirs that contain a hydrocarbon fluid other than oil, volatile or otherwise. It is theoretically possible to have a retrograde gas condensate reservoir with an initial producing gas/oil ratio as low as 2,000 scf/STB, but this would take a unique combination of very high discovery pressure (exceeding 8,000 psia), modest temperature, and high concentrations of intermediates: C_2 through C_6. Table 6-1 shows the mole compositions of several different types of typical reservoir fluids.

Table 6-1
Mole Composition of Typical Hydrocarbon Reservoir Fluids

Component	Dry Gas	Single-phase Wet Gas	Retrograde Gas Condensate	Volatile Oil
C_1	96.0	90.0	75.0	60.0
C_2	2.0	3.0	7.0	8.0
C_3	1.0	2.0	4.5	4.0
C_4	0.5	2.0	3.0	4.0
C_5	0.5	1.0	2.0	3.0
C_6	—	0.5	2.5	4.0
C_{7+}	—	1.5	6.0	17.0
	100.0	100.0	100.0	100.0
Mol. Wt. C_{7+}	—	115	125	180
GOR, scf/STB	High	26,000	7,000	2,000
Tank Gravity, °API	—	60	55	50
Liquid Color	—	Water White to Light Yellow	Light Yellow to Yellow	Amber to Darker Colors

DEFINITION OF RESERVOIR TYPE FROM PHASE DIAGRAMS

Figure 6-1 provides a pressure-temperature phase diagram for a reservoir fluid. Note that the volume is a constant as is the total composition of the fluid represented. The figure may be used as a guide to relationships between the several reservoir fluid types. However, it is not truly representative because only pressure and temperature are allowed to change (not composition), yet both a single-phase oil and a single-phase gas are shown. Still, it is useful for descriptive purposes.

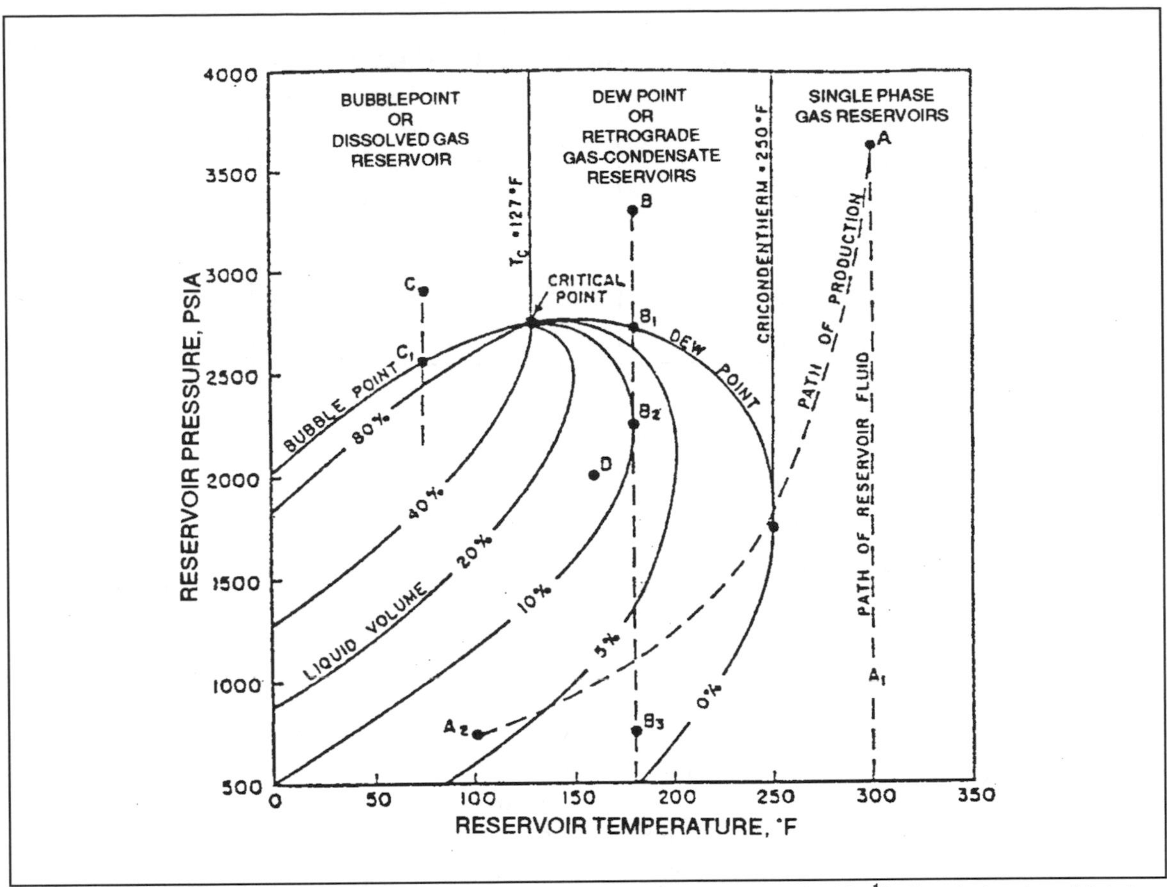

Fig. 6-1. Pressure-Temperature Diagram of a Reservoir Fluid (from Craft and Hawkins[1]). Permission to publish by Prentice-Hall.

Single-phase gas reservoirs are shown by point A upon discovery. Notice that this is to the right of the cricondentherm. Pressure depletion with production at constant reservoir temperature results in the reservoir fluid remaining as gas. However, cooling and pressure drop in the wellbore and surface facilities allow the condensing of some hydrocarbons along the line A to A_2.

Retrograde gas condensate reservoirs (sometimes called dew-point reservoirs) exist at pressures sufficient to be at or above the upper boundary of the two-phase envelope and at a temperature between the critical and cricondentherm values. Interestingly, we frequently find retrograde gas-condensate reservoirs with conditions on or very close to the dewpoint line at the time of discovery. This seems to be due to the presence of a large percentage of the intermediates (C_2 to C_6). It is also quite common to find a volatile oil rim, in which the gas cap would be exactly at the dewpoint. As a practical matter, should the observed dewpoint in the laboratory duplicate the discovery pressure within a few percentage points, we would expect possibly to find oil downdip.

Figure 6-2 presents descriptive phase diagrams for a gas-cap gas and oil zone fluid for the two cases of a retrograde gas-cap, and a nonretrograde gas-cap. Notice that the oil bubblepoint curve intersects the gas dewpoint curve at the time of discovery since phase equilibrium exists. Note also the relative positions of the critical points for the nonretrograde and retrograde cases.

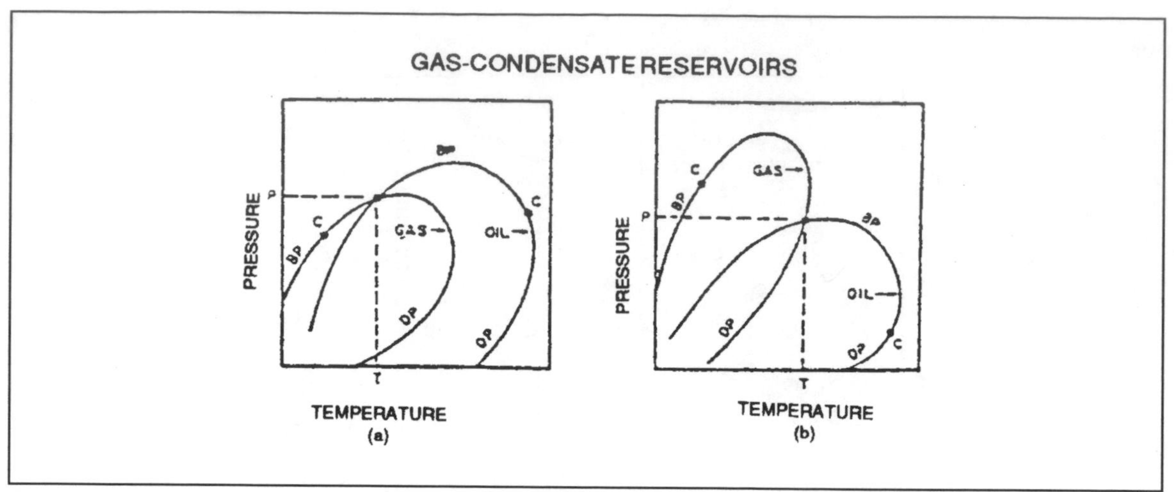

Fig. 6-2. Phase diagrams of a gas-cap gas and oil-zone liquid for (a) retrograde gas-cap gas, and (b) nonretrograde gas-cap gas (from Craft & Hawkins[1]). Permission to publish by Prentice-Hall.

CALCULATION OF IN-PLACE GAS AND OIL AND RESERVOIR PERFORMANCE

A. Laboratory PVT Study

A straight-forward way of predicting performance of a volumetric, retrograde gas-condensate reservoir is to duplicate reservoir depletion by laboratory studies of one or more representative reservoir fluid samples. The recommended method involves the continuous depletion of the gas phase in a constant-volume cell maintained at reservoir temperature. Normally, pressure control is obtained with the use of a mercury pump attached to the bottom of the high pressure cell. Beginning with the reservoir sample in the cell at the initial reservoir pressure, mercury is withdrawn and pressure permitted to drop several hundred psi. Assuming that the new pressure is below the dewpoint, after equilibrium has occurred and the condensed liquid has drained down to the bottom of the cell, the mercury is reinjected and gas is removed from the top of the cell at such a rate that a constant pressure is maintained in the cell. Gas is removed until the hydrocarbon volume (now two phase) is returned to the original cell volume. After the volume of gas removed and the volume of retrograde liquid in the cell are measured, then the cycle (pressure lowered, equilibrium obtained, gas produced, volumes measured) is repeated down to a selected abandonment pressure. To help obtain equilibrium, the cell is rocked. More details concerning this laboratory procedure are given in Craft and Hawkins[1].

An example from Craft and Hawkins[1], illustrates the calculation steps.

Problem 6-1:

Calculate the volumetric reservoir performance of a retrograde gas-condensate reservoir given the following data and the data of Table 6-2.

Initial pressure (dewpoint)	2960 psia
Abandonment pressure	500 psia
Reservoir temperature	195 °F
Connate water	30%

Porosity	25%
Standard conditions	14.7 psia & 60 °F
Initial cell volume	947.5 cc
Mol. Wt. of C_{7+} in initial fluid	114 lb/lb-mole
Sp. Gr. of C_{7+} in initial fluid	0.755 at 60 °F

Table 6-2
VOLUME, COMPOSITION, AND GAS DEVIATION FACTORS FOR A RETROGRADE CONDENSATE FLUID[1]

(1)	(2)	(3)	(4)	(5)	(6)	(7)	(8)	(9)	(10)	(11)	(12)
	Composition of Produced Gas Increments, Mole Fraction										
Pressure psia	C_1	C_2	C_3	C_4	C_5	C_6	C_{7+}	Produced Gas, cu cm at 195°F and Cell Pressure	Retro-grade Liquid Volume, cu cm	Retro-grade Volume, Percent of Hydrocarbon Volume	Gas Deviation Factor at 195°F and Cell Pressure
2960	0.752	0.077	0.044	0.031	0.022	0.022	0.052	0.0	0.0	0.0	0.771
2500	0.783	0.077	0.043	0.028	0.019	0.016	0.034	175.3	62.5	6.6	0.794
2000	0.795	0.078	0.042	0.027	0.017	0.014	0.027	227.0	77.7	8.2	0.805
1500	0.798	0.079	0.042	0.027	0.016	0.013	0.025	340.4	75.0	7.9	0.835
1000	0.793	0.080	0.043	0.028	0.017	0.013	0.026	544.7	67.2	7.1	0.875
500	0.768	0.082	0.048	0.033	0.021	0.015	0.033	1080.7	56.9	6.0	0.945

Assumptions:

(1) The same molecular weight and specific gravity for the C_7+ content for all produced gas.

(2) Liquid recovery from the gas is 25 percent of the butanes, 50 percent of the pentane, 75 percent of the hexane, and 100 percent of the heptanes and heavier.

Solution: (summary in Table 6-3)

(1) Calculate the increments of gross production in Mscf per acre-foot of net bulk reservoir rock, and enter in Column 2 of Table 6-3.

$$V_{HC} = 43{,}560 \times 0.25 \times (1 - 0.30) = 7623 \text{ cu ft/ac-ft}$$

For example, in the pressure increment between 2960 and 2500 psia,

$$\Delta V = 7623 \times (175.3 \text{ cu cm})/(947.5 \text{ cu cm}) = 1410 \text{ cu ft/ac-ft at 2500 psia and 195 °F}$$

$$\Delta G_p = (379.4\, p\Delta V)/(1000\, zRT) = (379.4 \times 2500 \times 1410)/(1000 \times 0.794 \times 10.73 \times 655)$$

$$= 240.1 \text{ Mscf/net ac-ft}$$

Column 3 contains the summation of gross gas production in Column 2.

(2) Calculate the Mscf of residue gas and the barrels of liquid obtained from each increment of gross gas production, and enter in Columns 4 and 6 using assumption (2). Then, the mole fraction recovery as liquid is:

$$\Delta n_L = 0.25 \times 0.028 + 0.50 \times 0.019 + 0.75 \times 0.016 + 0.034 = 0.0625 \text{ mole fraction}$$

Table 6-3
GAS AND LIQUID RECOVERIES IN PERCENT AND PER ACRE-FOOT FOR PROBLEM 6-1[1]

(1)	(2)	(3)	(4)	(5)	(6)	(7)	(8)	(9)	(10)	(11)
Pressure, psia	Incre-ments of Gross Gas Produc-tion, Mscf	Cumu-lative Gross Gas Produc-tion, Mscf, Σ(2)	Residue Gas in Each In-crement, Mscf	Cumu-lative Residue Gas Produc-tion Mscf Σ(4)	Liquid in Each In-crement, bbl	Cumu-lative Liquid Produc-tion, bbl Σ(6)	Avg Gas-Oil Ratio of Each Incre-ment, scf Residue Gas per bbl (4)+(4)	Cumu-lative Gross Recovery, % (3) X 100/1580	Cumu-lative Residue Gas Recovery, % (5) X 100/1441	Cumu-lative Liquid Recovery, % (7) X 100/143.2
2960							10,600			
2500	240.1	240.1	225.1	225.1	15.3	15.3	14,700	15.2	15.6	10.7
2000	245.2	485.3	232.3	457.4	13.1	28.4	17,730	30.7	31.7	10.8
1500	266.0	751.3	252.8	710.2	13.3	41.7	19,010	47.6	49.3	29.1
1000	270.9	1022.1	256.9	967.1	14.0	55.7	18,350	64.7	67.1	38.9
500	248.7	1270.8	233.0	1200.1	15.9	71.6	14,650	80.4	83.3	50.0

Assuming mole fraction is the same as volume fraction in gas, the Mscf recovered as liquid from 240.1 Mscf is

$$\Delta G_L = (0.0625)(240.1) = 15.006 \text{ Mscf}$$

This volume of gas can be converted to gallons of liquid using the gal/Mscf values of Table 3-3 of "Fluid Properties" chapter (see gas density) for C_4, C_5, and C_6. The average of the iso and normal compounds is used for C_4 and C_5.

For C_{7+}, (114 lb/lb-mole)/(379.4 scf/lb-mole x 0.8337 lb/gal x 0.755) = 47.71 gal/Mscf

Then the total liquid recovered from 240.1 Mscf is,

$$1.681 \times 32.04 + 2.281 \times 36.32 + 2.881 \times 41.03 + 8.163 \times 47.71$$

$$= 644.4 \text{ gal}$$

$$= 15.3 \text{ bbl}$$

The residue gas recovered from the 240.1 Mscf is,

$$240.1 \times (1 - 0.0625) = 225.1 \text{ Mscf}$$

Column 5 is the summation of Column 4, and Column 7 is the summation of Column 6.

(3) Calculate the gas/oil ratio for each increment of gross production in units of residue gas per barrel of liquid. Enter in Column 8 of Table 6-3. For instance,

$$(225.1 \times 1000)/15.3 = 14{,}700 \text{ scf/bbl}$$

(4) Calculate the cumulative percent recoveries of gross gas, residue gas, and liquid. Enter in Columns 9, 10, 11. The initial gross gas in place is:

(379.4 pV)/1000 zRT

= (379.4 x 2960 x 7623)/(1000 x 0.771 x 10.73 x 655)

= 1580 Mscf/ac-ft

The original liquid mole fraction is 0.088 and the total liquid recovery is 3.808 gal/Mscf of gross gas, which are calculated from the initial composition in the same manner as shown in part (2) above. Then,

G = (1 - 0.088) x 1580 = 1441 Mscf residue gas/ac-ft

N = (3.808 x 1580)/42 = 143.2 bbl/ac-ft

At 2500 psia, then

Gross gas recovery = (100 x 240.1)/1580 = 15.2 percent

Residue gas recovery = (100 x 225.1)/1441 = 15.6 percent

Liquid recovery = (100 x 15.3)/143.2 = 10.7 percent

Reference to Column 8 of Table 6-3 will show that the laboratory cell producing gas/oil ratio increases and then declines below 1500 psia due to revaporization of liquids. As a practical matter, field experience suggests that essentially all liquids are held to rock surfaces by capillary forces, and virtually none is revaporized. In an actual reservoir, the producing gas/oil ratio normally does not decline.

Reference to Columns 10 and 11 will indicate that 83.3 percent of the residue gas is calculated as recovered to an abandonment pressure of 500 psia. However, only 50 percent of the liquids are recovered, due to retrograde condensation. Gas cycling projects often are justified on the recovery of the liquids which would otherwise be lost due to capillary effects if production operations only permitted pressure depletion of the resource involved.

B. Using Equilibrium Constants (or so-called "K" Values)

While gas condensate reservoir behavior is best made using laboratory determinations as illustrated in Problem 6-1, calculations can be made with equilibrium constants, or K values. This method requires that a representative sample of the reservoir fluid has been analyzed for composition. Best results are obtained when sampling bottomhole pressure has not dropped below the dewpoint.

Then, volumetric depletion performance may be calculated using equilibrium ratios. An equilibrium ratio or K value is defined as the ratio of the mole fraction (y) of any component, i, in the vapor phase to the mole fraction (x) of that same component in the liquid phase, or:

$$K_i = y_i / x_i \tag{6-1}$$

These ratios depend on temperature and pressure and, unfortunately, also on the composition of the system. If a set of K values can be found which are applicable, then the composition of the vapor and liquid phases of a retrograde gas-condensate system can be calculated at any temperature and pressure. Also, it is possible to calculate the vapor and liquid volumes. However, this is a specialized technique, and published methods[4,16] must be used to adjust the K values to fit the composition of a particular retrograde system. Without some sort of laboratory check, Standing[7] states that the equilibrium ratio method, used by itself, can lead to considerable error. With time, the K values are changing because the pressure and composition of the system (remaining in the reservoir or lab cell) are changing. However, the calculation procedure has merit if a laboratory PVT study has not been made, or a sample is no longer available.

A retrograde gas condensate system is considered to be a complex hydrocarbon mixture, and will be characterized by being comprised of a number of pure components. The basic equations of the calculation procedure are:

$$\sum_{i=1}^{m} y_i = \sum_{i=1}^{m} \frac{z_i}{(L / K_i) + V} = 1 \tag{6-2}$$

and

$$\sum_{i=1}^{m} x_i = \frac{1}{V} \sum_{i=1}^{m} \frac{z_i}{(L / V) + K_i} = \sum_{i=1}^{m} \frac{z_i}{L + VK_i} = 1 \tag{6-3}$$

where:

- y_i = mole fraction of component i in the vapor phase,
- x_i = mole fraction of component i in the liquid phase
- z_i = mole fraction of component i in the total mixture,
- K_i = component i equilibrium ration, y_i/x_i,
- V = mole fraction of the vapor phase in the total mixture,
- L = mole fraction of the liquid phase in the total mixture, and
- m = number of pure components.

This is best illustrated with an example taken from Guerrero[6].

Problem 6-2:

A natural gas has a composition at initial reservoir conditions of 2500 psia and 150 °F as shown in Column 2 of Table 6-4. What are the mole fractions of liquid and gas at 2000 psia given isothermal pressure decline?

Solution:

(1) Guerrero used the method of Hadden[4] for determining K values. The method is based on convergence pressure of a system represented as a pseudobinary, or two-component, mixture. The lightest component, usually methane, is used as component one. Then component two is actually a pseudocomponent made up of the remaining components in the system. Table 6-4 shows the calculations using a convergence pressure of 3,000 psia. Note that three values of V were assumed before the liquid and vapor fractions individually added to one. The K values were taken from the *Engineering Data Book*[5].

(2) Then check the convergence pressure used. Convergence pressure is merely a parameter used to characterize the effect of system composition. Reference 5 indicates that convergence pressure is mainly a function of temperature and composition of the liquid phase. The procedure involves first calculating the weighted-average critical temperature (considering the liquid) as shown in Columns 12 through 15 of Table 6-4. Here, the critical temperature is 23,724.99 / 49.9 or 475 °F. On Figure 6-3, this being the critical loci of hydrocarbon systems, the critical temperature of the hypothetical heavy component lies to the right of hexane. The loci of critical points of all pseudobinary mixtures of methane and the hypothetical heavy component is sketched (interpolated) in basically parallel to the existing curves. To determine convergence pressure, enter Figure 6-3 at reservoir temperature (150 °F), move up to the appropriate curve, and read the indicted convergence pressure on the Y-axis. For this problem, the apparent convergence pressure is 3,100 psia, or sufficiently close to the 3,000 psia value used in the trial-and-error calculations.

In the calculation of Problem 6-2, both liquid and vapor are present. This suggests that slightly over 11 mol percent liquid (condensate) is already present at a reservoir pressure of 2,000 psia and 150 °F. Liquid drop-out was expected due to the substantial fractions of C_5, C_6, and C_{7+} present.

Since liquid drop-out has been calculated to be substantial, a reservoir gas of the composition given in Table 6-4 would be a candidate for cycling operations. Cycling would allow the exploitation of this hydrocarbon resource with minimum liquid losses. Since the initial reservoir pressure was reported as 2,500 psia, the reservoir fluid apparently was already at or near its dewpoint pressure. Liquid losses will be large unless cycling operations are initiated early in field life.

While Problem 6-2 involved a fluid of substantial liquid content having a dewpoint close to the discovery pressure, many retrograde gas-condensate reservoirs have dewpoint pressures which are substantially lower than the discovery pressure. If a good fluid analysis is available, it is possible to calculate estimated upper and lower dewpoints of natural gas, using K values.[6, 16] Such calculations are useful in preparing the reservoir development plan.

Table 6-4
CALCULATIONS FOR SELECTION OF K VALUES BASED ON CONVERGENCE PRESSURE[6]

(1)	(2)	(3)	(4)	(5)	(6)	(7)
		K values	Assume V = 0.875 L/V = 0.125/0.875 = 0.143		Assume V = 0.888 L/V = 0.112/0.888 = 0.126	
Component	Mole fraction, Z	at 2,000 psia and 150° F.*	K + 0.143	(2)÷(4)	K + 0.126	(2)÷(6)
CO_2	0.0050	†1.55	1.693	0.0030	1.676	0.0030
C_1	0.8255	2.05	2.193	0.3764	2.176	0.3794
C_2	0.0375	0.94	1.083	0.0346	1.066	0.0352
C_3	0.0170	0.60	0.743	0.0229	0.726	0.0234
iC_4	0.0061	0.42	0.563	0.0108	0.546	0.0112
nC_4	0.0081	0.375	0.518	0.0156	0.501	0.0162
C_5	0.0293	‡0.240	0.383	0.0765	0.366	0.0801
C_6	0.0218	0.137	0.280	0.0779	0.263	0.0829
C_7 +	0.0497	§0.069	0.212	0.2344	0.195	0.2549
				0.8521		0.8863

(8)	(9)	(10)	(11)	(12)	(13)	(14)	(15)
Assume V = 0.8893 L/V = 0.1107/0.8893 = 0.124		Mole fraction					
		Liquid X	Vapor Y	Molecular weight[3] M	Weight	Critical	Wt. × t_c
K+0.124	(2)÷(8)	(9)÷0.8903	(10)×(3)		(10)×(12)	Temp.[3], t_c, °F.	(13)×(14)
1.674	0.0030	0.0033	0.0051	44.01	0.145	87.5	12.69
2.174	0.3797	0.4266	0.8753		Omit lightest component		
1.064	0.0352	0.0395	0.0371	30.07	1.138	89.9	106.80
0.724	0.0235	0.0264	0.0158	44.09	1.164	206.0	239.78
0.544	0.0112	0.0125	0.0053	58.12	0.727	274.6	199.63
0.499	0.0162	0.0182	0.0068	58.12	1.058	305.7	323.43
0.364	0.0805	0.0904	0.0217	72.15	6.522	‡377.9	2,464.66
0.261	0.0835	0.0938	0.0129	86.17	8.083	454.2	3,671.30

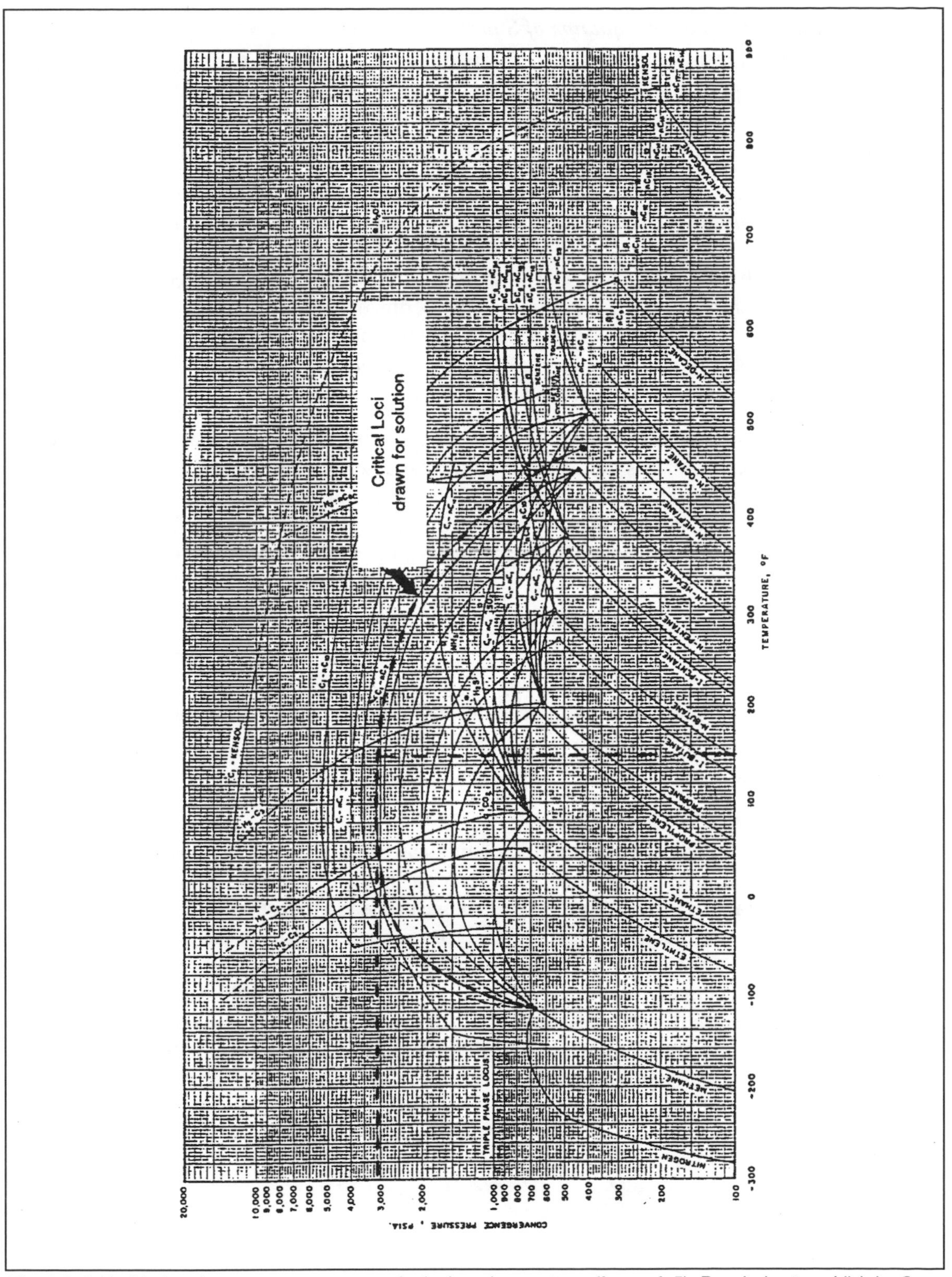

Fiig. 6-3. Critical loci and convergence pressures for hydrocarbon systems (from ref. 5). Permission to publish by Gas Processors Suppliers Association.

Phase Equilibria Using an Equation of State

The method of Haddon[4,5] illustrated in Problem 6-2, gives good results and can be used without the aid of a computer. The K-value charts[5], needed in this method, are based on laboratory experimentation of different hydrocarbon systems. Charts are given for individual hydrocarbon components through decane at various convergence pressures to 10,000 psia.

The trend today, if a computer is available, is to predict hydrocarbon phase equilibria using an equation of state (EOS). Farley *et al.*[10] have presented a method for predicting the dewpoints of gas-condensate systems and their subsequent normal and retrograde phase behavior with pressure decline. Use is made of a modified form of the Benedict-Webb-Rubin EOS based on regression calculations on experimental data. Much of the problem with predicting gas-condensate system behavior is due to the presence of the heavier molecular weight hydrocarbons. Farley *et al.,* chose to characterize the "heavies" by splitting the reported C_{10+} fraction into extended components C_{11} through C_{20}.

Coats[11] has discussed simulation of gas-condensate reservoir performance. He suggested that the properties of gas condensates above the dewpoint can be represented adequately using only two pseudocomponents. Below the dewpoint, especially when cycling, it was found necessary to use a minimum of six total components. The C_{7+} fraction was to be split into a number of components. Then, the compositional model could suitably match laboratory expansion data.

Coats presents a single generalized EOS that can be used to represent a number of the currently popular EOS equations such as that of Peng and Robinson[17]. Abhvani and Beaumont[16] present a procedure using the generalized EOS to calculate equilibrium constants. Although such computer procedures involve the simultaneous solution of nonlinear equations, and therefore require iterative techniques, the method is fairly stable and converges rapidly.

Equilibrium Constant Conclusion

The use of an equation of state for predicting gas-condensate performance can be tedious, but it can be an improvement over the Haddon[4] method based on convergence pressure. Since reservoir predictions from K values alone can lead to substantial errors, nothing takes the place of carefully collected and analyzed samples obtained early in the life of the reservoir. Figure 6-4 presents hypothetical phase relations for a gas-condensate system. The difficulty in accurately predicting phase behavior knowing only the system composition, reservoir temperature, discovery pressure, and perhaps current pressure is evident. Any change in system composition due to retrograde condensation substantially alters the shape and dimensions of the two-phase region.

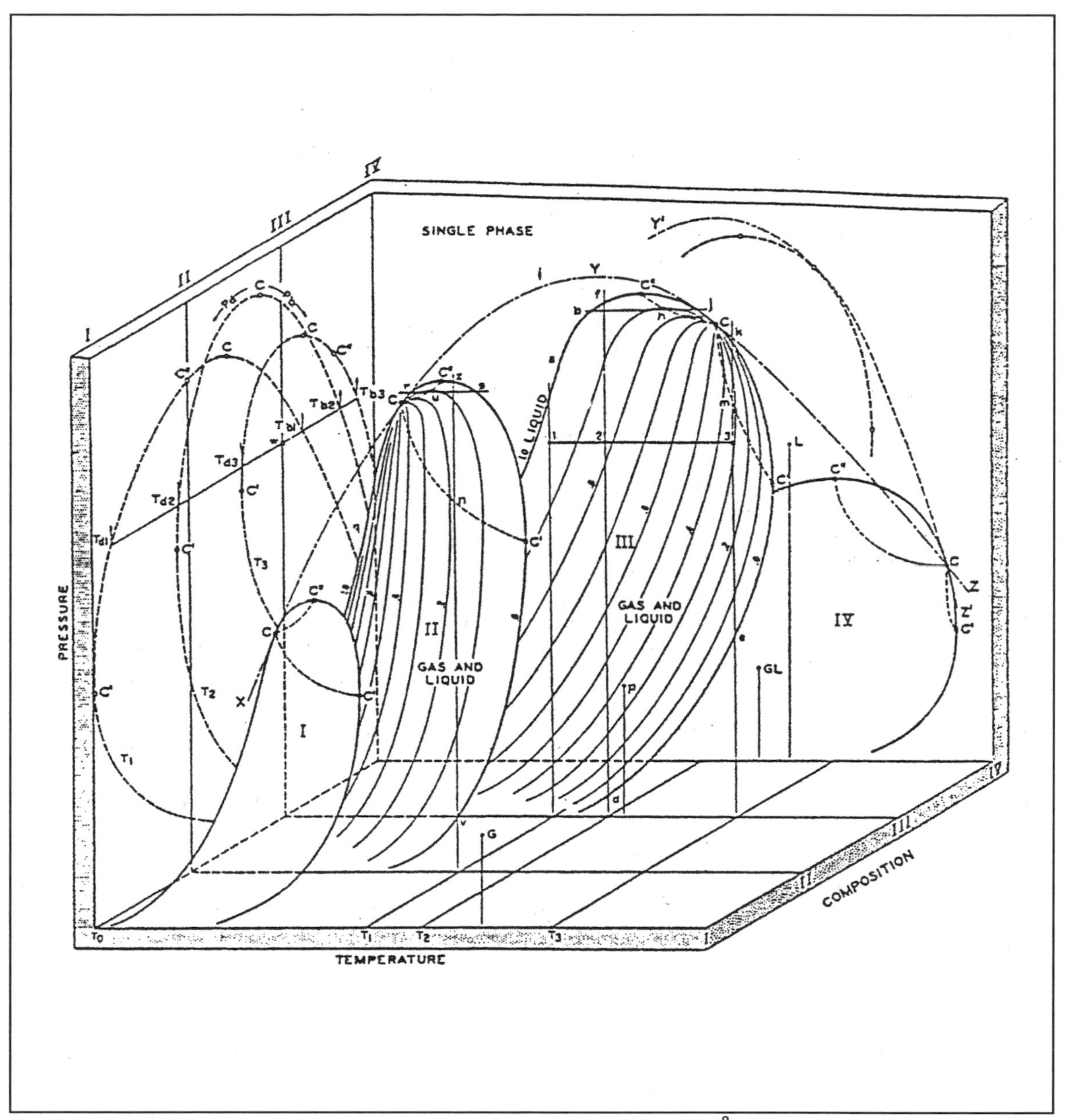

Fig. 6-4. Hypothetical phase relations for a gas-condensate system (from Eilerts et al.[8]). Permission to publish by American Gas Association.

C. Empirical Methods

Where laboratory studies of the reservoir fluid are not available, it is sometimes possible to use the published results of Jacoby, Koeller, and Berry[9]. This is often the case for small reservoirs. It is also a useful method where estimates of future condensate and gas production are needed in advance of a laboratory analysis, even if such work is ultimately planned. The authors studied several rich natural gas-condensate systems (gas/oil ratios of 3,600 to 60,000 scf/STB), one natural volatile oil system (gas/oil ratio of 2,363 scf/STB), and a series of related synthetic reservoir fluid mixtures (gas/oil ratios of 2,000 to 25,000 scf/STB). The correlations were prepared using regression analysis. It was found that the cumulative oil and separator gas in place recovered from the saturation pressure to an abandonment pressure of 500 psi could be correlated with initial GOR, initial tank oil gravity, and reservoir temperature. It was noted that depletion from saturation pressure to 500 psi would recover an average of 92 percent of the total separator gas in place.

The calculated ultimate oil recovery by depletion from the saturation pressure to 500 psia is correlated by the equation:

$$N_p = -0.061743 + (143.55/R_i) + 0.00012184T + 0.0010114(°API) \tag{6-4}$$

where:

N_p = cumulative stock-tank oil production from $p_{d,b}$ to 500 psia, stock-tank barrels of oil per barrel of hydrocarbon pore space,
R_i = initial separator gas/oil ratio, scf/STB,
T = reservoir temperature, °F,
$°API$ = initial stock-tank oil gravity,
$p_{d,b}$ = saturation pressure (dewpoint or bubblepoint), psia

In Equation 6-4 it is worth noting that pressure was not found to be an important correlating factor. This might not be the case for lean gases. Equation 6-4 was found to fit the available data with a standard deviation of the percentage errors of 11.3 percent.

Separator gas in place at the saturation pressure was correlated by the equation:

$$G = -229.4 + 148.43\,(R_i\ /\ 100)^{0.2} + 124{,}130\ /\ T + 21.831\,(°API) + 0.26356\,p_{d,b} \tag{6-5}$$

where:

G = total primary separator gas in place initially, scf/BBL HCPS.

Equation 6-5 fits the available data with a standard deviation of the percentage errors of 3.8 percent.

Figures 6-5 and 6-6 are convenient nomographs for the solution of Equations 6-4 and 6-5. The original reservoir pressure should be used in the nomographs, in the absence of information on the dewpoint or bubblepoint of the initial well stream fluid. In this instance, a check calculation can be made to suggest whether the original reservoir pressure is substantially above the saturation pressure of the original fluid in place. Divide the total initial separator gas in place (G) by the initial GOR (R_i) to obtain the most probable value of the stock tank oil in place (Bbl/Bbl HCPS). If this value is substantially larger than that shown by the oil-in-place curve of Figure 6-7, this would be an indication that reservoir pressure is above the fluid dewpoint pressure. However, such an instance could also be correlation error.

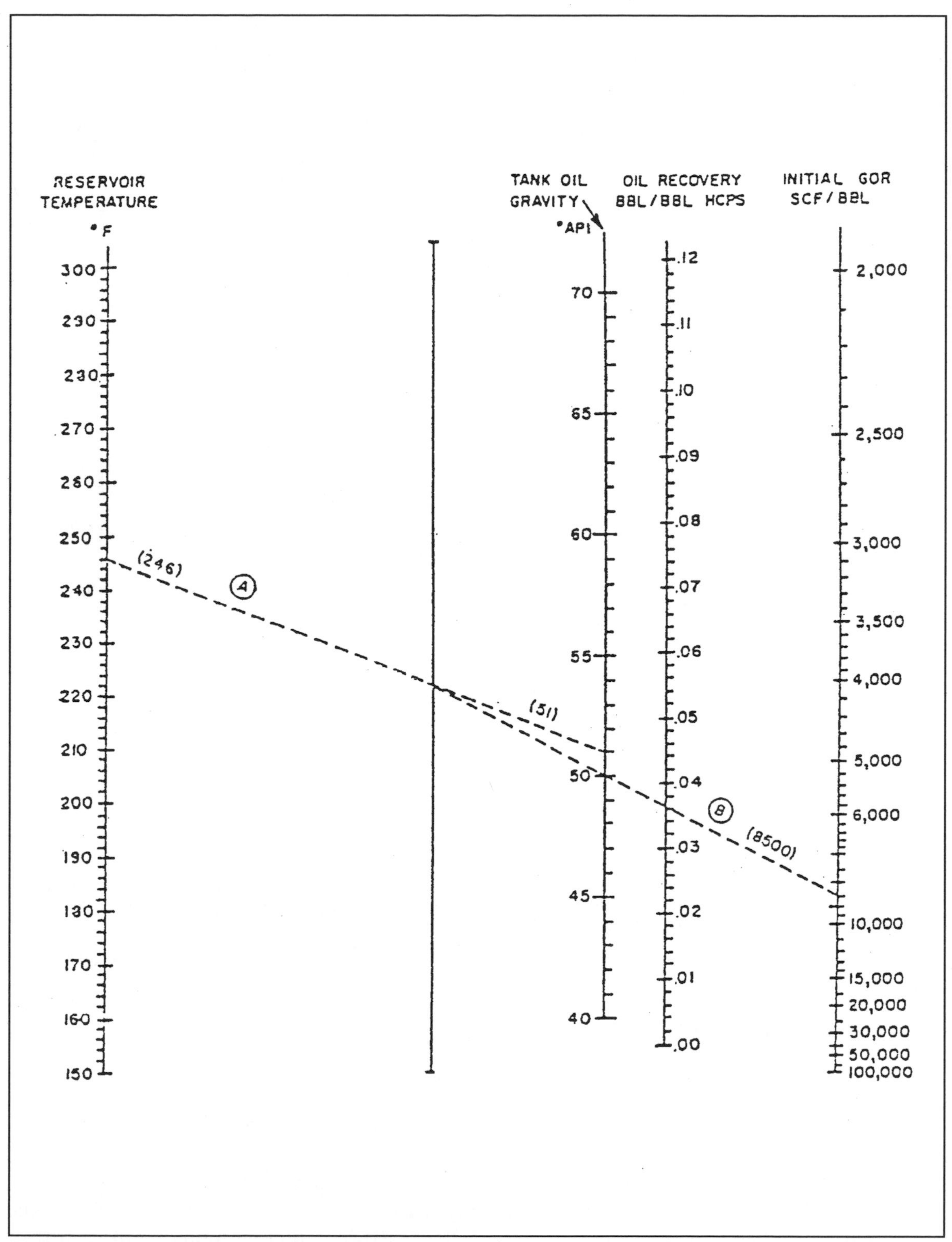

Fig. 6-5. Nomograph for predicting ultimate tank-oil recovery from very volatile oil and rich gas-condensate reservoirs (from Jacoby et al.[9]). Permission to publish by SPE.

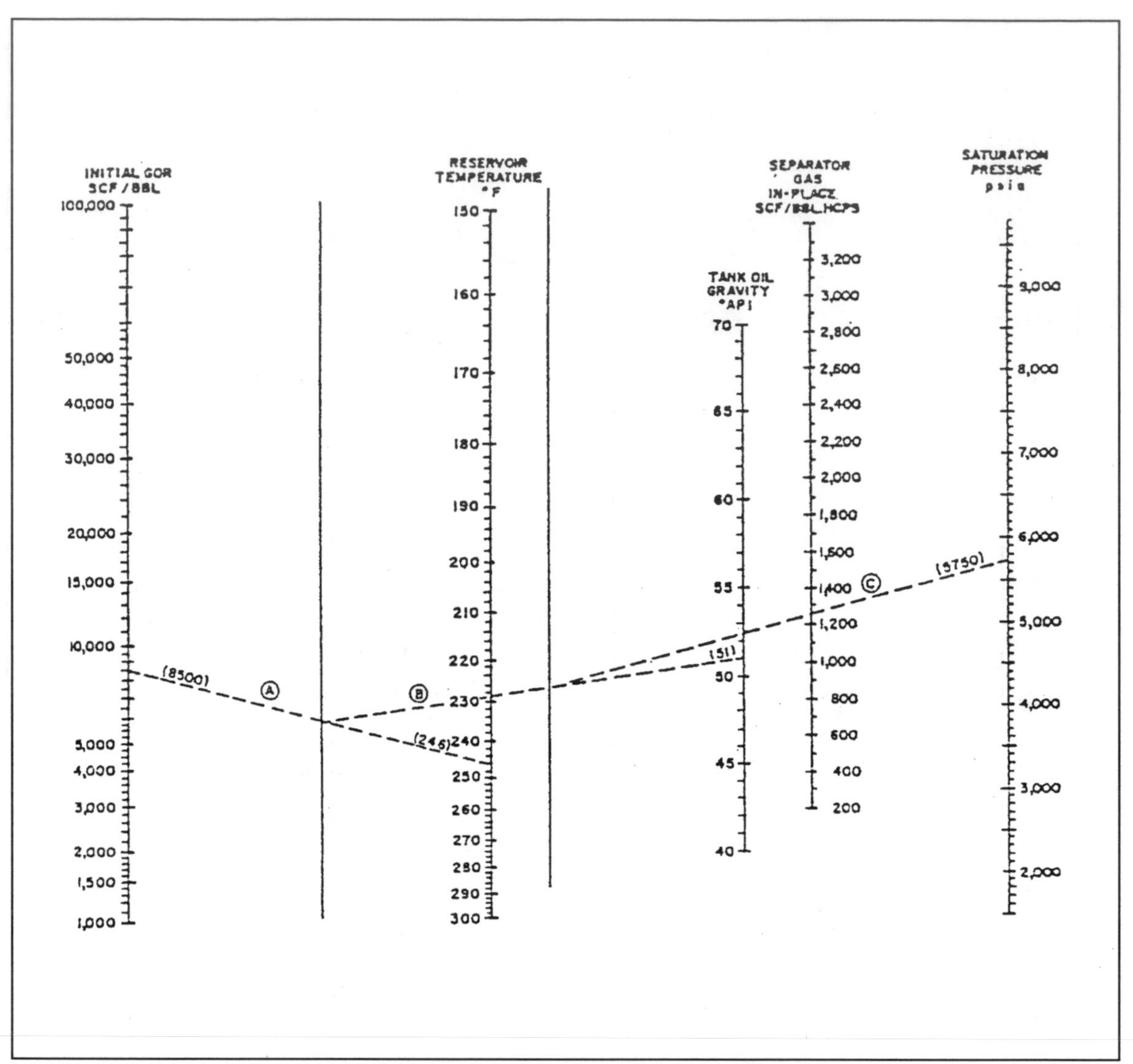

Fig. 6-6. Nomograph for predicting total separator gas initially in place in very volatile oil and rich gas-condensate reservoirs (from Jacoby et al.[9]). Permission to publish by SPE.

Where the initial pressure is indicated to be substantially above the dewpoint pressure, a revised value of oil recovered should be calculated with:

$$N_p\,(\,revised\,) \;=\; N_p\,(\,nomograph\,)\left[\frac{G\,(\,nomograph\,)\;/\;R_i}{oil\;\;in\;\;place\;\;(\,Fig.\,6\text{--}7\,)}\right] \tag{6-6}$$

A newer method by Eaton and Jacoby[15] for gas-condensate depletion performance estimation (empirical basis) was published in 1965. Twenty-five different gas condensates and two volatile oils were studied experimentally in the laboratory, and depletion performance of each fluid system was calculated. By multiple regression analysis, correlations were developed for: gas in

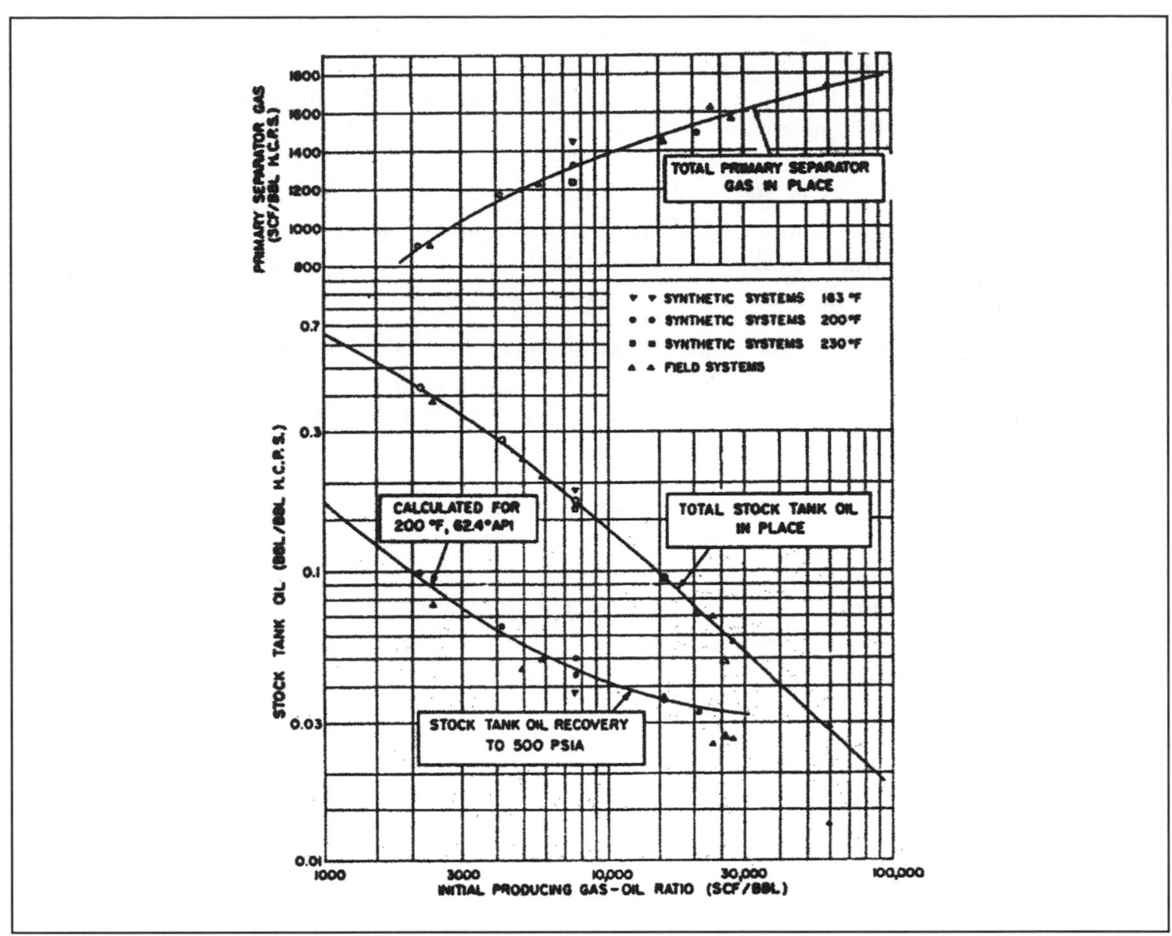

Fig. 6-7. Predicted oil recovery by depletion and total oil and gas in placed (reservoir fluid saturated at initial conditions) (from Jacoby et al.[9]). Permission to publish by SPE.

place, stock-tank oil in place, and stock tank oil production to 500 psia (abandonment pressure). Further, their results indicated an average gas recovery efficiency of 92.6% for production to 500 psia.

The three correlations are:

$$1n(G) = 4.5484 + 0.0831\ ln(R\ sub\ t) + 0.4265\ ln(p) - 0.3185\ ln(T) \quad (6\text{-}7)$$

$$ln(N) = 2.60977 - 0.90398\ ln(R\ sub\ t) + 0.48490\ ln(p) - 0.30084\ ln(T) + 0.29243\ ln(\&API) \quad (6\text{-}8)$$

$$ln(N\ sub\ p) = -20.243 - 0.65314\ ln(R\ sub\ t) + 1.3921\ ln(p) + 2.7958\ ln(\&API) \quad (6\text{-}9)$$

where:

G = total original surface gas in place, scf/bbl HCPS,
N = original stock-tank oil in place, STB/bbl HCPS,
N_p = ultimate oil production from original pressure to 500 psia, STB/bbl HCPS,

R_t = total scf of gas off the whole separator train (original conditions) per barrel of stock-tank oil, scf/STB,
p = original reservoir pressure (which may or may not be the saturation pressure), psia, and
T = reservoir temperature, °F

Unlike the earlier Jacoby *et al.*[9] results, the Eaton and Jacoby statistical studies indicated that stock tank oil gravity was not a significant parameter in the gas-in-place correlation (Equation 6-7). Similarly, the ultimate oil production relationship (Equation 6-9) determined by Eaton and Jacoby included pressure, but excluded temperature.

These three correlations are presented graphically in Figures 6-8, 6-9, and 6-10. The gas-in-place correlation had an error standard deviation of 3.57%, while the standard deviation of the errors of the ultimate oil produced relationship was 14.1% (9.3% if the two extreme errors were thrown out).

It is best not to use these correlations outside the range of data used in their preparation:

p : 4000 to 12,000 psia,
T : 160 to 190°F
R_t : 2500 to 60,000 scf/STB
°API : 45 to 65.

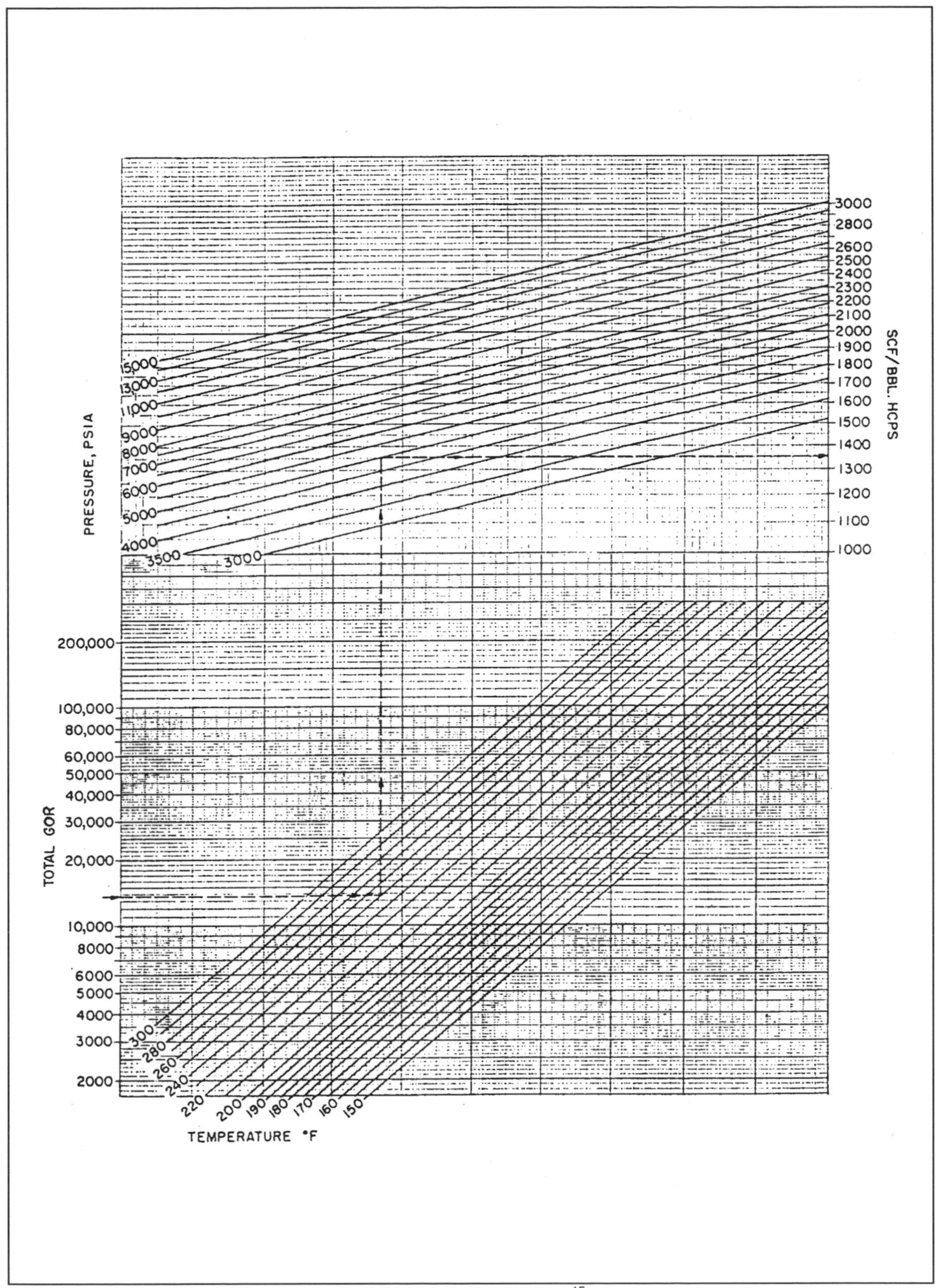

Fig. 6-8. Gas-condensate gas-in-place correlation (from Eaton and Jacoby[15]). Permission to publish by JPT.

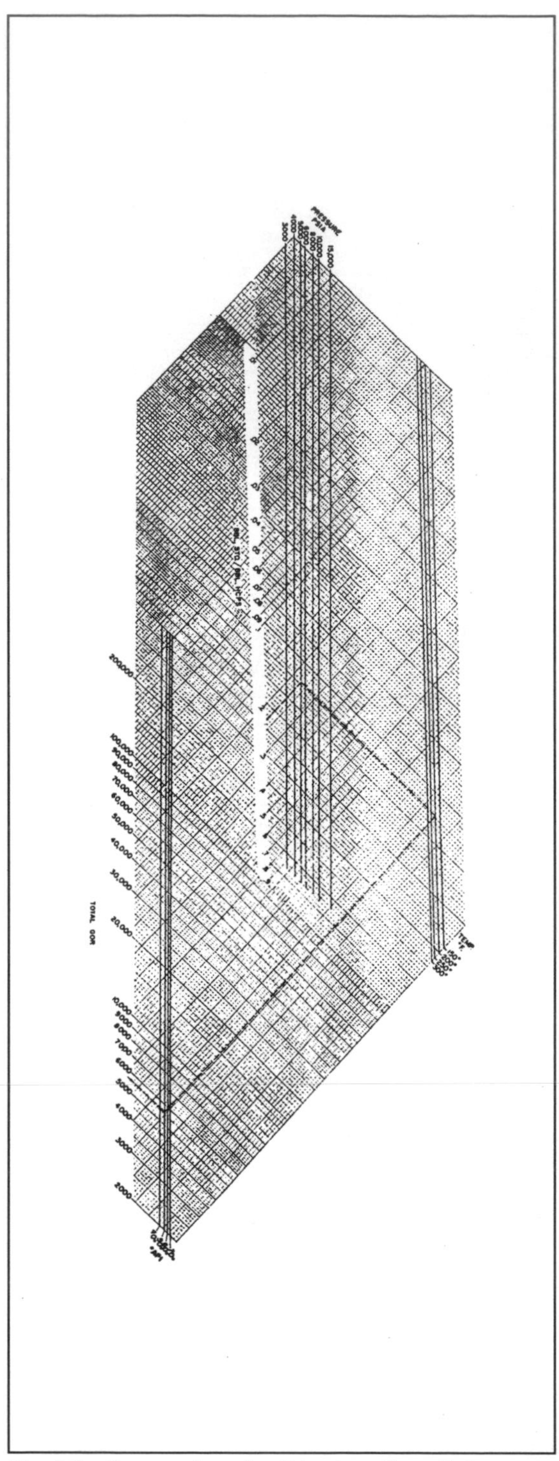

Fig. 6-9. Gas condensate oil in place (from Eaton and Jacoby[15]). Permission to publish by JPT.

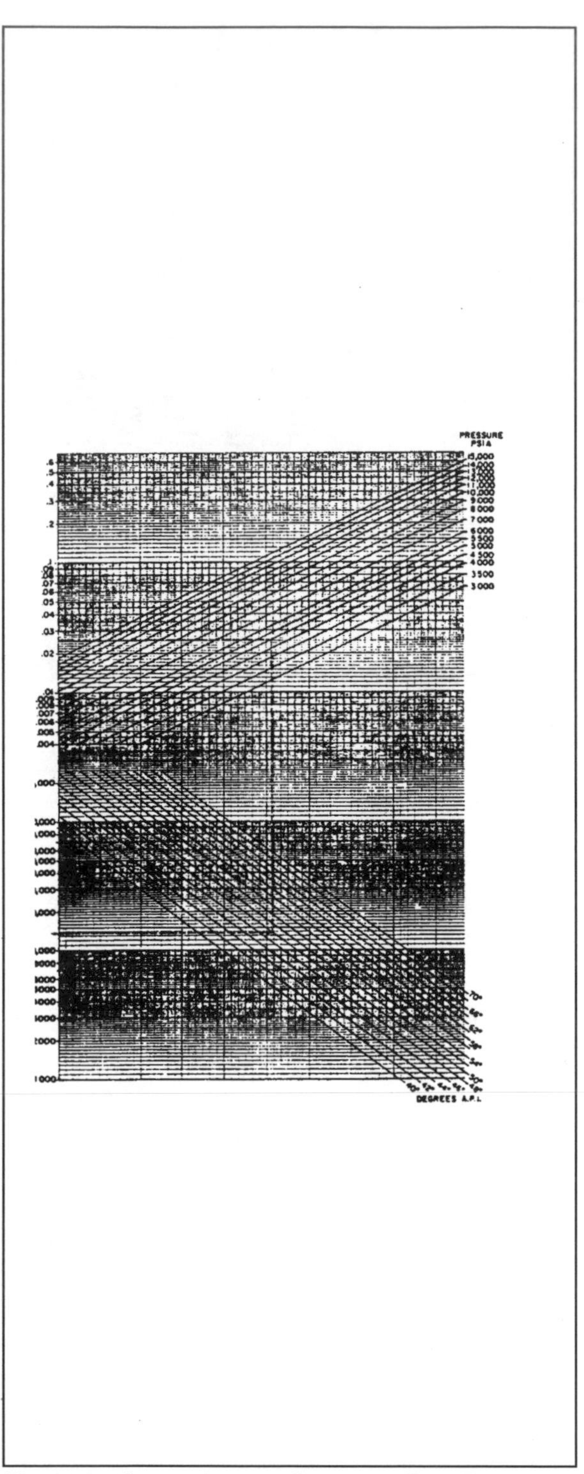

Fig. 6-10. Gas-condensate oil recovery (from Eaton and Jacoby[15]). Permission to publish by JPT.

MATERIAL BALANCE CALCULATIONS IN RETROGRADE CONDENSATE RESERVOIRS

If a retrograde condensate reservoir is behaving volumetrically and some production and corresponding reservoir pressure data are available, then the calculation method of Problem 6-1 can be used for material balance purposes. The accuracy of these computations is tied directly to sampling accuracy and to the degree that laboratory analysis matches field performance.

Gas-condensate reservoirs (whether exhibiting retrograde behavior or not) may perform volumetrically or may produce under a partial or total water drive. If pressure maintenance occurs, then the recovery will depend upon the stabilization pressure and displacement efficiency of the invading water (i.e., a frontal displacement mechanism). The liquid recovery for retrograde reservoirs will be less since the liquid will usually be immobile and will be trapped with some gas behind the invading water front. Unfortunately, recoveries are usually lower with water influx than with volumetric depletion.

If no oil zone is associated with the gas-condensate reservoir, the material balance equation is:

$$\frac{p_{sc} G_p}{T_{sc}} = \frac{p_i V_i}{z_i T} - \frac{p (V_i - W_e + B_w W_p)}{z T} \tag{6-10}$$

or

$$G (B_g - B_{gi}) + W_e = G_p B_g + B_w W_p \tag{6-11}$$

The terms are defined in the "Gas Reservoir" chapter. Equations 6-10 and 6-11 can be used to determine water influx, W_e, the original gas in place, G, or the initial reservoir gas volume, V_i. Care should be taken in the determination of z at the reservoir pressure of interest. It must include the condensate (or oil); i.e., it is a two-phase gas deviation factor.

CYCLING OF GAS-CONDENSATE RESERVOIRS

Incentive exists to cycle gas-condensate reservoirs in those instances where natural depletion of the resource will result in substantial loss of liquid hydrocarbons. This occurs in water drive fields where "wet" gas is trapped, or in volumetric-type reservoirs where retrograde behavior exists. Liquid hydrocarbons formed during pressure depletion are not normally revaporized at lower reservoir pressures and thus are trapped as a residual liquid saturation. Where the reservoir rock has favorable characteristics, cycling with "dry" gas should permit recovery of part of the liquids which otherwise would be lost.

There is evidence[12] that at least part of any residual liquid saturation that formed prior to the initiation of gas cycling operations will be revaporized into the dry gas. Figure 6-11 provides a schematic diagram of such a system. For maximum benefit, cycling should begin before the dewpoint of the reservoir hydrocarbon fluid is reached. Most cycling projects recover about 50 percent of the liquid hydrocarbons that would otherwise be lost.

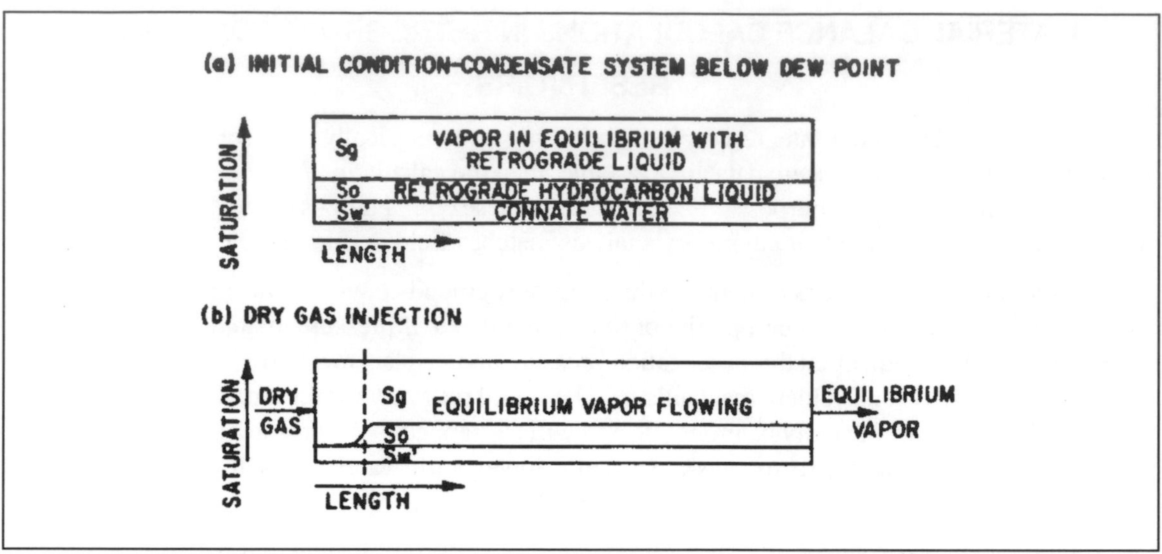

Fig. 6-11. Gas-condensate system subjected to dry gas injection (from Smith and Yarborough[12]). Permission to publish by SPE.

Field *et al.*[13] have reported on gas cycling operations in the Kaybob South Field, located 160 miles northwest of Edmonton, Alberta, Canada. The reservoir is at a depth of 10,500 ft and is totally underlain by water. It has an elongated shape, a productive area of 19,180 ac, and a length of 32 mi. The reservoir rock is a dolomitized reef complex with a gross productive thickness exceeding 75 ft. Figure 6-12 presents predicted cycling performance for the northwestern one-half of the reservoir. The displacement fronts represent areal coverages expected.

Not all cycling projects are successful. Sprinkle *et al.*[14] have reported the adverse influence of stratification on gas cycling operations. In this instance the presence of a high permeability layer was postulated as the reason for poor liquid recoveries and ultimate abandonment of cycling operations in a Texas Gulf Coast Frio sand reservoir.

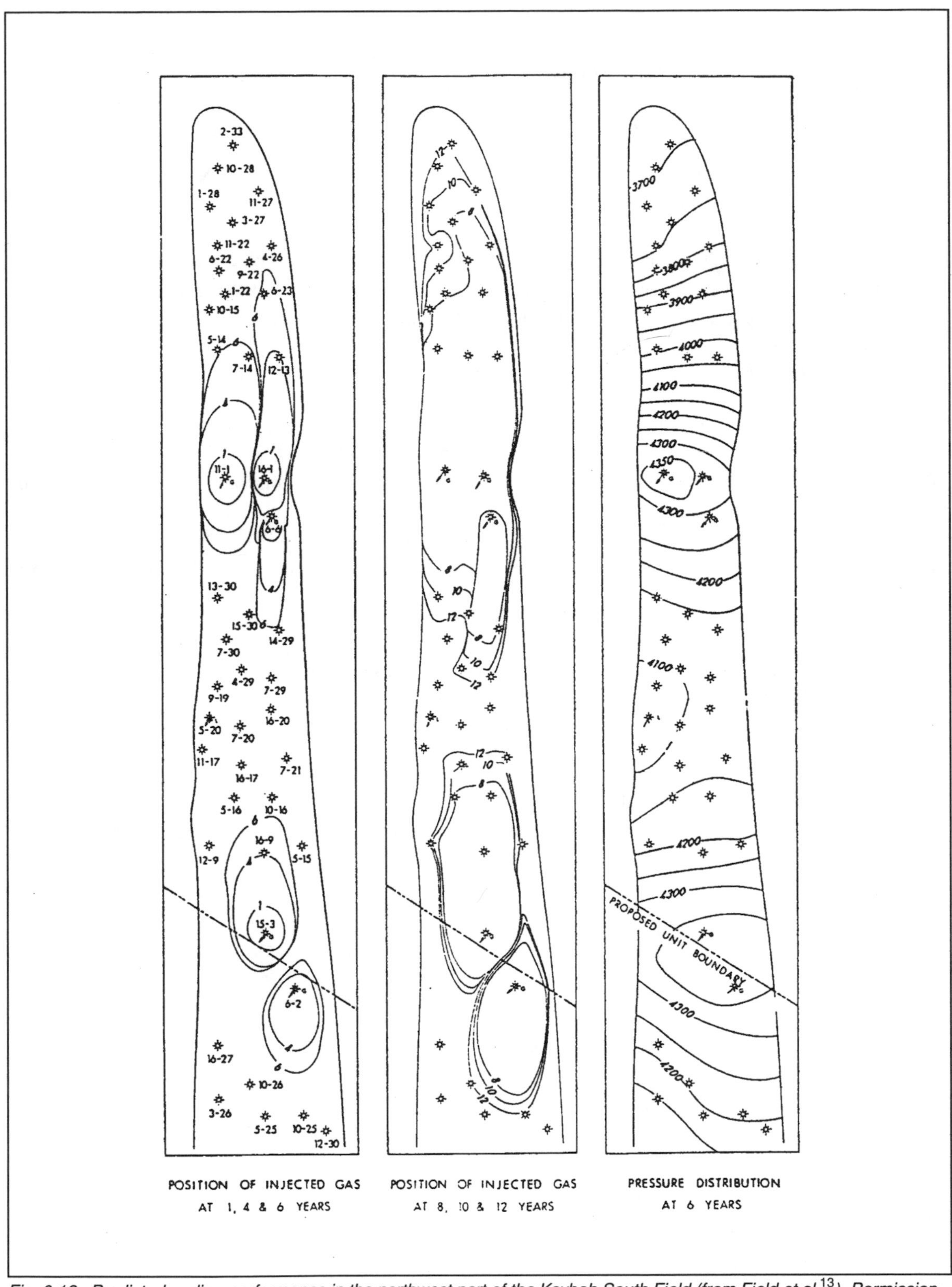

Fig. 6-12. Predicted cycling performance in the northwest part of the Kaybob South Field (from Field et al.[13]). Permission to publish by JPT.

REFERENCES

1. Craft, B. C. and Hawkins, M. F.: *Applied Petroleum Reservoir Engineering*, Prentice-Hall, Inc., Englewood Cliffs, N. J. (1959) Ch. 2.
2. Cragoe, C. S.: "Thermodynamic Properties of Petroleum Products," Bureau of Standards, U.S. Dept. of Commerce, Misc. Publ. No. 97 (1929) 22.
3. Mathews, T. A., Roland, C. H. and Katz, D. L.: "High Pressure Gas Measurement," *Proc*., NGAA (1942) 41.
4. Hadden, S. T.: "Convergence Pressure in Hydrocarbon Vapor-Liquid Equilibrium," *Chem. Eng. Prog*. Symp. Series 49. No. 7 (1953) 53.
5. *Engineering Data Book*, 9th Edition, Gas Processors Suppliers Association, Tulsa, (1972) 4th Rev. (1979).
6. Guerrero, E. T.: *Practical Reservoir Engineering*, Petroleum Publishing Co., Tulsa (1968) 225ff.
7. Standing, M. B.: *Volumetric and Phase Behavior of Oil Field Hydrocarbon Systems*, Reinhold Publishing Corp., New York (1952) Ch. 6.
8. Eilerts, C. K., *et al*.: *Phase Relations of Gas-Condensate Fluids - Test Results*," Monograph 10, U.S. Bureau of Mines, Published by Amer. Gas Assoc. (1957) 1, Ch. 2.
9. Jacoby, R. H., Koeller, R. C. and Berry, V. J., Jr.: "Effect of Composition and Temperature on Phase Behavior and Depletion Performance of Rich Gas-Condensate Systems," Petr. *Trans*., AIME (1958) 216, 406-411.
10. Farley, R. W., Weinaug, C. F. and Wolf, J. F.: "Predicting Depletion Behavior of Condensates," SPE Jour. (Sept. 1969) 343-350.
11. Coats, K. H.: "Simulation of Gas-Condensate Reservoir Performance," *JPT* (Oct. 1985) 1870 - 1886.
12. Smith, L. R. and Yarborough, L.: "Equilibrium Revaporization of Retrograde Condensate by Dry Gas Injection," SPE (March 1968) 87-94.
13. Field, M. B., Givens, J. W. and Paxman, D. S.: "Kaybob South - Reservoir Simulation of a Gas Cycling Project with Bottom Water Drive," *JPT* (April 1970) 481 - 492.
14. Sprinkle, T. L., Merrick, R. J. and Caudle, B. H.: "Adverse Influence of Stratification on a Gas Cycling Project," *JPT* (Feb. 1971) 191 - 194.
15. Eaton, B. A. and Jacoby, R. H.: "A New Depletion Performance Correlation for Gas-Condensate Reservoir Fluids," *JPT* (July 1965) 852 - 856.
16. Abhvani, A. S. and Beaumont, D. N.: "Development of an Efficient Algorithm for the Calculation of Two-Phase Flash Equilibria," SPE Reservoir Engineering (Nov. 1987) 695 - 702.
17. Peng, D. Y. and Robinson, D. B.: "A New Two Constant Equation of State," *Ind. & Eng. Chem.* (1976) 15, 59 - 64.

7 FLUID FLOW IN RESERVOIRS

DARCY'S LAW

The basic work on flow through porous materials was published in 1856 by Darcy[1], who was investigating the flow of water through sand filters for water purification. Figure 7-1 shows a schematic drawing of Darcy's experiment. Darcy interpreted the results in essentially the following equation form:

$$q = kA\,[\,(h_1 - h_2\,)\,/\,L]\tag{7-1}$$

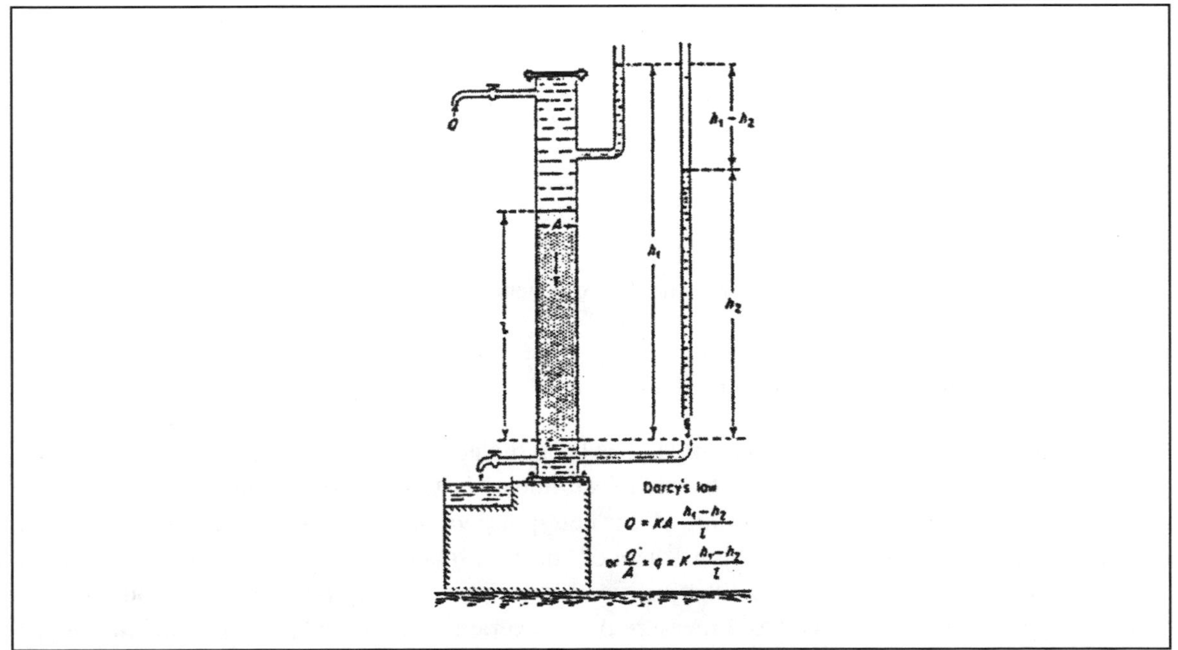

Fig. 7-1. Schematic drawing of Darcy's experiment on flow of water through sand (after Darcy[1]).

In this equation, A is the cross-sectional area (here a horizontal area normal to downward flow) and h_1 and h_2 are heights above the standard datum of the water in manometers at the input and output faces, respectively. L is length of the sand pack, and k is a constant of proportionality found to be characteristic of the porous media.

Darcy's experiment was limited to gravity flow under steady state conditions of a single fluid (water), 100 percent saturating the porous media. Later investigations found that other fluids flowing in porous media also can be modeled with the Darcy equation as long as they do not react with the porous media. Reacting normally means any type of interaction that changes the size or shape of the flow channels or changes the surface forces between the rock and fluid.

In reservoir engineering, it is necessary to modify Equation 7-1 to reflect differing fluid viscosities, dip angle for flow, and various flow geometries. Also note that flow must be laminar. This restriction is not a problem for most liquid flow situations, since flow rates normally are too small to cause turbulence.

Due to its low viscosity, gas generally has a higher mobility (k/μ) than liquid. Thus, modeling of gas flow in porous media frequently requires an adaptation of Darcy's law to account for the additional pressure drop due to turbulence. (See the section on deliverability testing in the "Gas Reservoirs" chapter.) Some high rate liquid wells, especially in fractured formations also require accounting for turbulence.

GENERALIZED FORM OF DARCY'S LAW

In the chapter on rock properties, the Darcy equation was presented for the linear horizontal system. It is:

$$q = \frac{-A\ k(\Delta p)}{\mu L} \tag{7-2}$$

where:

k = permeability, darcies,
q = outlet flow rate, cc/sec,
μ = fluid viscosity at temperature of the system, cp,
L = system length, cm,
A = system cross-sectional area, cm^2,
Δp = pressure differential across porous medium, atm.

Note that the proportionality constant, k, has been replaced by k/μ. Notice that the dimensions of permeability are determined by the dimensions of the other terms in Equation 7-2. A rock of one darcy permeability is one in which a fluid of one centipoise viscosity will move at a velocity of one centimeter per second under a pressure gradient of one atmosphere per centimeter. Reference to the "Rock Properties" chapter will show the justification for permeability having dimensions of area. Conceptually, this can be viewed as a measure of the average area of the "pore throats" (minimum areas) of the flow channels.

Darcy's law does not apply to flow within individual pore channels, but rather is a statistical relationship which averages the behavior of many pore channels. The actual velocity of flow within the pores is much higher than the apparent velocity, q/A. If the porous media is 100 percent saturated with one fluid, actual pore velocity may be approximated by $q/A\phi$, where ϕ is porosity expressed as a fraction.

Often, flow can be other than horizontal. A more generalized form of Equation 7-2 is:

$$v = \frac{q}{A} = -\frac{k}{\mu}\left[\frac{dp}{du} + \frac{\rho}{1033}\sin\alpha\right] \quad (7\text{-}3)$$

where

v = apparent velocity, cm/sec,

dp/du = pressure gradient, atm per cm in direction u,

ρ = fluid density, gm/cm^3,

α = angle of dip from the horizontal, $\alpha > 0$ if the flow is updip
$\alpha < 0$ if the flow is downdip, and

$\frac{\rho}{1033}\sin\alpha$ = gravity gradient, atm per cm in the direction of flow.

CLASSIFICATION OF FLUID FLOW IN POROUS MEDIA

There are several different bases for this classification: flow regime, dip angle, number of phases flowing, and system geometry.

I. FLOW REGIME

The term "flow regime" here signifies whether the conditions are steady state, semi-steady state, or transient flow.

A. Steady State Flow

Steady state flow exists when there is no change in density at any position within the reservoir as a function of time. Practically, this means that there will be no change in pressure at any position as well. Equations 7-1, 7-2, and 7-3 assume steady state conditions.

B. Semi-Steady State Flow

This is also called quasi- or pseudo- steady state flow. This flow regime exists when conditions are such that pressure is declining linearly with time at any reservoir position. Here, the rate of pressure decline is directly proportional to the reservoir withdrawal rate and inversely proportional to drainage volume. Semi-steady state is sometimes treated mathematically as a sequence of steady state conditions.

C. Transient Flow

Pressure in the reservoir is changing non-linearly with time. Most real reservoir flow problems are transient in nature, although to make it easier, they are often modeled as steady state or semi-steady state. To obtain the transient flow equation, three different relationships are needed: a mass balance, an equation of state, and Darcy's law. The derivation is given in the "Well Testing" chapter.

II. DIP ANGLE

A. Horizontal Flow

When the dip angle is zero, the force of gravity will not be a driving force in the fluid flow relationship.

B. Non-Horizontal Flow

When flow is not horizontal, the gravity gradient must be considered. In Equation 7-3, Darcy's law for nonhorizontal flow, the gravity gradient term is related to the fluid density and to dz/du (the change in elevation with distance along the flow path). With angles of dip exceeding 5 degrees, gravity usually makes a meaningful contribution to flow.

III. NUMBER OF FLOWING PHASES

A. Single-phase Flow

Here, one fluid only is flowing within the porous media. This is normally either oil, gas, or water. There may be an immobile second phase present such as connate water at the irreducible saturation.

B. Multiple-phase Flow

When two or more phases are flowing simultaneously in porous media, the mathematical description becomes quite complex. Relative permeability and viscosity considerations are normally used to control the relative amounts of each phase flowing at a particular point in the system. For multi-dimensional transient flow, a computer solution may be convenient. To date, no analytical solution has been found to the transient, multiple-phase flow problem in a heterogeneous reservoir. This lack of an analytical solution with which to compare the numerical solutions has led some reservoir analysts to question the validity of reservoir simulation results. There is a brief discussion of reservoir simulation in the last chapter.

IV. FLOW GEOMETRY

A number of different flow geometries have been considered in reservoir fluid flow. The three most common are: linear, radial, and 5-spot. For steady state flow, Equation 7-3 is used, and the geometry of the system is incorporated into the flow equation during the integration process.

HORIZONTAL STEADY-STATE SINGLE-PHASE FLOW OF FLUIDS

The basic flow equation is:

$$q = -A \frac{k}{\mu} \frac{dp}{dL} \tag{7-4}$$

A number of authors[2,3,4,5,6,7,8] have provided detailed derivations of flow equations for linear, radial, and other flow configurations by manipulation of Equation 7-4. Table 7-1 provides a brief summary of some of these equations both for incompressible and compressible flow. Note that these equations have been written using darcy units (not millidarcies) for permeability. Figures 7-2 and 7-3 provide schematic drawings for the linear- and radial-flow systems.

For compressible fluids (usually gases), the derivation includes Boyle's law, or with the same results, uses mass flow rates. Note that p_m, is the mean system pressure (arithmetic average of the inlet and outlet pressures). Actually, the incompressible flow equations can be used for compressible flow if the volume flow rate at the mean system pressure is used.

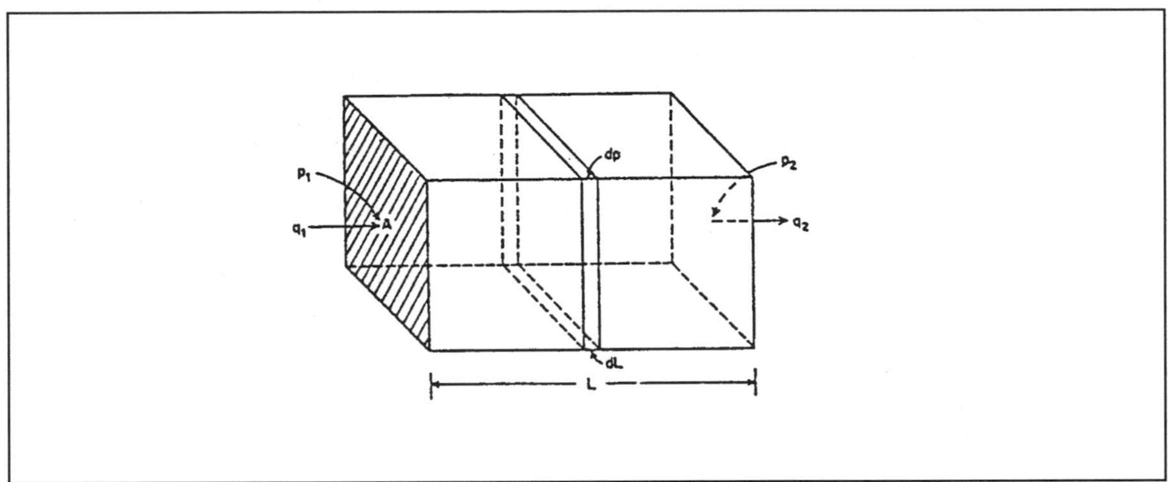

Fig. 7-2. Linear-flow system illustrating notations of Darcy's equation. For incompressible fluids, $q_1 = q_2$. (from Gatlin[9]) Permission to publish by Prentice-Hall.

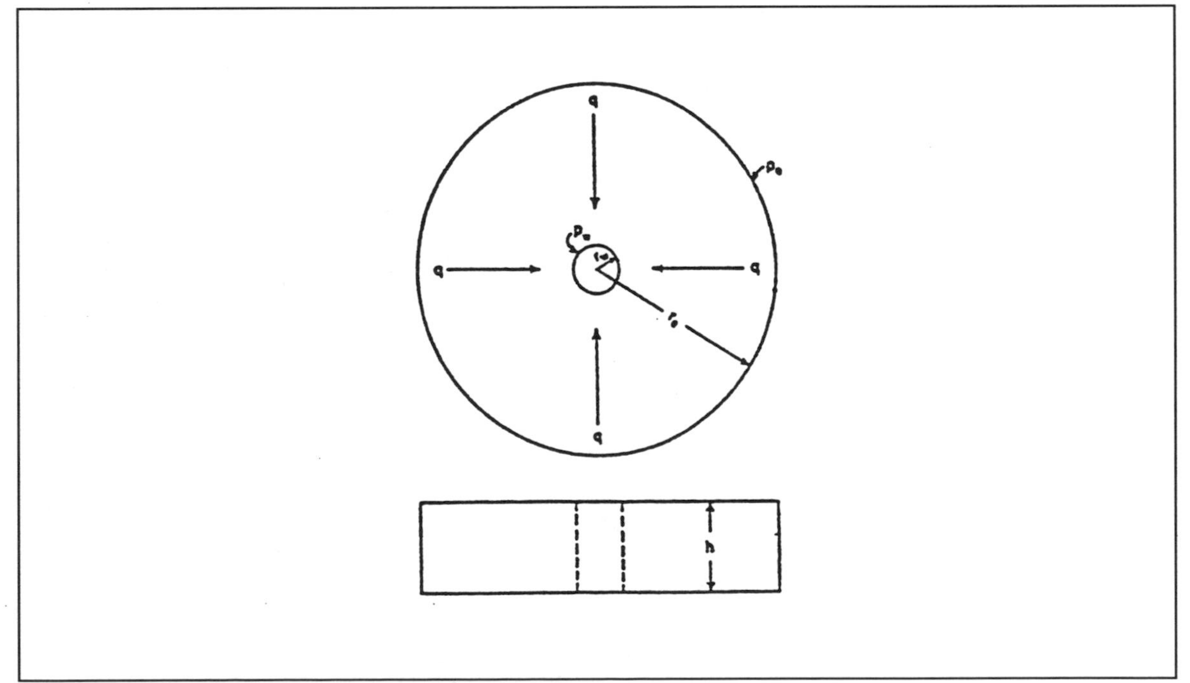

Fig. 7-3. Ideal radial-flow system (from Gatlin[9]) Permission to publish by Prentice-Hall.

Table 7-1
Steady-state Flow Equations For Homogeneous Fluids

	Darcy units	Field units	
Flow Rate	cc/sec	barrels per day	ft^3 per day
Pressure	atm	psi	psi
Length	cm	ft	ft
Viscosity	cp	cp	cp
Linear flow	$q = A \frac{k}{\mu} \frac{\Delta p}{L}$ *	$q = 1.127\, A \frac{k}{\mu} \frac{\Delta p}{L}$	$q = 6.33\, A \frac{k}{\mu} \frac{\Delta p}{L}$
	$q_a = A \frac{k}{\mu} \frac{p_m}{p_a} \frac{\Delta p}{L}$ †	$q_a = 1.127\, A \frac{k}{\mu} \frac{p_m}{p_a} \frac{\Delta p}{L}$	$q_a = 6.33\, A \frac{k}{\mu} \frac{p_m}{p_a} \frac{\Delta p}{L}$
Radial flow	$q = 2\Pi \frac{kh}{\mu} \frac{\Delta p}{Ln(r_e/r_w)}$ *	$q = 7.08 \frac{kh}{\mu} \frac{\Delta p}{Ln(r_e/r_w)}$ $q = 3.07 \frac{kh}{\mu} \frac{\Delta p}{Log(r_e/r_w)}$	$q = 39.8 \frac{kh}{\mu} \frac{\Delta p}{Ln(r_e/r_w)}$ $q = 17.3 \frac{kh}{\mu} \frac{\Delta p}{Log(r_e/r_w)}$
	$q_w = 2\Pi \frac{kh}{\mu} \frac{p_m}{p_w} \frac{\Delta p}{Ln(r_e/r_w)}$ †	$q_w = 7.08 \frac{kh}{\mu} \frac{p_m}{p_w} \frac{\Delta p}{Ln(r_e/r_w)}$ $q_w = 3.07 \frac{kh}{\mu} \frac{p_m}{p_w} \frac{\Delta p}{Log(r_e/r_w)}$	$q_w = 39.8 \frac{kh}{\mu} \frac{p_m}{p_w} \frac{\Delta p}{Ln(r_e/r_w)}$ $q_w = 17.3 \frac{kh}{\mu} \frac{p_m}{p_w} \frac{\Delta p}{Log(r_e/r_w)}$
Five-spot	$q = \Pi \frac{kh}{\mu} \frac{\Delta p}{Ln \frac{d}{r_w} - 0.619}$ *	$q = 3.54 \frac{kh}{\mu} \frac{\Delta p}{Ln \frac{d}{r_w} - 0.619}$ $q = 1.54 \frac{kh}{\mu} \frac{\Delta p}{Log \frac{d}{r_w} - 0.269}$	$q = 19.9 \frac{kh}{\mu} \frac{\Delta p}{Ln \frac{d}{r_w} - 0.619}$ $q = 8.63 \frac{kh}{\mu} \frac{\Delta p}{Log \frac{d}{r_w} - 0.269}$
	$q_w = \Pi \frac{kh}{\mu} \frac{p_m}{p_w} \frac{\Delta p}{Ln \frac{d}{r_w} - 0.619}$ †	$q_w = 3.54 \frac{kh}{\mu} \frac{p_m}{p_w} \frac{\Delta p}{Ln \frac{d}{r_w} - 0.619}$ $q_w = 1.54 \frac{kh}{\mu} \frac{p_m}{p_w} \frac{\Delta p}{Log \frac{d}{r_w} - 0.269}$	$q_w = 19.9 \frac{kh}{\mu} \frac{p_m}{p_w} \frac{\Delta p}{Ln \frac{d}{r_w} - 0.619}$ $q_w = 8.63 \frac{kh}{\mu} \frac{p_m}{p_w} \frac{\Delta p}{Log \frac{d}{r_w} - 0.269}$

* Incompressible flow

† Compressible flow

Note: Permeability (k) above has units of darcies.

I. LINEAR BEDS IN SERIES

On occasion, calculations are necessary where linear beds in series are thought to be present. Figure 7-4 illustrates the flow system. Notice that the pressure drops are additive:

$$(p_1 - p_4) = (p_1 - p_2) + (p_2 - p_3) + (p_3 - p_4)$$

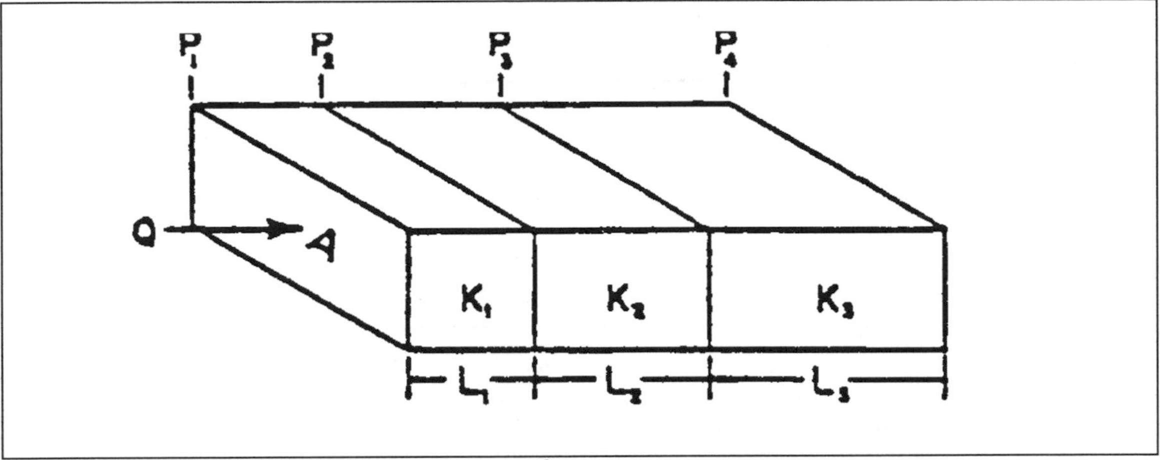

Fig. 7-4. Series flow in linear beds (from Craft & Hawkins[3]). Permission to publish by Prentice-Hall.

Writing the equivalent expressions from Darcy's equation for flow in linear beds,

$$\frac{q_t\, L_t\, \mu}{1.127\, k_{avg}\, A} = \frac{q_1\, L_1\, \mu}{1.127\, k_1\, A_1} + \frac{q_2\, L_2\, \mu}{1.127\, k_2\, A_2} + \frac{q_3\, L_3\, \mu}{1.127\, k_3\, A_3}$$

Since the flow rates, cross-sections and viscosities (assuming that modest change in pressure does not alter viscosity) are equal in each bed, then,

$$k_{avg} = \frac{L_t}{L_1 / k_1 + L_2 / k_2 + L_3 / k_3} = \frac{\sum L_i}{\sum (L_i / k_i)} \tag{7-5}$$

The average permeability k_{avg} is the value which can be used assuming a single bed and yield the same flow rate as for the series of beds.

Equation 7-5 also applies to compressible fluids. Sum the pressure drop terms:

$$(p_1^2 - p_4^2) = (p_1^2 - p_2^2) + (p_2^2 - p_3^2) + (p_3^2 - p_4^2)$$

and substitute in the equivalent expressions from Darcy's equation for flow, then Equation 7-5 results.

II. LINEAR BEDS IN PARALLEL

Often flow in the reservoir is through parallel strata having differing permeabilities. This is shown in Figure 7-5 for a linear system. The total flow is the sum of the individual flows in each zone, or:

$$q_t = q_1 + q_2 + q_3$$

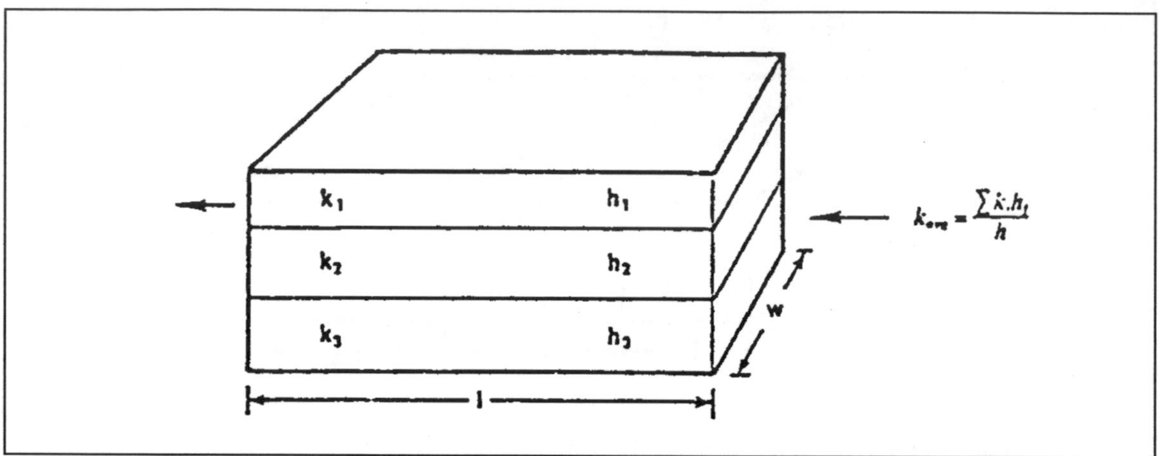

Fig. 7-5. Parallel flow in linear, horizontal beds (from Slider[7]). Permission to publish by Petroleum Publishing Company.

The equivalent expression from Darcy's equation is:

$$\frac{k_{avg} A_t (p_1 - p_2)}{L} = \frac{k_1 A_1 (p_1 - p_2)}{L} = \frac{k_2 A_2 (p_1 - p_2)}{L} + \frac{k_3 A_3 (p_1 - p_2)}{L}$$

Simplifying:

$$k_{avg} = \frac{k_1 A_1 + k_2 A_2 + k_3 A_3}{A_t} = \frac{\sum k_i A_i}{\sum A_i} = \frac{\sum k_i h_i}{h_t} \qquad (7\text{-}6)$$

If the average permeability of Equation 7-6 is used with the total area in the Darcy equation, the calculated flow rate will be equal to the total flow rate from the individual beds.

Equation 7-6 assumes that no cross flow occurs between the individual beds. If this is not true, then some error will result.

III. RADIAL BEDS IN SERIES

Depositionally, it is hard to imagine radial beds in series occurring in an actual reservoir. However, this consideration is needed due to the alteration of reservoir properties that can occur in the vicinity of wellbores (both producers and injectors) during drilling, production, and stimulation operations. Figure 7-6 shows an example of this type of a system.

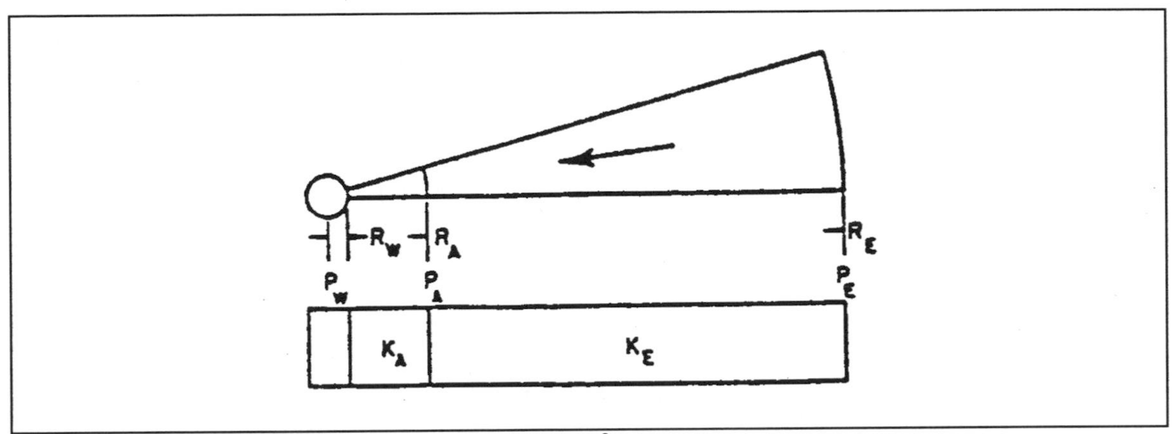

Fig. 7-6. Radial flow in beds in series (from Craft and Hawkins[3]). Permission to publish by Prentice-Hall.

Noting that the pressure drops are in series, then:

$$(p_e - p_w) = (p_e - p_a) + (p_a - p_w)$$

and from Darcy's equation for radial flow (see Table 7-1):

$$\frac{q\,\mu \ln(r_e / r_w)}{7.08\, k_{avg}\, h} = \frac{q\,\mu \ln(r_e / r_a)}{7.08\, k_e\, h} + \frac{q\,\mu \ln(r_a / r_w)}{7.08\, k_a\, h}$$

Solving for k_{avg}, then:

$$k_{avg} = \frac{k_a\, k_e \ln(r_e / r_w)}{k_a \ln(r_e / r_a) + k_e \ln(r_a / r_w)} \qquad (7\text{-}7)$$

Equation 7-7 has had wide usage in determining the impact of wellbore damage or stimulation due to acidizing and fracturing, and productivity increases due to heating near the wellbore.

IV. RADIAL BEDS IN PARALLEL

Most sedimentary reservoirs are comprised of strata of differing properties. Where calculation of the producing rate is desired, it is useful to determine the average permeability, or that permeability which will allow the system to be treated as a single radial bed with the total thickness. Figure 7-7 illustrates such a system. Since the flow rates are additive (assuming no crossflow),

$$q_t = q_1 + q_2 + q_3 + \ldots + q_n$$

Therefore:

$$\frac{7.08\, k_{avg}\, h_t\,(p_e - p_w)}{\mu \ln(r_e / r_w)} = \frac{7.08\, k_1\, h_1\,(p_e - p_w)}{\mu \ln(r_e / r_w)} + \frac{7.08\, k_2\, h_2\,(p_e - p_w)}{\mu \ln(r_e / r_w)} + etc.$$

Or,

$$k_{avg} = \frac{k_1 h_1 + k_2 h_2 + \ldots + k_n h_n}{h_t} \tag{7-8}$$

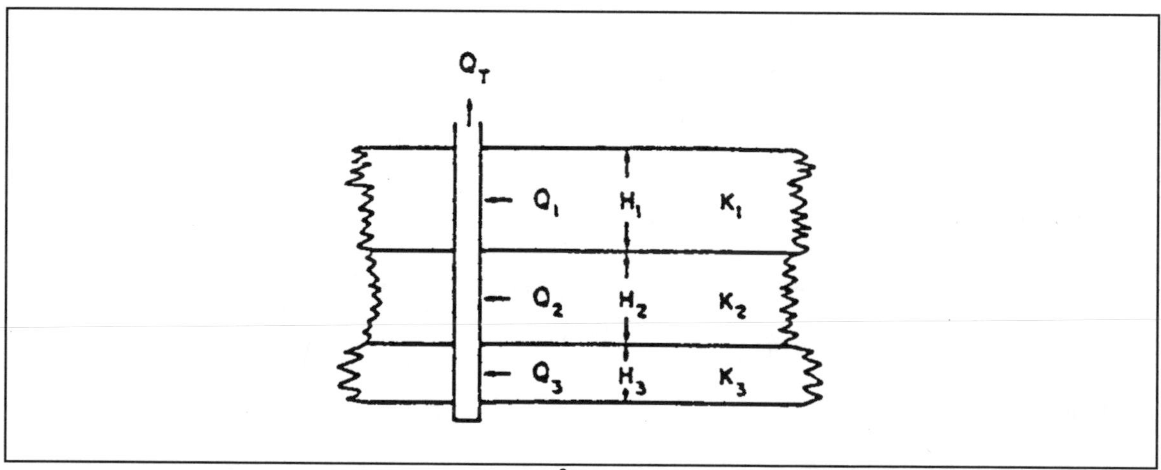

Fig. 7-7. Radial flow in parallel beds (from Craft & Hawkins[3]). Permission to publish by Prentice-Hall.

It is common industry usage to consider the group of terms, kh, as being the "flow capacity" of a bed in darcy-feet or millidarcy-feet. The summation of the capacities of the beds is usually termed the total capacity of the producing formation. Where the total capacity of a productive formation is calculated or determined by testing to be low, modern stimulation methods will often allow economic producing rates to be attained.

V. APPROXIMATING COMPLEX GEOMETRIES

Often the flow geometry will not be strictly linear or radial. In these instances, it is often possible to idealize the system sufficiently so that engineering answers can be obtained. A simple example is a rectangular drainage area with a well in the center, such that most of the area will be experiencing approximate radial flow. In this instance, an effective radius can be determined:

$$r_e = [\,(\,acres\,/\,well\,)\,(\,43{,}560\,)\,/\,\pi\,]^{1/2} \tag{7-9}$$

Equation 7-9 introduces only minor error. Near the drainage radius where the flow is not radial, the cross-sectional area normal to flow is very large, and the corresponding pressure drop quite small.

Sometimes, it is more convenient to correct the flow rate by recognizing that flow to the producing well(s) is from only part of a circle. Figure 7-8 illustrates this possibility.

In the example of Figure 7-8, we would treat the flow rate from the well(s) as representing 1/3 of that given by the radial equation from table 7-1. The pressure distribution would be the same in both instances since there would be no flow across the radial arms (or rays) of the system.

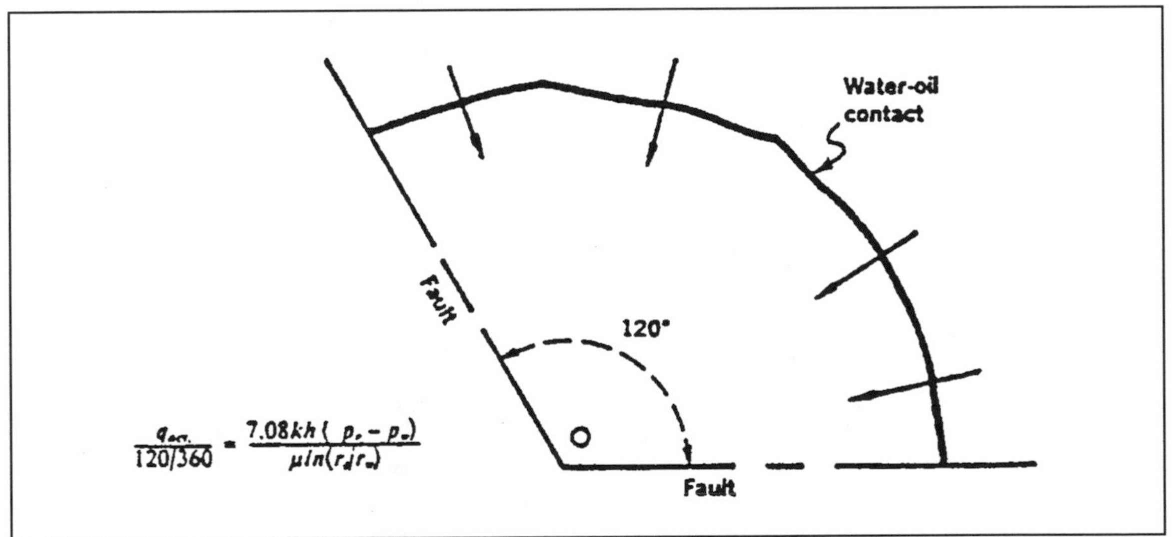

Fig. 7-8. Example application of a radial flow equation to flow from a portion of a circle (from Slider[7]). Permission to publish by Petroleum Publishing Company.

It is useful to consider a configuration where two Darcy flow equations are combined to represent the physical case believed to occur in the reservoir. An example is given in Figure 7-9. In this instance, it is appropriate to use a linear system and one-half of a radial system in series. It is necessary to make certain that the total model drainage area is equal to the total area of the actual reservoir system. The pressure drops in the model linear and radial portions are additive. Thus, the equation given in Figure 7-9 results. This could be solved for flow rate, q, if desired.

In the determination of flow from fault blocks, the radial equation sometimes can be used. It is possible to use two or more different drainage radii such as shown in Figure 7-10.

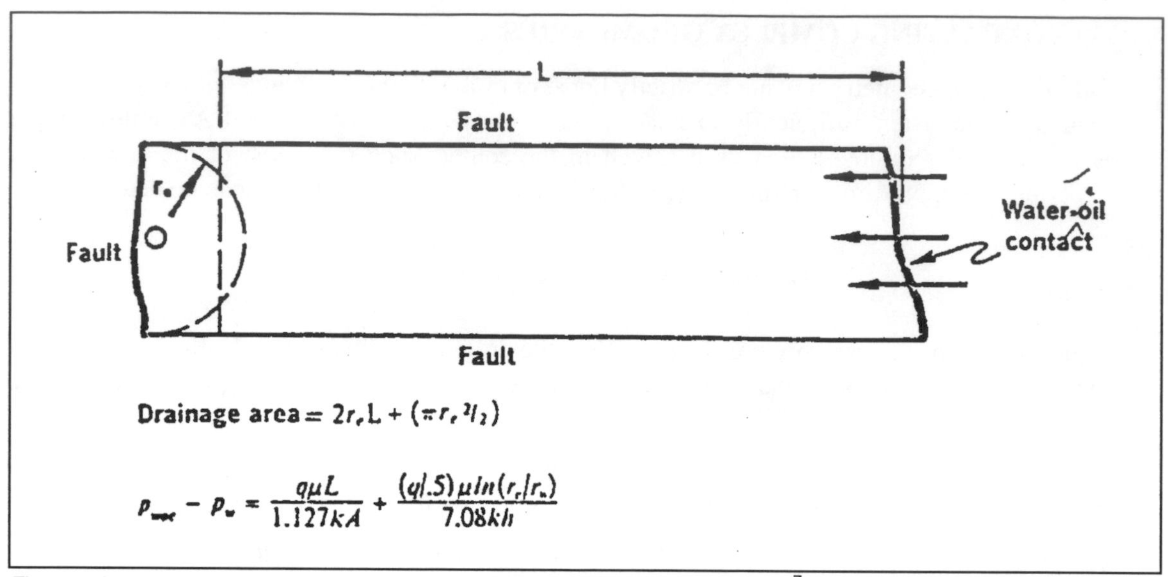

Fig. 7-9. Approximating flow equations with two simple systems in series (from Slider[7]). Permission to publish by Petroleum Publishing Company.

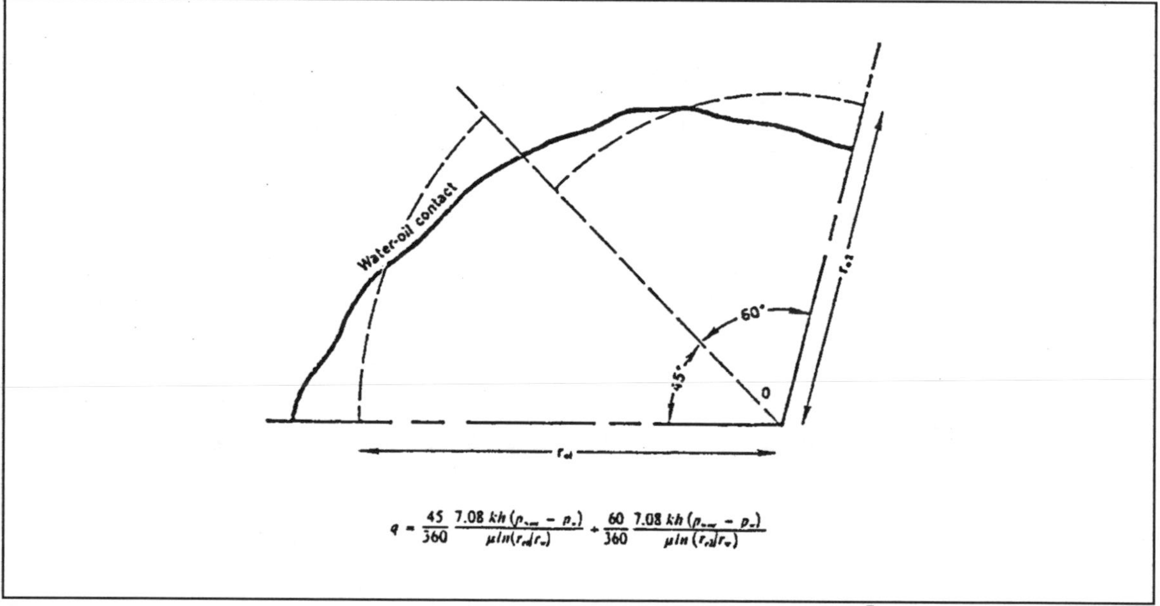

Fig. 7-10. Approximating flow equations with two simple systems in parallel (from Slider[7]). Permission to publish by Petroleum Publishing Company.

When considering the systems of Figures 7-8, 7-9, and 7-10, the calculated rates are valid only when the oil/water contact is stationary. Inward movement of the contact changes the calculation. Also, the presence of a contact would suggest a system with dip, a complexity not considered in this analysis. With a moving water-oil contact, the differences in fluid mobility behind and ahead of the contact should be taken into account. For flooding systems, this problem has been treated by Deppe,[10] Prats *et al.*,[11] and Smith.[4]

SEMI-STEADY STATE RADIAL FLOW OF COMPRESSIBLE LIQUIDS IN BOUNDED AREAS

Strictly speaking, the equations in Table 7-1 are for the case where there is flow across the external boundary to replace the fluid produced at the wellbore. This allows p_e to remain constant, as does the pressure at any radial position, assuming a constant rate of production. Where wells are draining the surrounding areas in volumetric (bounded) reservoirs; the external boundary is a no-flow boundary, not one of constant pressure. A no-flow boundary condition has a zero pressure gradient at the boundary. Brownscombe and Collins[12] have treated the problem for closed circular reservoirs. If the pressure is declining at a constant rate (semi-steady state), then the wellbore flow rate is:

$$q_w = c_e \, \pi \, r_e^2 \, h \, \phi \, \frac{dp}{dt}$$

where:

q_w = the reservoir flow rate at the well.

For semi-steady-state radial flow, the volumetric flow rate at any radial position r is proportional to the reservoir volume between r and r_e. So:

$$q = q_w \left[1 - \frac{r^2}{r_e^2} \right]$$

Substituting these last two relationships into Darcy's equation, and writing $A = 2\pi rh$, then:

$$q_w \left[1 - \frac{r^2}{r_e^2} \right] = 1.127 \, \frac{k}{\mu} \, (2 \, \pi \, r \, h) \, \frac{dp}{dr}$$

Separating variables and integrating from r_w to r_e, and from p_w to p_e results in:

$$q_{sc} = \frac{7.08 \, k \, h \, (p_e - p_w)}{\mu \, B_o \, [\ln(r_e / r_w) - (1/2)]} \qquad (7\text{-}10)$$

where:

$q_{sc} = q_w/B_o$ = surface flow rate, STB/D

Notice that the difference between flow and no flow across the external boundary is limited to the one-half which appears in the denominator of Equation 7-10. Instances where Equation 7-10 or its steady state counterpart in Table 7-1 strictly apply are not often met in petroleum reservoirs. However, they provide results that are often of adequate accuracy for engineering purposes. This is particularly true when comparative results are more important than absolute results. The equations have application for study of "before" and "after" conditions in wells. Figure 7-11 presents typical radial pressure distributions for flow and no-flow conditions at the external boundary. Notice that the steady-state case has a pressure profile that is linear with the logarithm of radial distance from the wellbore. Further, for distances reasonably close to the wellbore, the semi-steady-state pressure distribution is essentially the same as the one for steady state.

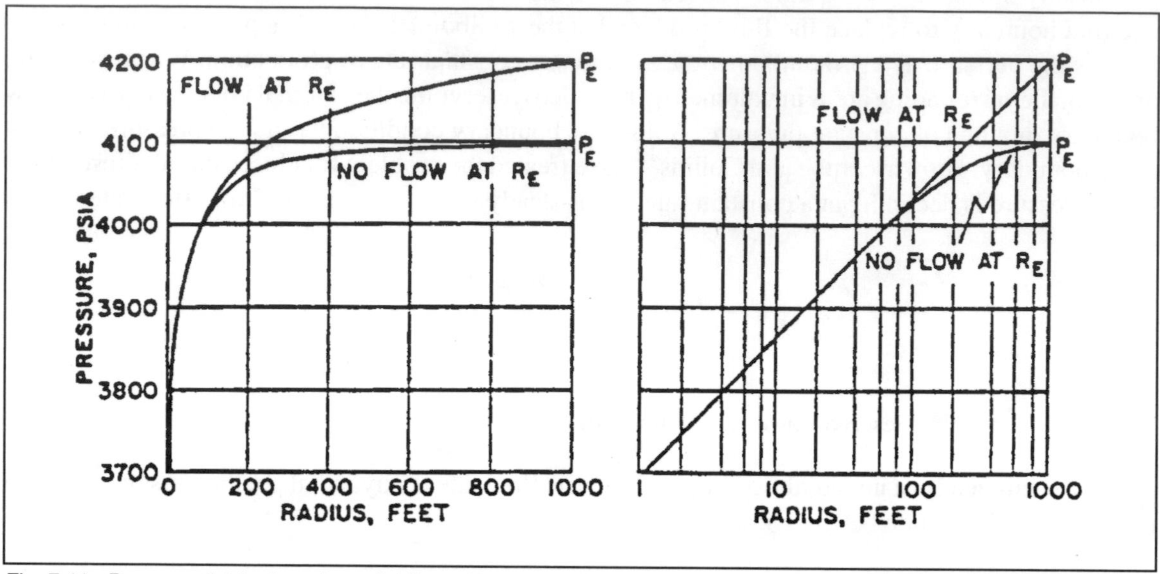

Fig. 7-11. Pressure distributions for radial flow of incompressible fluid and of compressible liquid with no flow across the external boundary at 1000 ft (from Craft & Hawkins[3]). Permission to publish by Prentice-Hall.

AVERAGE PRESSURE IN A RADIAL FLOW SYSTEM

The volumetric average pressure for a radial system may be expressed by the integral

$$p_{avg} = \frac{\int p\, dV}{V} = \frac{\int_{r_w}^{r_e} p \times 2\pi\, rh\,\phi\, dr}{\pi r_e^2 h\,\phi} = \frac{2\int_{r_w}^{r_e} pr\, dr}{r_e^2}$$

From Darcy's flow equation:

$$p = p_w + \frac{q_{sc} B_o\, \mu\, \ln(r/r_w)}{7.08\, k\, h}$$

or then:

$$p_{avg} = \frac{2p_w}{r_e^2}\int_{r_w}^{r_e} r\, dr + \frac{2\,\mu\, q_{sc} B_o}{7.08\, k\, h\, r_e^2}\int_{r_w}^{r_e} \ln(r/r_w)\, r\, dr$$

and carrying out the integration,

$$p_{avg} = \frac{2\,p_w}{r_e^2}\left[\frac{r_e^2}{2} - \frac{r_w^2}{2}\right] + \frac{2\,\mu\,q_{sc}\,B_o}{7.08\,k\,h\,r_e^2}\left[\frac{r_e^2}{2}\ln\left(r_e / r_w\right) - \frac{r_e^2}{4} + \frac{r_w^2}{4}\right]$$

Because r_w^2 is much smaller then r_e^2, the r_w^2 terms may be neglected, and the equation simplifies to:

$$p_{avg} = p_w + \frac{\mu\,q_{sc}\,B_o}{7.08\,k\,h}\left[\ln\left(r_e / r_w\right) - (1/2)\right] \tag{7-11}$$

Equation 7-11 represents the average pressure within the drainage area of a steady-state, radial, horizontal, single-phase flow distribution.

For a bounded radial-drainage system (horizontal, single-phase, semi-steady state), a similar result can be developed by using Equation 7-10. The result is:

$$p_{avg} = p_w + \frac{\mu\,q_{sc}\,B_o}{7.08\,k\,h}\left[\ln\left(r_e / r_w\right) - (3/4)\right] \tag{7-12}$$

As a matter of interest, the difference between Equations 7-11 and 7-12 is:

$$\Delta p_{avg} = \frac{\mu\,q_{sc}\,B_o}{28.32\,k\,h} \tag{7-13}$$

For the normal range of reservoir values, the difference between the results of Equations 7-11 and 7-12 is usually small.

READJUSTMENT TIME IN RADIAL FLOW SYSTEMS

It is useful to know the time necessary in a producing well for a logarithmic pressure distribution to be established out to the drainage radius, r_e. This will aid in determining the minimum time when a well may be expected to have reached some sort of stable producing rate where Darcy's law can be used. At any producing time and corresponding radius of investigation (distance out from the well that pressure has been disturbed), the amount of fluid removed is proportional to the effective compressibility, the volume of fluid contained in the affected area, and the drop in the average pressure, $(p_e - p_{avg})$.

$$\Delta V = c_e V \Delta p = c_e \left[\frac{\pi r_e^2 h \phi}{5.615} \right] (p_e - p_{avg}) \ barrels$$

Also, notice that:

$$p_e - p_{avg} = (p_e - p_w) - (p_{avg} - p_w)$$

Therefore:

$$p_e - p_{avg} = \frac{q_{sc} \mu B_o \ln(r_e / r_w)}{7.08 \, k h} - \frac{q_{sc} \mu B_o [\ln(r_e / r_w) - (1/2)]}{7.08 \, k h} = \frac{q_{sc} \mu B_o}{14.16 \, k h}$$

So, the total volume of fluid removed is:

$$\Delta V = c_e \left[\frac{\pi r_e^2 h \phi}{5.615} \right] \left[\frac{q_{sc} \mu B_o}{14.16 \, k h} \right]$$

If the reservoir production rate is $q_{sc}B_o$, then the readjustment time in days is:

$$t_r = \frac{\Delta V}{q_{sc} B_o} = \frac{0.04 \mu c_e \phi r_e^2}{k} \qquad (7\text{-}14)$$

For example, a well producing clean oil of 1 cp viscosity, with an effective compressibility of 10×10^{-6} psi^{-1}, porosity of 0.25, drainage radius of 660 feet, and permeability of 0.1 darcies, the readjustment time calculates to be 0.44 days. The times indicated by Equation 7-14 are intended to be order of magnitude values only since the required data may not be accurately known.

The readjustment time, calculated by Equation 7-14, is the time that it takes for the pressure disturbance, caused by placing a well on production, to just reach the external boundary, r_e. This corresponds directly to the equation given in the "Well Testing" chapter (under "Drawdown Pressure Behavior") in terms of dimensionless time (with respect to the drainage radius): $t_{de} = 0.25$.

WELL PRODUCTIVITY

I. PRODUCTIVITY INDEX

The productivity index concept is an attempt to find a simple function that relates the ability of a producing well to give up fluids. Simply defined, productivity index (PI) is the producing rate in stock tank barrels of oil per day divided by the pressure drawdown (psi) taken in the reservoir:

$$PI = Productivity\ Index = J = \frac{q_{sc}}{(\bar{p} - p_{wf})}\ STBO/D/psi \qquad (7\text{-}15)$$

The producing rate is measured directly at the surface at stock tank conditions. Static pressure, $\bar{p}$, is normally obtained from a pressure buildup. After a period of stable oil production, the producing bottomhole pressure, p_{wf}, is measured with a bottomhole pressure gauge or by determining the fluid level in the annulus (if the annulus is open).

Note, that Equation 7-15 involves stock tank oil rate only. This is the normal definition for productivity index; although at times this quantity is referred to as the "net oil productivity index." The gross liquid PI is defined as the gross liquid rate (oil and water, reservoir barrels per day) divided by the corresponding pressure drawdown. The gross PI is normally used in well pump sizing calculations.

Many people assume that productivity index is an all-rates, all-times constant for a given well. This is not always the case, but when producing rate is not excessively high and well bottomhole producing pressures are all above the bubblepoint pressure, then PI will normally remain relatively constant.

In an active water drive (where the pressure remains above the bubblepoint), a constant productivity index over a range of rates is a satisfactory assumption. For a solution-gas drive in which the flowing pressures are below the bubblepoint, the PI decreases with increasing rates. The two major reasons for this behavior include: (1) the buildup of a free gas saturation near the wellbore due to pressures less than the bubblepoint pressure (relative permeability effects), and (2) increased pressure drop due to turbulence at higher flow rates. Other factors are the increase in oil viscosity below the bubblepoint pressure and possible formation compaction (at higher rates with resulting low producing pressures) causing reduced absolute permeability.

Some states and countries have limits on producing rates of oil wells which include factors such as well spacing, market demand for oil, well depth, and producing gas/oil ratio. Table 7-2 provides an example calculation of producing rate where restrictions apply. Figure 7-12 provides a plot of the data of Table 7-2, showing typical pressure, production rate, and gas/oil ratio curves for a solution-gas drive type reservoir. For the well represented, production is limited first by a 100 BOPD well allowable (region A in Figure 7-12), second by producing gas-oil-ratios (region B), and finally by productivity (region C). In region B, where the actual GOR is above the penalty GOR of 2000, oil rate is restricted by a gas limit to a maximum of:

$$\frac{(\,Max.\ oil\ allowable\,) \times (\,penalty\ GOR\,)}{(\,actual\ GOR\,)}$$

Table 7-2
PRODUCING RATE FOR A WELL IN A DEPLETION-TYPE RESERVOIR
(from Craft and Hawkins[3]).

Well allowable = 100 STB/day - GOR penalty = 2000 scf/stb

(1) Cumultive Oil Production, STB	(2) Reservoir Pres-sure, psia	(3) Daily GOR, SCF/STB	(4) Productivity Index, STB/day/psi	(5) Maximum Physical Rate, $p_w = 0$, STB/day	(6) Maximum Rate at 2000 GOR Penalty, STB day	(7) Well Rate, STB/day
0	2200	470	0.100	220.0	425.5	100.0[a]
20,000	1600	480	0.102	163.2	416.7	100.0[a]
40,000	1480	530	0.100	148.0	377.4	100.0[a]
60,000	1400	705	0.095	133.0	283.7	100.0[a]
80,000	1320	1175	0.088	116.2	170.2	100.0[a]
100,000	1250	2000	0.082	102.5	100.0	100.0[a]
120,000	1120	3290	0.073	81.8	60.8	60.8[b]
140,000	890	4350	0.062	55.2	46.0	46.0[b]
160,000	650	4470	0.050	32.5	44.7	32.5[c]
180,000	350	4000	0.040	14.0	50.0	14.0[c]
200,000	200	1880	0.030	6.0	106.4	6.0[c]

Column (5) = Col. (2) x Col (4), assuming $p_w = 0$.
Column (6) = (100 STB/day x 2000 SCF/STB) / Col. (3).

[a]Limited by state regulations
[b]Limited by excessive gas/oil ratio.
[c]Limited by maximum physical rate.

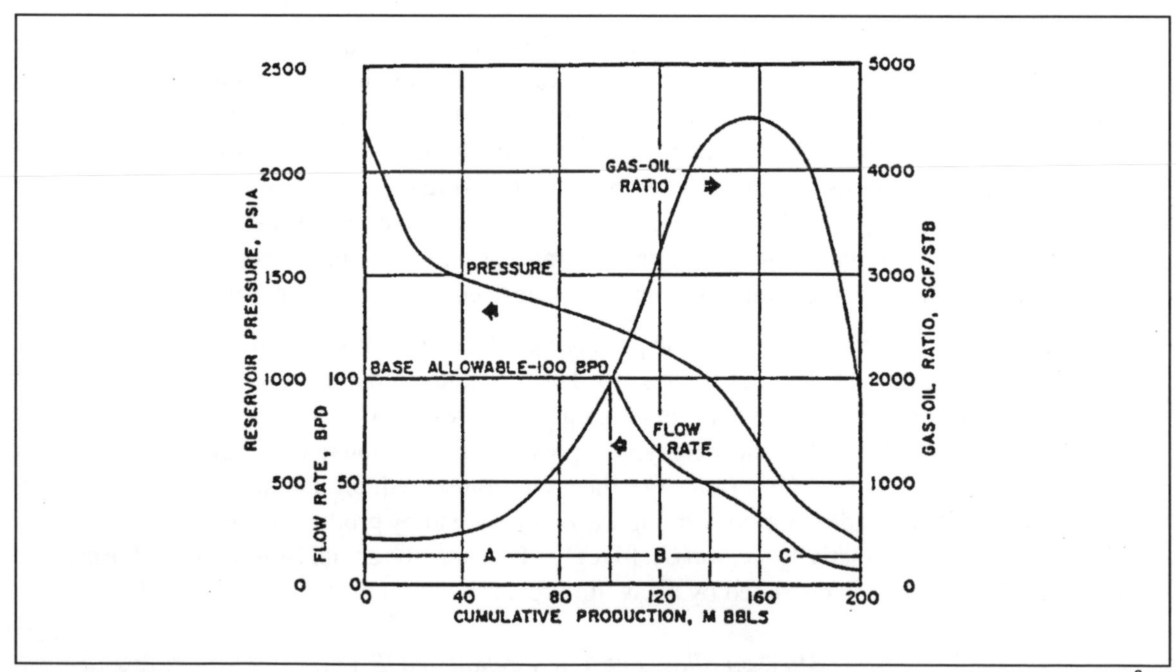

Fig. 7-12. Typical pressure, production rate, and gas/oil ratio curves for a depletion-type reservoir (from Craft & Hawkins[3]). Permission to publish by Prentice-Hall.

To consider the effect of reservoir properties, Darcy's radial-flow equation can be manipulated to solve for productivity index (J):

$$J = \frac{q_{sc}}{(\bar{p} - p_{wf})} = \frac{7.08\, k\, k_{ro}\, h}{\mu\, B_o\, \ln(r_e / r_w)} \tag{7-16}$$

Then, if we assume that 7.08 / [1n(r_e/r_w)] is approximately equal to one, then a PI estimate can be obtained:

$$J \approx \frac{k\, k_{ro}\, h}{\mu\, B_o} \tag{7-17}$$

It is important to note that permeability units here are darcies and the productivity index units are STBO/psi. According to Lewis and Horner,[12] field data indicate that a better approximation for PI is:

$$J \approx 0.6\, \frac{k\, k_{ro}\, h}{\mu\, B_o} \tag{7-18}$$

where units are the same as in equation 7-17.

The productivity index is normally calculated using actual well measurements of rate and drawdown rather than using 7-16, 7-17, or 7-18. However, if the well is not yet drilled or has not been tested, then Equations 7-16 through 7-18 may be helpful.

A. PRODUCTIVITY INDEX EXAMPLE PROBLEM

Problem 7-1:

Problem Statement: An independent in Wyoming has a pumping well in which he would like to maximize production with the existing beam pumping unit. Calculate the well productivity index,. and assuming that the PI is constant, estimate the maximum rate that this well can produce.

This example involves a mature well in a water drive reservoir in the Willow Draw Field, Wyoming. In this particular well, the producing formation is penetrated at a relatively low structural position. The approximate depth of the perforations is 4200' (from the ground level). The pump depth is also assumed to be 4200'. The well is currently making 5 BOPD and 180 BWPD. Other data include:

Tank oil gravity	= 15 °API
γ_g	= 0.80
Surface temperature	= 70 °F
Geothermal gradient	= 1.5°/100 ft
γ_w	= 1.0
Pumping fluid level	= 3200 ft GL
Static fluid level	= 2600 ft GL
Atmospheric pressure	= 11.8 psia = surface casing pressure

Solution:

(1) For the pumping case, find the pressure at the top of the liquid in the annulus. This will be equal to the pressure due to the weight of the gas in the annulus above the liquid plus the atmospheric pressure. To calculate this pressure, we use the "Bureau of Mines" equation 5-37 from the "Gas Reservoirs" chapter:

$$p_{\substack{bot \\ gas}} = p_{top} \exp\left(\frac{0.01875\ \gamma_g\ D}{z_{avg}\ T_{avg}}\right)$$

$$T_{avg} = 70 + \frac{(1.5)(32)}{2} = 94°F = 554\ °R$$

$$z_{avg} \approx 1.0 \ \ (\textit{because close } 14.7\ \textit{psia})$$

$$p_{\substack{bot \\ gas}} = 11.8 \exp\left[\frac{(0.01875)(0.8)(3200)}{(1)(554)}\right] = 12.9\ psia$$

(2) Now find the weight of the pumping annular liquid column (from 3200' to 4200'). To do this, it is assumed that the well has been pumping long enough for the annulus to approach gravitational equilibrium: in the annulus above the pump, the liquid is 100% oil. Although a small error will result, the weight of this oil column will be calculated using tank-oil specific gravity from the API gravity equation.

$$\textit{Weight of oil column} = \left[\frac{141.5}{131.5 + 15}\right][0.4335][4200 - 3200] = 418.7\ psi$$

(3) Calculate the bottomhole pumping pressure by adding the results of steps (1) and (2):

$$p_{wp} = 12.9 + 418.7 = 431.6\ psia$$

(4) Estimate the weight of the static annular gas column. Since the weight of the pumping gas column was small:

$$\Delta p_{\substack{pump \\ gas \\ column}} = 12.9 - 11.8 = 1.1\ psi$$

the weight of the static column will be estimated using a ratio of the height of the static gas column (2600') to the height of the pumping gas column (3200').

$$\Delta p_{\substack{static\\gas\\column}} = (1.1)\left[\frac{2600}{3200}\right] = 0.9 \; psi$$

$$p_{\substack{bot\\gas}} = 11.8 + 0.9 = 12.7 \; psi$$

(5) Now that the pressure at the top of the static liquid column has been computed, calculate the weight of the static liquid column. There are several assumptions involved in this step: (a) When the well is shut in at the surface, there continues to be fluid entry into the wellbore until pressure equalizes across the formation face. Thus, the fluid level in the annulus above the pump rises. It is assumed that this fluid that enters the annulus is comprised of oil and water at the same ratio that was being produced just before shut-in. (b) It is further assumed that the part of the static liquid column that was in the annulus before shut-in (the pumping annular liquid column), is still comprised of 100% oil. Thus, to calculate the weight of the static liquid column, it must be analyzed in two parts: the old pumping liquid column (100% oil) and that part of the column which entered after shut-in, which is considered to be made up of water and oil.

Static liquid column = 1000 ft oil + 600 ft (water & oil)

1000 ft oil column weight [from step (2)] = 418.7 psi

600 ft (water & oil) column weight =

$$\left[\left(\frac{180}{185}\right)(0.4335) + \left(\frac{5}{185}\right)\left(\frac{141.5}{131.5 + 15}\right)(0.4335)\right] 600 = 259.9 \; psi$$

So, static liquid column weight = 418.7 + 259.9 = 678.6 psi

(6) Calculate static bottomhole pressure by adding the results from (4) and (5).

$$p_{ws} = 12.7 + 678.6 = 691.3 \; psia$$

(7) Calculate the productivity index:

$$J = 5 / (691.3 - 431.6) = 0.0193 \; STBO / D / psi$$

(8) Calculate maximum rate. For this step, the height of the current pumping fluid level above the pump (1000 ft) is considered. For proper pump maintenance, there should actually be some liquid above the pump (say 100 ft). However, for the sake of this example problem, it will be assumed that the maximum rate possible from this well will occur when the pumping fluid level is reduced to the level of the pump. As it has been seen that a difference in gas column height of 600 ft made a difference in gas column weight of only 0.2 psi, in this part of the problem, gas column weight will be ignored. Therefore, if the pumping fluid level is reduced to a depth of 4200 ft, then pumping bottomhole pressure will be approximately equal to the pressure at the bottom of the gas column which was determined in step (4):

$p_{wp} = 12.9\ psia$

So, $\Delta p = 691.3 - 12.9 = 678.4\ psi$

and, $(q_o)_{max} = (J)(\Delta p) = (0.0193)(678.4) = 13.1\ STBO/D$

So, the total liquids associated with this oil rate would be (oil and water):

$$(q_L)_{max} = (13.1)(185/5) = 484.4\ B/D$$

The PI has been assumed to remain constant here even thought the pumping bottomhole pressure has been lowered considerably. This increased drawdown could lower the pressure in the reservoir around the wellbore to a point below the bubblepoint. If there is sufficient gas in the oil, the relative permeability effects could be considerable. It should also be kept in mind that a large drawdown could cause water coning which would yield an increased water cut. Thus, the resulting oil rate could be lower than the initial oil rate.

However, in this case (real example), success was complete. On the basis of similar calculations, the surface pump stroke was increased from 64" to 84" and the speed was increased from 8.5 to 13.5 strokes per minute. Resulting oil rate (10.4 BOPD) was exactly what a constant PI would predict:

$$(q_o)_{new} = (84/64)(13.5/8.5)(5) = 10.4\ STBO/D$$

B. PRODUCTIVITY INDEX CHANGE WITH TIME

When wells make increasing amounts of water, oil producing rates normally decline. Where attempts are being made to predict future oil producing rates using the productivity index, it is sometimes useful to include the water, and use the gross liquid productivity index. This PI is a total liquid productivity index relating total reservoir barrels per day of water and oil divided by the pressure drawdown.

The oil produced is derived from a companion plot of log of water-oil ratio versus cumulative oil production. It should be realized that this total liquid PI will also probably not remain completely constant as water saturation is increasing around the well. It may decrease (but not as much as oil PI) until a large percentage of water is being produced, and then begin to increase.

Especially with solution-gas drive and gas-cap expansion-drive reservoirs, PI decreases with cumulative recovery. Water drive reservoir well PI's tend to be more constant with time, but can also decline due to changes in relative permeabilities to oil and water, and when flow is below the bubblepoint.

From Equation 7-16, it can be seen that productivity index is a function of several variables that will change with time. Oil viscosity and formation volume factor are both functions of pressure; oil relative permeability is a function of saturation. One approach that may be used to consider the effects of time on PI is to use the concept of a relative PI. An initial well PI (J_i) is calculated. Then, at any later point in time the PI (J) may be estimated (if the k_{ro}, μ_o, and B_o values at that time and initially are known) by using the following equations:

$$J_{rel} = \left(\frac{k_{ro}}{\mu_o B_o} \right) / \left(\frac{k_{roi}}{\mu_{oi} B_{oi}} \right) \tag{7-19}$$

$$J = (J_i)\ (J_{rel}) \tag{7-20}$$

where the subscript "i" refers to initial conditions (or conditions when J_i was measured). Then, of course, $q_o = (J)(\Delta p)$.

Brown[13] suggests that for wells with pressures below the bubblepoint (where the PI changes with rate), the J and J_i in Equation 7-20 should be at the same Δp to have a valid basis for comparison.

II. INFLOW PERFORMANCE

Because PI varies with rate and time in solution-gas-drive wells producing below the bubblepoint pressure, the "inflow performance" relationship of Vogel[14] is useful. The different reservoir conditions investigated by Vogel resulted in curves of ($p_{wf} / \bar{p}$) vs. producing rate which exhibited similar shapes. The empirical equation for well rate in a solution gas drive reservoir is:

$$\left\{ \frac{q_o}{(q_o)_{max}} \right\} = 1 - 0.20 \left[\frac{p_{wf}}{\bar{p}} \right] - 0.80 \left[\frac{p_{wf}}{\bar{p}} \right]^2 \tag{7-21}$$

where:

q_o = oil producing rate with bottomhole pressure of p_{wf}, STBO/D,
$(q_o)_{max}$ = theoretical oil producing rate with $p_{wf} = 0$ psig, STBO/D,
p_{wf} = producing bottomhole pressure, psig,
$\bar{p}$ = static reservoir pressure, psig.

If one producing rate and corresponding bottomhole producing pressure is known, then $(q_o)_{max}$ may be calculated, and the equation may be used for any other rate. Equation 7-21 assumes that the well has a flow efficiency of 1.0 (the well is neither damaged nor stimulated). Further, the equation matches well performance better in the early stages of depletion than for later stages.

Standing[15,16] has extended Vogel's results to wells with flow efficiencies different than 1.0 and to prediction of future well performance. Brown[13] has discussed the use of inflow performance.

Fetkovich[17] noted that oil wells producing below the bubblepoint pressure behave much like a gas well. He proposed a generalized equation for these wells which also allows predicting performance into the future:

$$q_o = C\left(\frac{\bar{p}}{\bar{p}_i}\right)\left(\bar{p}^2 - p_{wf}^2\right)^n \tag{7-22}$$

where:

q_o = oil producing rate, STBO/D,
C = constant to be determined by testing the well,
n = constant to be determined by testing the well,
$\bar{p}_i$ = initial static pressure, psia,
$\bar{p}$ = current static pressure, psia,
p_{wf} = producing bottomhole pressure, psia.

The constants C and n are determined by initial testing of the oil well at different rates via a flow-after-flow or isochronal testing procedure. Brown[13] has presented example calculations.

III. INJECTIVITY INDEX

It is often convenient to use an injectivity index when considering water injection wells:

$$Injectivity\ Index = II = I = \frac{q_{sc}}{(p_w - \bar{p})} \tag{7-23}$$

where:

I = injectivity index, BPD/psi,
q_{sc} = water injection rate, STBW/D,
p_w = bottomhole injection pressure, psi,
$\bar{p}$ = static reservoir pressure, psi

With injection wells it is possible to measure all pressures at the surface. In this case, frictional losses in the tubing or casing strings are assumed not to be an important factor in the determinations. The injectivity index is useful in determining the condition of the sandface (whether remedial work is needed), and in noting the fill up of a disposal formation or a formation being subjected to secondary or tertiary recovery operations. The II is also helpful in determining the maximum amounts of water that can be injected (with certain limitations due to fracturing, etc.).

IV. SPECIFIC PRODUCTIVITY INDEX (or Specific Injectivity Index)

Sometimes it is helpful to determine the specific productivity index, J_s (or specific injectivity index, I_s). In such an instance, the index (J or I) is divided by the net productive thickness of the formation to give (STB/D/psi)/(Net ft). This might be useful in situations, where the specific index is known, to predict the future productivity (or injectivity) of wells to be drilled in an area of the field where a good estimate of the formation thickness is possible.

HORIZONTAL, STEADY-STATE, MULTIPLE-PHASE FLOW OF FLUIDS

When two or more fluid phases are flowing, the concept of relative permeability must be introduced. The chapter on "Rock Properties" has treated permeability concepts in detail.

Of substantial interest will be producing or reservoir gas/oil ratios and water/oil ratios.

I. GAS/OIL RATIOS (GOR)

In a linear system, the reservoir flow rates of gas and oil are,

$$q_g = \frac{6.33\, A\, k_g\,(p_i - p_o)}{\mu_g\, L} \tag{7-24}$$

$$q_o = \frac{1.127\, k_o\, A\,(p_i - p_o)}{\mu_o\, L} \tag{7-25}$$

where:

q_g = volumetric rate of gas flow at reservoir temperature and pressure, p_m, cu ft/day,
p_m = mean pressure, $(p_i + p_o)/2$, or the average of the inlet and outlet pressures, psia,
q_o = volumetric oil rate at reservoir conditions, barrels per day.

Dividing Equation 7-24 by Equation 7-25, then the reservoir gas/oil ratio in cubic feet per barrel results,

$$(GOR)_{reservoir} = 5.615\,\frac{\mu_o\, k_g}{\mu_g\, k_o} \tag{7-26}$$

If a unitless ratio is desired, then the 5.615 conversion factor can be deleted. Should the surface gas/oil ratio be needed, then it should be noted that:

1. Gas is liberated from solution in the oil. By definition, the solution gas in a unit of oil is R_s, with usual units of SCF of gas per STB of oil.
2. The "free" gas flowing in the reservoir expands in passing from mean reservoir pressure, p_m, to pressure at standard conditions, p_{sc}. Also, reservoir temperature of the gas, T_f, changes to temperature at standard conditions, T_{sc}. The real gas law is used to adjust to surface conditions; therefore, the temperatures should be in absolute units, usually °Rankine.

The surface gas-oil ratio, scf/STB, is:

$$(GOR)_{surf} = R_s + \left[\frac{B_o}{B_g}\right]\left[\frac{q_g}{q_o}\right]_{res} = R_s + \left[5.615\,\frac{\mu_o\, k_g}{\mu_g\, k_o}\right]\left[\frac{B_o\, p_m\, T_{sc}}{p_{sc}\, T_f\, z}\right] \tag{7-27}$$

where:

B_g = gas formation volume factor, cu ft/scf and
R_s = solution GOR @ p_m, scf/STB.

II. WATER/OIL RATIO (WOR)

In a linear system, the reservoir flow rate for water is:

$$q_w = \frac{1.127 \; k_w A \, (p_i - p_o)}{\mu_w L} \tag{7-28}$$

Division of Equation 7-28 by Equation 7-25 yields a reservoir WOR, provided it can be assumed that the capillary pressure is negligible compared to the total pressure differential across the system.

$$(WOR)_{reservoir} = \frac{\mu_o k_w}{\mu_w k_o} \tag{7-29}$$

Considering the formation volume factors allows the surface WOR to be calculated:

$$(WOR)_{surface} = \left[\frac{B_o}{B_w} \right] (WOR)_{reservoir} \tag{7-30}$$

Often, the water formation volume factor will be very nearly 1.0, and therefore is frequently neglected in reservoir engineering calculations.

HORIZONTAL, TRANSIENT, SINGLE-PHASE FLOW

This subject is covered for oil and gas wells in the "Well Testing" chapter, where the derivation is given of the partial differential equation that governs such flow. Transient aquifer behavior is treated in the "Water Drive" chapter.

REFERENCES

1. Darcy, H.: *Les fontaines publiques de la ville de Dijon*, Victor Dalmont (1856).
2. Pirson, S. J.: *Oil Reservoir Engineering*, 2nd Ed., McGraw-Hill Book Co., New York City (1958) Ch. 8.
3. Craft, B. C. and Hawkins, M. F.: *Applied Petroleum Reservoir Engineering*, PrenticeHall, Inc., Englewood Cliffs, N.J. (1959) Ch. 6.
4. Smith, Charles R.: *Mechanics of Secondary Oil Recovery*, Reinhold Publ. Corp., New York City (1966) Ch. 6.
5. Amyx, J. W.; Bass, D. M., Jr.; Whiting, R. I.: *Petroleum Reservoir Engineering*, McGraw-Hill Book Co., New York City (1960) 68 ff.
6. Frick, T. C.: "Petroleum Production Handbook," Vol. 11, *Reservoir Engineering*, McGraw-Hill Book Co., New York City (1962) Ch. 32.
7. Slider, H. C.: *Practical Petroleum Reservoir Engineering Methods*, Petroleum Publishing Co., Tulsa (1976) Ch. 1.
8. Calhoun, J. C.: *Fundamentals of Reservoir Engineering*, 2nd Printing, Univ. of Oklahoma Press, Norman, OK., (1955) Part II, 75 ff.
9. Gatlin, Carl: *Petroleum Engineering - Drilling and Well Completions*, Prentice-Hall, Englewood Cliffs, N.J. (1960) 24 ff.
10. Deppe, J. C.: AIME paper 1472G presented at the 4th Biennial Secondary Recovery Symposium, Wichita Falls, Texas (May 2-3, 1960).
11. Prats, M., Strickler, W. R. and Matthews, C. S.: *Trans.*, AIME (1955) **204**, 160ff.
12. Lewis, James A. and Horner, William L.: "Productivity Index and Measurable Reservoir Characteristics," AIME (March 1942).
13. Brown, Kermitt E.: *The Technology of Artificial Lift Methods*, Volume 1, Petroleum Publishing Company, Tulsa (1977) Ch. 1.
14. Vogel, J. V.: "Inflow Performance Relationship for Solution Gas Drive Wells," *JPT* (January 1968) 83-93.
15. Standing, M. B.: "Inflow Performance Relationships for Damaged Wells Producing by Solution Gas Drive," *JPT* (November 1970) 1399-1400.
16. Standing, M. B.: "Concerning the Calculation of Inflow Performance of Wells Producing from Solution Gas Drive Reservoirs," *JPT* (September 1971) 1141-1142.
17. Fetkovich, M. J.: "The Isochronal Testing of Oil Wells," SPE Paper 4529 presented at the 1973 Annual Fall Meeting of SPE of AIME, Las Vegas, Nevada (September 30 - October 3).

8 OIL RESERVOIR DRIVE MECHANISMS

INTRODUCTION

A reservoir drive mechanism may be defined as a distinct form of energy within a reservoir causing the expulsion (or production) of fluids. The normal drive mechanisms include solution gas drive, gas cap drive, water drive, segregation drive, and compaction drive. A particular reservoir may be producing by one or more drive mechanisms.

A convenient approach to the study of reservoir energy types is with the material balance equation which was first presented by Schilthuis.[1] Provided good data is available, material balance techniques allow dependable estimates of the initial hydrocarbons in place as well as prediction of future reservoir performance.

GENERAL MATERIAL BALANCE EQUATION

Although not necessary, the material balance equation is normally written on a reservoir volume basis. Very simply, one representation is: the initial hydrocarbon volume is equal to the remaining hydrocarbon volume (after a given amount of production) plus the volume of the water that has encroached into the reservoir. The reservoir is actually being treated as a large tank with no flow effects considered. For this reason, the material balance equation is sometimes described as a zero dimensional model. Although this basic method now somewhat has been superceded by the 1, 2, and 3 dimensional reservoir simulators (which include a mass balance, flow effects, heterogeneity, and sometimes interphase mass transfer), the reservoir analyst will obtain a much better feel for the reservoir if he begins a study with the simple tools. Then, if additional answers are needed, more sophisticated methods may be used. Material balance is not merely a topic of historical interest.

Figure 8-1A illustrates the "tank" representation of an oil reservoir with an initial gas cap at the discovery (initial) conditions. The total fluid volume in this diagram represents the hydrocarbon pore volume (HCPV) of the reservoir at the discovery pressure, p_i.

Figure 8-1B indicates the effect of fluid production with the resulting decrease in pressure. Notice that the original gas cap has expanded at the current static reservoir pressure, p. Also, illustrated is the net water influx that occurred between p_i and p. Pressure p is assumed below the bubblepoint pressure, so there has been some liberated free gas from the oil. It is assumed that some was produced and some remains in the reservoir. So, that depicted is the net liberated free gas; i.e., that remaining in the reservoir. Finally, the volume of the remaining oil in place is represented.

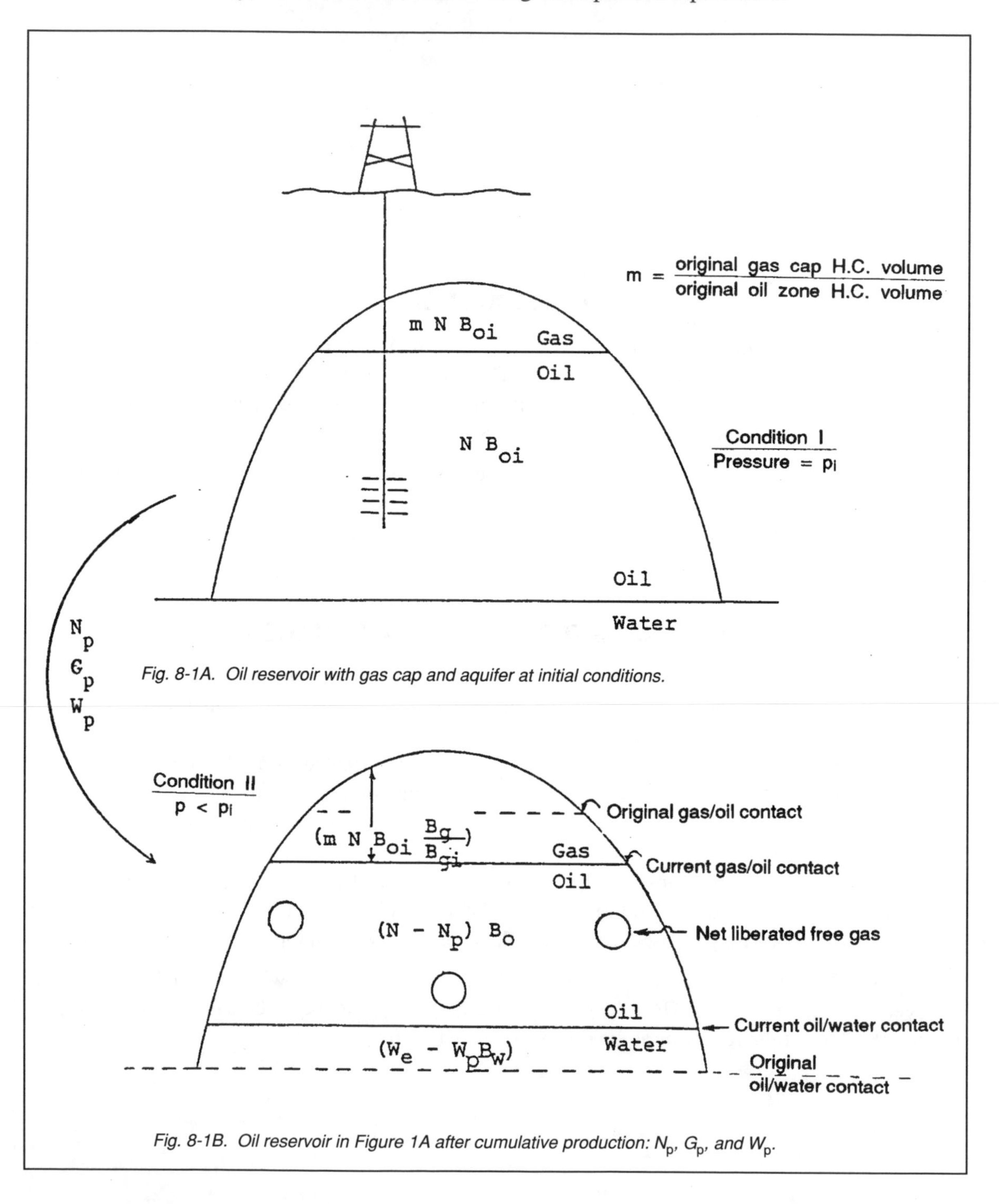

Fig. 8-1A. Oil reservoir with gas cap and aquifer at initial conditions.

Fig. 8-1B. Oil reservoir in Figure 1A after cumulative production: N_p, G_p, and W_p.

Not shown in Figure 8-1B is the decrease in hydrocarbon pore volume (HCPV) caused by rock and connate water expansion, which is often ignored, especially below the bubblepoint.

The general material balance equation may be written as:

Original hydrocarbon volume =
Remaining hydrocarbon volume + *volume of net water influx from the aquifer*
+ *reduction of HCPV due to rock and connate water expansion* (8-1)

Because we are dealing with compressible fluids (and rock) in a volume balance, each term of the equation on the right-hand side must be referred to the same pressure. It is simpler if the pressure basis is the static pressure, p, existing in the reservoir after a given amount of cumulative production: N_p, G_p, and W_p. Consequently, all volumes in Equation 8-1 are in reservoir barrels. The necessary terms are defined as:

Nomenclature:

N_p = cumulative oil production in stock-tank barrels,
N = initial oil in place in stock-tank barrels,
m = ratio of the initial hydrocarbon volume of the gas cap to the initial hydrocarbon volume of the oil zone (m is a constant), dimensionless,
B_o = oil-formation volume factor, rb/STB,
G_p = cumulative gas production, scf,
R_p = cumulative produced gas/oil ratio, (G_p / N_p), scf/STB,
R_s = solution gas/oil ratio, scf/STB,
B_g = gas formation volume factor, reservoir barrels / scf,
W_e = cumulative water influx from the aquifer into the reservoir, res. barrels,
W_p = cumulative amount of water produced, STB,
B_w = water-formation volume factor, rb/STB,
S_w = connate water saturation, fraction,
c_w = connate water isothermal compressibility, (1/psi),
c_f = pore volume isothermal compressibility, (1/psi),
p = static reservoir pressure, psia,
Δp = (p_i - p), psi, and
i = subscript indicating initial conditions.

It will be helpful to refer to Figure 8-1.

Original hydrocarbon volume =
original oil volume + *original gas volume*.

Original oil volume =
(initial stock tank oil in place) × *(initial oil-formation volume factor)*
$= NB_{oi}\ rb$ (8-2)

where "rb" signifies reservoir barrels. The constant "m" is defined in the nomenclature as:

$$m = \frac{\textit{original gas cap hydrocarbon volume}}{\textit{original oil zone hydrocarbon volume}} \tag{8-3}$$

"m" may be estimated from mud logs, logs, core data, well completion data, and well tests. At times, structure maps constructed from seismic data are also used. Therefore,

$$Original\ gas\ volume = m\,N\,B_{oi}\quad rb \tag{8-4}$$

$$Original\ hydrocarbon\ volume = N\,B_{oi}\,(1+m)\quad rb \tag{8-5}$$

Remaining hydrocarbon volume = expanded original gas cap volume + remaining oil volume + net liberated free gas rb, (8-6)

where: "net liberated free gas" is the liberated solution gas (in going from p_i to p) that was not produced and therefore remains in the reservoir.

$$Expanded\ original\ gas\ cap\ volume = m\,N\,B_{oi}\,(B_g/B_{gi})\quad rb \tag{8-7}$$

$$Remaining\ oil\ volume = (N - N_p)B_o\quad rb \tag{8-8}$$

Net liberated free gas = (gross liberated free gas) - (produced free gas) (8-9)

Equation 8-7 assumes that the gas cap is not being produced. However, even if some gas cap gas is being produced, it will be included in the produced free gas term. Thus, there would be offsetting errors in Equation 8-6. The "net liberated free gas" term would be smaller than it should be, but the "expanded gas cap volume" would be larger than actual by the same amount.

Earlier, it was mentioned that the current static reservoir pressure, p, is less than the bubblepoint pressure. Therefore, some gas has come out of solution in going from p_i to p. The amount of free gas liberated from one stock tank barrel of original oil would be $(R_{si} - R_s)$ scf, and the reservoir volume equals $(R_{si} - R_s)B_g$ reservoir barrels. The actual quantity of interest, however, is the reservoir volume of gas liberated from the total original oil in place. Therefore:

$$Gross\ liberated\ free\ gas = N\,B_g\,(R_{si} - R_s)\quad rb \tag{8-10}$$

All volumes concerning the right-hand side of Equation 8-1 are at reservoir conditions with static pressure, p. Considering the cumulative produced gas, G_p, solution gas is considered to be that part of the gas that would be in solution in the produced oil, N_p, if still in the reservoir. Thus, produced solution gas is equal to $(N_p)(R_s)$ scf. Therefore, the remainder of the produced gas, $(N_p)(R_p - R_s)$ scf, is called the "produced free gas." This produced free gas is made up of solution gas that was liberated from the oil in the reservoir and possibly some gas cap gas. So,

$$Produced\ free\ gas = N_p\,B_g\,(R_p - R_s)\quad rb \tag{8-11}$$

Therefore,

$$Net\ liberated\ free\ gas = N\,B_g\,(R_{si} - R_s) - N_p\,B_g\,(R_p - R_s)\quad rb \tag{8-12}$$

Substituting equations 8-7, 8-8 and 8-12 into equation 8-6:

$$\textit{Remaining hydrocarbon volume} = (N - N_p)B_o + m\,N\,B_{oi}\,(B_g/B_{gi}) + N\,B_g\,(R_{si} - R_s) - N_p\,B_g\,(R_p - R_s) \qquad rb \tag{8-13}$$

In Equation 8-1 there is a term representing the net water influx into the reservoir from the aquifer. The significance of this volume is depicted in Figure 8-1. Notice in part A that a container representing the original hydrocarbon volume is shown. Here, the bottom of the container is at the oil/water contact. In part B, which depicts the original hydrocarbon volume as well as the volume at the current pressure, some water has encroached into the reservoir across the original oil/water contact. This water entered the reservoir due to the pressure reduction caused by fluid withdrawals. The total amount of water that has encroached is represented as W_e reservoir barrels of water. W_e is never seen at the surface, so it may as well be kept in the units of Equation 8-1, reservoir barrels. W_p is the symbol for the cumulative water produced (STB) during this same time period. Therefore, the net water influx into the reservoir is:

$$\textit{Net water influx} = W_e - W_p\,B_w \qquad rb \tag{8-14}$$

According to Dake[2], the decrease in hydrocarbon pore volume (due to expansion of the connate water and the rock matrix), accompanying a decrease in pressure in the reservoir, may be expressed as:

$$-\,d\,(HCPV) = (1 + m)\,NB_{oi}\left(\frac{c_w\,S_w + c_f}{1 - S_w}\right)\Delta p \tag{8-15}$$

Inserting Equations 8-5, 8-13, 8-14, and 8-15 into Equation 8-1, the general material balance equation results:

$$\begin{aligned} NB_{oi}\,(1 + m) = {} & (N - N_p)\,B_o + mNB_{oi}\left(\frac{B_g}{B_{gi}}\right) + NB_g\,(R_{si} - R_s) \\ & - N_p\,B_g\,(R_p - R_s) + W_e - W_p\,B_w \\ & + (1 + m)\,NB_{oi}\left(\frac{c_w\,S_w + c_f}{1 - S_w}\right)\Delta p \end{aligned} \tag{8-16}$$

Equation 8-16 can be rearranged such that terms involving N are collected together and placed on the left-hand side, and other terms placed on the right-hand side.

$$\begin{aligned} & N\left\{ B_o - B_{oi} + mB_{oi}\left[\left(\frac{B_g}{B_{gi}}\right) - 1\right] + B_g\,(R_{si} - R_s) + B_{oi}\,(1 + m)\left(\frac{c_w\,S_w + c_f}{1 - S_w}\right)\Delta p \right\} \\ & \quad = N_p\,[B_o + B_g\,(R_p - R_s)] - (W_e - W_p\,B_w) \end{aligned} \tag{8-17}$$

Then, Equation 8-17 is solved for N:

$$N = \frac{N_p\,[B_o + B_g\,(R_p - R_s)] - (W_e - W_p\,B_w)}{B_o - B_{oi} + mB_{oi}\left[\left(\frac{B_g}{B_{gi}}\right) - 1\right] + B_g\,(R_{si} - R_s) + B_{oi}\,(1 + m)\left(\frac{c_w\,S_w + c_f}{1 - S_w}\right)\Delta p} \tag{8-18}$$

Discussion:

Equations 8-16 through 8-18, in general, lack any time dependence. Basically, the material balance is a function of cumulative fluid withdrawals; however, water influx (W_e) does usually have some time dependence. This will be discussed in the chapter on Water Drive.

The general material balance equation may be regarded as having three unknowns: N, W_e, and m. Although techniques exist[4] for more than one unknown, solution is much cleaner and more accurate if reasonable estimates exist for two of these quantities. The first use of the material balance equation is indicated in Equation 8-18: to solve for the original oil in place. Other uses include:

(1) to develop a reservoir fluid withdrawal versus pressure relationship, which is used to predict reservoir performance;

(2) to determine the existence and rate of water encroachment;

(3) to verify the existence of a gas cap; and

(4) to compare with the volumetrically calculated oil in place, N_{vol}. It is important to realize that N_{mb} (from material balance) should not necessarily be close to N_{vol}. N_{mb} requires pressure continuity; whereas N_{vol} does not. The material balance method will "feel" the presence of an impermeable boundary (perhaps a sealing fault between two reservoirs); whereas, the volumetric method can not necessarily distinguish between one or multiple reservoirs. Alternatively, perhaps there is an extension to the reservoir which can be sensed with the material balance method.

Unfortunately, for a reservoir with a very active water drive, the material balance technique is useless in determining N. With a very active water drive, (p_i - p) will remain small because the encroaching water maintains the pressure. Without a reasonable pressure drop occurring with fluid withdrawals, to attempt the calculation of N using Equation 8-18 is futile.

Similarly, material balance calculations should be regarded as highly suspect during the initial depletion of a reservoir (such as the first five percent of the reserves) because the denominator in Equation 8-18 will be quite small. Therefore, any errors in the data will be magnified. Additional initial problems reported by Amyx *et al.*[5] include: (1) produced water and gas often are not measured very carefully, and (2) early pressure data are sometimes meager or unknown. Confidence in material balance predictions should grow with reservoir maturity.

As Dake[2] points out, even though pressure only appears explicitly in the hydrocarbon pore volume reduction term, other parameters in the material balance relationship are directly dependent on pressure, such as: B_o, B_g, and R_s. The terms in the material balance equation must be evaluated at a single pressure: the reservoir static pressure existing after the cumulative fluid withdrawals. The static pressure for the drainage area of a particular well is normally determined from a pressure buildup test. The total reservoir static pressure then may be determined by volume-averaging the individual well static pressures. Well static pressure determination will be discussed in the chapter on well testing.

The reservoir analyst attempting to use one of the Equations 8-16 through 8-18 should ensure that the proper units are used for each parameter. In particular, the proper unit for the gas reservoir volume factors, B_g and B_{gi}, is reservoir barrels / standard cubic foot. Although, some may consider such a unit to be non-standard, its use is essential (when R_s, R_{si}, and R_p are in scf/STB) to maintain units consistency.

To perform a material balance study, a representative fluid sample is obtained and studied in the laboratory. Then, the hydrocarbon fluids are assumed to perform in the reservoir in a manner similar to the laboratory performance. Pirson[3] states that a situation termed "vaporization hysteresis" sometimes occurs in the reservoir causing more gas to remain in solution in the reservoir than in the laboratory pressure cell at the same pressure. According to this hypothesis, such a condition can occur particularly when permeability is small or with high drawdowns or excessive withdrawal rates. In fact, it may take several hundred psi lower pressure in the reservoir to attain vaporization conditions similar to that found in the laboratory.

According to Slider,[6] conventional material balance procedures assume a constant free gas composition for the life of the reservoir. In the case of a volatile oil reservoir, due to the presence of a substantial amount of intermediate hydrocarbons, the composition of the free gas in the reservoir does not remain constant. Free gas in such a reservoir makes a contribution to the production of stock-tank liquids. Therefore, conventional material balance methods will probably be quite inaccurate. Although special material balance techniques can still be used, it is suggested that this calculation be coupled with special laboratory PVT studies and analysis techniques similar to those that have been described in the literature.[7-13]

The general material balance equation, 8-17 or 8-18 can be rewritten to determine the different active drive mechanisms. In this way, a reservoir producing under a combination of drive mechanisms can be analyzed to find the relative magnitudes of the different forms of reservoir energy and relate this to ultimate recovery efficiency. (See the "Combination Drive" chapter.)

In a reservoir where the bubblepoint pressure is crossed, it is common practice to break the material balance calculation into two parts: (1) from initial pressure down to the bubblepoint and (2) from the bubblepoint down to the pressure of interest. Historically, this has been done for a variety of reasons including the convenient simplification of the equation for calculations entirely above the bubblepoint and because of the discontinuity in the oil-formation volume factor curve at the bubblepoint. However, Slider[6] points out that the general material balance equation contains no restrictions in application when crossing the bubblepoint. Slider has a good point, and there are times when it is convenient to use the general material balance equation in "one pass" from the initial pressure (above the bubblepoint) down to the current pressure (below the bubblepoint). There are also instances when the problem simplifies by making the calculation in two parts. Both approaches will be illustrated with an example.

The reservoir engineer involved in a material balance study should be aware of the decrease in hydrocarbon pore volume due to the expansion of connate water and reservoir rock that occurs with decreasing pressure. Due to the much higher compressibility of gas (roughly 100 times as great) than either water or rock, the effects of the water and rock usually are insignificant when a substantial gas saturation has developed. Many people simply ignore the rock-water compressibility term in the general material balance equation for calculations below the bubblepoint pressure. However, when above the bubblepoint, to ignore the contribution of the rock and connate water will likely result in a substantial overestimation of oil in place. Such effects will be illustrated in the following problem.

Problem 8-1

This problem has been adapted from Smith.[14]

Statement: Calculate the original oil in place for a reservoir whose discovery pressure was 5000 psia with a bubblepoint of 2750 psia. Initial data include:

B_{oi} = 1.305 rb/STB
S_w = 21.6%
ϕ = 10%
c_o = 15.3 x 10^{-6} psi^{-1}
c_w = 3.5 x 10^{-6} psi^{-1}
c_r = 3.0 x 10^{-7} psi^{-1}
R_{si} = R_{sb} = 500 scf/STB
T_f = 240 °F

Note that the oil compressibility given is the undersaturated oil (above the bubblepoint) compressibility. And it is quite important to realize that c_r, is the rock matrix (or grain) compressibility, not the pore volume compressibility.

At a static reservoir pressure of 3350 psia, the cumulative oil produced, N_p, was 1.510 MMSTBO with an oil formation volume factor of 1.338. Notice that this pressure is above the bubblepoint, so there was no free gas in the reservoir at that time.

At the bubblepoint, p_b = 2750 psia, and B_{ob} = 1.350 rb/STB.

At 1500 psia: N_p = 6.436 MMSTBO; G_p = 3,732 MMscf; R_s = 375 scf/STB; B_o = 1.250; and the z-factor for the gas, z = 0.90.

There has been no significant water encroachment or water production.

Find the indicated oil in place at 3350 psia and at 1500 psia. Perform the calculations with and without considering the effect of the connate water and rock expansion. Finally, for the below the bubblepoint case, work the problem two ways: (1) break the problem into two parts (above the bubblepoint and below) and (2) do not break the problem into two parts.

Discussion:

Notice in the general material balance Equation 8-18 that the reservoir compressibility that is required is not the rock matrix (or grain) compressibility, but the pore volume compressibility. Unfortunately, what is given in the problem statement is the rock matrix compressibility. So, a way to convert between the two is needed. Consider the unit bulk volume of rock as follows:

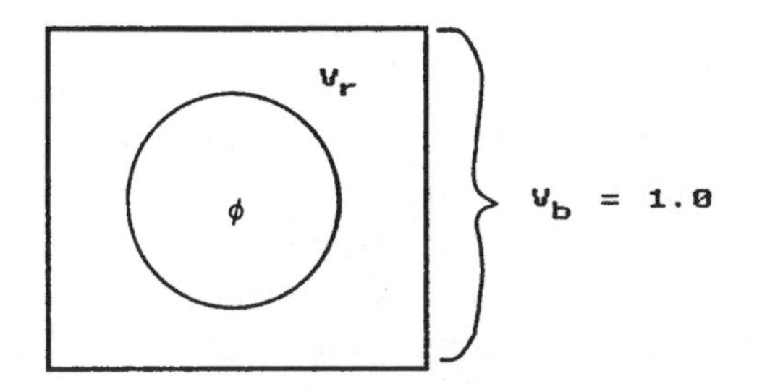

For this development, the outer rock boundaries in the simple picture above are considered to be constrained (held in fixed location), and the overburden is constant. Therefore, any reduction in pore pressure will result in the rock expanding into the pore volume shown. The beginning pressure is p_i. The rock matrix volume is V_r. Therefore, the porosity is:

$$\phi = 1 - V_r \tag{8-19}$$

Now, some fluid within the pore space is withdrawn leaving a pressure p such that $p < p_i$. The general isothermal compressibility equation is:

$$c = -\frac{1}{V}\frac{\partial V}{\partial p}\bigg|_T \tag{8-20}$$

where the "minus" sign is included in the definition because normal behavior for a substance is to increase in volume with a decrease in pressure. Therefore, the term $\partial V / \partial p$ is usually a negative quantity. The minus sign then allows the compressibility to be a positive number.

Writing the compressibility equation for the rock matrix,

$$c_r = -\frac{1}{V_r}\frac{\partial V_r}{\partial p}\bigg|_T \tag{8-21}$$

For the porosity, the equation is:

$$c_f = \frac{1}{\phi}\frac{\partial \phi}{\partial p}\bigg|_T \tag{8-22}$$

Note that the "minus" sign has been omitted (porosity decreases with a decrease in pressure.) Substituting Equation 8-19 into Equation 8-22, and simplifying we obtain:

$$c_f = -\frac{1}{\phi}\frac{\partial V_r}{\partial p} \tag{8-23}$$

Then, rearranging Equation 8-21 to yield an expression for $\partial V_r / \partial p$ and substituting this into Equation 8-23, the result is:

$$c_f = \frac{c_r V_r}{\phi} \tag{8-24}$$

With this expression, pore volume compressibility can be calculated if rock matrix compressibility and porosity are known. So, using Equation 8-19 again:

$$c_f = c_r\left(\frac{1-\phi}{\phi}\right) \tag{8-25}$$

In the development of Equation 8-25, it was assumed that the boundaries of the rock did not move. Actually, when there is a fluid pressure reduction within the pore space, then more of the overburden is being carried by the rock matrix. This in turn causes some compaction of the rock bulk volume downward. The net effect is a loss of pore volume which may be simplistically viewed as an expansion of the rock matrix into a small portion of the former pore volume.

Solution:

(A) Solve for N using the data at a reservoir static pressure of 3350 psia neglecting the effect of connate water and rock expansion.

For convenience, Equation 8-18 is rewritten here.

$$N = \frac{N_p\left[B_o + B_g\left(R_p - R_s\right)\right] - \left(W_e - W_p B_w\right)}{B_o - B_{oi} + mB_{oi}\left[\left(\frac{B_g}{B_{gi}}\right) - 1\right] + B_g\left(R_{si} - R_s\right) + B_{oi}\left(1 + m\right)\left(\frac{c_w S_w + c_f}{1 - S_w}\right)\Delta p} \quad (8\text{-}18)$$

Notice, that for this reservoir, above the bubblepoint, most terms in Equation 8-18 are equal to zero. As the oil is undersaturated, there is no gas cap. So, m = 0. Further, the only gas produced is solution gas. So, $R_p = R_s$. Of course, $R_s = R_{si}$. The remaining equation is:

$$N = \frac{N_p B_o - \left(W_e - W_p B_w\right)}{B_o - B_{oi} + B_{oi}\left(\frac{c_w S_w + c_f}{1 - S_w}\right)\Delta p} \quad (8\text{-}26)$$

For this particular reservoir, $W_e = W_p = 0$.

The connate water and rock expansion are not to be considered here. So, assuming $c_w = c_f = 0$ and substituting the pertinent data into Equation 8-26:

$$N = \frac{(1.510)(1.338)}{1.338 - 1.305} = 61.2\ MMSTBO$$

(B) Solve for N as in part (A), but now consider the effects of the expansions of connate water and the reservoir rock.

First, Equation 8-25 is used to calculate the pore volume compressibility:

$$c_f = \frac{(3)(10^{-7})(1 - 0.10)}{0.1} = 2.7 \times 10^{-6}\quad psi^{-1}$$

Using Equation 8-26:

$$N = \frac{(1.510)(1.338)}{1.338 - 1.305 + 1.305\left[\frac{(3.5)(10^{-6})(0.216) + (2.7)(10^{-6})}{(1 - 0.216)}\right](1650)}$$

$$= 47.5 MMSTBO$$

Notice that the answer for N obtained in part A, when neglecting the effect of the rock and water expansions, was 29% too high.

(C) Calculate N with the available data at initial conditions and at 3350 psia considering the effects of water and rock expansion, but using a formulation involving the isothermal undersaturated oil compressibility.

Notice that the denominator of Equation 8-26 contains $B_o - B_{oi}$.

From Smith[14] an equation involving oil compressibility and oil formation volume factors for conditions entirely above the bubblepoint is:

$$B_o = B_{oi}\,[1 + c_o(p_i - p)] \tag{8-27}$$

Rearranging Equation 8-27 and using it in Equation 8-26:

$$N = \frac{N_p B_o - (W_e - W_p B_w)}{B_{oi}\, c_o\, \Delta p + B_{oi}\left(\dfrac{c_w S_w + c_f}{1 - S_w}\right)\Delta p} \tag{8-28}$$

The drive mechanism for this reservoir, above the bubblepoint is actually a liquid and rock expansion drive. The oil, water, and rock are expanding causing the expulsion of fluids. To incorporate the effects of the oil, rock, and water into a single compressibility term, an "effective oil compressibility" is defined:

$$c_{oe} = c_o + \frac{c_w S_w + c_f}{1 - S_w} \tag{8-29}$$

Then, Equation 8-28 becomes:

$$N = \frac{N_p B_o - (W_e - W_p B_w)}{c_{oe} B_{oi}\, \Delta p} \tag{8-30}$$

Using Equation 8-29 to calculate effective oil compressibility:

$$c_{oe} = (15.3)(10^{-6}) + \frac{(3.5)(10^{-6})(0.216) + (2.7)(10^{-6})}{(1 - 0.216)}$$

$$= 19.71 \times 10^{-6}\ psi^{-1}$$

Now, using Equation 8-30:

$$N = \frac{(1.510)(1.338)}{(19.71)(10^{-6})(1.305)(1650)} = 47.6\ MMSTBO$$

This answer essentially agrees with that obtained in (B) except for some slight rounding effects.

(D) Using the general material balance Equation 8-18, calculate N with data available initially and at 1500 psia. In other words, do not break the calculation at the bubblepoint and perform in two parts; do it in one calculation. Also, consider the effects of the connate water and rock expansions.

Considering Equation 8-18, several quantities appear that are not yet available: B_g and R_p. As the current pressure, 1500 psia, is below the bubblepoint, free gas should be present in the reservoir. Considering 14.7 psia and 520°R to be standard conditions, gas reservoir volume factor may be calculated with the following equation:

$$B_g = 0.00503 \frac{T_f\, z}{p} = (0.00503)(460 + 240)(0.90) \ / \ 1500 \quad rb/SCF$$

$$B_g = 0.00211 \quad rb/scf$$

To calculate R_p:

$$R_p = G_p / N_p = \frac{(3{,}732)(10^6)}{(6.436)(10^6)} = 579.863 \quad scf/STB$$

Recall from part (B) that $c_f = 2.7 \times 10^{-6}\ psi^{1}$

Also, $W_e = W_p = m = 0$. Now, all values are available that are needed for Equation 8-18.

$$N = \frac{6.436\,[\,1.250 + 0.00211\,(579.863 - 375)\,]}{1.25 - 1.305 + 0.00211\,(500 - 375) + 1.305\left[\dfrac{(3.5)(0.216) + 2.7}{(1 - 0.216)}\right](10^{-6})(3500)}$$

= *47.3 MMSTBO*

Notice that this value is different than that obtained in (B) and (C), which were calculations above the bubblepoint. In this particular problem, the difference is only minor. Normally, there would probably be more of a discrepancy. The calculation that utilizes the greater amount of production data would usually be preferred.

(E) With data available initially, at the bubblepoint, and at 1500 psia, calculate N by considering the problem in two steps: (1) from initial conditions to the bubblepoint and (2) from the bubblepoint down to 1500 psia. Include the effects of rock and connate water expansion.

Part 1:

Equation 8-30 can be rearranged to permit a calculation of the cumulative oil produced down to the bubblepoint pressure. It is important to remember, however, that the resulting equation is only valid for pressures above (and down to) the bubblepoint. So, $p \geq p_b$.

$$N_p = \frac{N\, c_{oe}\, B_{oi}\,(p_i - p) + (W_e - W_p\, B_w)}{B_o} \tag{8-31}$$

Using N from part (C), N = 47.6 MMSTBO, and Equation 8-31:

$$N_{p@pb} = \frac{(47.6)(19.71)(10^{-6})(1.305)(5000 - 2750)}{1.350} = 2.04\ MMSTBO$$

The gas produced down to the bubblepoint is here assumed to be entirely solution gas.

So:

$$G_{p@pb} = N_p R_{sb} = (2.04)(10^6)(500)\ scf = 1{,}020\ MMscf$$

Part 2:

For this part of the calculation, beginning at the bubblepoint; i.e., the bubblepoint will be considered to be initial conditions. Consequently, the calculated N will be the oil in place at the bubblepoint, and will, therefore, neglect the oil production above the bubblepoint.

The cumulative oil produced from the bubblepoint, 2750 psia, down to 1500 psia is:

$$[N_p]_{below\ pb} = 6.436 - 2.040 = 4.396\ MMSTBO$$

Cumulative gas produced below the bubblepoint pressure is:

$$[G_p]_{below\ pb} = (3{,}732 - 1{,}020) = 2{,}712\ MMscf$$

And the below-the-bubblepoint, cumulative-produced gas/oil ratio:

$$R_p = \frac{2{,}712}{4{,}396} = 616.92\ scf/STBO$$

Now, using values of N_p and R_p from the bubblepoint and remembering that $B_{oi} = B_{ob}$, Equation 8-18 is used to calculate the oil in place that existed at the bubblepoint:

$$N_{@pb} = \frac{4.396\,[\,1.250 + 0.00211\,(616.92 - 375)\,]}{1.25 - 1.35 + 0.00211\,(500 - 375) + 1.35\left[\dfrac{(3.5)(0.216) + 2.7}{(1 - 0.216)}\right](10^{-6})(1250)}$$

$= 45.21\ MMSTBO$

Then, to get the true N, the production down to the bubblepoint must be added.

$$N = N_{@pb} + N_{p@pb} = 45.21 + 2.04 = 47.25\ MMSTB$$

Strictly speaking, the calculation should be done again, beginning with part 1. The value of N just calculated, 47.3 MMSTBO, should be used in Equation 8-31 to restart the calculation procedure. In this case, there is little difference between 47.3 and 47.6, and the re-calculation can be neglected.

Notice that the N obtained here in a two-part calculation, 47.3 MMSTBO, agrees with that computed in (D).

(F) Work the second part of (E) neglecting the effects of rock and connate water expansion.

Using the values of N_p and R_p below the bubblepoint and $B_{oi} = B_{ob}$, the oil in place existing at the bubblepoint is calculated just as in (E); however, here the water and pore volume compressibilities are assumed to be 0. Using Equation 8-18:

$$N_{@p_b} = \frac{4.396\,[\,1.250 + 0.00211\,(\,616.92 - 375\,)\,]}{1.250 - 1.350 - 0.00211\,(\,500 - 375\,)} = 47.26 \;\; MMSTBO$$

Then,

$$N = 2.04 + 47.26 = 49.3 \;\; MMSTBO$$

The neglecting of the rock and water expansion below the bubblepoint here produced an N that is 4% too high. This is certainly within the accuracy of data normally available to the reservoir engineer. If compressibility data (oil, pore volume, and/or water) is desired but not available, then methods presented in the "Rock Properties" and "Fluid Properties" chapters may be used to estimate them.

MATERIAL BALANCE EXPRESSED AS A LINEAR EQUATION

In a classic paper, Havlena and Odeh[4] described techniques that in certain situations allow the material balance Equation 8-18 to be written as a linear equation and even solved graphically. They further stated that with their analysis methods, the otherwise static material balance relation attained a dynamic meaning.

This discussion has been adapted from Dake's treatment[2] of the work of Havlena and Odeh. The following definitions are needed.

$$F = N_p\,[B_o + (R_p - R_s)\,B_g] + W_p\,B_w \quad rb \tag{8-32}$$

which represents underground withdrawal volume.

$$E_o = (B_o - B_{oi}) + (R_{si} - R_s)\,B_g \quad rb/STB \tag{8-33}$$

This term describes the expansion of the oil and liberated free gas.

$$E_g = B_{oi}\,[\,(B_g / B_{gi}\,) - 1\,] \quad rb/STB \tag{8-34}$$

which concerns the expansion of the gas cap gas.

$$E_{f,w} = B_{oi}\,(\,1 + m\,)\left(\frac{c_w\,S_w + c_f}{1 - S_w}\right)\Delta p \quad rb\,/\,STB \tag{8-35}$$

Here, the rock and connate water expansion is considered.

Then, the general material balance Equation 8-17 may be written as:

$$F = N\,(E_o + mE_g + E_{f,w}) + W_e \qquad (8\text{-}36)$$

The analysis method requires that each of the applicable functions, Equations 8-32 through 8-35, be evaluated at several different stages of cumulative production. Plotting and analysis are musts[4] when using this method as complete automation is not likely to be successful. For the reservoir with no initial gas cap, negligible water influx, and negligible compressibilities of connate water and rock, equation 8-36 reduces to:

$$F = NE_o \qquad (8\text{-}37)$$

In this case, a plot of F versus E_o should result in a straight line with slope N that passes through the origin as in Figure 8-2. Above p_b, water and pore volume compressibilities should be included, and figure 8-2 should be constructed with $(E_o + E_{f,w})$ as the x-axis, not simply E_o. If the plot is not linear, the curve shape may well be diagnostic in determining the actual reservoir drive mechanisms.

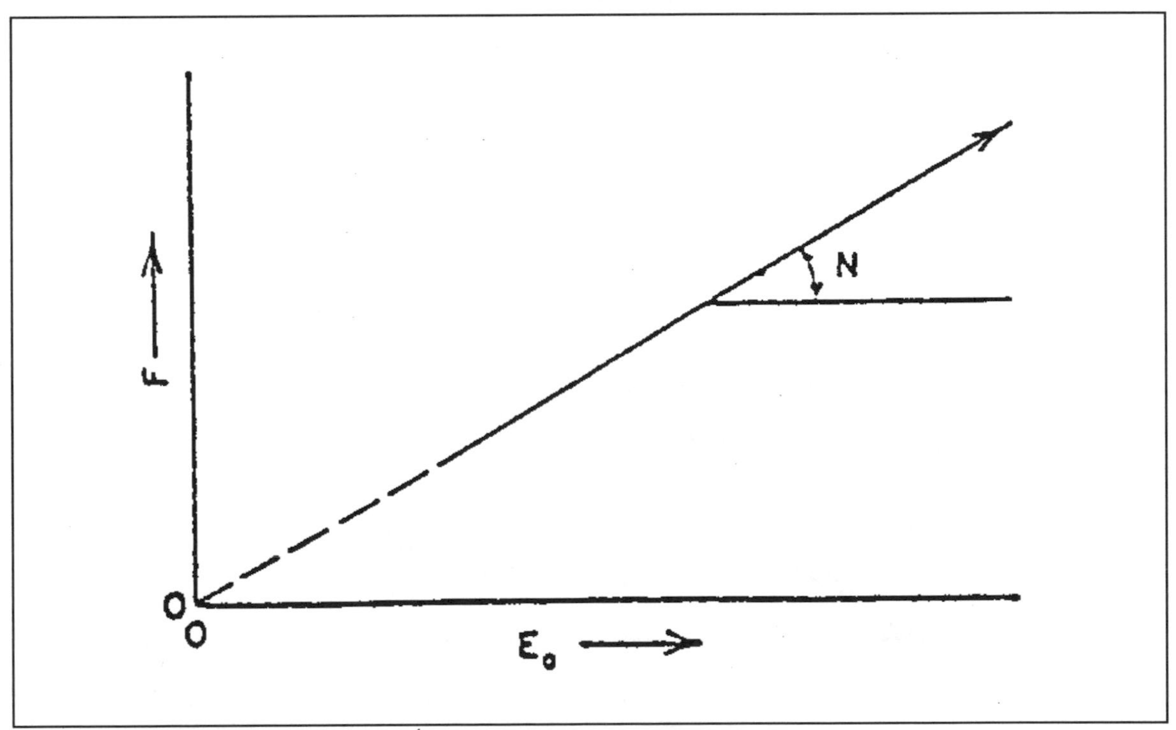

Fig. 8-2. F vs. E_o (from Havlena and Odeh[4]). Permission to publish by JPT.

Another case of interest is a reservoir with an initial gas cap but with negligible water encroachment. Because of the initial gas cap, the connate water and rock compressibilities may be neglected with little error incurred. Equation 8-36 becomes:

$$F = N\,(E_o + mE_g) \qquad (8\text{-}38)$$

In this case, a plot of F versus ($E_o + mE_g$) should result in a straight line through the origin with slope N. Figure 8-3 illustrates the plot. If m is not exactly known (the usual situation), then different m's may be assumed and used to make different plots on the same set of axes. The correct m will yield a straight line.

For the case of a water drive reservoir with no gas cap and negligible connate water and rock compressibilities, Equation 8-36 becomes:

$$F = NE_o + W_e \tag{8-39}$$

In this case, F/E_o should plot as a linear function of W_e / E_o. The y-intercept will be N. A suitable function describing the water influx is normally used in the analysis. Such functions and the corresponding analysis techniques are described in the "Water Drive" chapter.

More detailed information concerning the graphical techniques of Havlena and Odeh is available.[2,4] With any of these methods, once a straight line has been obtained by matching production and pressure data, then a "history match" has been achieved. At this point, the mathematical model may be used to predict future behavior of the reservoir.

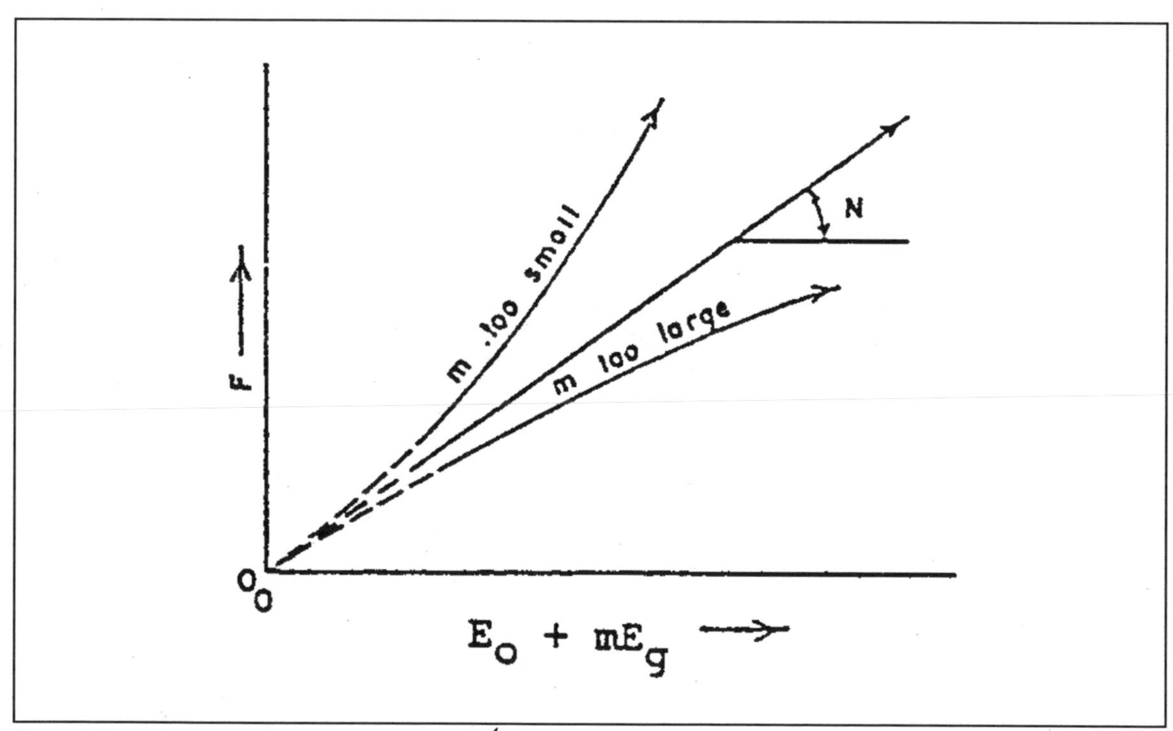

Fig. 8-3. F vs. ($E_o + mE_g$) (from Havlena and Odeh[4]). Permission to publish by JPT.

DRIVE MECHANISMS

Introduction

This discussion will be limited to reservoirs containing oil with an API gravity of 20 or greater. Bitumen, tar, and heavy oil-bearing reservoirs will not be considered.

The purpose is to discuss primary operations defined as reservoir operations where nothing is returned or added to the reservoir. So, pressure maintenance, which is sometimes considered as primary recovery will be viewed as secondary recovery. However, it is not a breach of this definition to have producing wells on artificial lift.

Reserves are defined to be the cumulative oil production at the economic limit, or:

$$N_{pa} = Reserves = (N)(RE)/100 \tag{8-40}$$

where:

N_{pa} = cumulative oil production at abandonment, STB, and
RE = recovery efficiency, %

Solution Gas Drive

Solution gas drive is sometimes called internal gas drive, dissolved gas drive, depletion drive, volumetric drive, or fluid expansion drive. This is the principal drive mechanism for approximately one-third of all oil reservoirs in the world.

In a solution gas drive reservoir, there is no gas cap or waterdrive. However, at least two-thirds of all reservoirs have water at the bottom. If the volume of water is not large (say less than 10 times the volume of the oil), then this water is often called a "water leg" and will not contribute much to the reservoir energy. There are many solution gas drive reservoirs with water legs. The average water saturation within the hydrocarbon pore volume is normally close to the irreducible value.

The initial reservoir pressure will either be above or equal to the bubblepoint pressure. Assuming for the moment that the initial pressure is above the bubblepoint, then with production, pressure will decline rapidly to the bubblepoint. During this period, all gas in the reservoir remains in solution. This operating regime down to the bubblepoint is often referred to as fluid-expansion drive. Pictorially, we have:

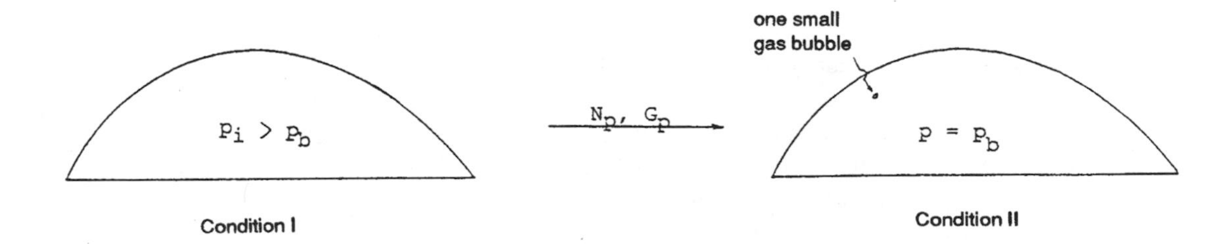

Condition I

Condition II

The drive mechanism in going from condition I to condition II is the expansion of oil, rock, and connate water. Beginning with the static reservoir pressure at the bubblepoint (condition II), further production will cause the reservoir pressure to be lowered below the bubblepoint with the consequent evolution of free gas in the reservoir. This may be viewed as:

The "*" superscripts for the production terms indicate cumulative production from p_b, not from p_i.

After gas saturation exceeds a critical value, it becomes mobile. The gas saturation in condition III is considered to be greater than the critical which is indicated above by the flow paths drawn between the gas pockets. In order for a gas cap not to develop, thus creating a gas cap drive, the vertical permeability must be small (say, less than 50 md.). At condition III, the free gas flow in the reservoir is evidenced by the increasing GOR's observed at the wells. In going from condition II to condition III, the drive mechanism consists mainly of gas drive and oil expansion. The effect of water and rock expansion is small compared to the energy of the highly expansible free gas.

Although the worldwide average is approximately 15 to 17%, solution gas drive recovery efficiencies vary from 5 to 30%. However, recovery from a fractured carbonate would likely be somewhat lower than 15%. Recovery efficiency prediction methods basically fall into two classes: (1) empirical or statistical methods and (2) mathematical models.

Arps *et al.*[15] developed the following statistical equation through multivariate regression analysis of case histories of 80 different solution gas drive reservoirs.

$$RE = 41.815 \left[\frac{\phi(1 - S_w)}{B_{ob}} \right]^{0.1611} \left[\frac{k}{\mu_{ob}} \right]^{0.0979} \left[S_w \right]^{0.3722} \left[\frac{p_b}{p_a} \right]^{0.1741} \tag{8-41}$$

Where:

RE = recovery efficiency, percent
ϕ = porosity, fraction,
S_w = connate water saturation, fraction,
B_{ob} = oil-formation volume factor at the bubblepoint, rb/STB
k = average formation permeability, darcies
μ_{ob} = viscosity of the reservoir oil at the bubblepoint, centipoise,
p_b = bubblepoint pressure, psig, and
p_a = abandonment pressure, psig

This equation represents the recovery efficiency from the bubblepoint pressure down to the abandonment pressure for a solution gas drive reservoir. Recovery above the bubblepoint would be in addition to that predicted by Equation 8-41. Recovery efficiency above the bubblepoint is normally in the range of 3% or less.

Equation 8-41 should be used only during the beginning stages of development of a reservoir when no production history is available. If enough information exists, then a mathematical model (such as material balance, decline analysis, or a reservoir simulator) is used. Equation 8-41 is useful in planning for the development of new discoveries. Because the Arps *et al.* equation was developed using data from reservoirs varying widely in lithology and geographical location, the efficiencies predicted by Equation 8-41 should be tempered with the engineer's knowledge of the performance of other reservoirs in the given geographical area. It should be mentioned that although the authors have enjoyed some success with Equation 8-41, Doscher *et al.*[19] have questioned its validity.

If reservoir fluid PVT data, rock property data, and (hopefully) some production data are available, then mathematical models may be used to predict performance. Material balance predictive models are discussed in the "Solution Gas Drive" chapter.

A generic graph illustrating the performance life of a solution gas drive reservoir is presented in Figure 8-4. Three variables (R = producing gas/oil ratio, oil-producing rate, and average reservoir pressure) are plotted versus cumulative oil production. In this figure, all wells are assumed to be placed on production at their maximum rate at time zero. No wells are ever shut-in until abandonment. Then, all wells are shut-in. With an actual reservoir, wells would begin producing at different times and may be restricted by market, regulatory, or engineering constraints.

In Figure 8-4, a reservoir with an initial pressure above the bubblepoint is shown. With production, the pressure declines somewhat linearly and rapidly to the bubblepoint. The pressure decline below the bubblepoint flattens some due to the presence of free gas (with its large compressibility) in the reservoir. With production, the pressure continues to decline. With pressure decrease, more and more gas comes out of solution. Until the critical gas saturation is reached (3 to 7 percent in sandstone reservoirs), this free gas is immobile. At this point, the free gas begins flowing to the wells. As the gas saturation builds, the gas mobility increases, more and more gas is produced, and the pressure declines more rapidly. Late in the life of the reservoir the slope of the pressure curve may flatten.

The reservoir producing gas/oil ratio, R, sometimes called the instantaneous producing gas/oil ratio, is equal to total gas (scf) produced from the reservoir in a given time interval (such as one day) divided by the total oil (STB) produced during the same interval. The equation that relates R to reservoir rock and fluid properties was derived in the "Rock Properties" chapter:

$$R = R_s + \frac{k_g}{k_o}\frac{\mu_o}{\mu_g}\frac{B_o}{B_g} \qquad (8\text{-}42)$$

From the initial pressure down to the bubblepoint pressure, since there is no free gas in the reservoir, the producing GOR normally remains constant and is equal to the solution GOR: $R = R_{si} = R_{sb}$. Notice in Figure 8-4 that there is a slight decrease seen in the producing GOR just below the bubblepoint for a short time, and then it begins to increase rapidly. This decrease results because free gas is coming out of solution in the reservoir but cannot flow until the critical gas saturation is reached.

Thus, for a while, the only gas that is produced is solution gas which is decreasing. Therefore, the producing GOR is also seen to decrease. This behavior is not always seen. In the vicinity of a producing well, a pressure sink exists which can cause a large (local) free gas saturation to build up. In this case, free gas will be able to flow sooner than if there were no such near well effects. As soon as the critical gas saturation is exceeded, the producing GOR rises. With larger gas saturations, higher GOR's are experienced.

Near the end of the reservoir primary producing life, the producing GOR curve may decline. Actually, on a reservoir volume basis, GOR continues to increase for the life of the solution gas drive reservoir. However, the standard industry definition of R, Equation 8-42, involves B_g. This relates the gas reservoir volume to its surface volume. B_g increases in size as reservoir pressure decreases. And at low pressures, B_g may increase more rapidly than the relative permeability and viscosity ratios. Therefore, since B_g is in the denominator, at low pressures, R may decline.

Notice in Figure 8-4 that, similar to pressure, the oil rate declines fairly rapidly down to the bubblepoint. Above the bubblepoint, where all free gas is in solution, the relative permeability to oil remains high since reservoir flow is one phase oil flow. The decrease in oil rate with production is due to the decrease in pressure. Below the bubblepoint, the rate of decline in q_o decreases because a free gas saturation is being formed which causes the pressure to decline more slowly. However, there is another cause for oil rate decline below the bubblepoint: relative permeability effects. As gas comes out of solution and builds a free gas saturation, the oil saturation decreases. This causes a decrease in the oil relative permeability.

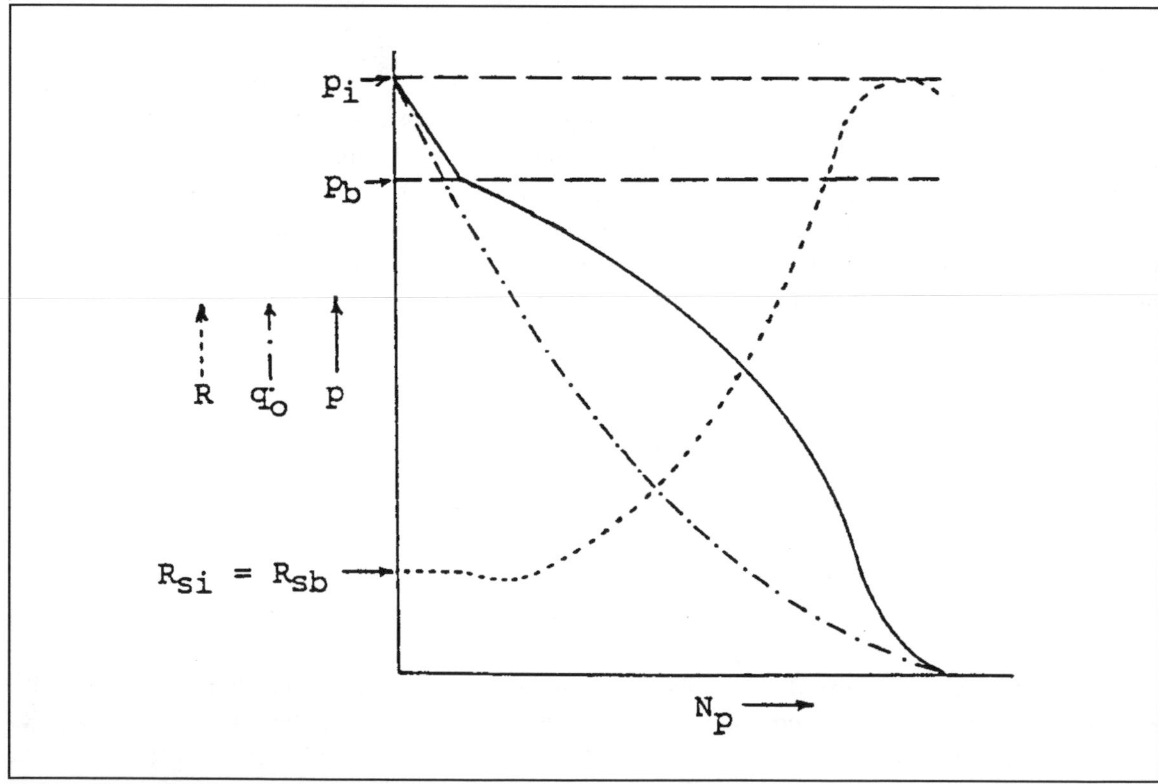

Fig. 8-4. Solution gas drive reservoir performance curves.

The low recovery from solution gas drive reservoirs emphasizes the need for additional energy to be added to the reservoir to produce part of the remaining 70+ percent of original oil that is still in place at the end of primary operations. This type of reservoir is usually an excellent candidate for secondary oil recovery.

Water Drive

Roughly two-thirds of all reservoirs have oil/water contacts. However, only about one-half of these develop a water drive. For a strict waterdrive, there is no gas cap. Therefore, the initial pressure is greater than the bubblepoint pressure. The vertical permeability is limited (at least in the upper part of the reservoir), or the water drive maintains pressure such that a gas cap does not form. Some of the different depletion stages that occur during the life of a water drive reservoir are presented below.

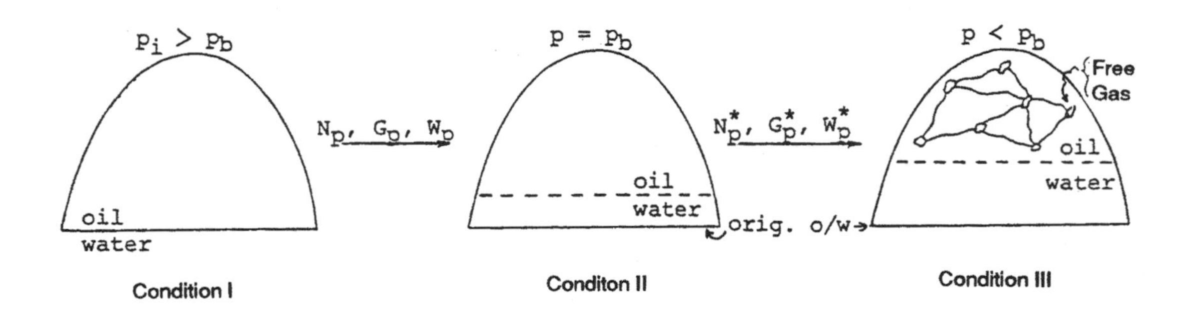

Note: The "*" notation between conditions II and III indicates cumulative production from p_b, not p_i.

Notice that at condition I (discovery) the level of the original oil/water contact is illustrated. After cumulative production N_p, G_p, and W_p, the reservoir pressure drops to the bubblepoint or condition II. Here, the oil/water contact has risen. The encroaching aquifer not only helps to maintain the pressure, but also immiscibly displaces much of the oil from the invaded portion of the reservoir. Further production brings the reservoir state to condition III where the pressure is well below the bubblepoint. The aquifer has invaded even more into the reservoir, and some free gas has come out of solution. Notice that the free gas does not form a gas cap. Rather, producing GOR increases after the critical gas saturation is reached.

The encroached water enters the reservoir from either of two sources:

(1) Appreciable expansion of the water (or aquifer) occurs where the water volume is very large (more than 10 times) compared to the oil volume. With pressure reduction, the water expands and partially replaces fluid withdrawals from the reservoir.

(2) The aquifer is part of an artesian system as shown in Figure 8-5. Note in this figure that the oil reservoir is situated within a formation that outcrops at a higher structural location than the reservoir. Therefore, the water underneath the oil reservoir is being recharged by water from the surface.

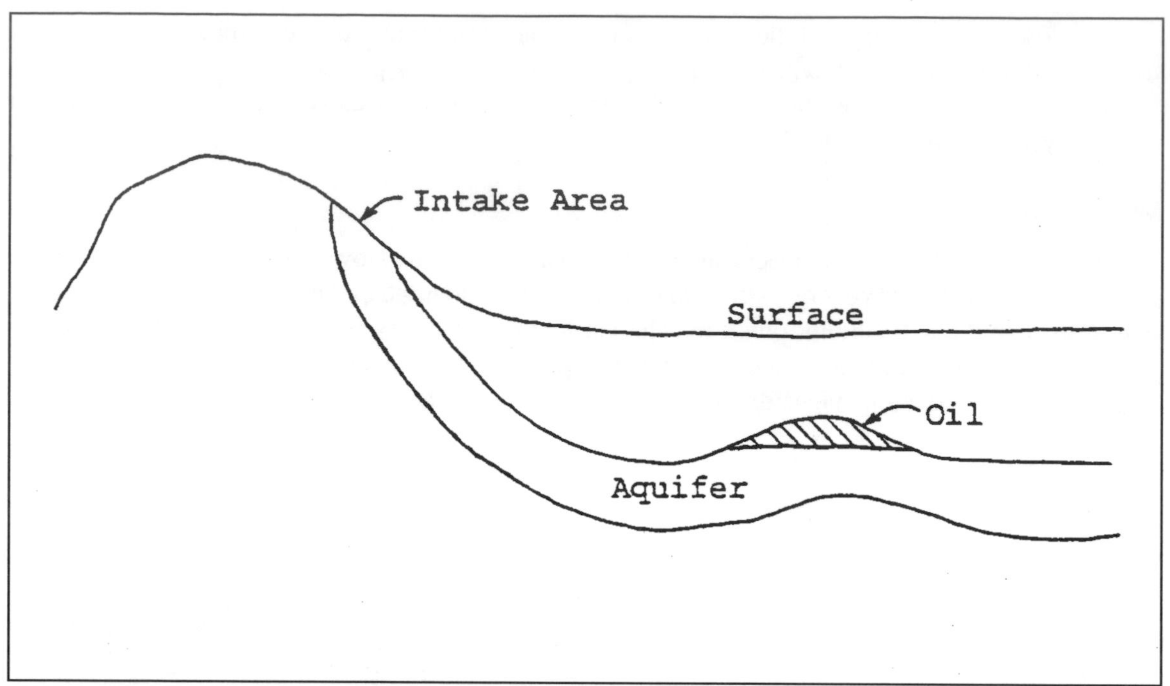

Fig. 8-5. Artesian aquifer - oil reservoir system.

Recovery efficiencies for waterdrive reservoirs range from 10 to 70 percent, with the average being 35 to 40 percent. The Arps *et al.*[15] statistical water drive recovery efficiency equation is:

$$RE = 54.898 \left[\frac{\phi(1 - S_w)}{B_{oi}} \right]^{0.0422} \left[\frac{k\,\mu_{wi}}{\mu_{oi}} \right]^{0.0770} \left[S_w \right]^{-0.1903} \left[\frac{p_i}{p_a} \right]^{-0.2159} \tag{8-43}$$

where:

RE = water drive recovery efficiency, percent
ϕ = porosity, fraction,
S_w = Average connate water saturation, fraction,
B_{oi} = initial oil formation volume factor, rb/STB
k = average absolute permeability, darcies
μ_{wi} = aquifer water viscosity at initial conditions, cp
μ_{oi} = reservoir oil viscosity at initial conditions, cp
p_i = initial reservoir pressure, psig, and
p_a = abandonment reservoir pressure, psig.

This equation is strictly for a natural water drive reservoir and is not meant to be used for waterflooding. As with the solution gas drive Arps equation, 8-43 is mainly for use early in reservoir life before enough data is available to use a mathematical model. It was developed[15] using multivariate regression analysis of actual case histories. Once again, the authors have had some success with the relationship; others[19] have not.

Material balance mathematical models for water drive reservoirs are discussed in the "Water drive" chapter.

Depending on the way that the water enters the reservoir, water drive reservoirs are sometimes characterized as either (1) bottomwater drive or (2) edgewater drive. In a bottomwater drive, the formation is usually thick with enough vertical permeability (at least in the lower part of the reservoir) so that the water can move vertically or perpendicular to the formation grain orientation. In this type of reservoir, since there is some vertical permeability, water coning can be a serious problem. In an edgewater drive, the water moves into the reservoir from the side (obviously from the downdip direction). Here, the formations may be thinner than those of bottomwater drive reservoirs. For material balance modeling, the equations are the same for both bottomwater and edgewater drives. However, the wells are handled differently. In an edgewater drive, an individual well often will water out more quickly than in a bottomwater drive.

The performance of a water drive reservoir is rate sensitive. Two cases will be considered.

Case 1: Large Permeable Aquifer, Low Relative Producing Rate

In this type of system, there is essentially 100% replacement of produced fluids with aquifer water, and often it is referred to as a total water drive. Possible performance curves for a total water drive are shown in Figure 8-6.

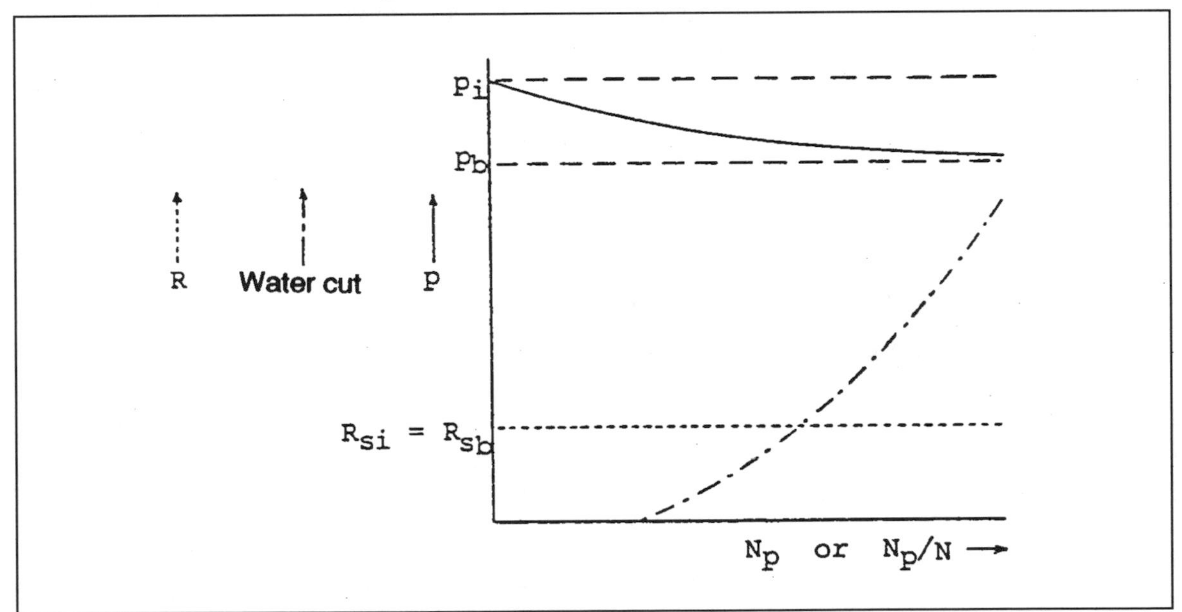

Fig. 8-6. Total water drive performance curves.

As was mentioned earlier, water drives are rate sensitive. The aquifer can deliver only so much water with a given pressure differential across the oil/water contact. If the reservoir is produced at a rate faster than the aquifer can deliver, then the pressure will decline more than shown in Figure 8-6. For a variety of reasons, such as the formation of free gas in the reservoir, as the pressure drops below the bubblepoint, recovery efficiency usually will decrease. Also, with both oil and gas flowing, performance calculations become more complex as the recovery mechanism is no longer a strict water drive, but a combination drive.

The producing gas/oil ratio should remain constant at R_{sb} for the life of the reservoir if the pressure stays above the bubblepoint.

A water drive reservoir may not produce water immediately; although some do. However, water production is inevitable as water encroaches into the reservoir. The water-cut curve of Figure 8-6 might be representative with a bottomwater drive and all wells continuing to produce.

The curve would be different with an edgewater drive and the wells being shut-in as they "water out." Indeed, it is the water cut that usually determines the economic limit of the operation. To be more precise, the economic limit is normally governed by the water lifting, handling, and disposal costs. The actual value of water cut at the economic limit is area dependent, but typical limiting values are in the range of 95 to 99 percent.

As far as actual reservoir behavior is concerned (as opposed to text book reservoirs), the total water drive reservoir is probably as close to true steady state behavior as the reservoir engineer will ever encounter.

Case II: High Relative Producing Rate

The performance curves (generic) for a partial water drive reservoir are illustrated in Figure 8-7. The assumptions include a constant producing rate with water production increasing steadily after it begins. As these relationships are rate sensitive, actual curves could look quite different, reflecting mechanical failures, market demand, etc.

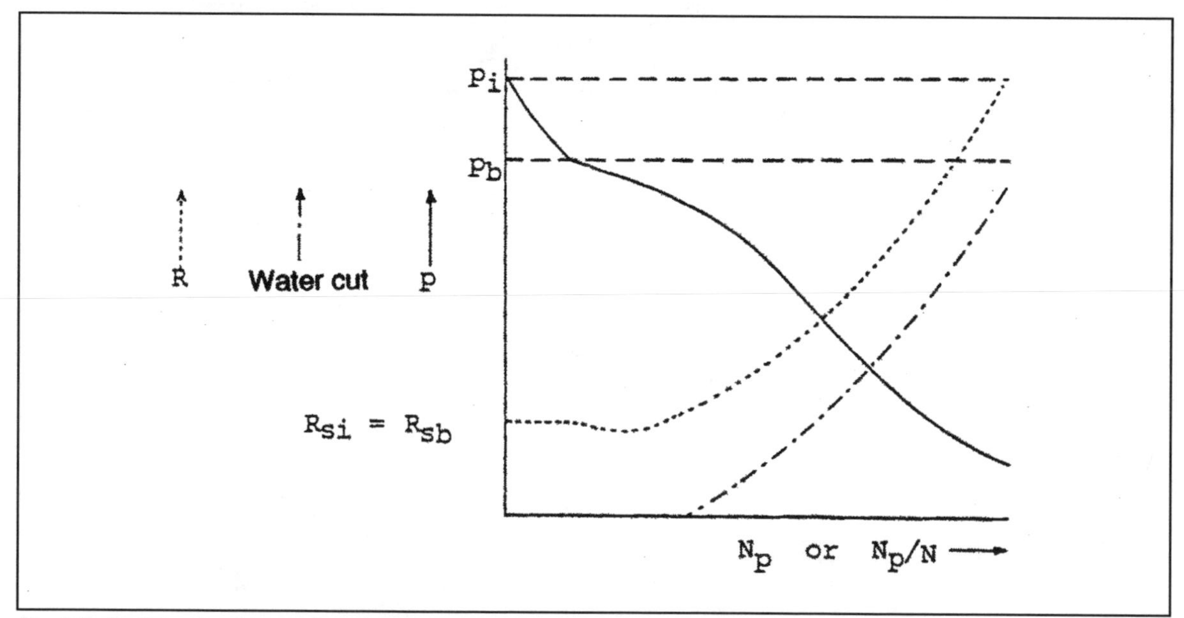

Fig. 8-7. Partial water drive performance curves.

In a partial water drive, the aquifer cannot keep up with the reservoir producing rate, so the pressure continually declines. Just below the bubblepoint, the pressure decline flattens slightly due to gas coming out of solution that cannot yet flow away from its liberation site. The critical gas saturation is eventually attained, and free gas flow starts. Consequently, the pressure begins to decline more quickly from this point. Later, the pressure curve may flatten as the aquifer influx rate increases (larger pressure differential) and the high water-cut wells are shut-in.

The partial water drive case is really a combination drive. In fact, if the reservoir is produced at such a high rate that relatively little water encroaches, the solution gas drive mechanism predominates; and oil recovery is lowered.

Gas-Cap Drive

For a conventional gas cap drive, one without gravity segregation of gas and oil, the reservoir must have an initial gas cap. Therefore, the initial reservoir pressure is exactly equal to the bubblepoint pressure of the oil and equal to the dewpoint pressure of the gas in the gas cap. This occurs because over the course of geologic time, equilibrium should exist between the oil and the gas. With an initial gas cap, the oil is holding the maximum amount of gas in solution that it can.

Although an oil/water contact may be present, no significant water drive develops. Vertical permeability is limited (less than 50 millidarcies); therefore, no segregation of the liberated gas and oil occurs.

The production phases of a gas-cap drive reservoir are shown in the following sketch. Notice that the well perforations are positioned so as to limit gas-cap gas production, thereby avoiding any unnecessary loss of reservoir energy. Condition II represents the reservoir after sufficient production has occurred to lower reservoir pressure and develop a mobile gas saturation within the oil zone. The gas cap has expanded causing an immiscible displacement of the oil downward. As with a water drive, there is frontal displacement of oil occurring. Frontal advance theory (discussed in the "Immiscible Displacement" chapter) indicates that, like water drive reservoirs, gas-cap drive reservoirs are rate sensitive.

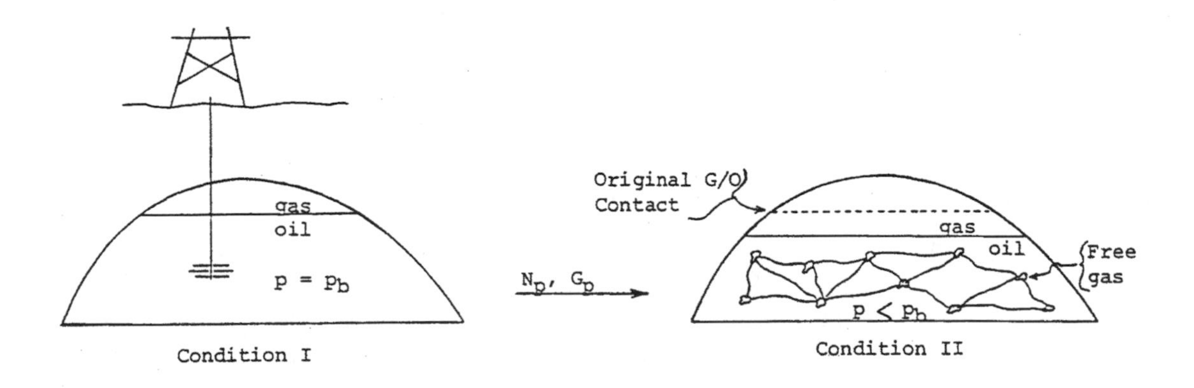

The average recovery efficiency for a gas cap drive reservoir is 30 to 40 percent of the original oil in place.

As defined earlier, "m" is equal to the original gas-cap hydrocarbon pore volume divided by the original oil zone hydrocarbon pore volume. The performance of a gas-cap drive reservoir is dependent on the gas-cap size. The gas-cap drive reservoir performance curves presented in Figure 8-8 demonstrate a large influence of "m." When m = 0.1 (original gas-cap size equal to 1/10 the size of the original oil zone), the curves are basically those of a solution gas drive reservoir, because the small gas cap provides only minor pressure support. However, when m = 10, the pressure declines quite slowly, and the producing GOR can remain quite constant. Because the pressure does not fall much below the bubblepoint, free gas in the oil zone is minimized. Consequently, the critical gas saturation may never be achieved; thus oil permeability remains high.

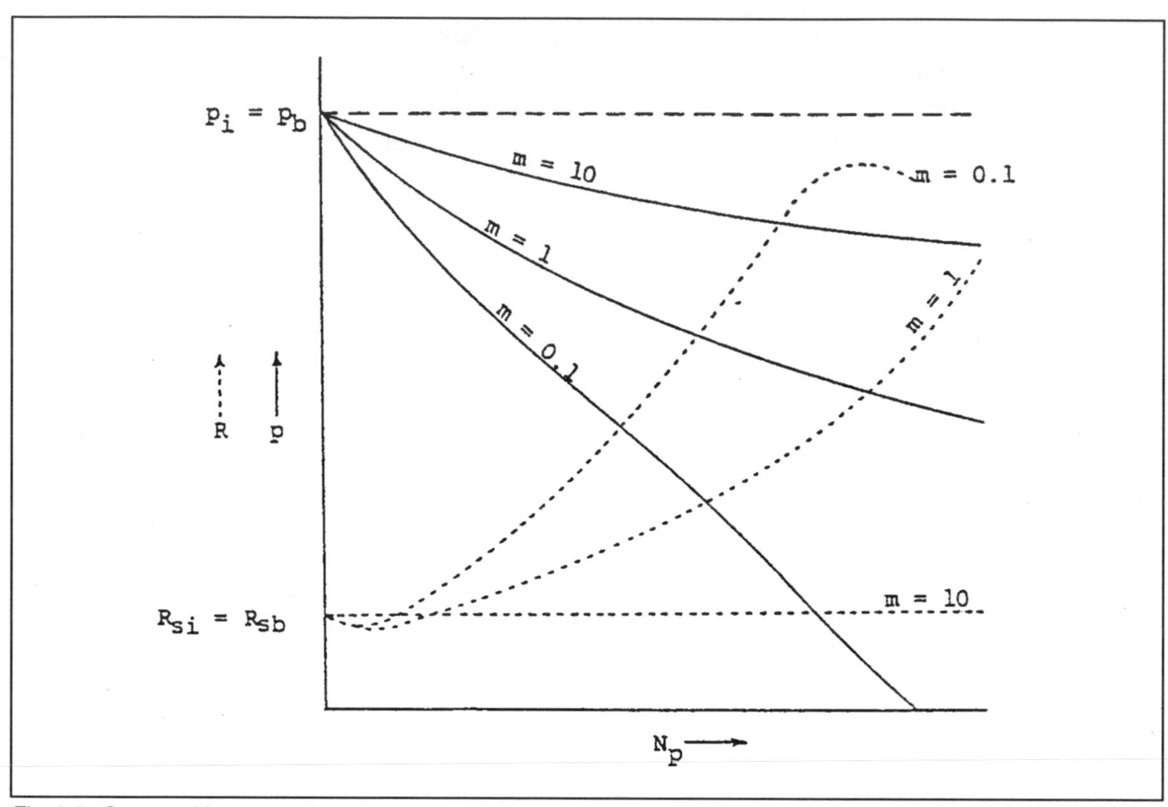

Fig. 8-8. Gas cap drive reservoir performance curves.

Segregation Drive

In a segregation drive reservoir, as free gas is liberated from the oil, using the vertical permeability, the gas moves up to the top of the reservoir, and the oil, moves downward. For this to occur, sufficient vertical permeability (at least 50 md.) must exist to allow the gravitational forces to overcome the viscous forces within the reservoir. Although some of these reservoirs do not have an initial gas cap, recovery will usually be greater if one exists. There may be an oil/water contact present, but no significant water drive develops. Some pressure depletion is essential for segregation drive to occur.

A similar mechanism, referred to as gravity drainage, often occurs in a reservoir with large dip. In this case, the oil is still moving downward and the gas upward, but the flow is parallel to the dip angle, instead of perpendicular to it. A large angle of dip allows the oil to be produced downdip without disturbing or coning the gas that is migrating to the top of the structure. In this section, gravity drainage and segregation drive will be considered to be the same mechanism.

This is the most efficient of the primary drive mechanisms provided economics are not considered. The rate of segregation normally is controlled by the oil mobility, rather than that of the gas, because after segregation begins, $(k_o / \mu_o) < (k_g / \mu_g)$. If the operator is patient, recovery should be higher from this kind of oil reservoir than from any of the other types. Recovery efficiencies from segregation drive reservoirs usually range between 40 to 80%.

Figure 8-9 shows possible pressure and producing GOR responses in a segregation drive reservoir. With a large initial gas cap, the pressure would decline less than shown. Notice that it is possible for the producing GOR (in downdip wells, at least) to decline with cumulative production. If not producing the wells at high rates, then most of the free gas can migrate to the top of the structure, leaving essentially only solution gas to be produced with the oil. Solution gas is declining with reservoir pressure. More often, the producing GOR remains relatively constant or increases slightly with time.This drive mechanism will be treated in more detail in the "Gas-cap Drive" chapter.

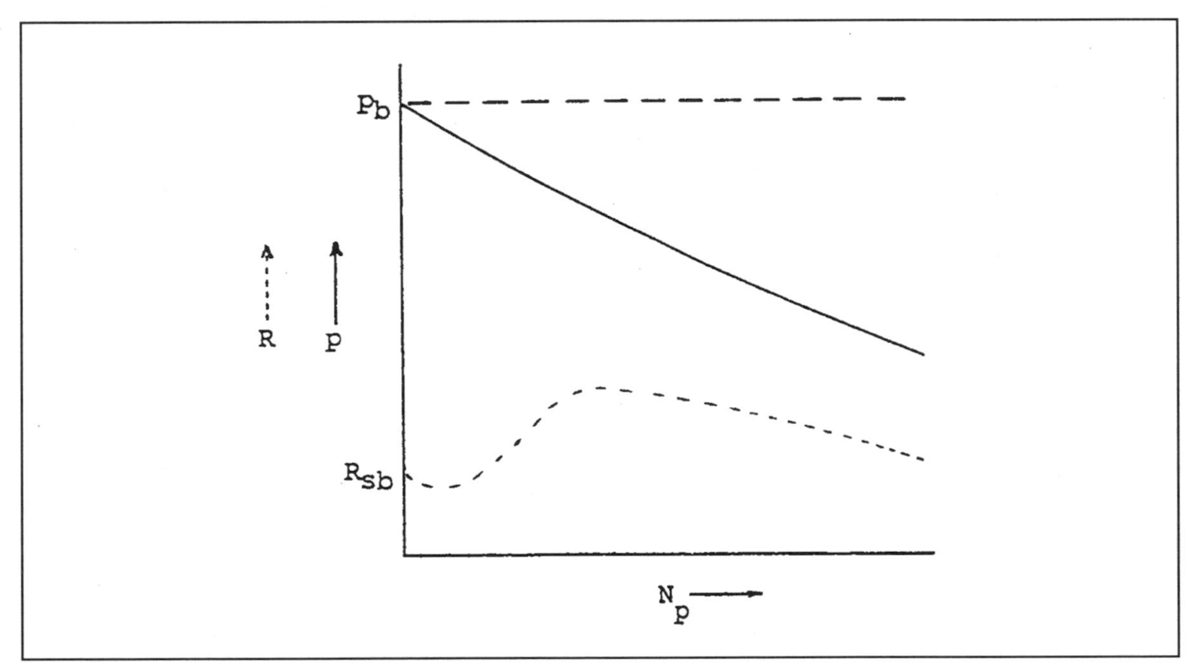

Fig. 8-9. Segregation-drive-reservoir performance curves.

Compaction Drive

At reservoir depth, overburden pressure refers to the combined weight of all overlying rock and fluids. In the majority of sedimentary basins, the overburden gradient is approximately 1 psi per foot of depth. Some of the weight of the overburden is carried by the rock grains, and the rest is supported by the fluid within the pore spaces. The portion of the overburden carried by the rock grains is called grain or matrix pressure.[2] In normally pressured regions, the fluid pressure gradient ranges from 0.433 to 0.465 psi per foot of depth. Therefore, the grain pressure will normally increase with depth at the rate of approximately 0.54 to 0.56 psi per foot.

At any reservoir depth, fluid withdrawal reduces the fluid pressure. Therefore, with production, more of the overburden is supported by the rock grains with a consequent increase in the grain pressure. Since rock is compressible, compaction of the rock grains results. This compaction causes a reduction in the rock pore volume and, in the extreme, can lead to surface subsidence.

Laboratory experiments[16] have shown that rock grain compressibility is not a constant, being higher at low grain pressures. This is significant because the higher the compressibility is, the more potential there is for compaction (drive).

Dake[2] defines compaction drive as the expulsion of reservoir fluids due to the dynamic reduction of pore volume. This is only likely to be a significant source of reservoir energy where the rock or pore volume compressibility is large such as in shallow reservoirs. In one Venezuelan reservoir, the compaction drive mechanism has been reported[17] to have accounted for more than 50% of the total oil recovery.

Overpressured reservoirs also have potential for a portion of the reservoir energy to be compaction drive. Since the fluid pressure is so high, the grain pressure is quite low (for that depth). Therefore, the pore volume compressibility is higher than it would be at the same depth with normal fluid pressures. Also, overpressured reservoirs often are associated with shale-imbedded, unconsolidated sand formations with pore volume compressibilities that can be ten times[20] greater than that of a consolidated rock. North Sea chalk reservoirs also can have significant formation compressibility.[21]

Loose, unconsolidated formations are often found at shallow depths. Compaction of this material is both inelastic and irreversible. Not only can the packing configuration be changed, but crushing of individual grains also occurs. Significant compaction usually contains at least some inelastic deformation. In this case, there may be a considerable time lag after the start of production before much compaction occurs (pressure threshold effect[17]).

Shallow, unconsolidated sands have the greatest potential for significant compaction drive. Appreciable compaction in this type of reservoir may be accompanied with visible surface subsidence. The inelastic deformation characteristics of unconsolidated formations make it difficult to predict in-situ performance based on laboratory compressibility tests.[2] However, the need is great, not only to be able to predict reservoir performance, but also to be able to investigate future surface subsidence, especially when the field surface location is next to a lake or sea.[18] After the subsidence has begun, there is evidence[17] that this subsidence can be analyzed using elastomechanical theory, production data, and a computer to relate the amount of subsidence with time to the degree of compaction. In this way, future compaction-subsidence-withdrawal-pressure relationships can be generated.

Combination Drive

Reservoir energy is often comprised of more than one drive mechanism. It is possible to have both solution gas drive and water drive acting in the same reservoir. Another possibility is water drive, gas-cap drive, and solution gas drive. The performance of these reservoirs is more difficult to predict than if a single drive mechanism is at work. Computational techniques are available to determine the relative magnitude of each of the individual drive mechanisms and the ultimate effect on recovery efficiency. This will be discussed in the "Combination Drive" chapter.

REFERENCES

1. Schilthuis, R. J.: "Active Oil and Reservoir Energy," *Trans*., AIME (1936) 33-52.
2. Dake, L. P.: *Fundamentals of Reservoir Engineering*, Elsevier Scientific Publishing Company, Amsterdam (1978).
3. Pirson, S. J.: *Oil Reservoir Engineering*, McGraw-Hill, Inc., New York City (1958).
4. Havlena, D. and Odeh, A. S.: "The Material Balance as an Equation of a Straight Line," *JPT* (August 1963) 896-900; *Trans*., AIME, 231.
5. Amyx, J. W., Bass, D. M. and Whiting, R. L.: *Petroleum Reservoir Engineering — Physical Properties*, McGraw-Hill, New York City (1960).
6. Slider, H. C.: *Practical Petroleum Reservoir Engineering Methods*, Petroleum Publishing Company, Tulsa (1976).
7. Cook, A. B., Spencer, G. B. and Bobrowski, F. P.: "Special Considerations in Predicting Reservoir Performance of Highly Volatile Type Oil Reservoirs," *Trans*., AIME (1951) **192**, 37-46.
8. Jacoby, R. H. and Berry, Jr., V. J.: "A Method for Predicting Depletion Performance of a Reservoir Producing Volatile Crude Oil," *Trans*., AIME (1957) **210**, 27-33.
9. Reudelhuber, F. O. and Hinds, R. F.: "A Compositional Material Balance Method for Prediction of Recovery from Volatile Oil Depletion Drive Reservoirs," *Trans*., AIME (1957) **210**, 19-26.
10. Abasov, M. T., *et al*.: "Physical and Mathematical Simulation for Development of Gas-Condensate Fields and Fields of Volatile Oils," *Proc*., Ninth World Pet. Conf., (1975) **4**, 369-376.
11. Brinkley, T. W.: "A Volumetric-Balance Applicable to the Spectrum of Reservoir Oils from Black Oils Through High Volatile Oils," *JPT*, (1963) **15**, 590-594.
12. Fussell, L. T. and Fussell, D. D.: "A New Mathematical Technique for Compositional Models incorporating the Redlich-Kwong Equation of State," paper SPE 6891 presented at the 1977 Annual Technical Conference and Exhibition, Denver.
13. Kazemi, H., Shank, G. D. and Vestal, C. R.: "An Efficient Multicomponent Numerical Simulator," paper SPE 6890 presented at the 1977 Annual Technical Conference and Exhibition, Denver.
14. Smith, C. R.: *Mechanics of Secondary Oil Recovery*, Reinhold Publishing Corporation, New York (1966).
15. Arps, J. J., Brons, F., van Everdingen, A. F., Buchwald, R. W. and Smith, A. E.: "A Statistical Study of Recovery Efficiency," *Bull. D14*, API, Washington, D. C. (October 1967).
16. Teeuw, D.: "Prediction of Formation Compaction from Laboratory Compressibility Data," *SPEJ* (September 1971) 263-271.
17. Merle, H. A., Kentie, C. J. P., van Opstal, G. H. C. and Schneider, G. M. G.: "The Bachaquero Study — A Composite Analysis of the Behavior of a Compaction Drive/Solution Gas Drive Reservoir," *JPT* (September 1976) 1107-1115.
18. Geertsma, J.: "Land Subsidence Above Compacting Reservoirs," *JPT* (June 1973) 734-744.
19. *Bull. D14*, "Statistical Analysis of Crude Oil Recovery and Recovery Efficiency," API, Dallas, 1984.
20. Tracy, G. W.: Private communication, 1986.
21. Ruddy, I., Anderson, M.A., Pattillo, P.D., Bishiawi, M. and Foged, N.: "Rock Compressibility, Compaction, and Subsidence in a High Porosity Chalk Reservoir," *JPT* (July 1989) 741-749.

9 SOLUTION GAS DRIVE RESERVOIRS

INTRODUCTION

It is suggested that the reader become familiar with the "Oil Reservoir Drive Mechanisms" chapter before attempting to understand and use the concepts of this chapter. Particularly important are the sections entitled "General Material Balance Equation" (including the example) and "Solution Gas Drive." It was shown in that chapter how to use material balance to calculate N, the original oil in place. The emphasis in this chapter will be the use of material balance to predict performance from a solution gas drive reservoir.

The simplest oil reservoir is generally thought to be the solution gas drive reservoir. With this type of reservoir, there is no initial gas cap. For a strict solution gas drive (not a combination drive), the vertical permeability is limited. Thus, no gas cap develops below the bubblepoint. Also, there is no significant encroached or produced water. Although below the bubblepoint pressure, there are two hydrocarbon phases (oil and gas) present in the reservoir; above the bubblepoint there is only one phase, oil. A solution gas drive reservoir will have a discovery pressure that is either equal to or greater than the bubblepoint pressure.

SOLUTION GAS DRIVE RESERVOIRS ABOVE THE BUBBLEPOINT

Since pressure is above the bubblepoint, no free gas exists. As pressure drops due to production, expansion of the oil is the principal force moving oil from the reservoir to the wellbore. Although not so compressible, water and reservoir rock also expand with reservoir pressure drop. A material balance Equation, 8-28, developed in the "Oil Reservoir Drive Mechanisms" chapter for solution gas drive reservoirs above the bubblepoint (but here, neglecting the water terms) is:

$$N = \frac{N_p B_o}{B_{oi} c_o \Delta p + B_{oi} \left(\dfrac{c_w S_w + c_f}{1 - S_w} \right) \Delta p} \tag{9-1}$$

where:

N = initial oil in place in stock-tank barrels,
N_p = cumulative oil production in stock-tank barrels,
B_o = oil formation volume factor, rb/STB,
c_o = isothermal, undersaturated (above p_b) oil compressibility, psi^{-1},
c_w = connate water isothermal compressibility, psi^{-1},
c_f = pore volume isothermal compressibility, psi^{-1},
S_w = connate water saturation, fraction,
p = static reservoir pressure, psia,
Δp = $(p_i - p)$, psi, and
i = subscript indicating initial conditions.

This equation includes the effects of the expanding oil, water, and rock which cause the expulsion of fluids. Above the bubblepoint, neglecting the effect of the water and rock may cause a substantial over-estimation of oil in place. So, the term: $B_{oi}\left(\frac{c_w S_w + c_f}{1 - S_w}\right)\Delta p$ in the denominator of 9-1 should not be neglected.

To simplify Equation 9-1 an effective (oil) compressibility may be defined:

$$c_{oe} = c_o + \frac{c_w S_w + c_f}{1 - S_w} \tag{9-2}$$

where:

c_{oe} = effective compressibility, psi^{-1}, which accounts for the compressibilities of oil, water, and reservoir rock.

Then, Equation 9-1 may be written as:

$$N = \frac{N_p B_o}{c_{oe} B_{oi} \Delta p} \tag{9-3}$$

As in the case of volumetric gas reservoirs, cumulative production from a solution gas drive reservoir above the bubblepoint is approximately a linear function of pressure.

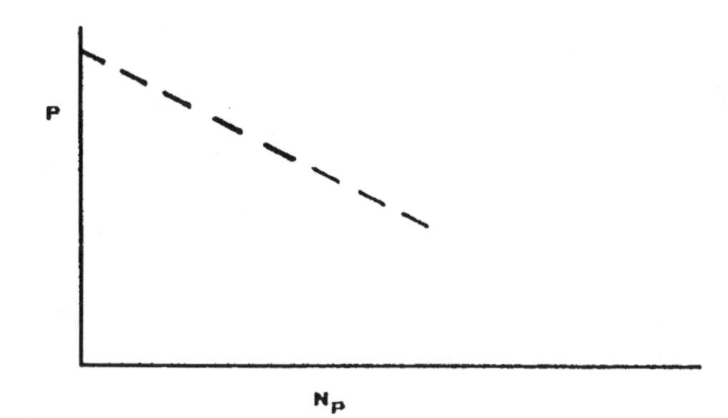

Estimates of original oil in place based on production and pressures above the bubblepoint often are not reliable because Equation 9-3 is quite sensitive to errors in the measured average reservoir pressure.

However, if a good estimate for N is available, then Equation 9-3 may be rearranged for prediction purposes:

$$N_p = \frac{N\, c_{oe}\, B_{oi}\,(p_i - p)}{B_o} \tag{9-4}$$

As with Equations 9-1 and 9-3, this equation applies only above the bubblepoint. To use 9-4, a static pressure, p (where $p_i > p \geq p_b$), is assumed. With knowledge of N, B_o, B_{oi}, c_{oe}, and p_i; the cumulative oil production obtained at p is estimated. In this way, cumulative oil production may be predicted between the initial and the bubblepoint pressures.

SOLUTION GAS DRIVE RESERVOIRS BELOW THE BUBBLEPOINT

As reservoir pressure drops below the bubblepoint pressure, free gas forms. At first, these bubbles of gas are small and discontinuous. As the pressure continues to drop, however, the bubbles coalesce, and at the critical gas saturation, a continuous free gas phase is formed. When this happens, gas begins to flow to wells along with the oil.

Below the bubblepoint there are two hydrocarbon phases in the reservoir: oil and free gas. Assuming no significant water encroachment or production, and assuming that the initial pressure is the bubblepoint pressure; then the general material balance equation (8-18) of the "Oil Reservoir Drive Mechanisms" chapter may be written as:

$$N_b = \frac{N_p\,[B_o + B_g(R_p - R_s)]}{B_o - B_{oi} + B_g(R_{si} - R_s) + B_{oi}\left(\dfrac{c_w S_w + c_f}{1 - S_w}\right)\Delta p} \tag{9-5}$$

where:

N_b = oil in place at the bubblepoint, stock-tank barrels
N_p = cumulative oil production from the bubblepoint, STB
B_o = oil formation volume factor, rb/STB
B_g = gas formation volume factor, reservoir barrels / scf
R_p = cumulative produced gas/oil ratio from p_b, (G_p/N_p), scf/STB
G_p = cumulative gas production from the bubblepoint, scf
p_b = bubblepoint pressure (also the initial pressure), psia
R_s = solution gas/oil ratio, scf/STB,
c_w = connate-water isothermal compressibility, psi^{-1}
S_w = connate-water saturation, fraction
c_f = pore volume isothermal compressibility, psi^{-1}
Δp = (p_b - p), psi
p = static reservoir pressure ($p < p_b$), psia, and
i = subscript indicating initial (bubblepoint) conditions.

Because the water and rock compressibilities are small compared to the compressibility of free gas; below the bubblepoint, a constant reservoir volume is often assumed. The consequence is that c_w and c_f are assumed to be zero. Negligible error usually results from such an approach, and Equation 9-5 reduces to:

$$N_b = \frac{N_p \left[B_o + B_g (R_p - R_s) \right]}{B_o - B_{oi} + B_g (R_{si} - R_s)} \tag{9-6}$$

The basic assumptions used in the material balance equation were discussed in the "Oil Reservoir Drive Mechanisms" chapter. The assumptions particular to Equation 9-6 are:

(1) Reservoir volume is constant.

(2) Reservoir temperature is constant.

(3) Pressure equilibrium exists at all times; i.e., no pressure gradients are found across the reservoir. Implications here are that fluid properties are fixed at any time. Saturation gradients are negligible.

(4) PVT data are applicable.

(5) There is no gas cap (initially or that develops later).

(6) Recovery is independent of rate.

(7) Production data are reliable.

(8) Initial conditions are those at the bubblepoint.

(9) There is negligible water encroachment or production.

Although some of these assumptions may be violated, Equation 9-6 is a powerful tool in the hands of an experienced petroleum engineer.

As it stands, Equation 9-6 may be used to calculate the oil in place at the bubblepoint. Material balance may also be used to predict future performance from a reservoir. Over the years there have been many prediction methods offered in the petroleum literature for solution gas drive reservoirs. Most are trial-and-error or iterative procedures. Two of these methods that are still being used today were developed by Tracy[2] and by Pirson.[5] Both are self-correcting (errors incurred in previous steps do not accumulate) and do converge to the correct solution.

TRACY MATERIAL BALANCE

Tracy[2] started with the Schilthuis form of the material balance equation:

$$N = \frac{N_p \left[U + (R_p - R_{si}) B_g \right] - (W_e - W_p B_w)}{(U - U_i) + m \dfrac{U_i}{B_{gi}} (B_g - B_{gi})} \tag{9-7}$$

where:

U = $B_o + (R_{si} - R_s)B_g$
U_i = B_{oi}
W_e = cumulative water influx from the aquifer into the reservoir, reservoir barrels
W_p = cumulative water produced, STB,
B_w = water formation volume factor, rb/STB,
B_g = gas formation volume factor, res. barrels/scf, and
m = ratio of the original hydrocarbon volume of the gas cap to the hydrocarbon volume of the original oil zone, dimensionless.

Note: There are terms that appear in this general equation that will not be present with a strict solution gas drive reservoir, such as: W_e, W_p, and m. It is preferable to start with this general equation, and then exclude terms as needed when applying the equation to a specific reservoir.

Substituting for U, U_i, and R_p:

$$N = \frac{N_p(B_o - R_s B_g) + G_p B_g - (W_e - W_p B_w)}{(B_o - B_{oi}) + (R_{si} - R_s)B_g + \frac{B_{oi}}{B_{gi}}(B_g - B_{gi})(m)}$$

An examination of this equation shows that cumulative oil production, N_p, is multiplied by a factor that is a function of pressure only. Cumulative gas production, G_p and net water influx, $(W_e - W_p B_w)$, are also multiplied by factors which are functions of pressure only. These "phi factors" are defined as:

$$\phi_o = \frac{(B_o - R_s B_g)}{(B_o - B_{oi}) + (R_{si} - R_s)B_g + m(B_{oi} / B_{gi})(B_g - B_{gi})} \quad (9\text{-}8)$$

$$\phi_g = \frac{B_g}{(B_o - B_{oi}) + (R_{si} - R_s)B_g + m(B_{oi} / B_{gi})(B_g - B_{gi})} \quad (9\text{-}9)$$

$$\phi_w = \frac{1}{(B_o - B_{oi}) + (R_{si} - R_s)B_g + m(B_{oi} / B_{gi})(B_g - B_{gi})} \quad (9\text{-}10)$$

Hence the Tracy material balance becomes:

$$N = N_p \phi_o + G_p \phi_g + (W_p B_w - W_e)\phi_w \quad (9\text{-}11)$$

Notice the units of the pressure factors:

ϕ_o = dimensionless;

ϕ_g = STB/scf (assuming G_p in scf);and ϕ_w = STB/res. barrel.

If there is no water production or influx:

$$N = N_p \phi_o + G_p \phi_g \quad (9\text{-}12)$$

Phi factors can be calculated at all desired pressures using data from a reservoir fluid analysis. Then a table or plot of these factors can be used to calculate oil in place or to predict future performance. Using phi factors makes material balance calculations simpler than with other forms of the material balance equation.

Phi factors are infinite at the bubblepoint and decline rapidly as pressure declines below the bubblepoint. Characteristic shapes of these pressure functions are shown in Figures 9-1 and 9-2. Their steep slope near the bubblepoint, plus the fact that pressure seldom is known precisely, graphically illustrate why it is futile to attempt to determine oil in place by material balance near the bubblepoint (initial pressure).

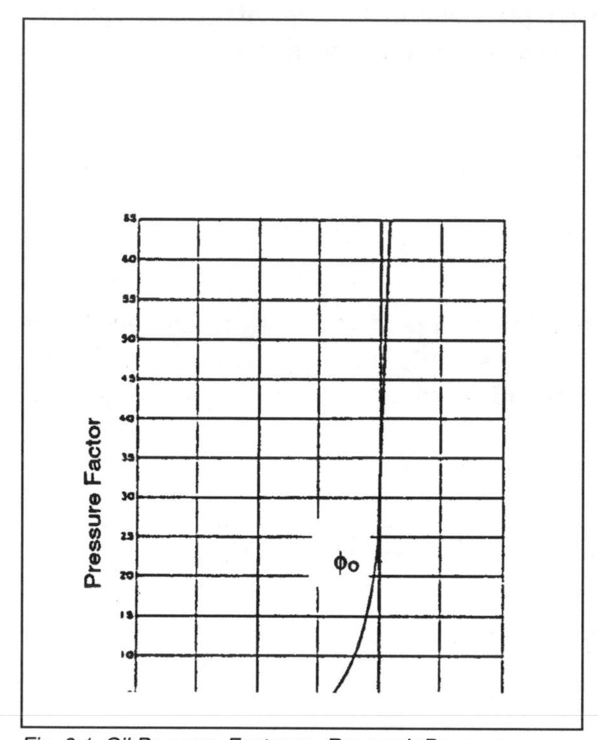

Fig. 9-1. Oil Pressure Factor vs. Reservoir Pressure.

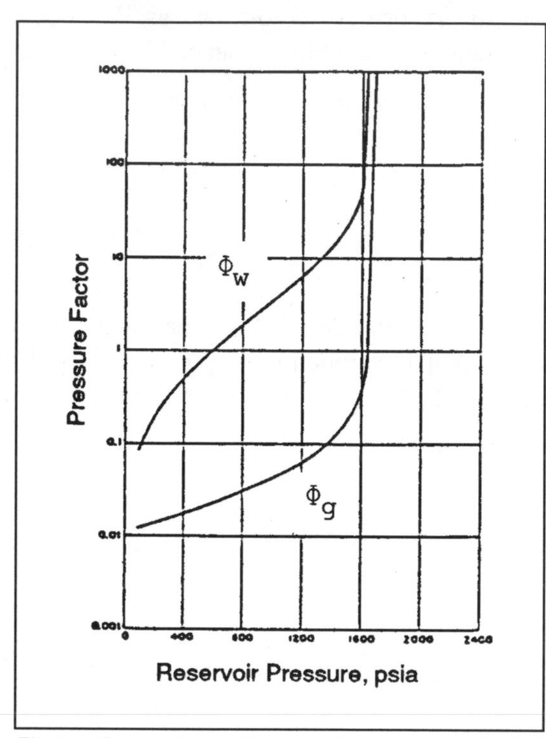

Fig. 9-2. Gas and Water Pressure Factors vs. Reservoir Pressure.

CALCULATING ORIGINAL OIL IN PLACE

Calculating original oil in place using Tracy's material balance can best be illustrated with an example. Phi factors were obtained from Figures 9-1 and 9-2.

Problem 9-1:

Production history for an originally saturated (discovered right at the bubblepoint) oil reservoir is as follows:

Table 9-1:

Pressure psia	ϕ_o	ϕ_g	Cumulative Oil Prod, STB	Cumulative Gas Prod, MCF
1690			0	0
1600	36.60	0.400	398,000	38,593
1500	14.30	0.179	1,570,000	155,797
1300	4.85	0.0813	3,175,000	424,043
1100	2.10	0.0508	4,470,000	803,028

Calculate oil in place with Equation 9-12:

$$N = N_p \phi_o + G_p \phi_g \tag{9-12}$$

$\underline{P = 1600:}$ $N = (398{,}000)(36.60) + (38{,}593{,}000)(0.400) = \underline{30.0 \times 10^6 \ STB}$

$\underline{P = 1500:}$ $N = (1{,}570{,}000)(14.30) + (155{,}797{,}000)(0.179) = \underline{50.3 \times 10^6 \ STB}$

$\underline{P = 1300:}$ $N = (3{,}175{,}000)(4.85) + (424{,}043{,}000)(0.0813) = \underline{49.9 \times 10^6 \ STB}$

$\underline{P = 1100:}$ $N = (4{,}470{,}000)(2.10) + (803{,}028{,}000)(0.0508) = \underline{50.2 \times 10^6 \ STB}$

All of the above calculations, except the first, indicate about 50 million barrels of original oil in place. The calculation at 1600 psia illustrates the sensitivity of oil in place calculations near bubblepoint pressure. If pressure had been 1620 psia instead of 1600 psia, oil in place would have calculated to be 50 million barrels. Reservoir pressure could easily be in error 20 psi or more.

Unexpected water influx into reservoirs thought to be solution gas drive will be indicated by calculated oil in place values that are increasing. This happens because the solution gas drive material balance Equation, 9-12 or 9-6, has neglected the water encroachment, W_e. If significant, this term should appear as a negative quantity in the numerator as in Equation 9-11 or 9-7.

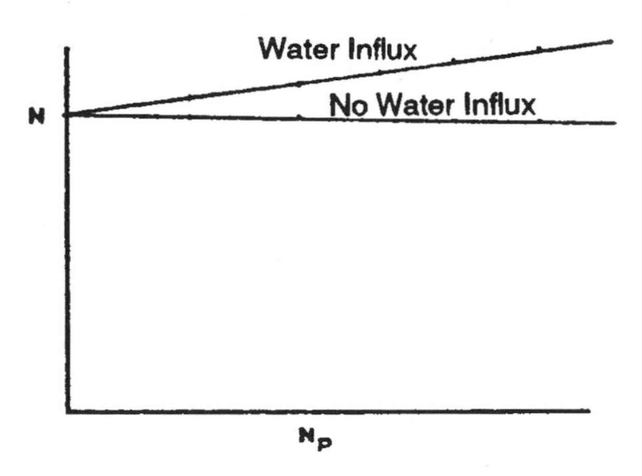

N calculated with Equation 9-6 or 9-12.

PREDICTING FUTURE PERFORMANCE

Above the bubblepoint, the Tracy model is not needed. Equation 9-4 is used down to p_b. The Tracy method is normally started at the bubblepoint pressure, but it may be begun below p_b.

To use the Tracy method for predicting future performance, it is necessary to choose the future pressures at which performance is desired. This normally means selecting the pressure step to be used. At each selected pressure; cumulative oil, cumulative gas, and producing GOR will be calculated. So, the goal is to determine a table of N_p, G_p, and R versus future reservoir static pressure such as:

n	Pressure	N_p	G_p	R
0	$p_o = p_i = p_b$	0	0	Rsb
1	p_1	N_{p1}	G_{p1}	R_1
2	p_2	N_{p2}	G_{p2}	R_2
3	p_3	N_{p3}	G_{p3}	R_3

The difference between successive pressures, the pressure decrement, is normally in the range of 25 to 300 psi, but should not be greater than 10% of the starting pressure. Notice that the calculated production figures do not include any production above the bubblepoint.

Calculating future performance of solution gas drive reservoirs involves the simultaneous solution of four equations:

1. Material balance equation:

$$N = N_p \phi_o + G_p \phi_g \tag{9-13}$$

2. Liquid saturation equation:

$$S_L = S_w + (1 - S_w)\left(\frac{N - N_p}{NB_{oi}}\right)(B_o) \tag{9-14}$$

This equation assumes: (1) the connate water saturation, S_w, is immobile and unchanging, (2) N is the oil in place at the bubblepoint, (3) N_p is cumulative production from the bubblepoint, and (4) $B_{oi} = B_{ob}$.

3. Gas/oil ratio equation:

$$R = R_s + \left(\frac{k_g}{k_o}\right)\left(\frac{\mu_o}{\mu_g}\right)\left(\frac{B_o}{B_g}\right) \tag{9-15}$$

4. Cumulative gas production:

$$G_{p,n} = G_{p,n-1} + \Delta N_p \left(\frac{R_n + R_{n-1}}{2} \right) \tag{9-16}$$

where:

n = index indicating value at the new pressure,
$n\text{-}1$ = index indicating value at the previous (higher) pressure, and
ΔN_p = incremental cumulative oil production between p_{n-1} and p_n, STB.

A material balance equation on a differential basis is needed to calculate incremental cumulative oil produced over a pressure decrement; i.e., from the previous pressure (p_{n-1}) to the new pressure (p_n). To start, we write difference equations for the gas and oil:

$$\Delta G_p = \Delta N_p \left(\frac{R_n + R_{n-1}}{2} \right) \tag{9-17}$$

$$N_{p,n} = N_{p,n-1} + \Delta N_p \tag{9-18}$$

$$G_{p,n} = G_{p,n-1} + \Delta G_p \tag{9-19}$$

where:

ΔG_p = incremental cumulative gas production between p_{n-1} and p_n, scf.

Now, these difference equations are substituted into the solution gas drive material balance equation (9-13) to obtain the material balance equation on a differential basis:

$$\Delta N_p = \frac{N - N_{p,n-1}\,\phi_o - G_{p,n-1}\,\phi_g}{\phi_o + \left(\frac{R_n + R_{n-1}}{2} \right) (\phi_g)} \tag{9-20}$$

This prediction equation relates the incremental oil produced over a pressure decrement to the cumulative produced oil and gas that existed at the beginning of the pressure step. It is also related to the average producing gas/oil ratio during the pressure step. Notice that the phi factors have no subscripts. These pressure functions, ϕ_o and ϕ_g, are evaluated at p_n, the pressure at the end of the pressure step. The procedure to calculate future performance is as follows:

Procedure:

1. Estimate R_n, the producing gas/oil ratio at the end of the pressure step (at p_n).

2. Calculate incremental oil production, ΔN_p, during the pressure decrement using the prediction Equation 9-20.

3. Calculate cumulative oil production to the end of the pressure step:

$$N_p = N_{p,n} = N_{p,n-1} + \Delta N_p$$

4. Calculate the average reservoir liquid saturation:

$$S_L = S_w + (1 - S_w)\left(\frac{N - N_p}{NB_{oi}}\right)(B_o)$$

5. Based on the average liquid saturation, S_L, determine k_g/k_o from the reservoir k_g/k_o relationship.

6. Calculate the instantaneous producing gas/oil ratio, R_n, at the end of the pressure step:

$$R_n = R_s + \left(\frac{k_g}{k_o}\right)\left(\frac{\mu_o}{\mu_g}\right)\left(\frac{B_o}{B_g}\right)$$

If the calculated R_n is the same as the estimated R_n (Step 1), then the estimate was correct. If the calculated R_n is within 10 percent of the estimated R_n, then use the calculated R_n and continue to step 7. If the calculated R_n differs from the estimated value by more than 10 percent, then use the calculated value, and go back to Step 2. (Since the trial-and-error solution converges, the calculated R_n will always be closer to the correct solution than the estimated one.)

7. Calculate the cumulative gas production to the end of the pressure decrement using incremental oil production and average GOR:

$$G_{p,n} = G_{p,n-1} + \Delta N_p\left(\frac{R_n + R_{n-1}}{2}\right)$$

8. Calculate the original oil in place (as a check):

$$N = N_p\,\phi_o + G_p\,\phi_g$$

This calculated N should be within 0.1% (with a computer) of the actual N.

$$\left|\frac{N_{actual} - (N)_{step\ 8}}{N_{actual}}\right| \leq 0.001$$

If the calculated N is within this tolerance, then convergence has been achieved. If not, then check for errors in the calculations, and go back to Step 2 using the last calculated R_n.

Prediction Based on a Unit Barrel:

For prediction purposes, the calculations can be made on the basis of a unit barrel; i.e., if both sides of Equation 9-12 are divided by N, then:

$$1 = N_p' \phi_o + G_p' \phi_g \tag{9-21}$$

where: $N_p' = N_p / N$ and $G_p' = G_p / N$

Notice that N_p' is actually fractional recovery efficiency below the bubblepoint.

The unit barrel formulation is convenient for cases where N is not accurately known or to test the effect of differing k_g/k_o relationships. If N is known, then when calculations are completed, N_p' and G_p' can be converted to actual barrels and scf simply by multiplying by N. Thus, the material balance calculations can be made independent of actual oil in place. The other equations become:

$$\Delta N_p' = \frac{1 - N_{p,n-1}' \phi_o - G_{p,n-1}' \phi_g}{\phi_o + \left(\frac{R_n + R_{n-1}}{2} \right) (\phi_g)} \tag{9-22}$$

$$N_{p,n}' = N_p' = N_{p,n-1}' + \Delta N_p' \tag{9-23}$$

$$S_L = S_w + (1 - S_w) \left(\frac{1 - N_p'}{B_{oi}} \right) B_o \tag{9-24}$$

$$R_n = R_s + \left(\frac{k_g}{k_o} \right) \left(\frac{\mu_o}{\mu_g} \right) \left(\frac{B_o}{B_g} \right) \tag{9-25}$$

$$G_{p,n}' = G_p' = G_{p,n-1}' + \Delta N_p' \left(\frac{R_n + R_{n-1}}{2} \right) \tag{9-26}$$

$$1 = N_p' \phi_o + G_p' \phi_g \tag{9-27}$$

For a particular pressure step, if the calculated result of Equation 9-27 is within the range 0.999 to 1.001, then convergence is achieved. Otherwise, another iteration is needed. (By hand, use 0.98 to 1.02).

Field-derived k_g/k_o Data

Reliable k_g/k_o data are possibly the most difficult of all reservoir data to obtain. Normally no k_g/k_o data are available for the reservoir being studied, and relative permeability data from a nearby or similar reservoir are often used. Laboratory k_g/k_o relationships are obtained on an altered sample of an infinitesimal fraction of the actual reservoir. Although the shape of the laboratory k_g/k_o data may be reasonable, its position relative to liquid or gas saturation is highly conjectural.

As described in the "Rock Properties" chapter, it is possible to use performance data, reservoir static pressure data, and PVT properties to generate a field-derived k_g/k_o relationship. Such data are helpful in positioning laboratory k_g/k_o curves with respect to the saturation axis. In the absence of laboratory data, the field-derived k_g/k_o points may be used to "calibrate" a k_g/k_o generalized correlation.

Two equations are needed: one to calculate k_g/k_o and the second to determine the corresponding reservoir saturation state.

$$k_g / k_o = (R - R_s)\left(\frac{\mu_g}{\mu_o}\right)\left(\frac{B_g}{B_o}\right) \tag{9-28}$$

This equation was obtained by rearranging Equation 9-15. The second equation, the solution gas drive liquid saturation equation, is:

$$S_L = S_w + (1 - S_w)\left(\frac{N - N_p}{NB_{oi}}\right)(B_o) \tag{9-29}$$

To use this equation properly, it should be remembered that initial conditions are those at the bubblepoint.

Converting Material Balance Prediction Results to Time

Material balance predictions of future performance so far have given cumulative production and gas/oil ratio performance as a function of reservoir pressure. If original oil in place and abandonment pressure are known, ultimate recovery can be calculated.

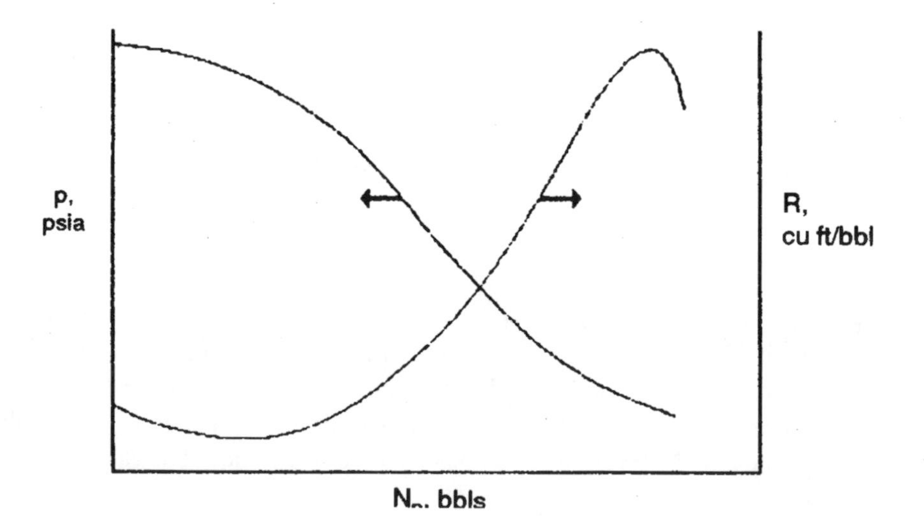

It is also important to be able to relate material balance prediction results to time as all economic evaluations are based on the time value of money. Since solution gas drive recovery is independent of time, sensitivity analyses can be made to justify well stimulation, infill drilling, etc.

To introduce time, the needed tool is a well equation relating static pressure and well pressure to flow rate. Most reservoir engineers seem to prefer Darcy's law in radial coordinates or the productivity index formulation. However, any other well equation such as the Vogel[6] or the Fetkovich[7] relationships (discussed in the "Fluid Flow" chapter) could be used.

The radial Darcy equation is:

$$q_o = 7.08 \frac{k_o \, h \, (p_e - p_w)}{\mu_o \, B_o \, \ln(r_e / r_w)} \tag{9-30}$$

and the productivity index equation is:

$$q_o = J \Delta p \tag{9-31}$$

where: J = productivity index, STBOPD/psi

Assuming that Equations 9-30 and 9-31 are equivalent, then

$$J = \frac{7.08 \, k_o \, h}{\mu_o \, B_o \, \ln(r_e / r_w)} \tag{9-32}$$

Normally, J is not calculated due to the uncertainties in the required data. To obtain J, a well is usually tested. Notice in Equation 9-32 that with production time, three things affecting productivity index are likely to change: B_o, μ_o and k_o. Hence, an equation that allows productivity index to change with time is:

$$J = (J_i)\,(J_{rel}) \tag{9-33}$$

where:

$$J_{rel} = \left[\frac{k_{ro}}{\mu_o B_o}\right] / \left[\frac{k_{ro}}{\mu_o B_o}\right]_i \tag{9-34}$$

and,

J_i = initial productivity index, and
J_{rel} = the relative productivity index, dimensionless.

Then, the well rate is: $q_o = (J)(\Delta p)$.

Notice that μ_o and B_o are functions of reservoir static pressure, but k_{ro} is usually considered to be a function of oil saturation. Recall that one of the calculations performed with a Tracy material balance model is liquid saturation at each new pressure. This can be related to oil saturation and k_{ro} determined.

Another of the results of the Tracy solution gas drive prediction method is ΔN_p or incremental cumulative oil production between pressures p_{n-1} and p_n. At pressure p_{n-1}, it is possible to calculate the total reservoir oil rate as:

$$Q_{field,n-1} = (\,no.\ wells\,)\ (J_i)\ (J_{rel,n-1})\ (p_{n-1} - p_w)$$

where:

J_i = the initial average single well productivity index, STOPD/psi,
p_w = the well bottomhole producing pressure, psia,

At pressure p_n:

$$Q_{field,\,n} = (\,no.\ wells\,)\ (J_i)\ (J_{rel,\,n})\ (p_n - p_w) \tag{9-36}$$

Thus, the average oil rate between p_{n-1} and p_n is:

$$Q_{field,\,avg} = (\,Q_{field,\,n-1} + Q_{field,\,n}\,)\,/\,2 \tag{9-37}$$

Then, the incremental time between p_{n-1} and p_n is:

$$\Delta t = \Delta N_p\ /\ Q_{field,\,avg} \tag{9-38}$$

Thus, the total time from p_o to p_n is:

$$t = \Sigma\,\Delta\,t \tag{9-39}$$

Naturally, if the calculated well rates exceed the statutory limits, then the statutory limiting rate should be used in the time calculations.

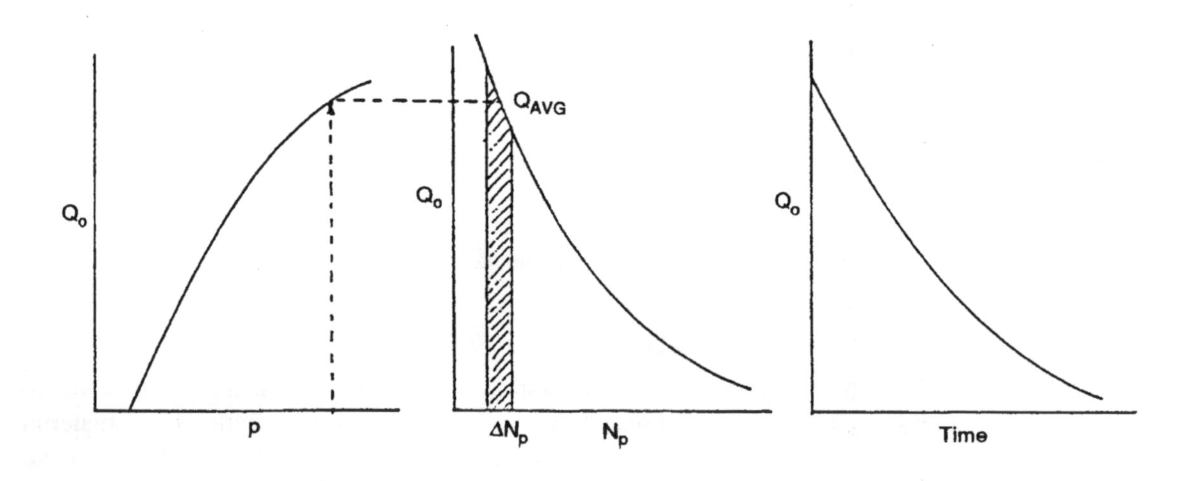

REFERENCES

1. Frick, T. C.: *Petroleum Production Handbook*, McGraw-Hill Book Co., Inc., New York City (1962) **Vol. II**, Ch. 25, 32, and 34.
2. Tracy, G. W.: Simplified Form of the Material Balance Equation, *Trans.*, AIME (1955) **204**, 243.
3. Craft, B. C., and Hawkins, M. F.: *Applied Reservoir Engineering*, Prentice-Hall, Inc. (1959) Ch. 3, 4.
4. Schilthuis, R. J.: "Active Oil and Reservoir Energy," *Trans.*, AIME (1936) **118**, 33.
5. Pirson, S. J.: *Oil Reservoir Engineering*, McGraw-Hill Book Co., Inc., New York City (1958) Ch. 10.
6. Vogel, J. V.: "Inflow Performance Relationship for Solution Gas Drive Wells," *JPT* (January 1968) 83-93.
7. Fetkovich, M. J.: "The Isochronal Testing of Oil Wells," SPE paper 4529 presented at the 1973 Annual Meeting, Las Vegas (September 30 - October 3).

10 MULTIZONE RESERVOIR PERFORMANCE

INTRODUCTION

When using classical material balance to describe producing behavior of an oil or gas reservoir, one of the basic assumptions made is that the reservoir consists of a single homogeneous "tank." Many reservoirs have been successfully analyzed as one zone. However, if the hydrocarbon trap is actually comprised of several zones or layers, classical material balance formulations may not closely predict actual performance. It is the intent of this chapter to describe techniques that can be used to analyze reservoirs made up of multiple producing zones. Both depletion drive oil reservoirs and gas reservoirs will be considered.

MULTIZONED SOLUTION GAS DRIVE OIL RESERVOIR

A. Analysis Method

To Start

To study performance from a multizoned SGD reservoir, a typical well producing from this reservoir is considered. Although more than two zones can be used, it is easier if the trap can be divided into two zones. This is the analysis that is presented here. Darcy's law in radial form is used to describe flow from each zone into the wellbore. With time, the higher permeability zone develops a lower average pressure (than exists in the tighter zone) because of its higher capacity per foot of thickness to flow fluids to the wellbore. Flow between zones is allowed and calculated with Darcy's law in linear form. This interzonal flow is controlled mainly by the vertical permeability and saturation conditions existing in the tighter zone.

Using the techniques of the "Solution Gas Drive" chapter, oil and gas recovery versus reservoir pressure is calculated. Each zone is considered separately, as if it were producing alone. Hence, a recovery versus pressure relationship such as that shown in Figure 10-1 is generated for each zone.

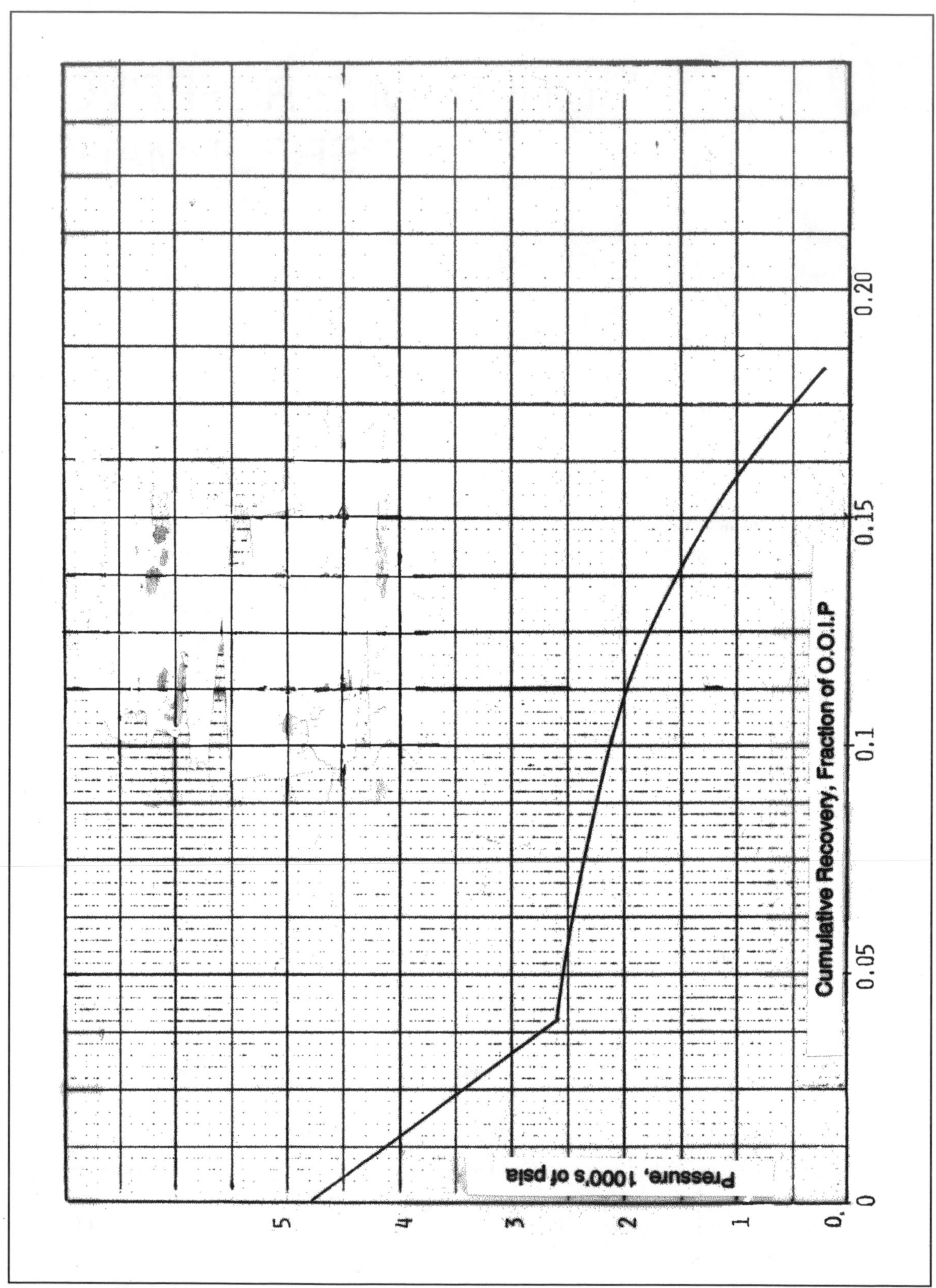

Fig. 10-1. Expected recovery vs. reservoir pressure (single producing zone).

The calculation procedure is best performed with a computer because conversion to time must be performed using relatively small time steps: probably a maximum of 10 days. This is necessary to maintain the stability of the iterative calculations that are involved for each time step.

No attempt will be made here to cover the details of any of the "outflow performance" models. That is, Darcy's law is used to model flow of fluids into the wellbore, but the calculation is really not complete at this point. Some method is needed to determine whether the fluids that have mathematically flowed into the wellbore can be flowed out of the wellbore with the existing well bottomhole pressure. Other sources[2,3] may be consulted for wellbore multiphase flow pressure-drop calculation techniques. Unless artificial lift is installed, a well's initial constant rate will have to begin declining after the bottomhole pressure has fallen to a certain point. This situation can be handled somewhat simplistically by letting the well flow at a constant rate until the bottomhole pressure falls to a given value, and then holding the well pressure constant from this point on. Since the average zone pressures are declining, the Darcy-predicted well rates will also decline.

Conversion to Time

For each time step, the following iterative procedure is used.

(1) Beginning with the average zone pressures existing at the end of the last time step, calculate the interzonal flow (linear) from the higher pressure zone to the lower pressure zone.

(2) Using Darcy's radial law, calculate the well pressure. Initially, use the well rate that existed for the last time step. (For the first time step, use the maximum sales rate.)

To keep this model simple, it is necessary to assume that the wells continue to produce until abandonment conditions are reached; i.e., there are no shut-ins that last for an excessive amount of time. Then, the interzonal flow can be assumed to occur from the higher pressure zone to the lower pressure zone without altering the saturations in the lower pressure zone. This assumption is justified in that the average pressure in both zones decreases continually throughout the entire producing history. To simplify the calculations, the interzonal flow is allocated directly to the wellbore. Thus, the total well rate is:

$$(q_o)_{well} = (q_o)_{rad,1} + (q_o)_{rad,2} + (q_o)_{inter} \tag{10-1}$$

or

$$(q_o)_{well} = (q_o)_{tot\ rad} + (q_o)_{inter}$$

where:

$(q_o)_{well}$ = oil production rate from the well, STB/D
$(q_o)_{rad,1}$ = oil flow rate from zone 1 into wellbore, STB/D
$(q_o)_{rad,2}$ = oil flow rate from zone 2 into wellbore, STB/D
$(q_o)_{tot\ rad}$ = $(q_o)_{rad,1} + (q_o)_{rad,2}$
$(q_o)_{inter}$ = linear interzonal flow rate, STB/D

Assuming two radial zones flowing in parallel with a common wellbore pressure, p_w, and since the individual zone rates are additive, we can use Darcy's law and solve for p_w:

$$p_w = \frac{c_1 \bar{p}_1 + c_2 \bar{p}_2 - (q_o)_{tot\ rad}}{c_1 + c_2} \tag{10-2}$$

where:

$$c_1 = \frac{7.08\ k_{o1}\ h_1}{\mu_{o_1} B_{o_1} \left[\ln \frac{r_e}{r_w} \right]} = J_1$$

$$c_2 = \frac{7.08\ k_{o2}\ h_2}{\mu_{o_2} B_{o_2} \left[\ln \frac{r_e}{r_w} \right]} = J_2$$

$\bar{p}_1$ = average pressure in zone 1, psia

$\bar{p}_2$ = average pressure in zone 2, psia

The indicated permeabilities are effective oil permeabilities in darcies.

If the wells were tested at the initial conditions, then, it would be better to use the productivity index formulation, rather than Darcy's law. At the initial conditions; i.e., when $p_i = \bar{p}_1 = \bar{p}_2$, then $J_i = (J_1 + J_2)_i = J_{i,1} + J_{i,2}$. Knowing the initial total well P.I., the initial zonal P.I.'s can be calculated. Also, they can be allowed to change with time as discussed in the "Solution Gas Drive" chapter. For example, the zone 1 equations are:

$$J_{rel,1} = \left[\frac{k_{ro}}{\mu_o B_o} \right]_1 / \left[\frac{k_{ro}}{\mu_o B_o} \right]_{i,1} \tag{10-3}$$

$$J_1 = (J_{i,1})(J_{rel,1}) \tag{10-4}$$

If the calculated wellbore pressure, p_w, is less than the minimum wellbore pressure (discussed earlier), then for this time step, and all future time steps, the wellbore pressure is set equal to the minimum value.

(3) Using the p_w from Step 2 and Darcy's radial law (or the productivity index formulation), calculate the individual zone producing rates. For example, the zone 1 rate is:

$$(q_o)_{rad,1} = \left[\frac{7.08\ k_{o1}\ h_1}{\mu_{o1} B_{o1} \left(\ln \frac{r_e}{r_w} \right)} \right] (\bar{p}_1 - p_w) = (J_1)(\bar{p}_1 - p_w) \tag{10-5}$$

(4) Calculate the incremental cumulative production (N_p) from each zone for the time step. Recall that for the higher pressured zone, this involves the zone (radial, Step 3) producing rate plus the interzonal rate.

(5) Evaluate total cumulative recovery from each zone. From the individual zone recovery versus pressure relationships (generated before the conversion to time computations began), determine the likely zone average pressures for the end of the time step.

(6) The interzonal flow rate is recalculated for the time step using the zone pressures from Step 5. This interzonal flow rate should be an average rate for the time step. Thus, the pressure difference (between Zone 1 and Zone 2) used in the linear Darcy equation should be an average of the pressure difference existing at the beginning of the time step and that existing at the end of the time step. If the new interzonal flow rate is more than 10% different than the one calculated in Step 1, then repeat Steps 2 through 6 using the new interzonal rate and average zone pressures (for the time step). If small time steps are used (maximum of 10 days), then these calculations should remain stable. Normally a maximum of two iterations is needed.

(7) Upon convergence, proceed to the next time step. This process is continued until economic depletion occurs.

An abandonment pressure chosen before a reservoir study such as this is likely to be erroneous. By considering calculated zone pressures and fluid production rate with time, an economic forecast can be determined. Abandonment conditions (pressure and time) occur at the point when production revenue drops to the level of production costs.

B. Example

Problem 10-1

To illustrate the results that can be obtained with the model just described, an example will be considered. Although the calculations were done with a computer, the input data and results will be given and discussed.

This reservoir consisted of two horizontal zones: a thin, high-permeability zone, and a thick, tight one. Rock and fluid properties follow in Table 10-1.

Table 10-1

Basic Data:

Absolute permeability, radial (Zone 1)	100	md
(Zone 2)	1	md
Porosity, fraction	0.15	
Connate-water saturation, fraction	0.25	
Pay thickness (Zone 1)	5	ft
(Zone 2)	50	f
Reservoir temperature	140	°F
Initial reservoir pressure	4800	psia
Bubblepoint pressure	2600	psia
Undersaturated oil compressibility	12.0×10^{-6}	psi-1
Water compressibility	3.0×10^{-6}	psi-1
Formation (pore volume) compressibility	4.0×10^{-6}	psi-1
Well drainage area	40	Ac
Sales-controlled maximum oil rate	200	STB/D
Minimum well pressure	200	psia
Reservoir abandonment pressure	200	psia
Wellbore radius	0.25	ft

Relative Permeability Data:

Gas Saturation, % Pore Space	k_g/k_o Ratio	k_{ro} Frac.
0	0	1.00
5	0.027	0.75
10	0.110	0.46
15	0.330	0.24
20	0.62	0.125
25	1.10	0.067
30	2.3	0.033
35	8.0	0.017
40	20.0	0.0131
45	100.0	0.0120

Fluid Properties:

Pressure Psia	Oil Formation Vol. Factor B/STB	Solution Gas/Oil Ratio Mscf/STB	Gas Deviation Factor Fraction	Oil Viscosity cp	Gas Viscosity cp
4800	1.623	1.0	0.85	0.50	0.029
2600	1.65	1.0	0.90	0.30	0.021
2000	1.52	0.76	0.915	0.35	0.0189
1500	1.395	0.57	0.933	0.40	0.0172
1000	1.28	0.38	0.954	0.47	0.0155
700	1.21	0.27	0.967	0.53	0.0144
300	1.118	0.12	0.986	0.64	0.0133
15	1.05	0.0	1.0	0.80	0.0120

Original Oil in Place (per well):

Zone 1	107,571 STB
Zone 2	1,075,707 STB
Total Reservoir	1,183,278 STB

Problem 10-1 Results and Discussion

The first step is to use the techniques of the "Solution Gas Drive" chapter to determine oil recovery versus zone pressure. This was performed for each of the two zones. Recall that each zone is considered as if it were producing alone. Considering zone recovery as a fraction of the OOIP (original oil in place) within that zone, the results for both zones are presented in Figure 10-1. If the abandonment pressure is 200 psia, then the recovery should be 18.06% of the original oil in place. And this is true for both zones. However, in a real reservoir, it is quite possible that the two zones could have different average pressures at abandonment conditions.

If ultimate production, or reserves, is assumed to be the amount of oil produced with abandonment pressure in both zones equal to 200 psia, then:

$$\text{Ultimate production} = (0.1806)(1{,}183{,}278) = 213{,}700 \text{ STB}$$

If the interzonal permeability between the two zones is set at 0.001 md, then Figure 10-2 gives the zone pressures versus percent recovery of the ultimate production. Notice that the pressure in the high permeability (100 md) layer decreases fairly rapidly at the start of production. Early production is primarily from this thin, permeable zone. However, as production proceeds, fluid movement (radial to the well plus interzonal flow) out of the tighter zone increases, and flow out of the more permeable zone decreases. For most of the life of the well, pressure declines at about the same rate in the two zones. Near the end, production is almost entirely from the tighter zone as the pressure difference between the permeable zone and the well has become insignificant.

The flow of fluids from the tighter, higher-pressured zone to the more permeable, lower-pressured zone is controlled by the interzonal permeability. In most instances, this parameter relates mainly to the vertical permeability in the lower permeability zone. Interzonal permeability is normally substantially lower than the radial or horizontal permeability. For this example, with an interzonal permeability of 0.001 md, the individual zone pressures are never the same except at the start and at abandonment.

Figure 10-3 contains the zone pressures versus percent recovery of the ultimate production for an interzonal permeability of 0.01 md. Once again, the early separation of the two curves occurs, but much less so than in Figure 10-2. In fact, the relationship is only shown out to 15% of recovery (of ultimate production) because after about 20% recovery, the individual zone pressures are only different by a few psi.

For cases with larger interzonal permeabilities than 0.01 md, the early divergence between zone pressures was even less, and less production was required before the zone pressures became essentially the same. For the situation where the interzonal permeability is the same magnitude as the tight zone radial permeability, the two zones deplete as one zone.

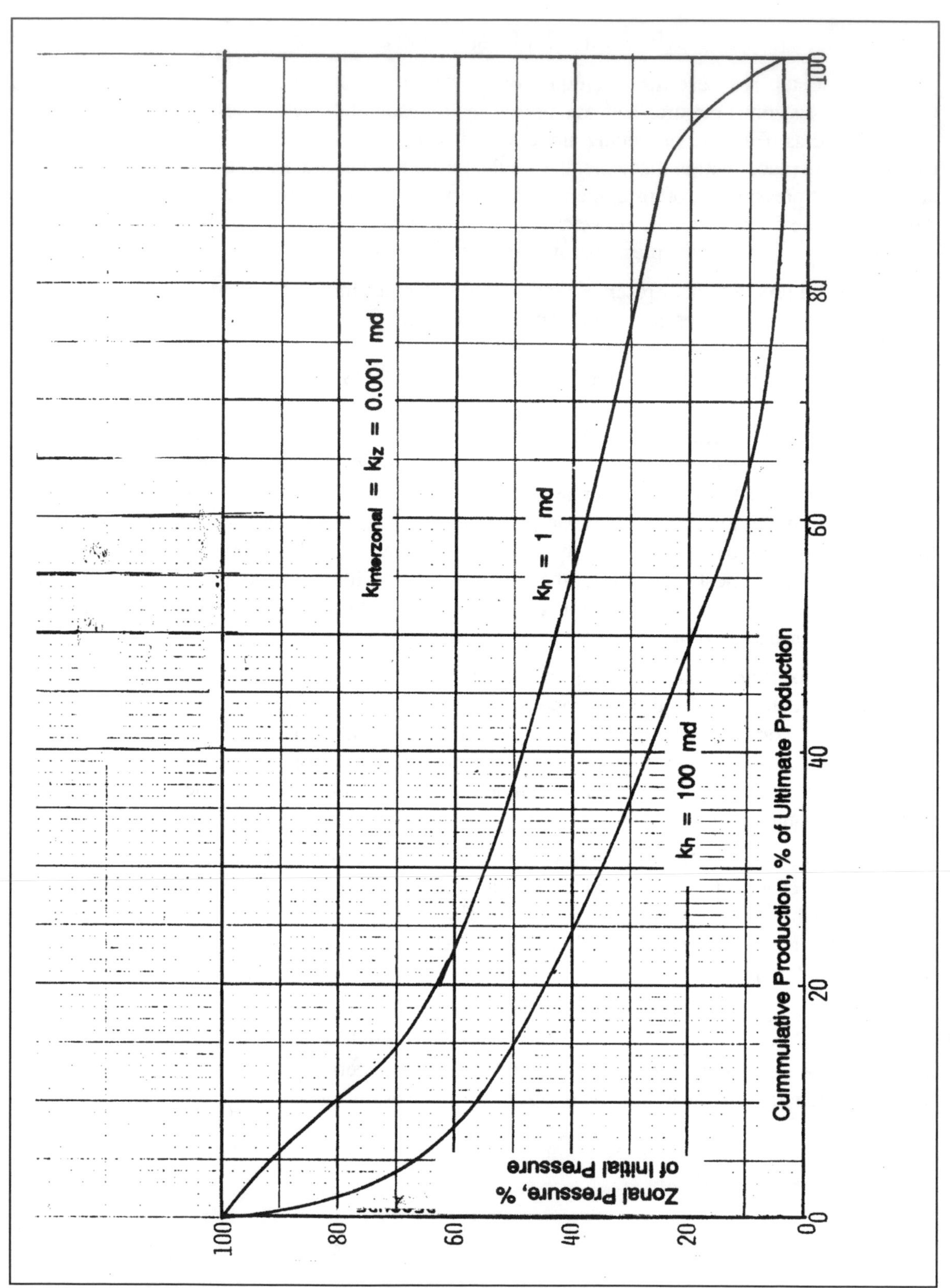

Figure 10-2. Zonal pressures vs. cumulative production.

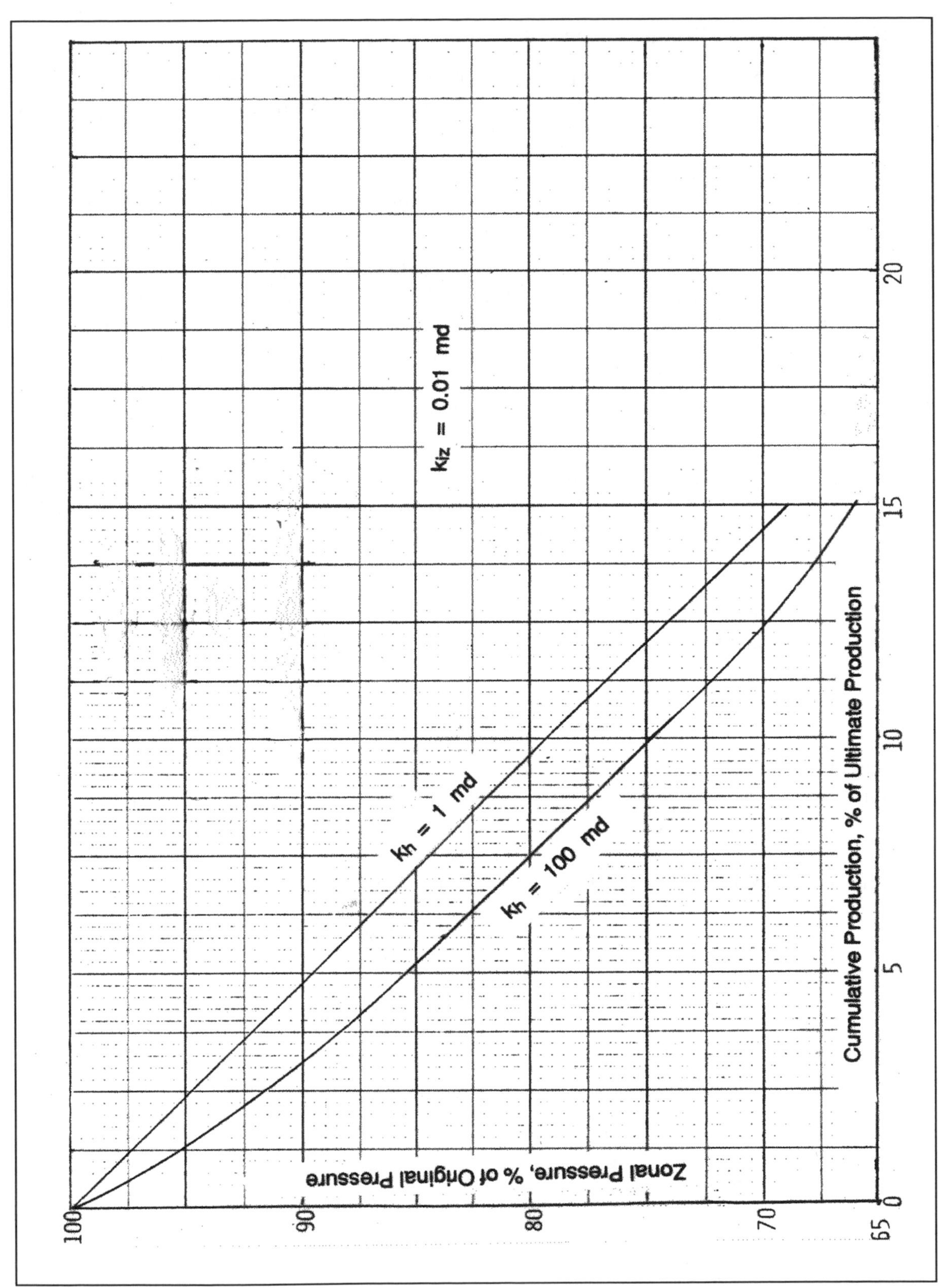

Fig. 10-3. Zonal pressures vs. cumulative production.

C. Effect of Multizone Production on Calculation of OOIP

The second use of material balance in a given reservoir is to predict future performance. As we have seen, prediction of future recovery is more complicated in a reservoir with multiple zones.

The first use of material balance in an oil reservoir is to calculate the OOIP. For "tank-type" reservoirs, the computational procedures were discussed in the "Oil Reservoir Drive Mechanisms" chapter. As with performance calculations, a multizone reservoir has additional problems in the material balance determination of OOIP.

Average reservoir pressures frequently are predicted using buildup tests. Such procedures are treated in the "Well Testing" chapter. Quite often, buildups are not designed to last long enough. In this instance, with a multizone reservoir, the average drainage area pressure determined from the test is not truly a "static" pressure, but actually closer to the average pressure of the lower pressure zone. If the well had been shut-in long enough for pressure equalization between the different zones to begin, then extrapolation of the shut-in pressures to a higher, more accurate value might have been possible.

If material balance calculations are made using an average pressure that is too low, then the predicted OOIP is also too low. Assuming a two-zone reservoir, Figure 10-4 can be used to determine what percent of reserves must be produced before material balance (using the average pressure in the more permeable zone) will yield reasonable estimates of OOIP. Four different permeability ratios (tight-zone-horizontal / interzonal) have been considered. Notice that with a permeability ratio of one, only 0.5% of the reserves need to be produced before an accurate OOIP can be predicted. This is true because a high interzonal permeability never allows the pressures within the two zones to become much different. On the other hand, if the permeability ratio is equal to 100, then 20% of the reserves must be produced. If the permeability ratio is 1000, then material balance will never yield a good estimate of OOIP (until abandonment is reached).

Actually, the results of Figure 10-4 were determined using the data of the earlier Problem 10-1 and considering different interzonal permeabilities. So, these results may not be directly applicable to other multizone reservoirs. However, at least the two zone effect on the calculation of OOIP has been illustrated. And using the described model, the amount of production needed to determine accurately the OOIP in a particular reservoir can be estimated.

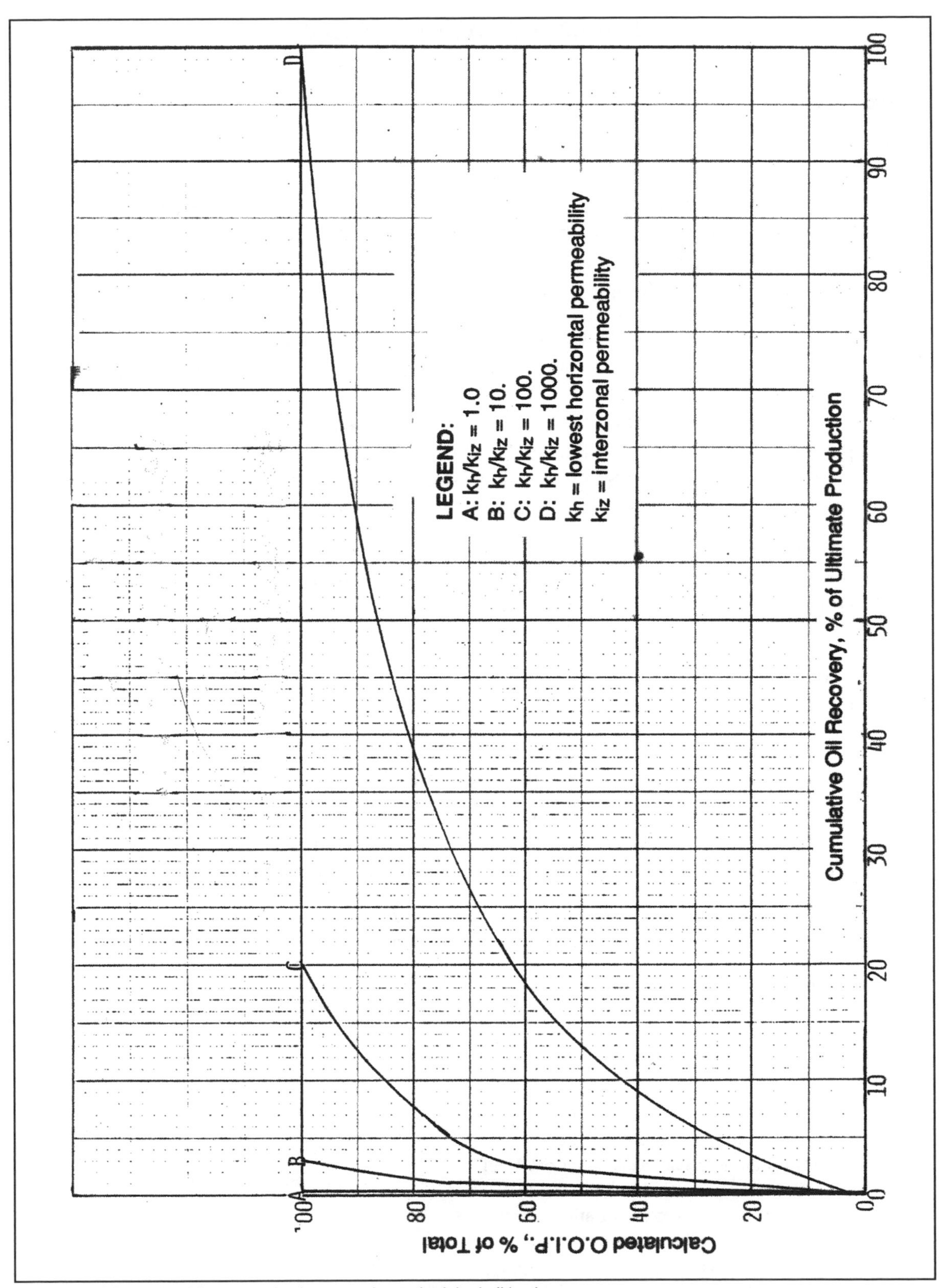

Fig. 10-4. Effect of multizone production on estimated original oil in place.

MULTIZONE GAS RESERVOIRS

A. Analysis Method

Introduction

The analysis of gas reservoir performance normally is handled by correlating (p/z) as a linear function of cumulative gas production. This treatment is quite adequate when the reservoir exists as a uniform zone and the hydrocarbon pore volume does not change significantly. Unfortunately, when the reservoir consists of more than a single zone and the interzonal permeability is limited, then this analysis is insufficient.

There are two major ways that hydrocarbon pore volume can decrease with cumulative production: (1) water encroachment, and (2) high formation (pore volume) compressibility. Water encroachment is considered in the "Water Drive" chapter. Particularly in overpressured reservoirs associated with unconsolidated formations and shale, pore volume compressibility can reach levels that are 10 times that of typical consolidated formations. If this is the case, then the effects of pore volume decrease and connate-water expansion should be considered in gas in place (OGIP) and performance calculations.

It is the purpose of this section to discuss techniques for considering multiple zones and high formation compressibility (say greater than 20×10^{-6} psi^{-1}) in material balance performance and OGIP calculations. The basic methods for handling multiple zones were illustrated earlier for a solution gas drive oil reservoir.

For a multizone reservoir, performance predictions begin with material balance calculations on each individual zone as if that zone were producing by itself. The purpose here is to determine a pressure versus cumulative-gas-produced relationship.

Material Balance With High Formation Compressibility

As mentioned earlier, the material balance formulation to be introduced here is especially needed in reservoirs that are overpressured and unconsolidated. This analysis is appropriate for a single zone or "tank." The material balance equation will be written as:

(original hydrocarbon volume) = (remaining hydrocarbon volume t some lower pressure) + (hydrocarbon pore volume decrease)

Expressing this relationship in mathematical terms:

$$G\,B_{gi} = (G - G_p)\,B_g + \frac{G\,B_{gi}}{1 - S_w}\left[(S_w\,c_w + c_f)\,\Delta p\right] \tag{10-6}$$

where:

G = initial gas in place, Mscf
G_p = cumulative gas production, Mscf
B_g = gas formation volume factor, B/Mscf, ($B_g = 5.034\ T\ z/p$),
S_w = connate water saturation, fraction
c_w = water compressibility, psi^{-1}, and
c_f = pore volume compressibility, psi^{-1}, and
Δp = pressure drop from initial pressure, or (p_i - p), psi.

Rearranging Equation 10-6:

$$G\left[1-\left(\frac{B_{gi}}{B_g}\right)(1-c_e\Delta p)\right]=G_p \tag{10-7}$$

where:

$$c_e=\frac{S_w c_w + c_f}{(1-S_w)} \tag{10-8}$$

Since $B_{gi}/B_g = (p/z)(z_i/p_i)$, Equation 10-7 can be written as:

$$G\left[1-\left(\frac{z_i}{p_i}\right)\left(\frac{p}{z}\right)(1-c_e\,\Delta p)\right]=G_p \tag{10-9}$$

It should be apparent from Equation 10-9 that if c_e is quite small, then the classical linear relationship between (p/z) and G_p results. However, if c_e is not insignificant, then (p/z) decreases nonlinearly with G_p. If a linear relationship is desired (especially for extrapolation purposes), Equation 10-9 suggests that (p/z) x $[1-(c_e)(\Delta p)]$ should be plotted versus G_p. These effects are illustrated in Figure 10-5. Notice that the (p/z) plot is nonlinear and curving downward; while the plot involving c_e is linear.

Multizone Performance Model

As with a SGD oil reservoir, the model here involves the drainage volume of a typical well that penetrates the different layers. For each zone, Equation 10-9 is used to develop a (p/z) x $[1 - (c_e)(p_i - p)]$ versus G_p relationship. From this, determine pressure versus recovery as illustrated in Figure 10-6. Darcy's radial equation (for a gas) is used to represent flow from each zone into the wellbore. Interzonal flow (from one layer to another) is calculated with the linear form of Darcy's law:

$$q_{iz}=\frac{1.127\;k_{iz}\;A\;(\bar{p}_1-\bar{p}_2)}{\mu_1\;L_{iz}\;B_g} \tag{10-10}$$

where:

q_{iz} = interzonal flow rate (Zone 1 to Zone 2), Mscf/D
k_{iz} = interzonal permeability, darcies
A = well's drainage area, ft^2
$\bar{p}_1$ = average pressure in Zone 1 (the tight zone), psia
$\bar{p}_2$ = average pressure in Zone 2 (more permeable zone), psia
μ_1 = gas viscosity (at $\bar{p}_1$), cp
L_{iz} = interzonal flow path length, ft, and
B_g = gas formation volume factor, $(5.034\,T\,z\,/\,\bar{p})$, B/Mscf

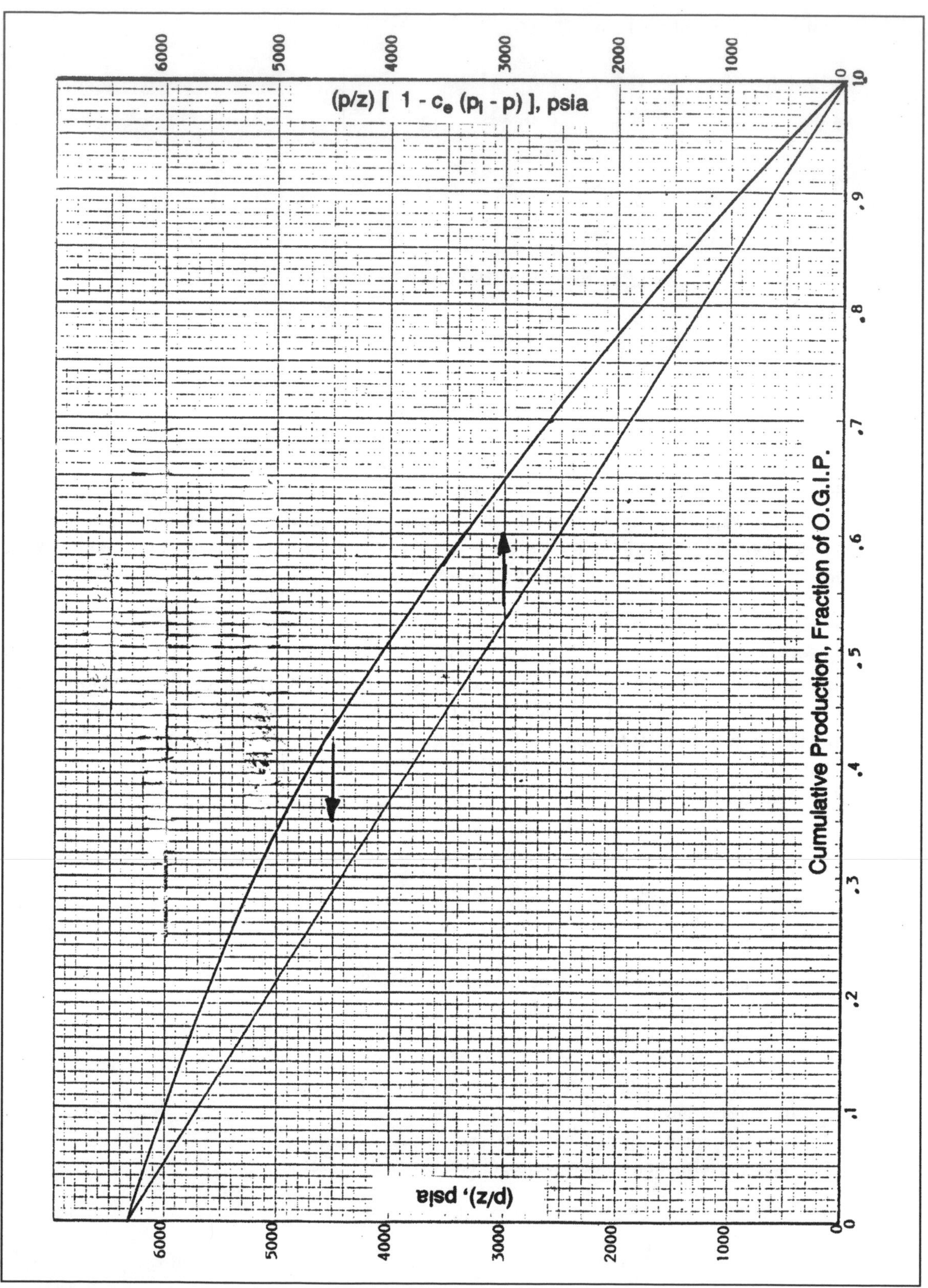

Fig. 10-5. Classical analysis of gas reservoirs including effect of formation and water expansion.

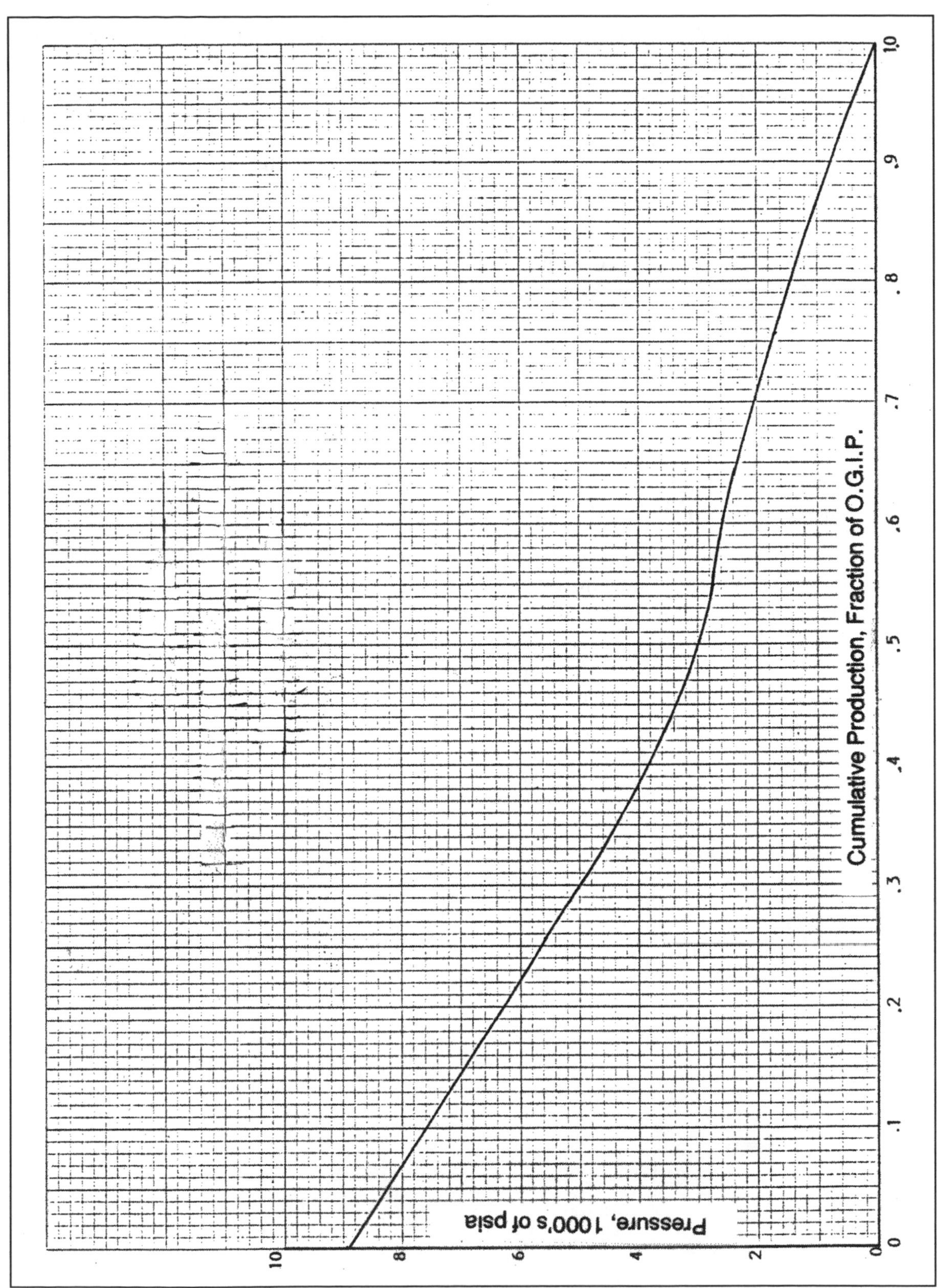

Fig. 10-6. Pressure vs. cumulative gas production for a single zone.

Assuming no thin permeability reduction layer existing between the two zones, then there are two common approaches for determining the values of k_{iz} and L_{iz} to be used in the interzonal flow equation.

(1) Assume the flow path to be from the center of the tight zone to a point barely within the permeable zone. In this case the flow length is simply one half the thickness of the tight zone, and k_{iz} is the vertical permeability of the tight zone.

(2) The flow path is assumed to be from the center of tight zone to the center of the permeable zone. So, L_{iz} is one-half of the combined thickness of both zones. Then, k_{iz} must be a linear flow, series average value ("Fluid Flow" chapter), or:

$$k_{iz} = \frac{(h_1 + h_2)/2}{\dfrac{(h_1/2)}{k_{v1}} + \dfrac{(h_2/2)}{k_{v2}}} = \frac{L_{iz}}{\dfrac{(h_1/2)}{k_{v1}} + \dfrac{(h_2/2)}{k_{v2}}} \tag{10-11}$$

In the example problems of this chapter, method 2 was used. However, there is usually not much difference in the (k_{iz}/L_{iz}) calculated by the two methods.

The conversion to time process is essentially the same as that presented for the multizone SGD oil reservoir. The hydrocarbon fluid is a gas this time, so the equation to determine well pressure, p_w, and the equations to determine zonal radial flow rates must be altered slightly. Again, to achieve a stable solution, the iterative calculation process required for each time step causes a time step size restriction. For the gas example problem to be presented shortly, the initial time steps were only one day, while the ending time steps (also the maximum) were 100 days.

B. Example

Problem 10-2

The following data were used to study a two-layer gas reservoir.

Table 10-2

Basic Data:

Absolute permeability, radial (Zone 1)	1	md
(Zone 2)	100	md
Pay thickness (Zone 1)	50	Ft
(Zone 2)	5	Ft
Porosity, fraction	0.20	
Connate-water saturation, fraction	0.30	
Reservoir temperature	210	°F
Formation depth	12,500	Ft
Initial reservoir pressure	8900	psia
Water compressibility	3.5×10^{-6}	psi-1
Formation (pore volume) compressibility	25.0×10^{-6}	psi-1
Approximate gas molecular weight	24.91	lb/mole
Total reservoir drainage area	2456.5	Ac
Number of wells	4	Producers
Maximum producing rate per well	10.0	MMscf/D
Minimum well pressure	200	psia
Reservoir abandonment pressure	200	psia
Wellbore radius	0.25	Ft

Relative Permeability (Imbibition) Data:

Water Saturation Percent	Relative Permeability, k_{rg}, Fraction
30	1.00
35	0.76
40	0.55
45	0.35
50	0.19
55	0.07
60	0.00

Fluid Properties

Pressure Psia	Gas Deviation Factor Dimensionless	Gas Viscosity cp
8900	1.395	0.0456
8500	1.370	0.0435
8000	1.305	0.0426
7000	1.208	0.0376
6000	1.080	0.0366
5000	0.965	0.0333
4000	0.848	0.0297
3000	0.791	0.0239
2000	0.833	0.0178
1000	0.877	0.0145
15	1.000	0.0127

Original Gas in Place (Total reservoir):

Using Equation (4-2) of the "Reservoir Volumetrics" chapter, the volumetric estimate of the original gas in place is:

Zone 1	252.3 Bscf
Zone 2	25.2 Bscf
Total Reservoir	277.5 Bscf

Problem 10-2 Results and Discussion

The first step is to use material balance techniques discussed earlier on each of the zones separately. Because the original reservoir pressure was significantly higher than the depth times the area water gradient (0.465 psi/ft), this reservoir is considered to be overpressured. Calculating the effective compressibility (Equation 10-8):

$$c_e = \frac{S_w\ c_w + c_f}{(1 - S_w)} = 37.2 \times 10^{-6}\ psi^{-1}$$

It can be seen that c_e is not insignificant; in fact, this is quite high. Thus, the material balance relationship used was equation 10-9. For each zone, a plot of $(p/z)[1 - c_e(p_i - p)]$ versus recovery was determined (Figure 10-5). This was converted to pressure versus recovery as shown in Figure 10-6.

To consider total reservoir performance, a typical well draining both zones was considered. Individual zone pressures versus recovery are given in Figure 10-7 for the case of interzonal permeability equal to 0.0001 md. Performance of a multizoned gas reservoir is quite similar to that of a multizoned oil reservoir. The early time zonal production rates are strongly a function of the zonal (k)(h) products. How much hydrocarbon is stored in a particular zone is directly related to the (ϕ))(h) product. In this problem, Zone 1 had 10 times the amount of gas initially stored there as did Zone 2. However, Zone 2 had an initial productivity 10 times that of Zone 1. Hence, the pressure in Zone 2 initially falls much faster than the pressure within Zone 1 (Figure 10-7).

Eventually, however, a pressure difference between the two zones develops that is large enough to cause interzonal flow. As long as interzonal flow is not significant, the pressure difference continues to increase. Increasing interzonal pressure difference causes a larger interzonal flow rate. Eventually, the system adjusts itself such that the pressure within each zone is declining at the same rate. The significance of this is that total production from each zone (when semisteady state is reached) is proportional to its (ϕ)(h) product, not its (k)(h) product. Note that total production from the low permeability or higher pressured zone includes both its radial flow rate to the well plus the interzonal flow rate.

Near the end of the producing life, the pressure in the more permeable zone approaches the minimum producing wellbore pressure (200 psia for this example). Given this situation, well production is almost entirely from the less permeable zone or zones. The more permeable zone has become only a transmission member, carrying the interzonal flow to the wellbore.

Figure 10-7 relates performance for this problem with an interzonal permeability (k_{iz}) equal to 1/10,000 of the horizontal permeability in the less permeable zone. For the cases studied with larger k_{iz} values, the pressure differences that developed between the zones were somewhat less. Although the number of different cases studied (gas and SGD oil) was not meant to be exhaustive, it appeared that a lower level of interzonal permeability is needed in a multizone gas reservoir (than for an oil reservoir) for a significant pressure difference between zones to develop. Gas normally has quite low viscosity (compared to oil), and gas reservoirs typically are developed with large well spacing. Notice in Equation 10-10 that low viscosity, coupled with a large drainage area, could yield interzonal flow rates that are significant even with a low value of interzonal permeability.

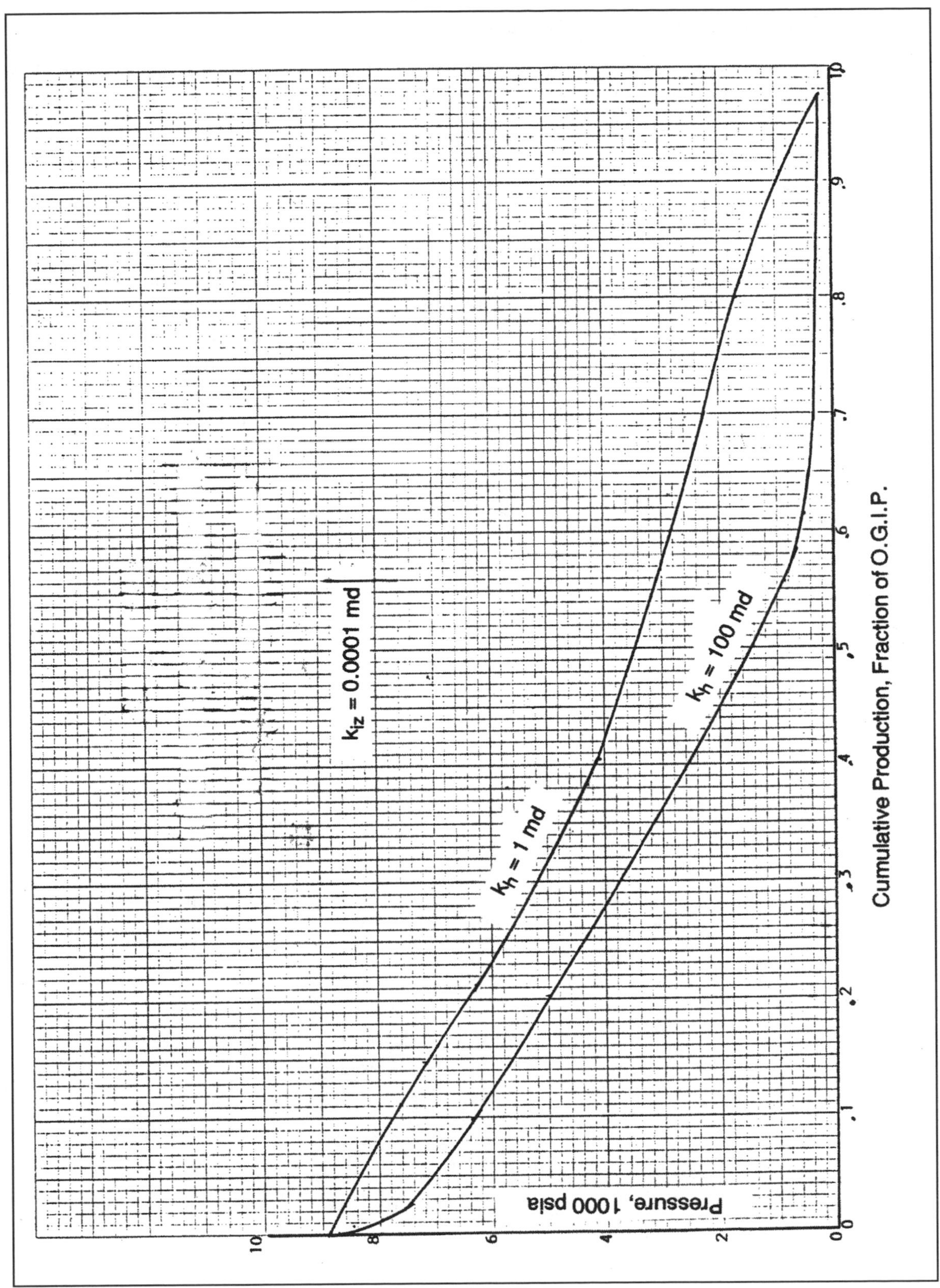

Fig. 10-7. Pressure vs. cumulative production for each zone individually.

C. Effect of Multiple Zones on Calculation of OGIP

Using pressure buildup estimated "static" pressures to calculate the original hydrocarbon in place in a multizoned reservoir is likely to yield poor results. This was discussed earlier in this chapter after the oil example. Because the pressure that is assigned to the well is too low, in a gas reservoir, the material balance calculated (OGIP) will also be too low.

Figure 10-8 gives a correlation for a two zone reservoir developed with the data of the earlier gas example problem and considering different interzonal permeabilities. Initially, with production, the pressure falls faster in the high (k)(h) zone. After a certain amount of production, then the two-zone pressures begin falling at about the same rate. Eventually, depending on the value of the interzonal permeability, the pressure difference between the two zones becomes minimal. At this point, pressure buildup "static" pressures will yield a good estimate of OGIP. Figure 10-8 indicates that if the interzonal permeability is at least one-tenth the value of the tight zone horizontal permeability, then only a minimal amount of production is needed to calculate an accurate value of OGIP. On the other hand, if the interzonal permeability is less than or equal to 1/10,000 the k_h of the tight zone, then material balance cannot be used to determine OGIP.

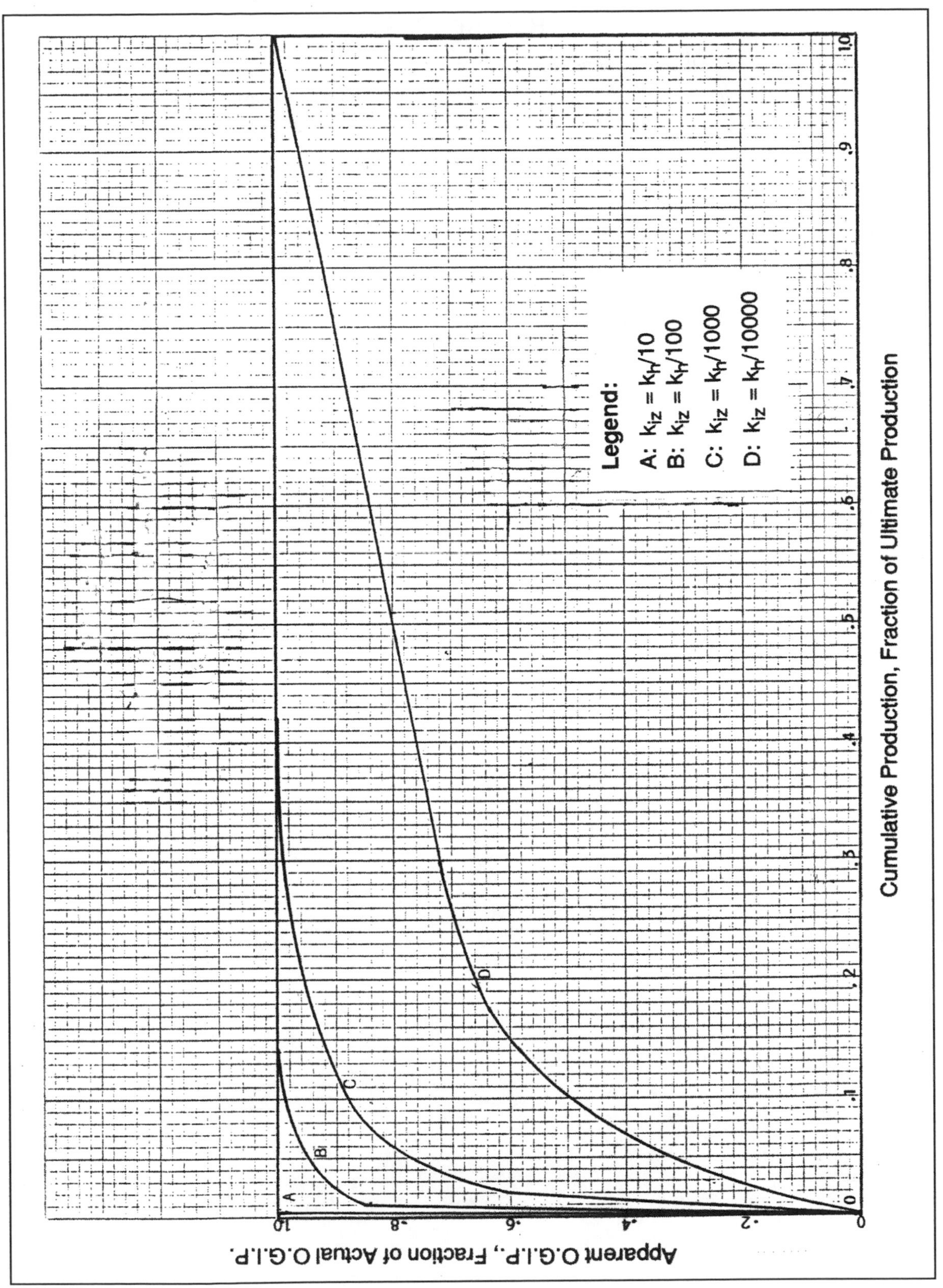

Fig. 10-8. Apparent original gas in place vs. cumulative production.

REFERENCES

1. Tracy, G. W.: "Fun with Material Balance," (Unpublished work), 1987.
2. Brown, K. E.: *The Technology of Artificial Lift Methods*, Petroleum Publishing Company, Tulsa (1977) **1**.
3. Beggs, H. D. and Brill, J. P.: "A Study of Two-Phase Flow in Inclined Pipes," *JPT* (May 1973).

11 IMMISCIBLE FLUID DISPLACEMENT MECHANISMS

INTRODUCTION

In a water drive oil reservoir there is an actual gradual displacement of the oil by advancing aquifer water which is immiscible with the oil. Production of fluids from the reservoir results in a pressure gradient across the oil/water contact which then causes water encroachment into the reservoir.

A similar situation occurs in a gas-cap drive-oil reservoir. With production from the reservoir, pressure is lowered resulting in an expansion of the gas-cap volume. The result is a displacement of the oil by the immiscible gas. Other immiscible displacements occur in secondary recovery operations such as water or immiscible gas injection.

The Buckley-Leverett[1] approach to the displacement of a fluid in porous media by an immiscible fluid was first presented in 1942. The theoretical development assumes that an immiscible displacement can be modeled mathematically using the concepts of relative permeability and a "leaky" piston. While the displacement is considered to be piston-like, some movable oil is bypassed. This is largely due to the effects of viscosity contrast, relative permeability, and capillary pressure. Capillary pressure is discussed in the "Rock Properties" chapter.

The theory permits a determination of the average pore-to-pore displacement efficiency in a linear system. The Welge[2] extension of this work permits a detailing of the displacement efficiencies that can be obtained after breakthrough of the displacing phase at the outlet end. A major limitation of the Buckley-Leverett theory is that it applies to a linear system only. Although there are a number of situations that can be modelled as linear displacements (edgewater drive, peripheral waterflood, gas-cap expansion), many of the injection-producing well patterns common to secondary recovery operations are not even close to linear systems. Nevertheless, the theory does permit the determination of a microscopic displacement efficiency. Then, using the concept of "sweep" efficiency, there are techniques[18] that can be used to extend the results to common nonlinear systems.

With immiscible displacement, it is generally assumed that a "front" develops. This is the place in the porous medium where there is a rapid saturation change from mostly displacing fluid to mainly displaced fluid. The front may be viewed as a kind of moving boundary, ahead of which oil flows, but behind which both oil and the displacing-phase flow. It is also possible to have two immiscible phases (such as oil and free gas) flowing ahead of the displacement front.

FRACTIONAL FLOW EQUATION

Leverett[3] is credited with the development of the fractional flow equation. This relationship allows the estimation of the fraction of the total flow rate that is displacing phase flowing at a given point in a linear system. Consider the linear reservoir or core segment below:

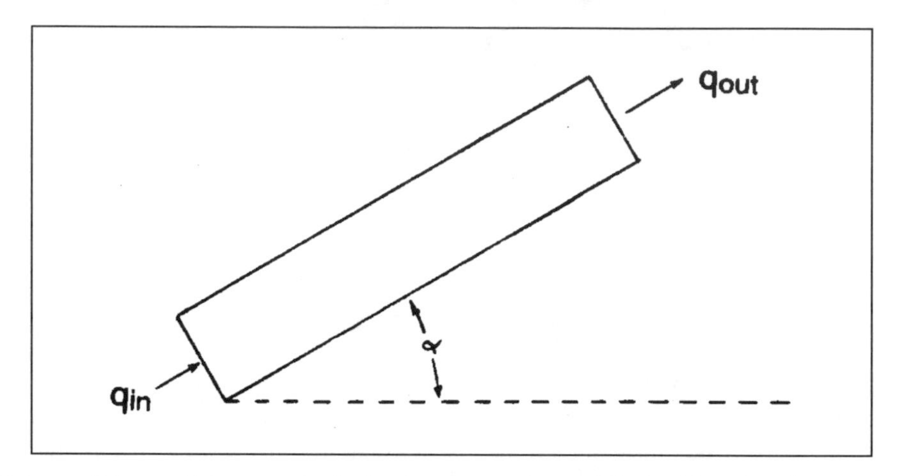

The initial conditions of the core include the pore volume filled with oil and connate water. The initial water saturation is at the irreducible level. The dip angle of the core is " α." At time zero, injection of 100% water is initiated at the lower end of the core with constant rate, q_{in}. Assumptions intrinsic to the development include: incompressible resident and injected fluids, constant system average pressure, and steady-state flow.

Therefore:

$$q_{in} = q_{out} = q_o + q_D = q_t \tag{11-1}$$

where:

q_{in} = injection rate per unit cross-sectional area,
q_{out} = fluid flow rate out of the core (at the outlet end) per unit cross-sectional area,
q_o = oil flow rate per unit cross-sectional area, at any point along the core,
q_D = displacing phase (here = water) flow rate per unit cross-sectional area, at the same point in the core as q_o,
q_t = total flow rate per unit cross-sectional area, at any point in the system,
D = subscript indicating the displacing phase (here, it is water, but it could also be gas), and
o = subscript indicating the displaced phase (oil).

Then, fractional flow may be defined as:

$$f_D = q_D \ / \ q_t = q_D \ / \ (q_D + q_o) \tag{11-2}$$

where:

f_D = fraction of the total flow rate at a given point that is the displacing phase at reservoir conditions of temperature and pressure.

Here, Equations 11-1 and 11-2 have been explained in terms of an oil/water system; however, the equations are also valid for gas displacing oil.

Since the flow of two immiscible fluids through a porous medium (with no reaction between the fluids and the medium) is to be modelled, the logical approach is to write Darcy's equation for each of the fluids:

$$\frac{q_o \mu_o}{k_o} = -\left[\frac{\partial p_o}{\partial u} + g \rho_o \sin \alpha\right] \tag{11-3}$$

$$\frac{q_D \mu_D}{k_D} = -\left[\frac{\partial p_D}{\partial u} + g \rho_D \sin \alpha\right] \tag{11-4}$$

where:

u = the linear direction of flow (measured from the inlet end),
$\partial p_o / \partial u$ = the pressure gradient in the oil phase,
$\partial p_D / \partial u$ = the pressure gradient in the displacing phase,
α = the angle of the fluid flow with respect to the horizontal (updip flow is assigned to be "+"; downdip is "-"),
q_o = displaced fluid (oil) flow rate per unit of cross-sectional area normal to u,
q_D = displacing fluid flow rate per unit of cross-sectional area normal to u,
ρ = fluid density,
μ = fluid viscosity,
$_o$ = subscript indicating displaced or oil phase, and
$_D$ = subscript indicating displacing phase

The reader attempting to gain a thorough understanding of the derivation and significance of the fractional flow equation should refer to the capillary pressure section of the "Rock Properties" chapter. There is a pressure difference across the interface between the displacing phase and the displaced phase with immiscible fluids. This pressure difference is defined to be capillary pressure:

$$p_c = p_D - p_o \tag{11-5}$$

Then:

$$\frac{\partial p_c}{\partial u} = \frac{\partial p_D}{\partial u} - \frac{\partial p_o}{\partial u} \tag{11-6}$$

Equation 11-5 may be confusing to those who recognize that capillary pressure is normally defined to be the pressure within the nonwetting phase minus the pressure in the wetting phase. Here, the definition 11-5 involves displacing and displaced phase pressures because this approach simplifies the development of the fractional flow equation.

Equation 11-4 may be subtracted from Equation 11-3, and then using Equation 11-6, the following equation results:

$$\frac{q_o \mu_o}{k_o} - \frac{q_D \mu_D}{k_D} = \frac{\partial p_c}{\partial u} + g\ (\Delta\rho)\ \sin\ \alpha \qquad (11\text{-}7)$$

Here, the density difference, $\Delta\rho$, has been defined to be:

$$\Delta\rho = \rho_D - \rho_o \qquad (11\text{-}8)$$

As discussed earlier, the two fluids are considered to be incompressible, and the average system pressure is assumed to be held constant. Therefore, steady-state flow results with continuity considerations requiring that:

$$q_t = q_o + q_D \qquad (11\text{-}1)$$

Manipulating Equations 11-1 and 11-2, it is possible to obtain:

$$q_D = f_D\, q_t \qquad (11\text{-}9)$$

and

$$q_o = (1 - f_D)\, q_t \qquad (11\text{-}10)$$

Substituting 11-9 and 11-10 into 11-7, rearranging, and then solving for f_D, results in the fractional flow equation in fundamental units:

$$f_D = \frac{1 - \dfrac{k_o}{q_t \mu_o}\left[\dfrac{\partial p_c}{\partial u} + g\ (\Delta\rho)\ \sin\ \alpha\right]}{1 + \dfrac{k_o}{k_D} \cdot \dfrac{\mu_D}{\mu_o}} \qquad (11\text{-}11)$$

where f_D is the fraction of the displacing phase in the total flow stream at a given point in the system. This equation is dimensionally correct, so long as the fundamental units of darcies, centipoise, cc/sec/cm^2, atmospheres and cm are used for permeability, viscosity, total flow rate per unit cross-sectional area, pressure and distance, respectively. If the difference in fluid density, $\Delta\rho$, has the units of gram/cc, then the gravitational constant, g, may be replaced by the constant 1/1033 to result in consistent units of atmospheres per cm in the direction of u for the terms $[\,(\Delta\rho)\sin\alpha\,] / 1033$ and $\partial p_c / \partial u$.

In common field units, Equation 11-11 becomes:

$$f_D = \frac{1 - \dfrac{1.127\, k_o}{\mu_o\, q_t}\left[\dfrac{\partial p_c}{\partial u} + 0.434\ (\Delta\gamma)\ \sin\ \alpha\right]}{1 + \dfrac{k_o}{k_D} \cdot \dfrac{\mu_D}{\mu_o}} \tag{11-12}$$

The units here are darcies, centipoise, barrels/day/square foot, psi, and feet for permeability, viscosity, total flow rate per unit cross-sectional area, pressure and distance, respectively. The term $\Delta\gamma$, indicates differential specific gravity or : $\Delta\gamma = \gamma_D - \gamma_o$.

As Pirson[4] has noted, the fractional flow equation is truly fundamental to the understanding and to the representation of the flow of two immiscible, insoluble fluids in porous media. Examination of the various terms of Equation 11-11 or 11-12 reveals the following factors influencing the fraction of the displacing fluid flowing at a given point in the system:

(1) Due to the assumptions made in the mathematical development, the displacement takes place at constant temperature and pressure, constant phase compositions, and at constant total flow rate. Where there is a partial miscibility of fluids involved, resulting in changing phase compositions and interfacial tensions, modification of the equations can be made.[15,16,17]

(2) The explicit fluid properties included are: μ_o, μ_D, γ_o, and γ_D . Due to the inclusion of the capillary pressure term, various system factors implicitly influence fractional flow. These include: S_o, S_w, wettability, interfacial tensions, and the geometries of the fluid saturations.

(3) The rock properties are represented through the effective permeability term, k_o, and the effective permeability ratio k_o/k_D. Since grain size, petro-fabric, composition, structure and cementing materials influence permeability, they also affect fractional flow.

Notice that the $\partial p_c / \partial u$ term is the partial derivative of capillary pressure with respect to distance in the direction of flow. This term cannot be evaluated directly. To consider the effect of this gradient, it is sometimes expressed as: $\partial p_c / \partial u = (\partial p_c / \partial S_D)(\partial S_D / \partial u)$. Then, with p_c defined to be $p_D - p_o$, $[\partial p_c / \partial S_D]$ is always a positive function over the saturation ranges found in the reservoir. (The interested reader should be able to see this after reading the capillary pressure section of the "Rock Properties" chapter. However, notice that capillary pressure there is defined in terms of wetting and nonwetting phases; whereas, here it is defined in terms of displacing and displaced phases). The expression $\partial S_D / \partial u$ is always negative. Therefore, the $\partial p_c / \partial u$ term is always negative; consequently, its effect in Equation 11-12 is always to increase the fractional flow of the displacing phase.

The mobility of the displacing and the displaced phases at any given point is defined as:

$$\lambda_D = k_D / \mu_D = k\, k_{rD} / \mu_D \tag{11-13}$$

and

$$\lambda_o = k_o / \mu_o = k\, k_{ro} / \mu_o \tag{11-14}$$

where:

λ_D = mobility of the displacing phase, darcy/cp,
k_D = effective permeability of the displacing phase, darcies,
μ_D = viscosity of the displacing phase, cp,
λ_o = mobility of the displaced phase (oil), darcy/cp,
k_o = effective permeability of the displaced phase, darcies, and
μ_o = viscosity of the displaced phase, cp.

Phase mobility is an indication of how easily that phase moves in the reservoir at the given point.

Mobility ratio, M, is usually defined[18] as the ratio of the mobility of the displacing phase to the mobility of the displaced phase. If there is a displacing-phase saturation gradient behind the front (the usual case), then the displacing-phase mobility should be evaluated at the average displacing-phase saturation behind the front.

$$M = \frac{(\lambda_D)_{behind}}{(\lambda_o)_{ahead}} = \frac{k_D(\bar{S}_D)\mu_o}{k_o(S_{o,OB})\mu_D} = \frac{k_{rD}(\bar{S}_D)\ \mu_o}{k_{ro}(S_{o,OB})\ \mu_D} \tag{11-15}$$

where:

$k_{rD}(\bar{S}_D)$ = the relative permeability of the displacing phase evaluated at the average displacing-phase saturation behind the front, and
$k_{ro}(S_{o,OB})$ = relative permeability to oil evaluated at the oil saturation in the oil bank.

Notice that the denominator of the fractional flow Equation 11-12 has a combination of permeabilities and viscosities that resembles the reciprocal of the mobility ratio. The $(k_o/k_D)(\mu_D/\mu_o)$ appearing in the fractional flow equation refers to conditions at the same point in the flow system. On the other hand, the mobility ratio compares fluid mobilities at two different points in the system: displacing phase behind the front and displaced phase ahead of the front.

An "unfavorable displacement" is one where $M > 1$; the displacing phase moves more easily in the porous medium than does the oil. When this situation exits, the more mobile displacing phase bypasses much of the oil. The most common reason for an unfavorable mobility ratio is high oil viscosity. A "favorable displacement" ($M < 1$) occurs when the oil has greater mobility than does the displacing phase. In this case, because the oil can move relatively easily ahead of the front, less oil is bypassed.

It is important to emphasize that the fractional flow equation includes in one relatively simple relationship, all the factors that affect immiscible fluid displacement in porous media. Because of the incompressible assumption, when compressible fluids such as oil or gas are involved, a design pressure should be specified before a particular displacement can be investigated with the fractional flow equation. If the pressure is expected to change much, then the displacement should be restudied every 100 psi or so over the expected pressure range. Then, for a given displacement pressure, a number of parameters in the fractional flow equation $(\Delta\gamma, \mu_o, \mu_D)$ become fixed. The only independent variable is the displacing-phase saturation which, in turn, specifies the relative permeability relationship. The section on permeabilities ("Rock Properties" chapter) suggests techniques by which representative values usually can be obtained for the specific reservoir being studied.

If the injected fluid wets the formation, then a suction capacity of the rock for the injected fluid results. In this case, the porous medium actually imbibes the injected fluid. This is an especially important mechanism if any of the oil is contained in the small-to-medium size pores. The rock will tend to imbibe the wetting phase into these tiny spaces thereby displacing oil into the main flow channels. A case in point is that of water injection into a water-wet system, which is the most common situation in present secondary recovery operations.

If the displacing fluid is heavier than the displaced fluid (as with water displacing oil), it should be obvious that the most efficient process will occur with the injection at the lowest elevation in the system. Inspection of Equation 11-12 will also indicate this is the case because injecting in the updip direction with a heavier displacing fluid will yield a smaller numerator in the fractional flow equation. Therefore, the fraction of the displacing fluid at any point is smaller which results in a larger percentage of displaced fluid (oil) flowing. The result is a higher displacement efficiency.

For a water/oil displacement process, Equation 11-12 may be written:

$$f_w = \frac{1 - \dfrac{1.127\, k_o}{q_t\, \mu_o}\left[\dfrac{\partial p_c}{\partial u} + 0.434\,(\Delta\gamma)\,\sin\,\alpha\right]}{1 + \dfrac{k_o}{k_w}\cdot\dfrac{\mu_w}{\mu_o}} \tag{11-16}$$

It should be evident that if the fraction of water flowing is kept to a minimum, then the maximum displacement of the oil should result. Further examination of Equation 11-16 indicates that forcing the term:

$$\frac{1.127\, k_o}{q_t\, \mu_o}\left[\frac{\partial p_c}{\partial u} + 0.434\,(\Delta\gamma)\,\sin\,\alpha\right]$$

to be as large a positive number as possible would normally be beneficial when water is displacing oil. This could be done when flooding updip by decreasing the water injection rate, by injecting the heaviest water available (perhaps a brine), by injecting the water as low on the structure as possible, or by altering the oil/water interfacial tension to create as favorable of $\partial p_c\,/\,\partial u$ as possible.

Notice in Equation 11-12 that the frontal displacement process is rate sensitive if either or both of the terms $\partial p_c\,/\,\partial u$ and $0.434\,(\Delta\gamma)\,(\sin\,\alpha)$ are not zero in a given system. Where immiscible fluids are being used in the displacement process, then the $\partial p_c\,/\,\partial u$ term cannot be neglected. The relative influence of the terms in the numerator of Equation 11-12 excluding the whole number one, can be controlled by means of the factor, q_t, the total flow rate per unit cross-sectional area. In secondary recovery, this is accomplished by controlling the fluid injection rates, whereas in the frontal displacement process of a natural water or gas drive, the fluid producing rates can be controlled.

Figure 11-1 illustrates the relationship between the fractional flow of the displacing phase and the displacing-phase saturation, as flow rate and dip are changed. Here, the displacing phase is wetting and more dense than the displaced phase.

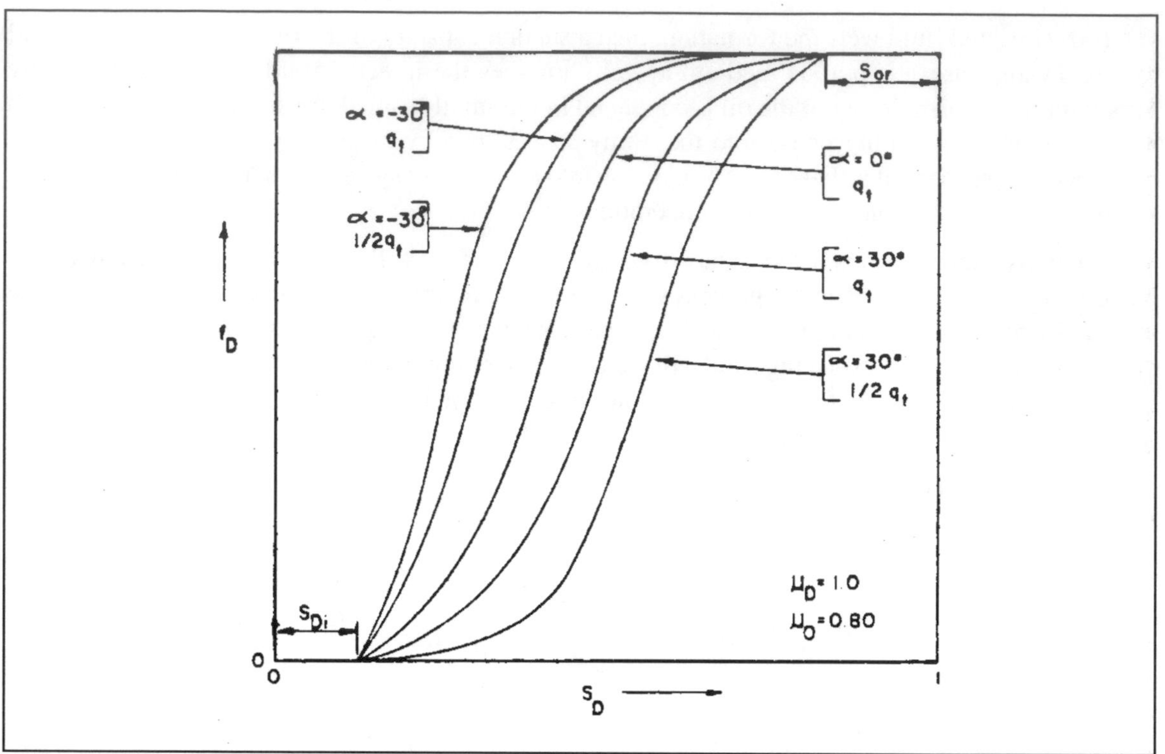

Fig. 11-1. Sketch of fractional flow relationships of the injected wetting phase as a function of reservoir dip and flow rate.

If it is necessary to inject in the downdip direction with a heavier displacing fluid, then Equation 11-12 indicates that the lowest displacing-phase fractions (and thus the highest displacement efficiency) will be obtained with high injection rates. Normally, care should be exercised to prevent injection rates high enough to fracture the formation as this would likely cause bypassing of some oil.

Where the injected fluid is gas, then Equation 11-12 becomes:

$$f_g = \frac{1 - \dfrac{1.127\, k_o}{\mu_o\, q_t}\left[\dfrac{\partial p_c}{\partial u} + 0.434\ (\Delta\gamma)\ \sin\ \alpha\right]}{1 + \dfrac{k_o}{k_g} \cdot \dfrac{\mu_g}{\mu_o}} \tag{11-17}$$

where:

$p_c = p_g - p_o$
$\Delta\gamma = \gamma_g - \gamma_o$
γ_g = specific gravity of the gas (related to water; i.e., water = 1),
γ_o = specific gravity of the oil (related to water; i.e., water = 1),
$_g$ = subscript referring to the gas phase, and
$_o$ = subscript referring to the oil phase

The water saturation present must be treated as being at the irreducible level. Otherwise, the problem would involve 3-phase relative permeabilities, which were not treated in the Buckley-Leverett development of the fractional flow equation. At the irreducible level, the water saturation may be considered as a part of the rock matrix.

Figure 11-2 is a typical diagram of flowing gas fraction versus gas saturation. Note that the abscissa represents the gas saturation based on the hydrocarbon pore volume rather than the total pore volume. Illustrated in Figure 11-2 is a family of curves which show the influence of injection rate, q_t, and dip upon the fraction of gas flowing. To construct the fractional flow curve, the capillary pressure gradient term (involving $\partial p_c / \partial u$) is usually ignored. The negative (dashed) portions of the fractional flow curves in Figure 11-2 occur because the capillary pressure term has been neglected.

If the capillary pressure gradient, $(\partial p_c / \partial u)$, and the gravity term, $0.434(\Delta\gamma) \sin \alpha$, can be validly ignored; then the fractional flow equation 11-12 takes the following form:

$$f_D = \frac{1}{1 + \dfrac{k_o}{k_D} \dfrac{\mu_D}{\mu_o}} \quad (11\text{-}18)$$

where the o and D refer to the displaced and displacing phases. This form of the fractional flow equation indicates that the fraction is dependent only on relative permeability and viscosity ratios if the capillary and gravity terms can be neglected.

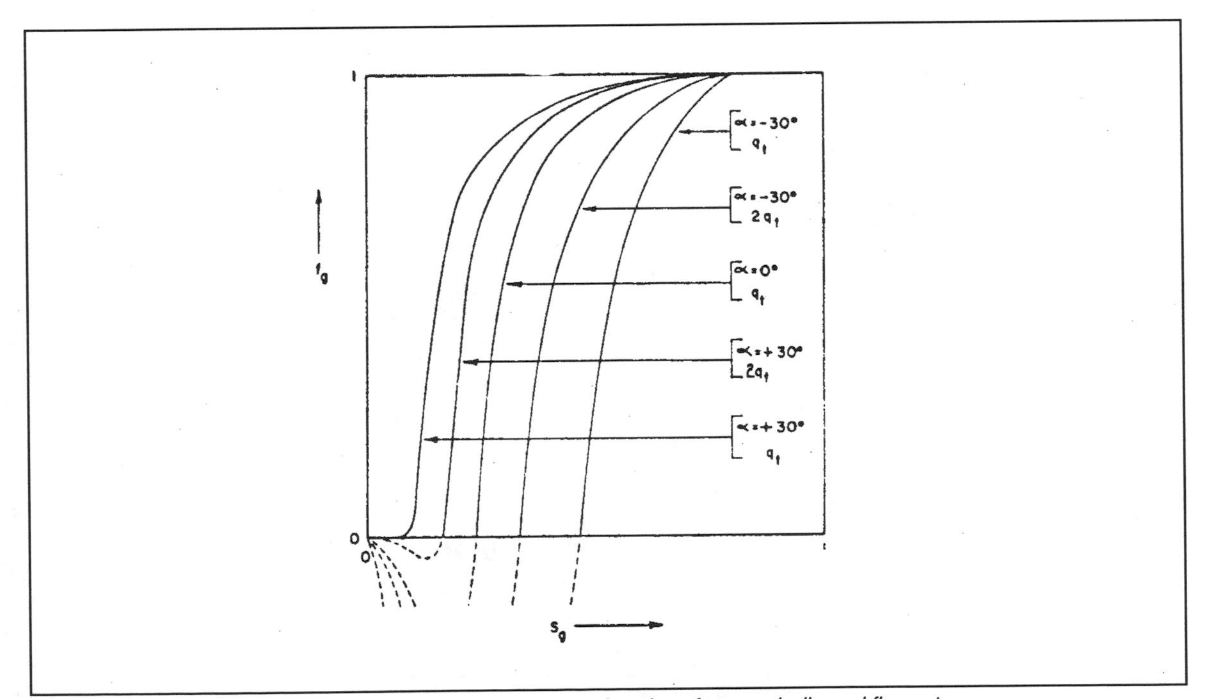

Fig. 11-2. Hypothetical plot of the fractional flow of gas as a function of reservoir dip and flow rate.

SOLVING THE FRACTIONAL FLOW EQUATION

I. Assemble the needed data.

A. PVT data: μ_o , μ_D , γ_o , and γ_D at displacement temperature and pressure. Generally, these data are desired at the conditions of the front as this is "where the action is."

B. Rock properties: absolute formation permeability, k, and displacing phase and oil relative permeability data (k_{rD} and k_{ro}). The relative permeability data should span the movable saturation range. Considering Equation 11-12, it would seem that capillary pressure data is also needed. However, even if p_c versus S_D data were available, at this point $\partial p_c / \partial u$ information will not be available.

C. Geologic property: α. Recall that this is the angle of the flow direction with respect to the horizontal.

D. Operational property: q_t, which depends on the water influx rate and/or well rates.

II. Ignore the term, $\partial p_c / \partial u$, in the fractional flow equation (for the time being), and calculate fractional flow, f_D, versus displacing-phase saturations (for the movable saturation range).

$$f_D = \frac{1 - \dfrac{0.488\ k_o\,(\Delta\gamma)\ \sin\ \alpha}{\mu_o\ q_t}}{1 + \dfrac{k_o\ \mu_D}{k_D\ \mu_o}} \tag{11-19}$$

III. Graph f_D versus S_D on Cartesian coordinate paper as shown below.

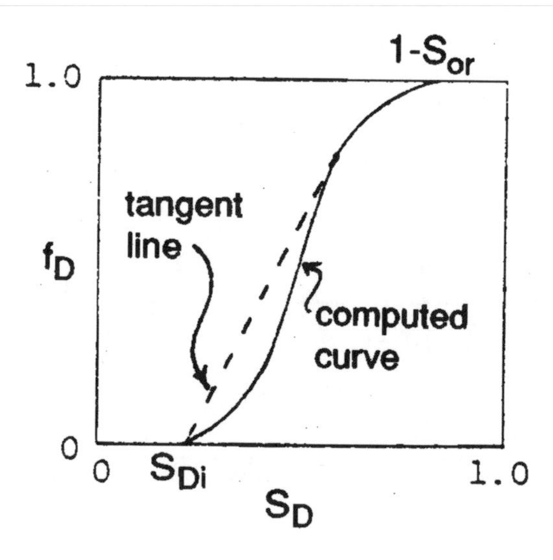

IV. **From the point on the f_D curve at the irreducible displacing-phase saturation, draw a tangent to the curve as shown.** Now, for the lower displacing-phase saturations, the correct f_D curve is the tangent segment up to the point of tangency with the original curve. However, for displacing-phase saturations higher than the tangent point, the correct f_D curve is the original curve, not the tangent line. It will be shown in the next section that the linear portion of the curve has the effect of restoring the neglected capillary pressure gradient term.

THE RATE OF FRONTAL ADVANCE EQUATION

Buckley and Leverett[1] first presented the rate of frontal advance equation in 1942. Consider an elemental volume from a linear porous medium, as shown in Figure 11-3. The medium contains two incompressible fluids: the displaced fluid (oil) and the displacing fluid which is usually water or gas. For the steady-state case, where pressure and temperature are constant, then the entering and exiting flow rates must be equal.

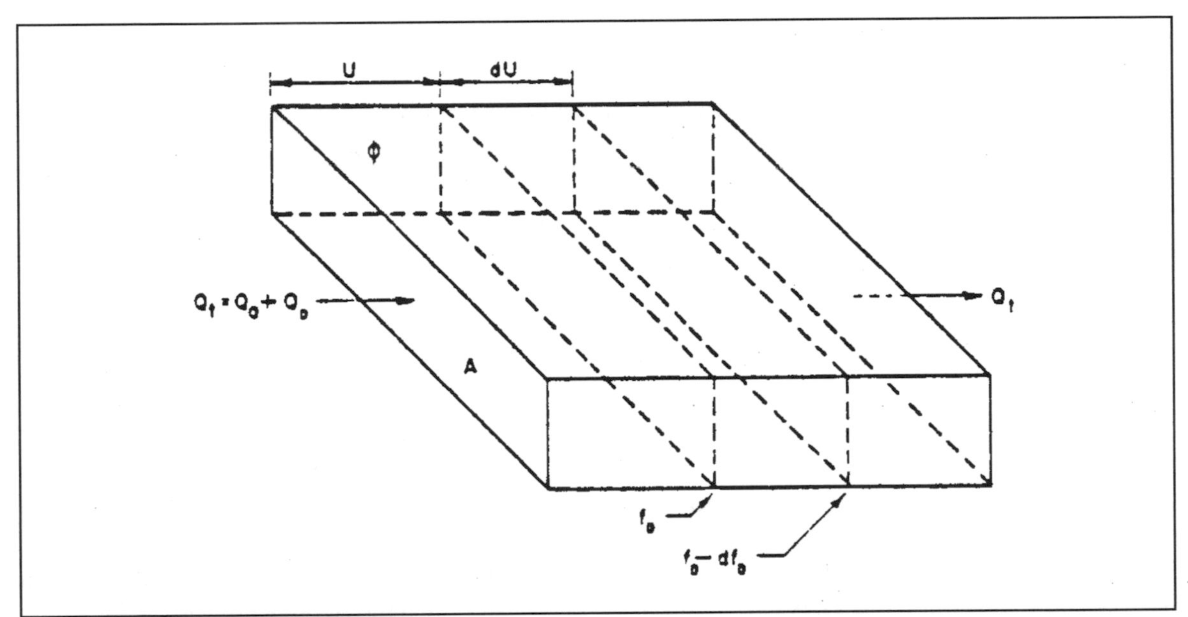

Fig. 11-3. Elemental reservoir volume containing displaced and displacing fluid phases.

The volume element in Figure 11-3 has a cross-sectional area, A, and a porosity, ϕ. If two phases are flowing in the element and an effective displacement of the oil is occurring, then the displacing-phase saturation in the element should be increasing with the passing of time. Thus, if the average displacing-phase saturation is S_D at time t, then at time t + dt, the average saturation is equal to $S_D + dS_D$. If the entering displacing-phase fraction is represented by f_D, then the exiting fluid displacing-phase fraction, which is leaving at a distance of du from the inlet face, may be considered to be $f_D - df_D$. Were this not the case, the mechanism would have little meaning, because a displacement of the oil would not be occurring. As the composition of each phase is assumed to be constant, continuity considerations (material balance) indicate that if oil is being displaced, the fractional flow of the displacing phase must be decreasing in the direction of the displacement. According to Pirson[4], the law of conservation of matter can be applied to the volume element of Figure 11-3. The resulting point or local equation is:

$$(\phi A\, du)\, dS_D = Q_t\, dt\, df_D \tag{11-20}$$

The left-hand side of 11-20 represents the increase in the displacing-phase volume in the element pore space, $\phi A\, du$ during the time dt. The right-hand side expresses the difference in the amount of displacing phase entering the volume element and the amount leaving over the time period, dt. The increase of the displacing phase within the pore volume is equal to the difference between the amount entering and that leaving the element.

Equation 11-20 may be rewritten in terms of the only independent variable (S_D) as:

$$du = \frac{Q_t}{A\phi}\left(\frac{\partial f_D}{\partial S_D}\right) dt \tag{11-21}$$

In this point or differential form, the equation represents the advance in a linear system of a plane of constant saturation, S_D, a distance du during the time interval, dt. Equation 11-21 affords a relation by which the distance that a plane of constant saturation has advanced may be determined. This is true because the distance is directly proportional to time and to the value of the partial derivative, $\partial f_D / \partial S_D$.

Integrating 11-21 allows the final form of the rate of frontal advance equation to be written:

$$u = \frac{Q_t}{A\phi}\left(\frac{\partial f_D}{\partial S_D}\right) t \tag{11-22}$$

Thus, at time t, a plane of constant saturation, S_D, has advanced a distance, u. At breakthrough of the displacing phase at the system outlet, the distance u would be equal to the length of the system, L, where time would be the breakthrough value. Any consistent set of units may be used. If barrels per day, square feet, feet, and days are used for Q_t, A, u, and t, then the right-hand side of Equations 11-21 and 11-22 must be multiplied by 5.615.

STABILIZED ZONE CONCEPT

A number of authors[5,6,7,8,9] have published treatments dealing with the stabilized zone between the displacing and displaced fluids in porous media. This concept was probably recognized first in the laboratory while observing core floods. It was seen at breakthrough that there is a drastic change in saturations; i.e., there is a so-called "shock front" in which there are planes of different saturations quite close together, apparently traveling at the same velocity.

Figure 11-4 illustrates the stabilized-zone concept. For a given porous medium, a stabilized zone will develop. At time, t_1, the displacing phase extends a distance into the linear system, as shown by the saturation profile. If the displacing phase already exists as an irreducible value, S_{Di}, in the porous media; then the displacement-phase saturation will range from this irreducible value up to a limiting maximum of 1 - S_{or}. Between these two saturation limits, two zones exist: the stabilized and the nonstabilized zones. The stabilized zone is at the front and is characterized by the saturation profile illustrated in Figure 11-4 at times t_1, t_2, and at breakthrough, t_B. The saturation profile in the stabilized zone is parallel for all times up to breakthrough. Again, this is mathematically equivalent to a constant velocity, $\partial u / \partial t$, for all displacing-phase saturations, S_D, in the stabilized zone.

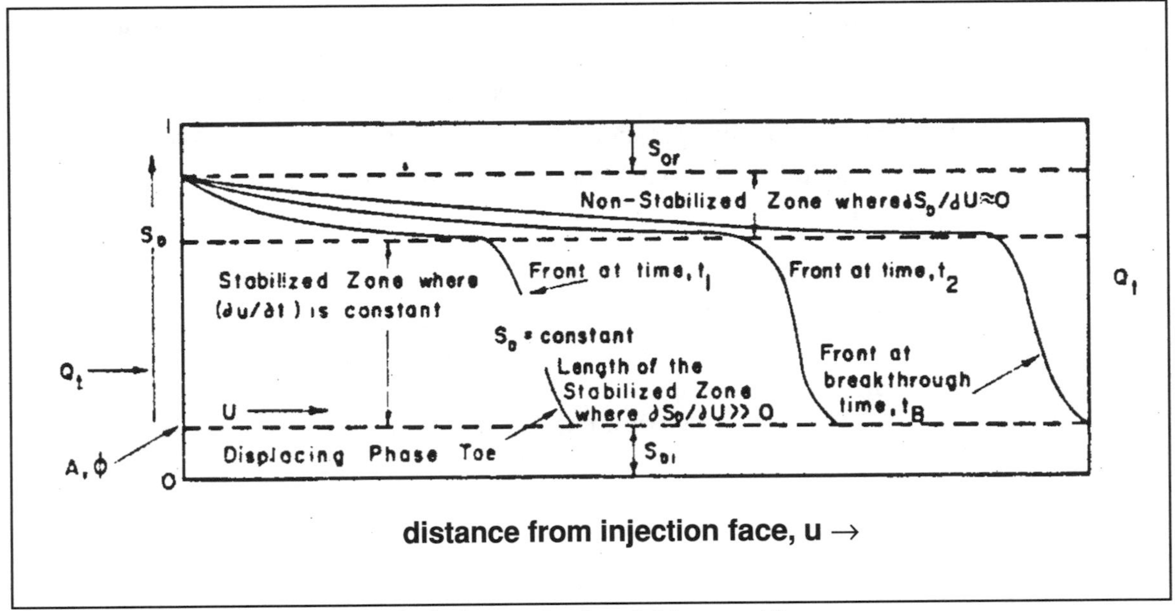

Fig. 11-4. Displacing-phase saturation distribution as a function of distance and time.

The nonstabilized zone, also called the "drag" zone, exists behind the front. In Figure 11-4, this saturation profile is seen to flatten with time as the front moves on through the system. It is the drag zone that contributes the "after-breakthrough" production of displaced oil.

Considering the displacement Equation 11-21, the velocity, $\partial u / \partial t$, of a given saturation plane in a linear immiscible flood (with constant injection rate, constant cross-sectional area, and constant porosity), is directly proportional to df_D/dS_D. But since laboratory core floods indicates the saturation planes at the front to be traveling at the same velocity, this suggests that df_D/dS_D has a constant value for the displacement saturations found at the front.

Consider the general fractional flow Equation 11-12:

$$f_D = \frac{1 - \dfrac{1.127\, k_o}{\mu_o\, q_t}\left[\dfrac{\partial p_c}{\partial u} + 0.434\ (\Delta\gamma)\ \sin\ \alpha\right]}{1 + \dfrac{k_o}{k_D} \cdot \dfrac{\mu_D}{\mu_o}} \tag{11-12}$$

The term, $\partial p_c / \partial u$, is of particular concern since data are seldom, if ever, available to properly define it. The usual approach is to neglect this term and then to correct for it later. If this is done (assume that $\partial p_c / \partial u = 0$), then for a particular system, a fractional flow curve similar to that in Figure 11-5 may be determined.

By constructing tangents at various points along this curve in Figure 11-5, the relationship of $\partial f_D / \partial S_D$ versus S_D may be generated. Figure 11-6 shows a graph of the derivative of the fractional flow curve of Figure 11-5 (which neglected capillary pressure effects). Notice that it is not a constant function anywhere. Thus, neglecting capillary pressure has generated a functional relationship, $\partial f_D / \partial S_D$ versus S_D, that does not conform to the experimental observation that the saturation planes in the stabilized zone are traveling at the same constant velocity (indicative that $\partial f_D / \partial S_D$ should be a constant for saturation values in the front). The implication is that the capillary pressure term, $\partial p_c / \partial u$, cannot be neglected.

At this point, the development will be continued neglecting capillary pressure effects. At a given time, say t_1, the frontal advance Equation 11-22 may be used with the $\partial f_D / \partial S_D$ data, generated as in Figures 11-5 and 11-6, to calculate the profile of displacing-phase saturation versus distance for a particular system. Then, at time t_1, the saturation profile might resemble Figure 11-7, where porosity, ϕ, cross-sectional area, A, and total flow rate, Q_t, are problem constants. The flow rate does not have to remain constant over all times, but if it does not, then displacement efficiencies may change due to rate sensitivity.

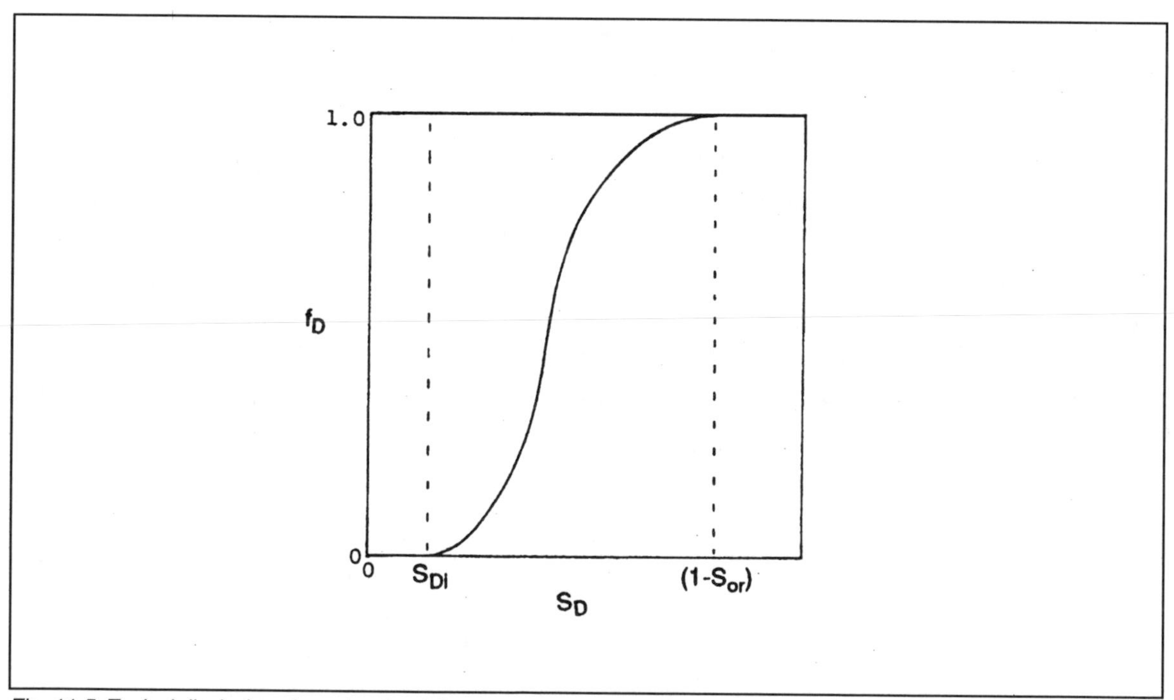

Fig. 11-5. Typical displacing-phase fractional flow curve constructed assuming that $\partial p_c / \partial u$ is equal to zero.

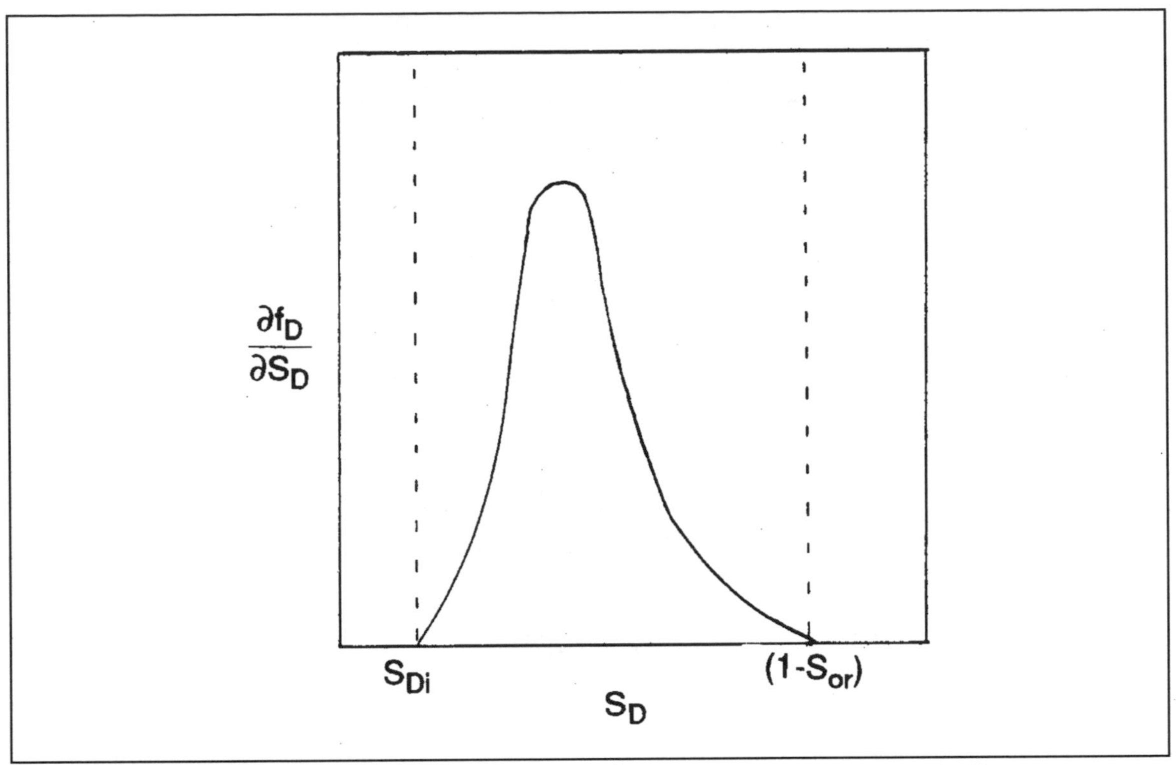

Fig. 11-6. Plot of the change of slope of f_D shown in Figure 11-5 which neglected capillary pressure effects.

Consider Figure 11-7. The saturation profile, as shown by curve "abce," is obtained by plotting the derived values of u (at time t_1) as the distance from the line representing the initial displacing-phase saturation, "aef." It should be apparent, at this point, that a physical difficulty occurs. Notice that for most distances, u, double saturation values are shown. Craft and Hawkins[10] illustrate a case where triple saturation values occur at a single point. How can multiple saturation values exist at a single point? This impossibility has resulted because the capillary pressure term in the fractional flow equation was neglected.

The simplest way to resolve this problem is to draw a vertical line somewhere on the profile (Figure 11-7), and let this represent the flood front. This would represent correctly the experimentally- observed fact that at the front, planes of different saturations are moving together with the same velocity. By balancing the cross-hatched areas, "A," when drawing the vertical line (as shown in Figure 11-7), material balance is preserved. It can be shown that this corresponds to replacing the capillary pressure term that was neglected in the fractional flow equation.

The method suggested by Buckley and Leverett,[1] consists of first drawing the fractional flow and fractional flow derivative diagrams, as shown in Figures 11-8 and 11-9. Then, by drawing a horizontal line, balance the areas, "A," on the derivative curve as in Figure 11-9. The balancing of the areas maintains material balance, while the curve value along "ghi" is the constant derivative value of the displacing-phase fractional flow for saturations at the front. The "ijk" portion of the curve represents the fractional flow derivative values for saturations behind the front.

According to Pirson,[4] this balancing the areas under the derivative curve is equivalent to

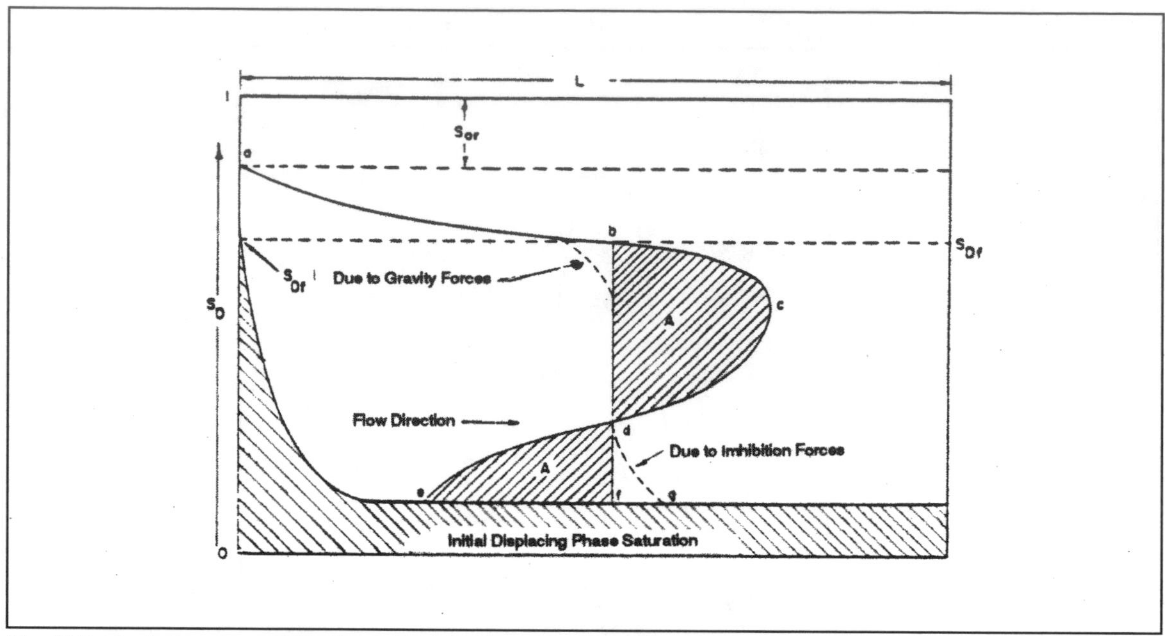

Fig. 11-7. Displacing-phase saturations after injection at a constant rate over time interval t_1.

redrawing the corresponding part of the f_D curve as a straight line beginning at the initial (irreducible) displacing-phase saturation on the original curve and then drawn tangent to the original curve as shown in Figure 11-8. Pirson further shows that the values, f_D and S_D, at the point of tangency are the displacing-phase fractional flow and saturation values at the front. Further, the tangent line graphically includes the neglected capillary pressure term. Therefore, the capillary pressure term contribution is represented by the distance between the straight line "ac" and the curved line segment "abc" in Figure 11-8.

An additional implication of Pirson's work is that it is correct to draw a vertical line (by balancing the areas "A") to represent the front on the saturation profile as shown in Figure 11-7. It is usually easier to draw the tangent to the fractional flow curve to find the displacing-phase saturation at the front, S_{Df}. Now, go to the saturation profile diagram and come down the curve until reaching S_{Df} (point "b" in Figure 11-7). From this point, draw the vertical line representing the front.

As discussed in the Fractional Flow section, the capillary pressure term of Equation 11-12 may be written as: $\partial p_c / \partial u = (\partial p_c / \partial S_D)(\partial S_D / \partial u)$. So, the change in capillary pressure with distance is controlled by both (1) the partial derivative of capillary pressure with respect to displacing-phase saturation and (2) the change in the displacing-phase saturation with distance. Figure 11-4 illustrates the stabilized zone concept and how it is affected by the capillary pressure term, $\partial p_c / \partial u$. Notice that in the stabilized zone (saturation values at the front that are moving at a constant velocity), $\partial S_D / \partial u$ is not zero, and large changes in capillary pressure over relatively short distances can occur. In fact, capillary forces in the stabilized zone normally are sufficiently high to permit saturation readjustment faster than the imposed displacement by the external drive. Thus, the saturation profile at the front remains parallel to itself over all distances and times up to breakthrough at the outlet end.

Notice in Figure 11-4 that the gradient, $\partial S_D / \partial u$, approaches zero in the "drag" or nonstabilized zone. Therefore, the capillary pressure term of the fractional flow equation may be neglected in this region (behind the front).

So, considering the fractional flow versus displacing-phase saturation diagram, the tangent line (shown in Figure 11-8) is used to represent fractional flow up to the tangent point. For higher displacing-phase saturations, the original curve, constructed neglecting capillary pressure, is used.

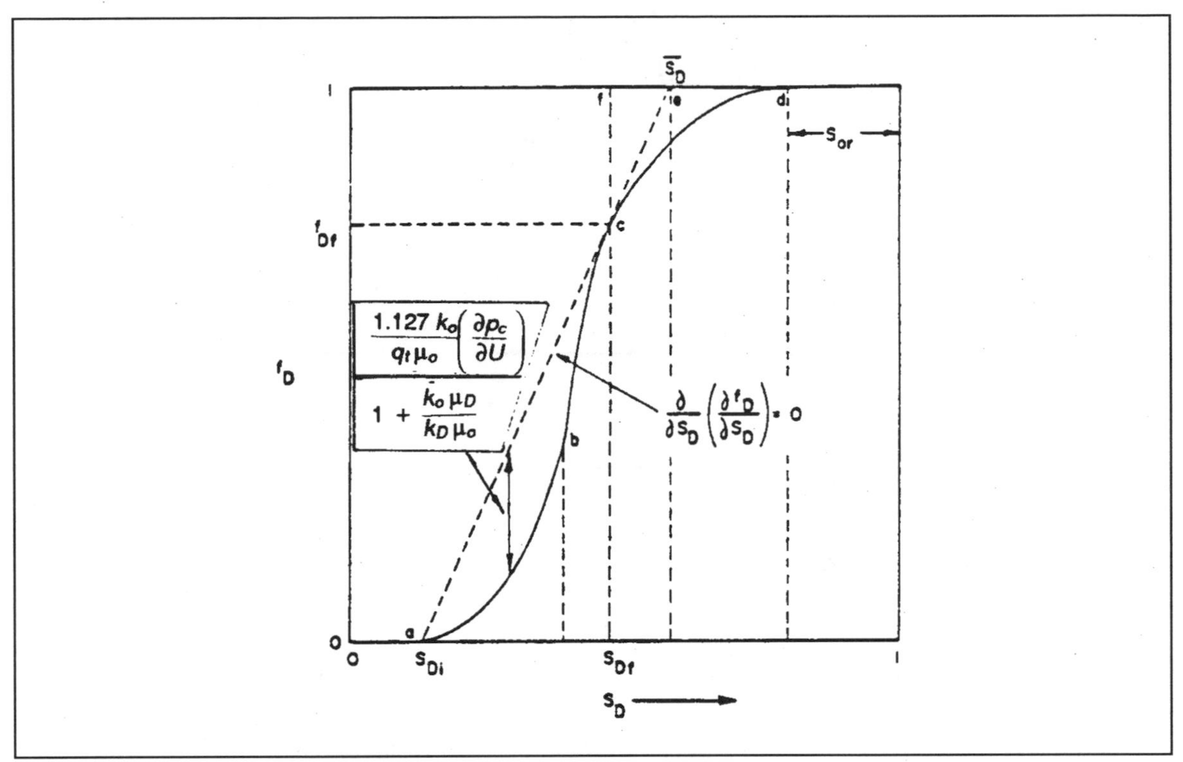

Fig. 11-8. Displacing-phase saturation distribution and graphical method of determining displacement efficiencies at breakthrough of the displacing phase.

In Figure 11-7 the front is represented by a vertical line drawn at point "b." If the displacing fluid (water) is wetting and denser than the displaced fluid (oil), the front will not be perfectly vertical, but will have a "toe" at the bottom due to the imbibition forces and a "slump" at the top of the front due to the gravity forces. With these modifications to Figure 11-7, good agreement is obtained with laboratory-demonstrated profiles as shown in Figure 11-4. For a gas/oil case, where the gas is nonwetting and less dense than the oil, then there would be no toe (because the gas is nonwetting) and no slump (because the gas is lighter than the oil). In fact, there would be a tendency for the gas to override the oil.

To determine displacement efficiency, the average saturations behind the front must be known. Welge[2] has shown that the mean displacing-phase saturation, $\overline{S_D}$, behind the front at breakthrough (or any time before breakthrough) may be determined by extending the tangent line on the fractional flow curve (Figure 11-8) to the point of intersection with the horizontal line: $f_D = 1.0$. Before the tangent line method was devised, the practice was to draw a vertical line on the fractional flow

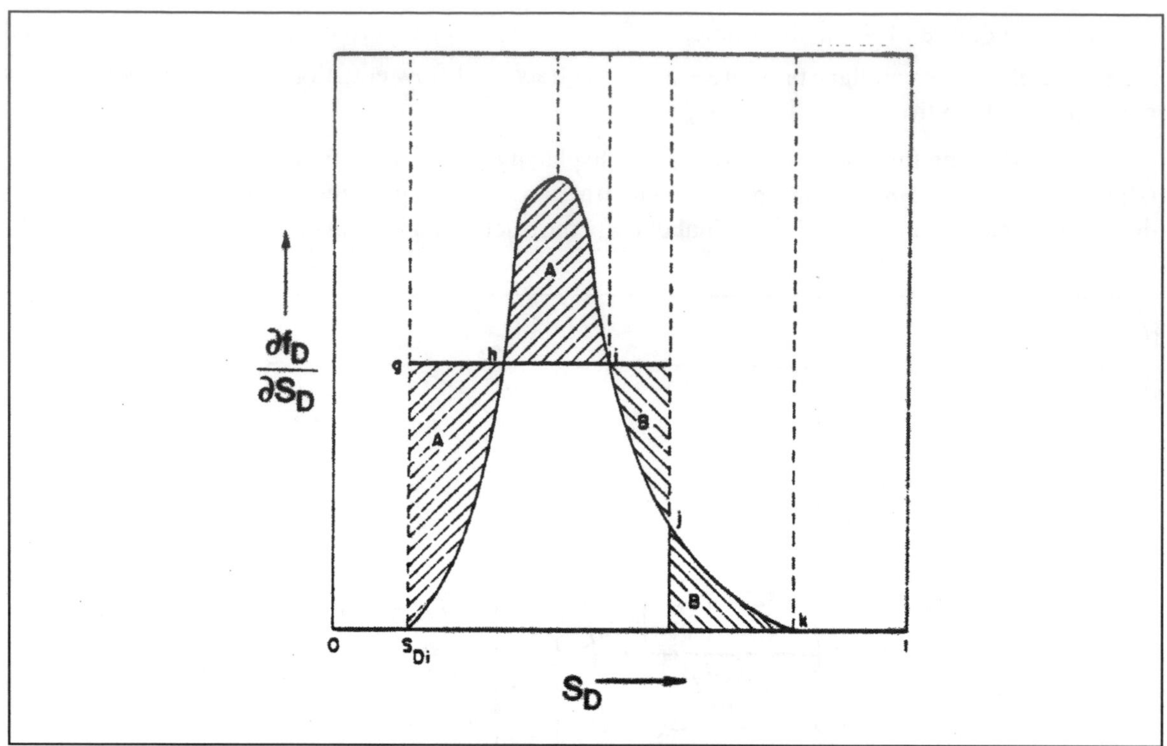

Fig. 11-9. Plot of the change of slope of fD as shown in Figure 11-8 versus SD and the construction for the breakthrough recovery.

derivative curve (Figure 11-9) such that the areas "B" balanced. Pirson[4] has shown these two methods to be equivalent. Figure 11-10 illustrates graphically the averaging that is accomplished (areas "A") by extending the tangent to the f_D curve to a value of one.

For a system initially saturated with oil to 1 - S_{Di} of the pore space, at breakthrough, the fraction of the original oil in place that is recovered is:

$$R.E. = \frac{\overline{S_D} - S_{Di}}{1 - S_{Di}} \tag{11-23}$$

where:

$R.E.$ = displacement efficiency or fractional recovery of the total original oil in place, fraction,
$\overline{S_D}$ = mean displacing-phase saturation behind the front at breakthrough (found by extending tangent of f_D curve to f_D = 1.0), fraction, and
S_{Di} = initial (irreducible) displacing-phase saturation, fraction.

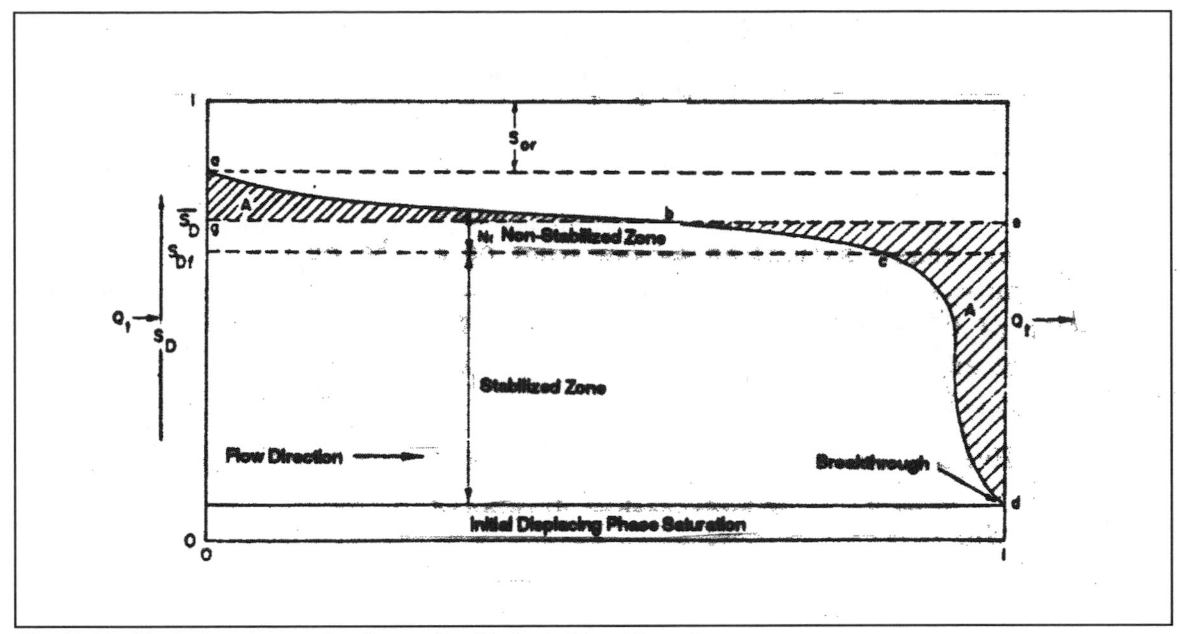

Fig. 11-10. Displacing-phase saturation profile at instant of breakthrough.

Notice that this formula gives recovery as a fraction of the total original oil in place. To find the recovery in terms of movable oil, the equation is:

$$R.E. = \frac{\overline{S_D} - S_{Di}}{1 - S_{Di} - S_{or}} \tag{11-24}$$

where:

$R.E.$ = displacement efficiency or fractional recovery of the movable original oil in place, fraction,

S_{or} = residual oil saturation, fraction, and

$1 - S_{Di} - S_{or}$ = fraction of the total pore volume that is filled with oil capable of moving.

After Breakthrough:

Typically, a significant portion of the movable oil has been produced at the outlet end at breakthrough. However, much remains behind the front in the drag zone, and is still mobile. Therefore, waterflood operators continue flooding after breakthrough. To predict recovery after breakthrough,[2] first the displacing-phase saturation or fractional flow at the outflow face must be determined. Then, on the fractional flow diagram, from the point on the curve equal to the outflow face saturation, S_{Dc}, a tangent line is extended to an f_D value equal to one as illustrated in Figure 11-11. For instance, assume that the economic limit reservoir water cut is 95%. Then, enter the fractional flow curve at an f_{Dc} value of 0.95. (The subscript "c" indicates that the value is taken at the outflow face after breakthrough.) At this point on the curve, construct a tangent. Extrapolate the tangent to f_D = 1.0, where the average displacing-phase saturation in the reservoir is found to be ($\overline{S_{Dc}}$). Then, the recovery is equal to:

$$R.E. = \frac{\overline{S_{Dc}} - S_{Di}}{1 - S_{Di}} \tag{11-25}$$

where:

$R.E.$ = displacement efficiency of the total original oil in place for continued injection past breakthrough, fraction,

$\overline{S_{Dc}}$ = mean displacing-phase saturation after breakthrough between the injector and producer, fraction.

$_c$ = subscript indicating conditions at the outflow face after breakthrough.

Again, if it is desired to base the after-breakthrough displacement efficiency on movable oil

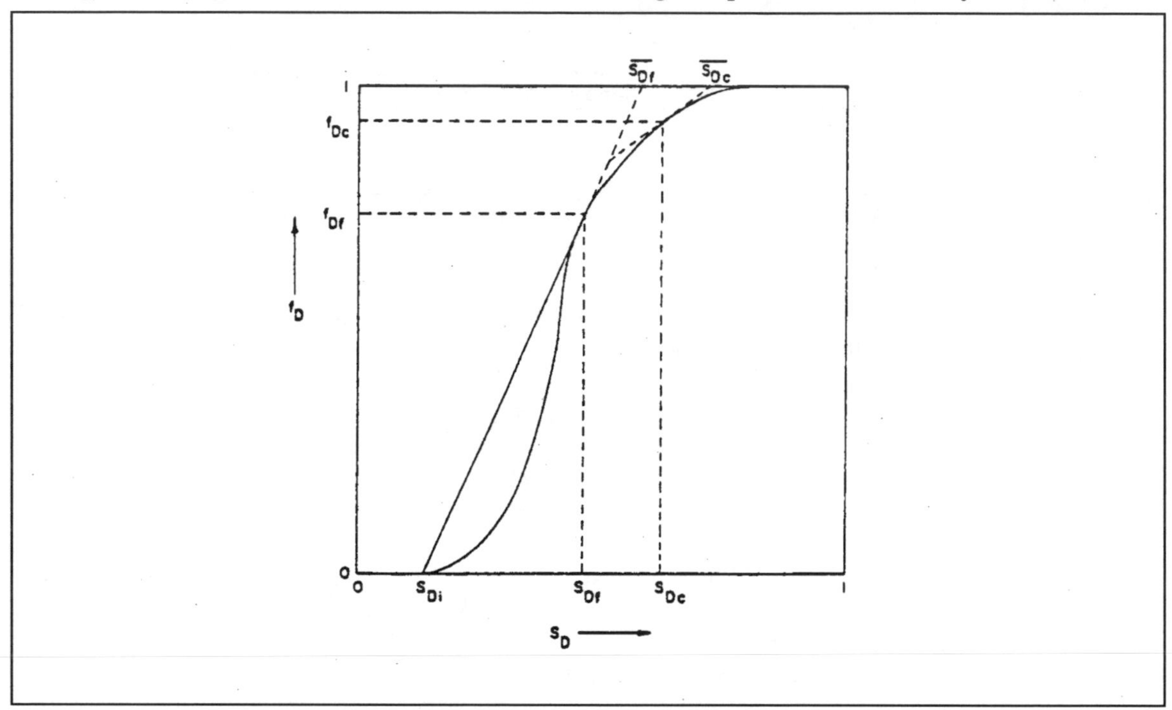

Fig. 11-11. Fractional flow curve showing graphical construction for average displacing-phase saturations for continued injection after breakthrough.

instead of total oil, then:

$$R.E. = \frac{\overline{S_{Dc}} - S_{Di}}{1 - S_{Di} - S_{or}} \tag{11-26}$$

where:

$R.E.$ = displacement efficiency of the movable original oil in place for continued injection past breakthrough, fraction.

Welge[2] has shown that after breakthrough:

$$Q_i = \frac{1}{\left[\dfrac{\partial f_D}{\partial S_D}\right]_{S_{Dc}}} \qquad (11\text{-}27)$$

where:

Q_i= cumulative water injected, pore volumes, and

$\left[\dfrac{\partial f_D}{\partial S_D}\right]_{S_{Dc}}$ = derivative of the fractional flow curve at the displacing-phase saturation existing at the outlet end (after breakthrough).

Equation 11-27 is used to estimate the cumulative displacing-phase injection that is needed to obtain a given displacing-phase saturation at the outlet end. For instance, assume that the economic reservoir water-cut limit is 95%. So, enter the fractional flow curve at $f_{Dc} = 0.95$. At this point on the curve, measure the derivative, [$(\partial f_D / \partial S_D)$]. Then, equation 11-27 indicates that the reciprocal of the derivative is equal to the number of pore volumes of water that will need to be injected to reach the economic water-cut limit.

EFFECTS OF WETTABILITY AND INITIAL SATURATION OF DISPLACING PHASE

If a reservoir exists that already has a displacing-phase saturation greater than the irreducible value, then the Buckley-Leverett analysis is slightly different. This situation might arise when waterflooding a reservoir that has already been naturally water driven. If the water saturation is initially too high, a front or oil bank will not develop. So, this situation is not solely of academic interest.

Water Displacing Oil in a Water-wet System

To begin with, construct the fractional flow curve in the fashion already discussed, i.e., neglecting capillary forces. As before, the curve will span water saturations of S_{Di} to $(1 - S_{or})$. Then, to predict the initial water cut at the outlet end, enter the fractional flow curve at a saturation equal to the average initial water saturation in the reservoir, S_{Dinit}. This is illustrated in Figure 11-12. This point on the fractional flow curve is also the beginning point for the tangent line that is to be drawn (Figure 11-12) to restore the capillary pressure term, $\partial p_c / \partial u$, in the fractional flow Equation 11-12. If the tangent line can be drawn, then an oil bank will form. The saturation and fractional flow values at the front are determined by the tangent point as shown in Figure 11-12. As before, extend this tangent line to $f_D = 1.0$ to find ($\overline{S_D}$), the average displacing-phase saturation behind the front.

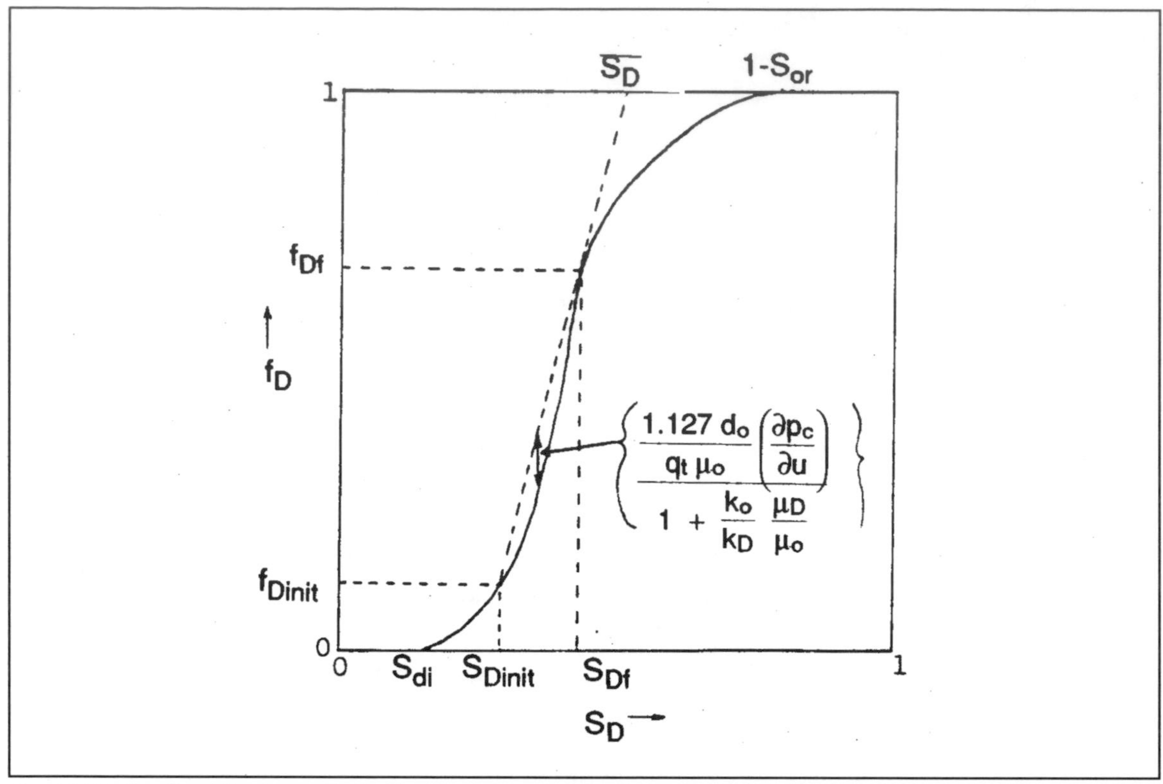

Fig. 11-12. Fractional flow curve illustrating the tangent construction and graphical analysis for the case where the initial displacing-phase saturation is higher than the irreducible value.

By inspection, one can see that with this situation, initial $S_d > S_{Di}$, the capillary pressure term is reduced in value from the $S_D = S_{Di}$ starting point case. In a normal water-wet formation, the pressure in the oil phase is higher than the pressure in the water phase, and this difference becomes larger as the oil saturation increases. The implication is not that the oil pressure is so high ahead of the front, but that the water pressure is low, especially when the water saturation ahead of the front is at the irreducible level. This is so because the water at the irreducible saturation exists in a pendular state (water rings around the grain contact points). Because these rings do not touch each other, no pressure differentials can be transmitted directly from one pendular ring to another. The oil phase, however, exists in a funicular state on both sides of the front, and thus forms continuous paths such that pressure differentials can be transmitted within the oil phase over the length of the system. Therefore, a positive pressure differential exists in the direction of flow ahead of the front only within the oil phase, which results in the beneficial effect of more oil production.

As the initial water saturation is increased above the irreducible value, then the pendular saturation state changes to a funicular one, which allows a pressure differential to be transmitted through the mobile water phase. This causes the pressure difference between the oil and water phases to be lessened (compared to the case where $S_D = S_{Di}$). Thus, the capillary pressure gradient term in the fractional flow equation is reduced. So, the pressure differential across the reservoir is transmitted through both the water and oil phases, and both water and oil are moving. At initial water saturations that are not much above the irreducible saturation, capillary pressure is still holding most of the water in place, and only that amount of the water saturation above the irreducible value can move. However, as the initial water saturation continues to increase, more and more of the water saturation is capable of flowing. Finally, at (or above) the critical water saturation (Figure 11-13), an oil bank or front does not develop. At this point, the capillary forces that allow a front to form are small. In this case, the entire displacement will take place in the nonstabilized zone. The oil will not be pushed (or banked), but will be dragged along with the water.

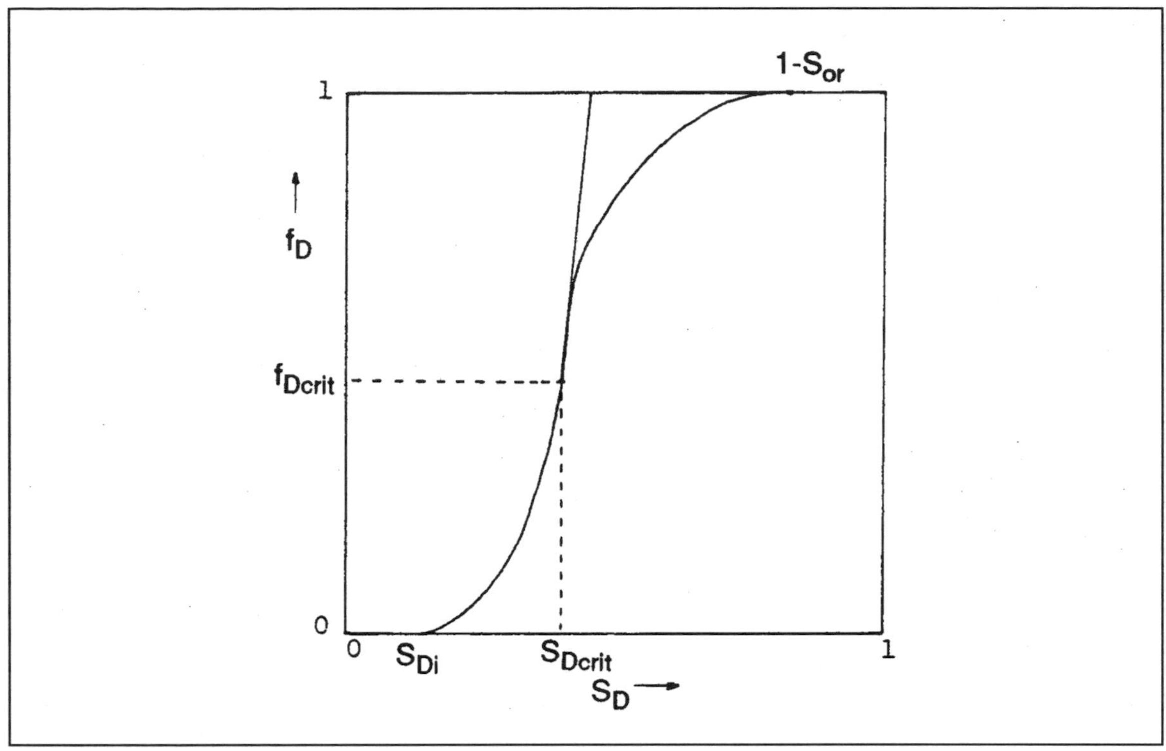

Fig. 11-13. Fractional flow curve illustrating the critical displacing-phase saturation; i.e., that initial displacing-phase saturation at (or above) which no front develops.

Another effect shown in Figure 11-12 (for water displacing oil in a water-wet system) is that the tangent line is steeper as the beginning displacing-phase saturation becomes higher. Thus, the average water saturation behind the front is lowered causing more oil to be left behind.

The highest displacing-phase saturation at the front occurs when the initial displacing-phase saturation is at the irreducible value. This can be seen by considering the tangent construction on the fractional flow curves in Figures 11-11 and 11-12. It should be apparent that as the initial displacing-phase saturation increases above the irreducible value, the displacing-phase saturation at the front (tangent point) decreases. Continuing this process (of increasing the initial displacing-phase saturation and drawing the tangent line) will cause the initial saturation and the corresponding flood front saturation to merge at the critical point. So, the critical point is the limiting lower value for the displacing-phase saturation at the flood front. Figure 11-14 serves to illustrate that for a beginning displacement phase saturation greater than the critical value, the displacement will have to be a "drag" zone displacement (entirely within the nonstabilized zone). Thus, the complete flood will take place as though breakthrough had already occurred. This situation illustrates very well the importance of capillary pressure in immiscible displacement operations. It is the capillary pressure effects that cause the oil bank to form.

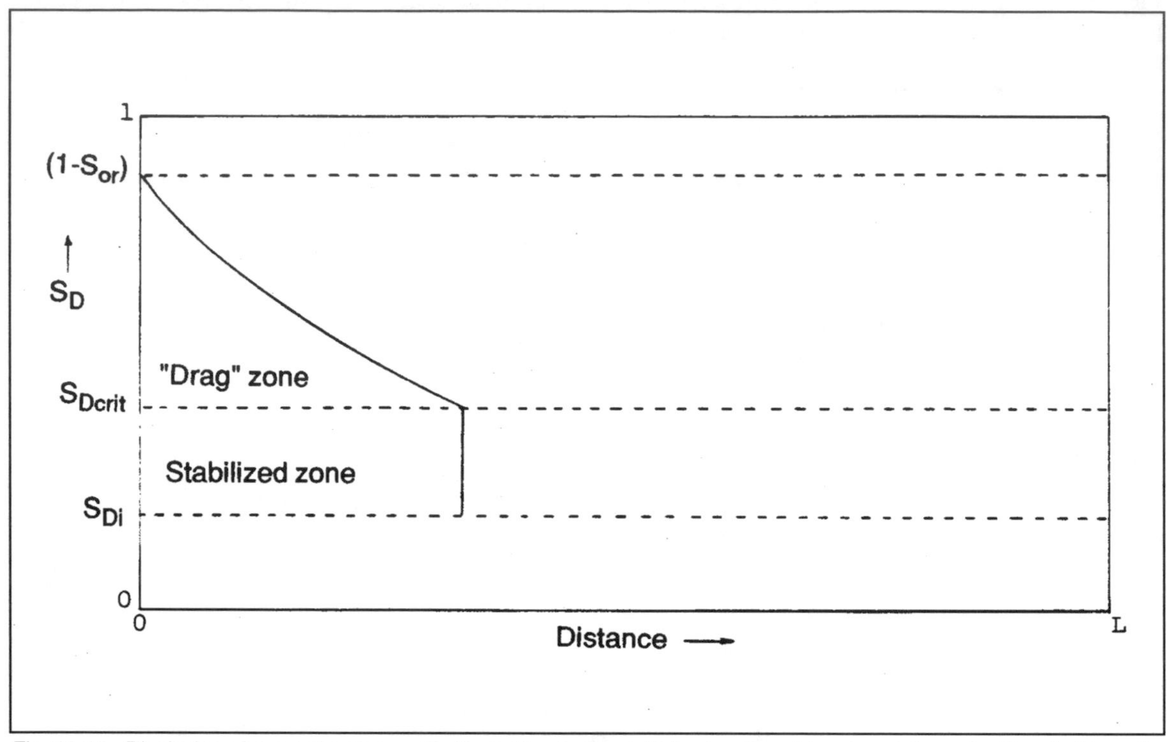

Fig. 11-14. Displacing-phase saturation profile illustrating that an initial saturation at (or above) the critical level will result in a "drag" zone displacement.

Water Displacing Oil in an Oil-wet System

In this case, the rock is more attracted to the oil than it is to the water. The oil resides next to the pore walls and within the smaller pores, while the water is normally in the middle of the larger pore spaces.

Water displacement in an oil-wet formation is not as efficient as in a water-wet formation. The reason for this is simple: the rock is "holding" on to the oil phase, which leaves the water in the middle of the flow channels. Thus, the relative permeability to water is higher at any given saturation than it is in a water-wet formation.

When the initial water saturation is at the irreducible level, it exists in an insular state. Because the water phase is discontinuous, no pressure differential can be transmitted through it. If a pressure differential is applied through the wetting oil phase, then a nonwetting globule can move, but only as far as the first capillary restriction which offers enough resistance to stop it. Thus, when water is displacing oil in an oil-wet reservoir with the initial oil saturation at $1 - S_{wi}$, there will be a front with only oil flowing ahead of it. But because the rock is not "holding" the water (but instead is attracted to the oil), the displacement is less efficient.

The capillary pressure (difference in pressure between the water and oil phases) ahead of the front is small due to the high wetting phase (oil) saturation. However, the capillary pressure behind the front is high with the oil-phase pressure being considerably less than the water-phase pressure. The result is that the oil phase pressure behind the front can theoretically be less than that ahead of the front. Thus, the oil tends to imbibe back across the front! The result is to increase the values of f_w and displace the f_w curve toward the low water saturations and lower oil recoveries.

If the initial water saturation is increased above the irreducible level, then a continuous water phase is established. Thus, the water phase is mobile and moves easily in the center of the flow channels. Because of this and because the oil tends to imbibe backwards, it may not take much increase in the initial water saturation (above the irreducible value) to cause a drag zone displacement.

Water-displacement oil recovery from an oil-wet reservoir depends on the degree of oil wetness of the formation, but it is usually reduced from that obtained in a similar water-wet reservoir.

Water Displacing Oil in a System with Heterogeneous Wettability

The comments thus far have concerned uniformly wetted porous media. Anderson[19] has presented an excellent discussion of the effects of fractional or non-uniform wettability, where portions of the rock surface are water-wet with the remainder being oil-wet. Where fractional wettability exists, with the sizes of the water-wet and oil-wet surfaces being on the order of a single pore, then waterflood behavior is similar to of uniformly wetted systems. Recovery should be between that of a 100% water-wet system and that of a 100% oil-wet system.

According to Anderson,[19] "mixed" wettability is a special type of fractional wettability in which the smaller pores are water-wet, while the larger pores (or the continuous flow paths) are oil-wet. In such a system, during a water displacement, there is no oil to be trapped in the smaller pores because they are filled entirely with connate water. The continuous oil-wet flow channels allow oil to drain in films along the pore wells which permits finite oil permeability to exist down to low oil saturations. The remaining oil saturation after water displacement in a mixed wettability system can be much lower than that obtained in uniformly wetted formations. Phenomenal recoveries have been obtained from the East Texas Woodbine formation that are believed to be due to the porous system's mixed wettability.

GAS DISPLACING OIL

Due to the wetting characteristics of oil and gas, it is unlikely that gas could exist as the wetting phase. So, oil should always be considered to be the wetting phase (with respect to the gas). In most reservoirs, water is also present. As already mentioned, it is necessary in Buckley-Leverett gas displacing oil calculations to assume that the water saturation does not exceed its irreducible value. Then, the water saturation is considered to be immobile and serves merely to reduce the effective pore space available to the gas and oil.

Since oil is the "wetting" phase, a gas displacement is similar to the case of a water displacement in an oil-wet reservoir which was just discussed. Of course, there are differences. The gas/oil fractional flow Equation 11-17 indicates that f_g is decreased (and therefore, recovery is increased) if the displacement is performed in a downdip direction. One of the major problems with a gas displacement is that the mobility ratio is unfavorable. (See Equations 11-13, 11-14, and 11-15.) This means that the mobility of gas, k_g/μ_g, is greater than the oil mobility, k_o/μ_o. Because the gas is more nonwetting than the oil, the gas is situated in the middle of the flow channels.

A significant density difference exists between free gas and oil. Thus, if gas displaces oil downward the force of gravity may reduce the influence of the adverse mobility ratio. If the rate of advance is not too high, the high gas mobility can be of benefit in a downdip displacement with variable cross-sectional area because it allows nearly 100 percent of the reservoir to be contacted by the gas.

Horizontal, immiscible displacements of oil with gas are usually not very efficient, especially with a significant beginning gas saturation. The usual result is early gas breakthrough.

PRACTICAL USE OF IMMISCIBLE DISPLACEMENT CONCEPTS

A number of examples of the practical application of the immiscible displacement concept as conceived by Buckley and Leverett have appeared in the literature.[4,10,11,12,13] Specific examples of immiscible displacement for the cases of water displacing oil and gas displacing oil are presented in Chapters 7 and 9, respectively, of Smith.[14] Details permitting the development of production curves on a time basis are also given in those chapters. While the examples given are only for two immiscible displacement cases, any other case where one fluid is displacing another insoluble fluid in porous media may be treated. Of course, a knowledge of the factors entering Equations 11-12 and 11-22 must be known. The method assumes that valid relative permeability information is at hand and the system being modelled is linear.

APPLICATION OF IMMISCIBLE DISPLACEMENT CONCEPTS TO NONLINEAR SYSTEMS

The most common secondary recovery well pattern is the five-spot: square pattern with an injector on each corner and a producing well in the middle. The method by which linear flooding concepts are extended to nonlinear systems will be illustrated in terms of the five-spot pattern. A quadrant of a five-spot is shown in Figure 11-15. Also shown in Figure 11-15 is a linear system approximation[7] for the five-spot pattern. If the streamlines for this pattern (not the approximation) were shown, it would be evident that the constant flow rate assumed for the linear flood is not the case for all locations of the 5-spot network. Obviously the flow rate per unit cross-sectional area will be smaller away from the wells than it is near a well. The sensitivity of the displacement may be determined by considering the probable injection rates attainable in the actual field case under consideration. Then, Equation 11-12 may be solved for a range of q_t values, and the plots of f_D versus displacing-phase saturation prepared. If considerable rate sensitivity is apparent, engineering judgment will have to be used in the choosing of a rate to properly model the displacement efficiency. It will also be necessary to apply an areal coverage factor (sweep efficiency), since the injected fluid will not contact 100% of the injection pattern area. Smith[14] presents techniques to estimate the effect of mobility ratio and flooding pattern on sweep efficiency. Also included[14] are the details of how to convert linear flooding predicted results to nonlinear systems.

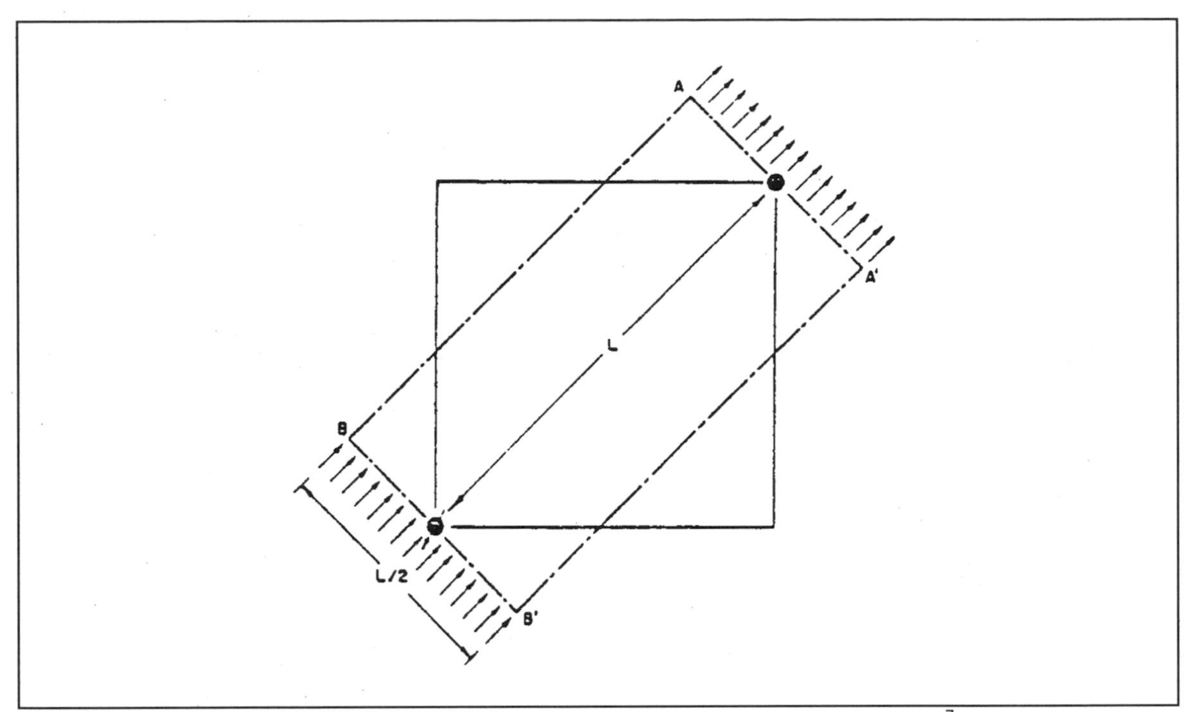

Fig. 11-15. Linear system approximation to the five-spot flooding pattern (after Rapoport and Leas[7]).

REFERENCES

1. Buckley, S. E. and Leverett, M. C.: *Trans.*, AIME (1942) **146** 107-116.
2. Welge, H. J.: *Trans.*, AIME (1952) **195**, 91.
3. Leverett, M. C.: *Trans.*, AIME (1941) **142**, 152-169.
4. Pirson, S. J.: *Oil Reservoir Engineering*, McGraw-Hill Book Co., Inc., New York City, (1958).
5. Jones-Parra, J. and Calhoun, J. C., Jr.: *Trans.*, AIME (1953) **198**, 335.
6. Terwilliger, P. L., *et al.*: *Trans.*, AIME (1951) **192**, 285.
7. Rapoport, L. A. and Leas, W. J.: *Trans.*, AIME (1953) **198**, 139.
8. Sheldon, J. W., Zondek, B. and Cardwell, W. T., Jr.: *Trans.*, AIME (1959) **216**, 290.
9. Rose, W. D. and Chanmapragada, R.: paper 1549G presented at the 1960 SPE Annual Meeting, Denver.
10. Craft, B. C. and Hawkins, M. F.: *Applied Petroleum Reservoir Engineering*, Prentice-Hall, Inc., Englewood Cliffs, N.J. (1959) 361.
11. Higgins, R. V. and Leighton, A. J.: "USBM RI 5618" (1960).
12. Higgins, R. V.: "USBM RI 5568" (1960).
13. Joslin, W. J.: *JPT* (January 1964) 87.
14. Smith, Charles R.: *Mechanics of Secondary Oil Recovery*, Reinhold Publ. Co., New York City (1966).
15. Collins, R. E.: *Flow of Fluids Through Porous Media*, The Petroleum Publishing Co., Tulsa (1976) 201-215,.
16. Nunge, R. J., *et al.*: "Flow Through Porous Media," American Chemical Soc. Publ., Washington D.C. (1970).
17. Pope, G. A.: "The Application of Fractional Flow Theory to Enhanced Oil Recovery," *SPEJ* (June 1980).
18. Craig, F. F., Jr.: "The Reservoir Engineering Aspects of Waterflooding," Monograph 3, SPE of AIME, Dallas (1971), (Third Printing, 1980).
19. Anderson, W. G.: "Wettability Literature Survey—Part 6: The Effects of Wettability on Waterflooding," *JPT* (December 1987) 1605-1622.

NOMENCLATURE

A = area

A = dimensionless time conversion constant (van Everdingen & Hurst aquifer model)

B = Fetkovich aquifer model constant

B_g = gas formation volume factor (reservoir volume divided by volume at standard conditions)

B_o = oil formation volume factor (volume at reservoir conditions divided by stock tank oil volume)

B_w = water formation volume factor

c = isothermal compressibility

c_e = effective (total system) compressibility

c_f = formation (pore volume) compressibility

c_g = gas compressibility

c_o = oil compressibility

c_{oe} = effective oil compressibility (applied to oil pore volume, undersaturated oil)

c_{pr} = pseudoreduced compressibility

c_w = water compressibility

c_{we} = effective (total) aquifer compressibility

C = coefficient of gas well back-pressure equation

C = wellbore storage coefficient

$\overline{C}$ = dimensionless wellbore storage coefficient

C_A = shape factor (well drainage geometry)

C_d = dimensionless wellbore storage coefficient

C_s = Schilthuis aquifer constant

C_v = van Everdingen & Hurst aquifer coefficient

d = diameter

D = non-Darcy flow coefficient

D = completion interval thickness

D = distance to closest well

D = distance to no-flow barrier

e = 2.7182 . . .

$\exp$ = e

$E_{f,w}$ = term relating expansion of connate water and rock in the material balance equation

E_g = term in the material balance equation relating to the expansion of the gas in the gas cap

E_o= term in the material balance equation relating to the expansion of the original oil including liberated solution gas

f= friction factor

f_g= fraction of the total flowing stream composed of gas

f_o= fraction of the total flowing stream composed of oil

f_w= fraction of the total flowing stream composed of water

F= cumulative production term in the material balance equation

F= wellbore storage factor of the McKinley type curve method

g= acceleration of gravity

g_c= conversion factor in Newton's Second Motion Law

G= original gas in place (OGIP)

G_a= apparent gas in place

G_p= cumulative gas production

G_{pc}= cumulative gas production from the gas cap

G_{poz}= cumulative gas production from the oil zone

h= formation thickness

I= injectivity index

I_s= specific injectivity index

J= J-function (capillary pressure) of Leverett

J= productivity index

J_s= specific productivity index

k= absolute permeability

k_f= fracture permeability

k_g= effective permeability to gas

k_m= matrix permeability

k_o= effective permeability to oil

k_{rg}= relative permeability to gas

k_{ro}= relative permeability to oil

k_{rw}= relative permeability to water

k_w= effective permeability to water

k_v= vertical permeability

L= length

L_f= fracture length (tip to tip)

Log= logarithm, base 10

Ln= logarithm, base e

m= mass

m= ratio of the original free gas reservoir volume to the original oil reservoir volume

m= slope

M= mobility ratio

M=molecular weight

n= number of samples

n= reciprocal of the slope of the gas well deliverability plot

n= total number of moles

N= stock tank oil originally in place

N_a= apparent oil in place

N_p= cumulative stock tank oil production

p= pressure

p_a= abandonment pressure

p_a= average aquifer pressure

p_b= bubble point pressure

p_c= capillary pressure

p_c= critical pressure

p_d= dew point pressure

p_e= external boundary pressure

p_g= pressure in the gas phase

p_i= initial pressure

p_o= pressure in the oil phase

p_{pc}= pseudocritical pressure

p_{pr}= pseudoreduced pressure

p_r= reduced pressure

p_{sc}= pressure at standard conditions

p_{sep}= separator pressure

p_w= bottom hole well pressure

p_w= pressure in the water phase

p_{wf}= flowing bottom hole pressure

p_{wp}= producing bottomhole pressure

p_{ws}= shut-in bottomhole pressure

p_{1hr}= pressure on straight line portion of semilog plot corresponding to a test time of one hour

$\bar{p}$= average (static) reservoir pressure

p*= "false" pressure obtained by extrapolating on the Horner plot to a Horner time ratio of one

Δp= pressure change, drawdown, or pressure drop

ΔP_d= dimensionless pressure

q = flow rate or production rate

q= fluid flow rate per unit of cross-sectional area

q_g= gas production rate

q_{gt}= net gas cap expansion rate

q_o= oil production rate

q_w= water production rate

q_{wt}= net water influx rate

Q= gas flow rate

Q= van Everdingen & Hurst cumulative influx function

Q_i= pore volumes of water injected

Q_{gin}= cumulative gas injection

r= radial distance

r_e= external boundary radius

r_i= radius of investigation

r_w= well radius

R= producing gas-oil ratio

R= universal gas constant

R_d= dimensionless radius (r / r_w)

R_p= cumulative produced gas-oil ratio

R_{poz}= cumulative produced GOR from the oil zone

R_s= solution gas-oil ratio

R_{sb}= solution gas-oil ratio at the bubble point

R_{si}= solution gas-oil ratio at original conditions

R_{sw}= solution gas-water ratio

s= skin effect or skin factor

S= saturation

S_D= displacing phase saturation

S_{Df}= displacing phase saturation at the front

$\bar{S}_D$= average displacing phase saturation behind the front

S_g= gas saturation

S_{gc}= critical gas saturation

S_o= oil saturation

S_{or}= residual oil saturation

S_w= water saturation

S_{wc}= connate water saturation

S_{wi}= irreducible water saturation

t= time

t= effective producing time

t_d= dimensionless time

t_s= producing time to stabilization

Δt= shut-in time

T= temperature

T= transmissibility $(k\,h / \mu)$

T_c= critical temperature

T_{pc}= pseudocritical temperature

T_{pr}= pseudoreduced temperature

T_{sc}= temperature at standard conditions

T_{sep}= separator temperature

u= linear flow direction

v= velocity

V= moles of vapor phase

V= volume

V_b= bulk volume

V_p= pore volume

V_w= wellbore volume

w= width

W_e= cumulative water influx

W_{in}= cumulative water injected

W_p= cumulative water produced

x= horizontal dimension

x= mole fraction of a component in the liquid phase

y= mole fraction of a component in the vapor phase

z= gas deviation factor

z= mole fraction of a component in a mixture

z= vertical dimension

α= angle of formation dip (from horizontal)

α

= angle of flow direction (from horizontal)

γ= specific gravity

γ_g= gas specific gravity

γ_o= oil specific gravity

γ_w= water specific gravity

Δ= difference

λ= mobility (k / μ)

λ= inter-porosity flow coefficient (double porosity reservoirs)

λ_g= gas mobility

λ_o= oil mobility

λ_w= water mobility

μ= viscosity

μ_g= gas viscosity

μ_o= oil viscosity

μ_w= water viscosity

ρ= density

ρ_g= gas density

ρ_o= oil density

ρ_w= water density

σ= surface tension

σ= interfacial tension

ψ= real gas potential (pseudopressure)

τ= pseudo-time

ϕ= porosity

ω= fracture to total system storativity ratio (double porosity reservoirs)

INDEX